Fundamentals of carrier transport

Second edition

Fundamentals of carrier transport is an accessible introduction to the behavior of charged carriers in semiconductors and semiconductor devices. It is written specifically for engineers and students without an extensive background in quantum mechanics and solid-state physics. This second edition contains many new and updated sections, including a completely new chapter on transport in ultrasmall devices.

The author begins by covering a range of essential physical principles. He then goes on to cover both low- and high-field transport, scattering, transport in devices, and transport in mesoscopic systems. The use of Monte Carlo simulation methods is explained in detail.

Many homework exercises are provided and there is a variety of worked examples. The book will be of great interest to graduate students of electrical engineering and applied physics. It will also be invaluable to practicing engineers working on semiconductor device research and development.

Mark Lundstrom is Professor of Electrical and Computer Engineering at Purdue University. He has published more than 200 scientific papers in the fields of semiconductor devices and device physics. He is a Fellow of the IEEE.

Fundamentals of carrier transport

Second Edition

Mark Lundstrom

School of Electrical and Computer Engineering
Purdue University, West Lafayette, Indiana, USA

CAMBRIDGE
UNIVERSITY PRESS

CAMBRIDGE UNIVERSITY PRESS
Cambridge, New York, Melbourne, Madrid, Cape Town, Singapore, São Paulo, Delhi

Cambridge University Press
The Edinburgh Building, Cambridge CB2 8RU, UK

Published in the United States of America by Cambridge University Press, New York

www.cambridge.org
Information on this title: www.cambridge.org/9780521637244

First edition published 1990
Second edition published 2000
This digitally printed version 2009

A catalogue record for this publication is available from the British Library

ISBN 978-0-521-63134-1 hardback
ISBN 978-0-521-63724-4 paperback

Contents

Preface to the second edition

The first edition of this book was written in a period when the drift–diffusion-based description of semiconductor devices was beginning to lose validity and many kinds of interesting transport effects (e.g. velocity overshoot, ballistic transport, real-space transfer, etc.) and their implications for devices were being explored. Since that time, semiconductor devices have continued to shrink in size, so that engineers and device researchers now face these issues daily. When the first edition was written, quantum transport in mesoscopic structures was also an active research field, with many uncertainties being debated. In the intervening years, this field has matured; the general principles are now understood and are becoming relevant to semiconductor technologists as devices continue their relentless march to microscopic dimensions.

The goals of the second edition are much like those of the first. The book is an attempt to help students with little formal training in quantum mechanics or solid state physics (i.e., the typical graduate of an undergraduate electrical engineering program) understand the fundamental concepts of carrier transport in semiconductors. Writing the second edition was an opportunity to update and clarify material in the first edition and to treat new topics. The most significant change in the second edition is the addition of Chapter 9 on transport in mesoscopic structures, a topic that device engineers now deal with.

Two classes of graduate students worked through early versions of this text and helped me to clarify the presentation and reduce the number of typos and errors. I am grateful to them for their willingness to suffer through those rough drafts and want to thank Muhammad A. Alam, Kausar Banoo, and Jung-Hoon Rhew in particular for their help in clarifying my thinking. Thanks also to my colleagues, Professors Supriyo Datta, Karl Hess, David K. Ferry, Robert W. Dutton, and Robert S. Eisenberg who helped to make this a better book by sharing their insights with me. The work of Professor Chihiro Hamaguchi's group (T. Kunikiyo, et al. *Journal of Applied Physics*, 75, pp. 297–312, 1994.) was the inspiration for the cover illustration. I hope that the result of these efforts will help device engineers and researchers acquire the understanding of transport fundamentals that is essential for device research and engineering.

Preface to the first edition

The operation of semiconductor devices is controlled by how electrons and holes respond to applied, built-in, and scattering potentials. Electrical engineers are used to treating transport phenomenologically – carriers drift in electric fields and diffuse in concentration gradients. For much of the past 40 years during which semiconductor technology advanced from point-contact transistors to megabit memories, drift–diffusion equations have served as the backbone of device analysis. As devices continue to decrease in size and increase in sophistication, however, this simple picture of carrier transport is beginning to lose validity. Device engineers now need a clear understanding of the physics of carrier transport in a variety of semiconductors as well as an understanding of the nature of transport in modern, small devices. The focus of this book is on carrier transport fundamentals beginning at the microscopic level and progressing to the macroscopic effects relevant to devices. The reader should acquire an understanding of the general features of low- and high-field transport in common semiconductors as well as of the characteristics of transport in small devices. He or she should learn how to evaluate scattering rates and mobilities from the semiconductor's material properties and should understand the various approaches commonly used to analyze and simulate devices.

The book is directed at electrical engineering graduate students or practicing device engineers who typically possess a mature understanding of semiconductor fundamentals and devices but only an acquaintance with the basics of quantum mechanics and solid-state physics. In addition to discussing physical principles, one objective is to familiarize the reader with commonly used theoretical approaches. Although this necessarily involves some amount of mathematics, the derivations have been written with the student in mind; intermediate steps and tricks have been displayed, so that students can learn the art of performing such calculations and develop confidence in their problem-solving skills. The reader is assumed to understand semiconductor fundamentals at the level of R.F. Pierret's *Advanced Semiconductor Fundamentals* (Vol. VI of the Modular Series) and to be familiar with devices at the level of S.M. Sze's *Physics of Semiconductor Devices*, 2nd Ed., (Wiley-Interscience, New York, 1981). The reader should also be acquainted with the rudiments of quantum mechanics and solid-state physics. Although a sophisticated understanding is not essential, readers seeking a thorough grounding in microscopic physics are encouraged to consult Supriyo Datta's *Quantum Phenomena* (Vol. VIII of the Modular Series). There is some overlap between Datta's text and this one (both treat the Boltzmann Transport Equation, carrier scattering and balance equations, for example), but *Quantum Phenomena* is an excellent companion text. The difference is that in this book the focus is on applications, while Datta's centers on the underlying physical principles.

In accordance with the philosophy of the Modular Series, this book is designed to be a self-contained treatment of the carrier transport fundamentals essential for modern device research and development. The book is not comprehensive, nor is it a review of current research; it represents the basics that I believe a device engineer needs to understand. Each chapter ends with references which the reader is encouraged to explore to deepen and broaden his understanding. Because the focus is on transport fundamentals which underlie the operation of devices in general, relatively little emphasis is placed on specific devices. The discussion is also restricted to electron transport in silicon and gallium arsenide, which I justify by the technological importance of these two semiconductors and because the results are representative of transport in most covalent or polar semiconductors. The text can be covered at a comfortable pace in one semester, which allows the instructor to supplement it with current research topics or with subjects that I've overlooked or slighted. After working through this text, students should be prepared to follow current device research and to actively participate in developing future devices.

For the most part, I've tried to stick with conventional notation, but a few deviations should be noted. A set of brackets, $\langle \cdot \rangle$, is used to denote an average over the distribution function. The double brackets, $\langle\langle \cdot \rangle\rangle$, denote the specially-defined average that appears in transport theory. Following Pierret in Vol. VI of the Modular Series, I use 'degrees' Kelvin (K) to measure temperature. Finally, the reader will note words and groups of words italicized from time to time. I use this device to call the reader's attention to the fact that an important concept, term, or name is being introduced.

While teaching my first graduate level course on these topics, the need for this text became apparent. To introduce students to classical, semiconductor transport as well as to the effects now occurring in modern devices, I compiled a collection of book excerpts, monograph chapters, and research papers. By the end of the semester it was clear that a systematic, consistent treatment of these topics required that a book be written. I'm grateful to the students in that class and to those in subsequent semesters who struggled through successive versions of the manuscript and pointed out errors, inconsistencies, and confusions.

Professor Christine Maziar tested the text in her course at the University of Texas, Austin and provided a wealth of suggestions and insights. Dr. Peter Blakey's careful review of the manuscript and his numerous suggestions on Chapters 6–8 were also a great help. Professor Robert Trew of North Carolina State University filled me in on the interesting mystery regarding the field-dependent diffusion coefficient for electrons in GaAs, and Professor Robert Pierret's meticulous review of the manuscript kept me on my toes.

My understanding of the subject of this text has been evolving during the course of my research on advanced semiconductor devices. I feel fortunate to work in a stimulating environment and owe a special debt to my colleague, Supriyo Datta, from whom I've learned much and to my research students who also taught me a lot. Special thanks go to Christine Maziar and Martin Klausmeier-Brown, for developing a Monte Carlo simulation program, and to Amitava Das and Mark Stettler, for extending and enhancing it and performing many of the simulations which illustrate the text. Finally, I now understand how demanding the writing of a book is — not only upon the author but upon his family as well. I'm especially thankful to my wife, Mary, who encouraged me to undertake the project and helped me find the quiet time needed to complete it.

An overview

Until recently, a deep understanding of electron transport in semiconductor devices has not been essential to the device engineer because the familiar drift–diffusion equation,

$$\mathbf{J}_n = nq\mu_n\mathscr{E} + qD_n\nabla n \tag{0.1}$$

described devices well, and the transport parameters, μ_n and D_n did not need to be computed from first principles because they had been carefully measured. For advanced devices, however, the situation is much different. In small devices, μ_n and D_n are no longer material- and field-dependent parameters; they depend on microscopic physics, on the structure of the device, and even on the applied bias. A variety of materials is now being investigated, and with modern epitaxial technology it is even possible to engineer material properties. This book begins at the microscopic level and progresses towards the macroscopic level of devices.

At the most basic level, electrons in semiconductors are quantum mechanical waves propagating through the device under the influence of the crystal, applied, and scattering potentials as indicated in Fig. 0.1a. *Chapter 1* begins at this level and shows that when the scale of the device is large enough, the electron can be treated much as a classical particle as indicated in Fig. 0.1b. Electron scattering, however, is the result of short-range forces and must be treated quantum mechanically. Calculation of scattering probabilities per unit time for the perturbing potentials encountered in common semiconductors is the subject of *Chapter 2*.

At the macroscopic level, equations like eq. (0.1) result by averaging an enormous number of nearly chaotic trajectories like that displayed in Fig. 0.1b. Alternatively, one may simply ask: what is the probability of finding an electron at $\mathbf{r}$ with momentum $\mathbf{p}$? The answer is $f(\mathbf{r},\mathbf{p},t)$, the distribution function, which defines the state of the device. As illustrated in Fig. 0.1c, the distribution function is a probability density function, which completely defines the average state of the system. In *Chapter 3*, we formulate the Boltzmann Transport Equation (BTE), an equation that determines $f(\mathbf{r},\mathbf{p},t)$, and show how to solve it under

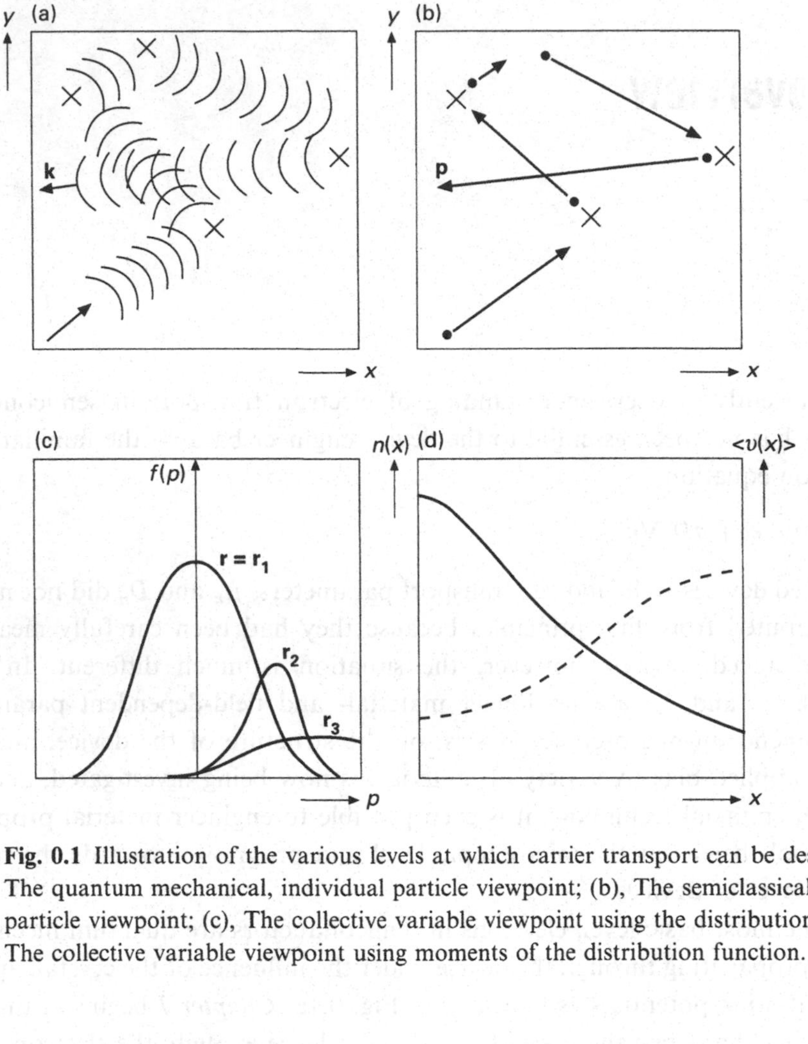

Fig. 0.1 Illustration of the various levels at which carrier transport can be described. (a), The quantum mechanical, individual particle viewpoint; (b), The semiclassical, individual particle viewpoint; (c), The collective variable viewpoint using the distribution function; (d), The collective variable viewpoint using moments of the distribution function.

simple circumstances. A systematic treatment of low-field transport based on solving the BTE is the subject of *Chapter 4*.

Device engineers rarely need to know the state of the system in detail as described by the distribution function. The quantities of interest are typically the average carrier density, velocity, and, perhaps, energy, which are the zeroth, first and second moments of the distribution function. A formal procedure for generating balance equations for these quantities is described in *Chapter 5*. In the process, we will discuss the derivation of the drift–diffusion equation from the BTE. At the drift–diffusion level, the quantities of interest are the position-dependent carrier and velocity profiles as illustrated in Fig. 0.1d. Another approach for computing macroscopic transport properties is to simulate a large number (typically several thousand) trajectories like that in Fig. 0.1b on a computer to average the results. The technique, known as Monte Carlo simulation, is the subject of *Chapter 6*. In *Chapter 7*, both the balance equation approach and Monte Carlo simulation are applied to the problem of analyzing high-field carrier transport in bulk semiconductors. Several of the interesting transport effects that occur in modern devices are examined in *Chapter 8*. Devices contain both low- and high-field regions, but the spatial variations are strong, so some qualitatively new transport features arise. Finally, the book concludes in *Chapter 9* by examining transport in *mesoscopic* devices whose size lies between the macroscopic and atomic regimes.

1 The quantum foundation

Conventional device analysis begins by assuming that carriers behave as classical particles which obey Newton's laws. A more fundamental treatment describes the electron by its wave function, $\Psi(\mathbf{r}, t)$, which is obtained by solving the Schrödinger equation,

$$i\hbar\frac{\partial\Psi}{\partial t} = -\frac{\hbar^2}{2m_0}\nabla^2\Psi + [E_{C0}(\mathbf{r}) + U_C(\mathbf{r}) + U_S(\mathbf{r}, t)]\Psi(\mathbf{r}, t). \tag{1.1}$$

The quantity $\Psi^*(\mathbf{r}, t)\Psi(\mathbf{r}, t)d\mathbf{r}$ is the probability of finding the electron between $\mathbf{r}$ and $\mathbf{r} + d\mathbf{r}$. Three different potential energies appear in the wave equation; the first, $E_{C0}(\mathbf{r})$, describes potentials that are built-in or applied to the device. (The energy band diagram of a semiconductor device is just a plot of this potential versus position. Device engineers usually refer to this potential as $E_C(\mathbf{r})$, but in this text E_C will refer to the position and momentum-dependent conduction band potential; it contains a potential energy component, $E_{C0}(\mathbf{r})$, and a kinetic energy component.) The second potential is the *crystal potential*, $U_C(\mathbf{r})$, which describes the electrostatic potential due to the atoms. (Since eq. (1.1) is a wave equation for a single electron, $U_C(\mathbf{r})$ also includes the average potential due to the other electrons in the solid.) Finally, U_S is a scattering potential due to random deviations in potential caused by ionized impurities or by lattice vibrations. Device analysis is usually based on an approximate solution to eq. (1.1) known as the *semiclassical* treatment which describes carrier dynamics in the applied and built-in potentials by Newton's laws without explicitly treating the crystal potential. The influence of the crystal potential is treated indirectly by the use of an effective mass or an energy band structure. Carrier scattering is treated quantum mechanically.

This chapter reviews techniques for treating the three different potentials in the wave equation. The emphasis is on justifying the semiclassical approach to carrier transport because it serves as the basis for most of conventional device analysis and for most of this text. It is important to understand the underlying approximations because they can be violated in advanced, ultra-small devices. For the most part, this chapter should be a review of introductory quantum mechanics and solid-state physics; results are stated, not derived, and their significance and relevance to device analysis is noted. For a thorough treatment of the fundamentals surveyed in this introductory chapter, the reader is referred to *Quantum Phenomena*, by Supriyo Datta [1.1].

1.1 Electrons in a nonuniform potential, $E_{C0}(\mathbf{r})$

Let us first review the nature of solutions to the wave equation in the absence of the crystal and scattering potentials; we further simplify the problem by reducing it to one spatial dimension. Application of the technique of separation of variables to the wave equation then shows that the solutions are of the form

$$\psi(z, t) = \psi(z)e^{-iEt/\hbar} = \psi(z)e^{-i\omega t}, \tag{1.2}$$

which oscillate in time with a frequency of $\omega = E/\hbar$. When eq. (1.2) is inserted in the wave equation, we obtain the time-independent wave equation,

$$\frac{d^2\psi}{dz^2} + k^2\psi = 0, \tag{1.3a}$$

where

$$k^2 = \frac{2m_0}{\hbar^2}[E - E_{C0}(z)]. \tag{1.3b}$$

The nature of the solutions is determined by whether k^2 is greater or less than zero.

General features of the solutions can be illustrated by a few very simple examples (see Fig. 1.1). First, we let $E_{C0}(z)$ be constant and set the constant to zero. Since $k^2 \geq 0$, the solutions to eq. (1.3) are of the form

$$\psi(z) = a_k e^{\pm ikz} \tag{1.4}$$

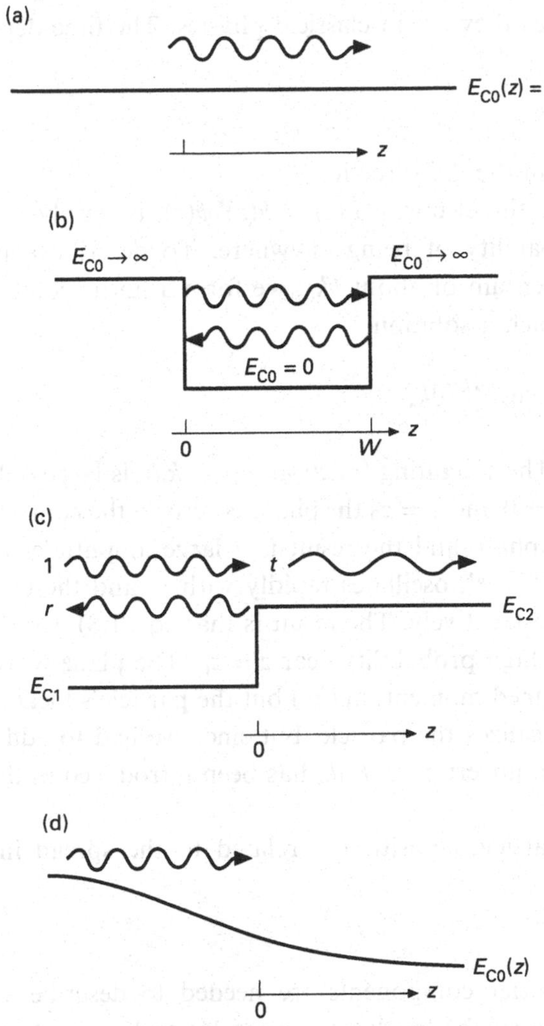

Fig. 1.1 Example potential profiles. (a) free electron, (b) infinite potential well, (c) finite potential step, and (d) slowly varying potential.

(or, equivalently, sin kz or cos kz), where

$$\hbar k = \sqrt{2m_0 E}, \tag{1.5}$$

and a_k is an arbitrary constant. According to eq. (1.5),

$$E(k) = \frac{\hbar^2 k^2}{2m_0} \tag{1.6}$$

is the relation between the electron's energy and its wave vector. Since $\hbar k$ can be shown to be the electron's momentum [1.2], the energy and momentum of a free electron are related exactly as they are in classical physics. The time-dependent solution,

$$\Psi(z, t) = a_k e^{i(\pm kz - \omega t)}, \tag{1.7}$$

represents a wave traveling in the $\pm\hat{z}$ direction.

The probability of finding the electron, $P(z) = \psi(z)^* \psi(z)$, is simply $|a_k|^2$; the electron has an equal probability of being anywhere. To describe a particle located near z_0 with a momentum of about $\hbar k_0$, we form a linear combination of the solutions, eq. (1.4). Such a solution,

$$\Psi(z, t) = \int_{-\infty}^{+\infty} a(k - k_0) e^{ik(z-z_0)} e^{-i\omega(k)t} dk, \tag{1.8}$$

is known as a *wave packet*. The weighting function, $a(k - k_0)$, is large only near k_0 as shown in Fig. 1.2. At $t = 0$ and $z = z_0$ the phase is zero so the contributions for all wave vectors add in phase and the result is a large amplitude. But for $|z| \gg |z_0|$, the exponential, $e^{ik(z-z_0)}$, oscillates rapidly with k, and the contributions from different k add destructively. The result is that eq. (1.8) describes an electron that is located with high probability near $z = z_0$. The plane wave solution, eq. (1.7), had a well-defined momentum ($\hbar k$) but the particle's location was undefined. Equation (1.8) localizes the particle, but since we had to add waves with different momentum, an uncertainty, $\hbar \Delta k$, has been introduced in the electron's momentum.

The uncertainty in the particle's position is related to the spread in wave vectors by [1.2]

$$\Delta z \Delta k \simeq 1, \tag{1.9}$$

which states that many Fourier components are needed to describe a small particle. Similarly, a particle can be localized in time by adding contributions with different frequencies such that

$$\Delta \omega \Delta t \simeq 1. \tag{1.10}$$

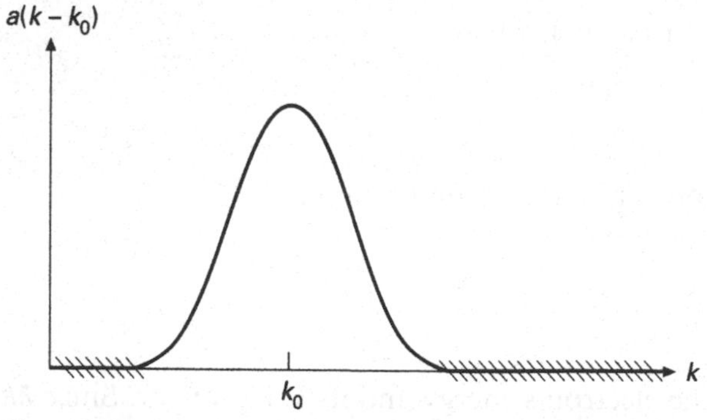

Fig. 1.2 Weighting function $a(k - k_0)$ used to localize an electron.

When a change of variables to momentum and energy is made in eq. (1.9) and eq. (1.10), we find the *uncertainty relations*:

$$\Delta z \Delta p \geq \hbar \tag{1.11}$$

and

$$\Delta E \Delta t \geq \hbar \tag{1.12}$$

which state that we cannot know both a carrier's position and momentum exactly and that we cannot determine its exact energy in a finite time.

Each of the individual components of the wave packet, eq. (1.8), travels at its own *phase velocity*

$$v_p = \frac{\omega(k)}{k}. \tag{1.13}$$

Since each of the components travels at a different velocity, both the center of the wave packet and its shape change with time. The center, which occurs where the components add constructively, moves at the *group velocity*, v_g, where

$$\boxed{v_g(k_0) = \left.\frac{d\omega(k)}{dk}\right|_{k=k_0} = \left.\frac{1}{\hbar}\frac{dE(k)}{dk}\right|_{k=k_0}}\;. \tag{1.14}$$

According to eq. (1.14) and eq. (1.6), for free electrons

$$v_g = \frac{\hbar k_0}{m_0} = \frac{\langle p \rangle}{m_0},$$

which shows that the group velocity of the electron wave packet is simply its average momentum divided by its mass – just what classical physics would give.

For electrons constrained within a potential well, the solutions are much different. Consider a second example (Fig. 1.1b) for which $E_{C0} = 0$ between 0 and W, but assume now that W is small and that $E_{C0} \to \infty$ at $x = 0$ and W so that the electron is *bound* – it must remain between 0 and W. The solutions are still given by eq. (1.4) but it is more convenient to use linear combinations of these, or $\sin(kz)$ and $\cos(kz)$. Because $\psi(z) = 0$ at $z = 0$, we find

$$\psi(z) = a_k \sin kz. \tag{1.15}$$

But $\psi(z)$ must also be zero at $z = W$, so the wave vector must be restricted to

$$k_n = \frac{n\pi}{W} \quad n = 1, 2, \ldots \tag{1.16}$$

In the first example, the electron was free, and we found a continuous distribution of wave vectors given by eq. (1.6), but in this example, the electron is bound and only certain k's, specified by eq. (1.16) are permitted. From eq. (1.6) we find that the energy is also quantized according to

$$\varepsilon_n = E(k_n) = \frac{\hbar^2 k_n^2}{2m_0} = \frac{\hbar^2 n^2 \pi^2}{2m_0 W^2} \qquad n = 1, 2, \ldots \tag{1.17}$$

With modern microfabrication technology, potential wells can be engineered into devices by appropriate variations of doping or composition. In a silicon metal-oxide-semiconductor field effect transistor (MOSFET), inversion layer carriers are confined in a potential well at the oxide–silicon interface. For such structures, the carriers are confined in only one direction; in the orthogonal plane, the potential is constant and the wave functions have the plane wave character of the first example. Electrons confined in the so-called *quantum wells* are discussed in Section 1.3.

As a third example, we consider an electron, which may be propagating through a device, when it encounters the potential step sketched in Fig. 1.1c. For the conditions shown, $E > E_{C0}(z)$ everywhere so the solutions are traveling waves of the form

$$\psi(z) = e^{ik_1 z} + re^{-ik_1 z} \qquad z < 0 \tag{1.18a}$$

and

$$\psi(z) = te^{ik_2 z} \qquad z > 0 \tag{1.18b}$$

where

$$k_1 = \frac{\sqrt{2m_0(E - E_{C1})}}{\hbar} \tag{1.19a}$$

and

$$k_2 = \frac{\sqrt{2m_0(E - E_{C2})}}{\hbar}. \tag{1.19b}$$

Both $\psi(z)$ and $d\psi/dz$ must be continuous everywhere otherwise $d^2\psi/dz^2$ would be infinite and eq. (1.3a) could not be satisfied (unless $E_{C0}(z)$ goes to infinity as it did for the second example). Applying these continuity conditions at $z = 0$, we find

$$r = \frac{k_1 - k_2}{k_1 + k_2} \tag{1.20a}$$

and

$$t = \frac{2k_1}{k_1 + k_2}. \tag{1.20b}$$

From a classical perspective, we expect that when the electron's energy is greater than the top of the step, it will simply transmit across. The finite probability of reflection at a step is a quantum mechanical effect due to the wave nature of carriers. Such reflections occur when the potential changes rapidly (meaning that it varies significantly on a distance comparable to the electron's wavelength).

The final example, illustrated in Fig. 1.1d, shows an electron moving through a slowly varying potential (the electron's energy is assumed to be greater than the potential energy). For such potentials, reflections do not occur and the solution can be obtained using the Wentzel–Kramers–Brillouin approximation as in [1.2]

$$\psi(z) = \frac{1}{\sqrt{k}} e^{i\int^{z} k(\eta)d\eta} \tag{1.21}$$

where $k(\eta)$ is given by eq. (1.3b). The absence of reflections can be understood if the slowly varying potential is approximated by a large number of small potential steps. The small reflections that occur at each interface add destructively so that no net reflected wave occurs. Electron motion in a slowly varying potential can be described classically because reflections do not occur.

1.1.1 Probability current

The goal of device analysis is often to compute the current through the device. Since quantum mechanics is based on probability, we evaluate the flow of probability,

$$\frac{\partial P(z)}{\partial t} = \frac{\partial}{\partial t}(\Psi^* \Psi) = \frac{\partial \Psi^*}{\partial t}\Psi + \Psi^* \frac{\partial \Psi}{\partial t}. \tag{1.22}$$

The derivatives can be evaluated from the wave equation, eq. (1.1), and its complex conjugate to find

$$\frac{\partial P}{\partial t} + \frac{\partial J}{\partial z} = 0 \tag{1.23}$$

where

$$J = \frac{\hbar}{2m_0 i}\left[\Psi^* \frac{\partial \Psi}{\partial z} - \Psi \frac{\partial \Psi^*}{\partial z}\right] \tag{1.24}$$

Equation (1.23) is a continuity equation – the first term is the rate of increase of probability and the second term, the divergence of a vector J, represents the flow of probability away from z. The vector J, which describes the flow of probability, is termed the *probability current*. For an ensemble of electrons, we interpret $\Psi(z)^* \Psi(z)$ as $n(z)$, the electron density and $J(z)$ as the electron flux.

For the plane wave solutions found in the first example, the probability current works out to be

$$J = \Psi^* \Psi \frac{\hbar k}{m_0}. \tag{1.25}$$

Since $\Psi^* \Psi$ is the electron density and $\hbar k$ the electron momentum, eq. (1.25) simply states that $J = nv$. For the second example, the states were bound, and $\psi(z)$ was proportional to $\sin kz$. For such states, eq. (1.24) shows that $J = 0$ which is consistent with the fact that the electron is localized and not going anywhere. For the potential step considered in the third example, the wave functions, eq. (1.18), produce a current

$$J = \frac{\hbar k_1}{m_0}(1 - |r|^2) \quad z < 0 \tag{1.26a}$$

and

$$J = \frac{\hbar k_2}{m_0}|t|^2 \quad z > 0, \tag{1.26b}$$

from which the incident, reflected, and transmitted currents J_i, J_r, J_t, are apparent. Transmission and reflection coefficients for the currents can be defined as

$$T \equiv \frac{J_t}{J_i} = \frac{k_2}{k_1}|t|^2 \tag{1.27}$$

and

$$R \equiv \frac{J_r}{J_i} = |r|^2. \tag{1.28}$$

Notice that T and R are real numbers unlike t and r and that $T + R = 1$ as expected.

For electrons moving in a slowly varying potential, the wave function is given by eq. (1.21). For electrons in such a potential,

$$J = \psi(z)^* \psi(z) \frac{\hbar k(z)}{m_0}. \tag{1.29}$$

Since $\psi^* \psi \sim 1/k(z)$, the current is constant – as it should be. When $\psi^* \psi$ is interpreted as the electron density, eq. (1.29) is seen to correspond to the classical current density, $J = n(z)v(z)$. The electron's momentum is $\hbar k(z)$, so from eq. (1.29) its velocity is

$$v(z) = \frac{\sqrt{2m_0(E - E_{C0}(z))}}{m_0}. \tag{1.30}$$

E is the electron's total energy and $E_{C0}(z)$ its potential energy, so eq. (1.30) is just the expression for the velocity of a classical particle.

These examples show that the wave nature of carriers is important when the potential changes rapidly, but when the potential varies gradually, reflections don't occur and electrons can be treated as classical particles which obey Newton's Laws. The electrons in the conduction band of a semiconductor always see at least one rapidly varying potential – it is the crystal potential due to the nucleus and core electrons.

1.2 Electrons in a periodic potential, $U_C(\mathbf{r})$

Electrons in a semiconductor crystal respond when fields are applied, but a crystal potential due to the lattice atoms and other electrons is always present. As sketched in Fig. 1.3a, the crystal potential, $U_C(z)$, displays the periodicity of the lattice. To find the wave functions we solve the one-electron wave equation

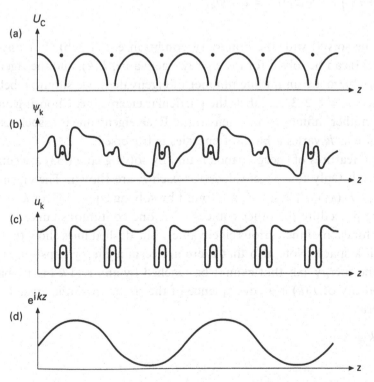

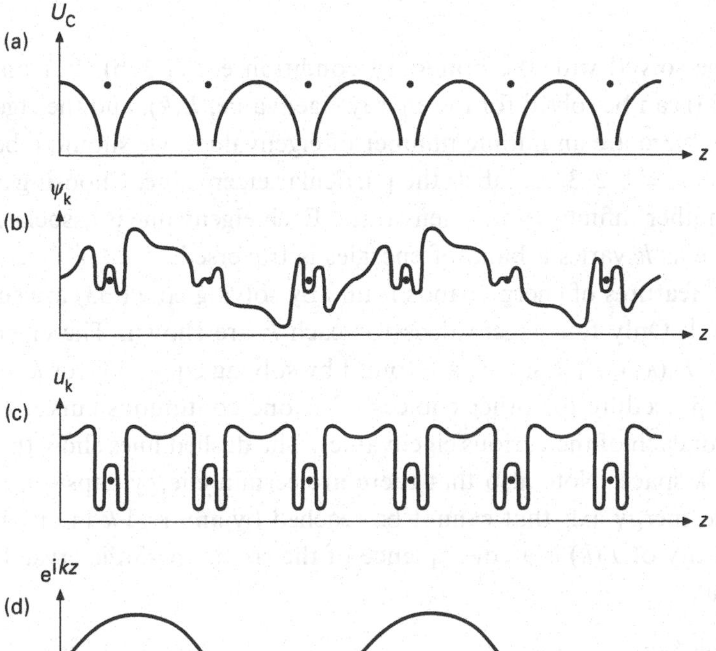

Fig. 1.3 Illustration of wave functions in a periodic potential. (a) $U_C(z)$, the crystal potential, (b) ψ_k, the eigenfunction, (c) u_k, the Bloch function, (d) $e^{ik_t \cdot z}$, a plane wave. From Harrison [1.4]. (Reproduced with permission from Dover, New York.)

$$\left[-\frac{\hbar^2}{2m_0}\frac{d^2}{dz^2}+U_C(z)\right]\psi(z)=E\psi(z). \tag{1.31}$$

The solutions sketched in Fig. 1.3b also reflect the periodicity of the lattice. The solutions for a periodic are called *Bloch waves* and consist of a function with the periodicity of the lattice multiplied by a plane wave. Figures 1.3c and 1.3d display the two components of a Bloch wave, which is mathematically defined by

$$\psi_k = u_k e^{ikz}, \tag{1.32a}$$

where

$$u_k(z+a)=u_k(z). \tag{1.32b}$$

The electron's momentum varies with position because the crystal potential alternately speeds up and slows down electrons. Nevertheless, the quantity $\hbar k$, termed the *crystal momentum*, often acts like the carrier's momentum.

To find $u_k(z)$ we insert eq. (1.32a) in eq. (1.31) and find

$$\left[\frac{1}{2m_0}\left(\frac{\hbar}{i}\frac{\partial}{\partial z}+\hbar k\right)^2+U_C(z)\right]u_k=E(k)u_k, \tag{1.33}$$

which must be solved with the boundary condition eq. (1.32b). For any k we select, eq. (1.33) can be solved for the energy eigenvalue, $E(k)$, and the eigenfunction, u_k. Since there are an infinite number of eigenvalues, we should label them as $E_n(k)$, where $n = 1, 2, 3, \ldots$ labels the particular eigenvalue. Choosing another k results in another infinite set of eigenvalues. Each eigenvalue is associated with a *band* because as k varies a band of energies is traversed.

The general features of energy bands found by solving eq. (1.33) are summarized in Fig. 1.4. Only four eigenvalues for each k are shown. The eigenvalues $E_1(k_1)$, $E_2(k_1)$, $E_3(k_1)$, $E_4(k_1)$, etc., are found by solving eq. (1.33) for $k = k_1$. By repeating the procedure for other choices of k, one continuous curve, $E_n(k)$, is mapped out for each of the various eigenvalues. The dashed lines show that $E_n(k)$ is periodic in **k**-space. Note also that there are certain energy gaps – forbidden regions on the energy axis that cannot be reached by any real k in any band.

The periodicity of $E(k)$ is a consequence of the spatial periodic crystal potential. In general,

$$E_n(k)=E_n(k+K_j), \tag{1.34}$$

where

$$K_j = j\frac{2\pi}{a} \quad j=1,2,\ldots \tag{1.35}$$

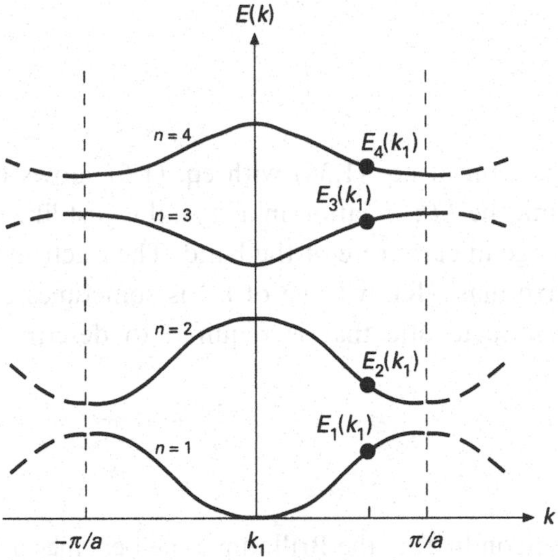

Fig. 1.4 Electron energy versus wave vector, $E(k)$.

is a *reciprocal lattice* vector, and a is the lattice constant. Because $E_n(k)$ is periodic, all information is available in one period or *Brillouin zone.* It is convenient to use a period centered about the origin (the so-called *reduced zone* representation).

The band structures of common semiconductors are well known from various experiments and from numerical solutions to the wave equation. For semiconductor work, approximate solutions to eq. (1.33), accurate near a band minimum, are often adequate. The so-called $\mathbf{k} \cdot \mathbf{p}$ method for obtaining such approximate solutions is discussed in the texts by Datta [1.1] and by Singh [1.6]. To treat very energetic carriers, however, a full, numerical tabulation of $E(\mathbf{k})$ throughout the Brillouin zone is essential [1.7].

If the band structure is known, $E(k)$ can always be expanded in a Taylor series as

$$E(k) = E(0) + \left.\frac{\partial E(k)}{\partial k}\right|_{k=0} k + \frac{1}{2}\left.\frac{\partial^2 E(k)}{\partial k^2}\right|_{k=0} k^2 + \ldots$$

When the band minimum occurs at $k = 0$, the gradient of $E(k)$ is zero at $k = 0$, so, to the lowest order,

$$E(k) = E(0) + \frac{\hbar^2 k^2}{2m^*}, \tag{1.36}$$

where

$$\boxed{\frac{1}{m^*} \equiv \frac{1}{\hbar^2}\frac{\partial^2 E(k)}{\partial k^2}} \tag{1.37}$$

is the *effective mass*. A comparison of eq. (1.36) with eq. (1.6) shows that for electrons near a band minimum, the $E(k)$ relation in a crystal is just like that for free electrons except for a change in curvature of the band. The electron mass is simply replaced by the effective mass. Knowledge of m^* is sometimes the only information from the $E(k)$ characteristic that is required to describe carrier transport.

1.2.1 Model band structure

In real, three-dimensional, semiconductors the Brillouin zone becomes a volume and $E(\mathbf{k})$ generally depends on the direction of $\mathbf{k}$. The simple $E(\mathbf{k})$ model shown in Fig. 1.5a describes both diamond and zincblende crystals such as silicon and GaAs. The conduction band has three minima; one at $\mathbf{k} = 0$ (called the Γ point), another along $\langle 111 \rangle$ directions at the boundary of the first Brillouin zone (called L) and a third near the zone boundary along $\langle 100 \rangle$ directions (the Δ-line). (The standard notation for labeling lines and points in the Brillouin zone is displayed in Fig. 1.5b.) The model semiconductor has three valence bands – each has a maximum at $\mathbf{k} = 0$. Two of the valence bands are degenerate at $\mathbf{k} = 0$; the third is *split-off* by the spin-orbit interaction.

Table 1.1 lists several band parameters, defined in Fig. 1.5, for a variety of semiconductors. For silicon and germanium, the lowest conduction band minima are along Δ and at L respectively. When the conduction band minimum doesn't occur at the same point as the valence band maximum, the semiconductor is called *indirect*. GaAs is seen to be *direct*, but it also has a minimum at L that is only a few tenths of an electron volt higher. This upper minimum plays an important role for electron transport in GaAs.

When the conduction band minimum or valence band maximum lies at $\mathbf{k} = 0$, $E(\mathbf{k})$ may be approximated as

$$E(\mathbf{k}) = \pm\frac{\hbar^2 k^2}{2m^*}, \tag{1.38}$$

where the positive sign is for the conduction band and the negative sign for the valence band. The energy zero is taken at the band extrema, so $E(\mathbf{k})$ represents the carrier's kinetic energy. Equation (1.38) describes a band whose constant energy surface in $\mathbf{k}$-space is a sphere; the effective mass is isotropic. This simple model is appropriate for the conduction band at Γ and for the split-off valence band, but is often used more widely – whenever rough estimates suffice.

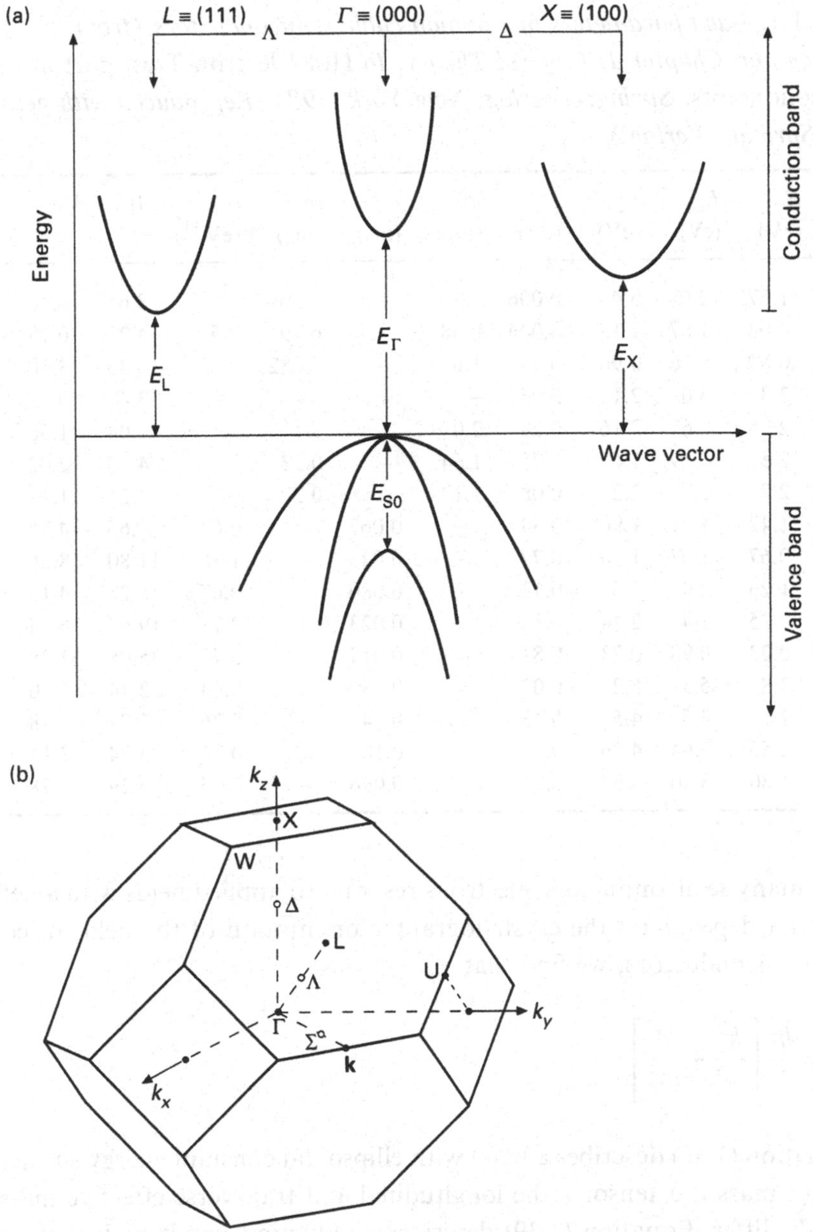

Fig. 1.5 (a) Band structure of the model semiconductor (from Reggiani, L., Chapter 1: General Theory. In *Hot Electron Transport in Semiconductors*. Springer-Verlag, New York, 1985.) (b) Standard notation for labeling high symmetry lines and points in the Brillouin zone for diamond and zincblende crystals. (Reproduced with permission from Springer-Verlag.)

Table 1.1. *Band parameters of common cubic semiconductors (from L. Reggiani, Chapter 1: General Theory. In* Hot Electron Transport in Semiconductors. *Springer-Verlag, New York, 1985. Reproduced with permission from Springer-Verlag.)*

	E_Γ (eV)	E_L (eV)	E_Δ (eV)	E_{so} (eV)	m_ℓ^* (m_0)	m^* (m_0)	m_t^* (m_0)	α (eV^{-1})	$\|A\|$	$\|B\|$	$\|C\|$
C	11.67	12.67	5.45	0.006	1.4	–	0.36	–	3.61	0.18	3.76
Si	4.08	1.87	1.13	0.044	0.98	–	0.19	0.5	4.22	0.78	4.80
Ge	0.89	0.76	0.96	0.29	1.64	–	0.082	0.65	13.35	8.50	13.11
AlP	3.3	3.0	2.1	0.05	–	–	–	–	3.47	0.12	3.98
AlAs	2.95	2.67	2.16	0.28	2.0	–	–	–	4.04	1.56	4.71
AlSb	2.5	2.39	1.6	0.75	1.64	–	0.23	–	4.15	2.02	4.95
GaP	2.7	2.7	2.2	0.08	1.12	–	0.22	–	4.20	1.96	4.65
GaAs	1.42	1.71	1.90	0.34	–	0.067	–	0.64	7.65	4.82	7.71
GaSb	0.67	1.07	1.30	0.77	–	0.045	–	1.36	11.80	8.06	11.71
InP	1.26	2.0	2.3	0.13	–	0.080	–	0.67	6.28	4.16	6.35
InAs	0.35	1.45	2.14	0.38	–	0.023	–	2.73	19.67	16.74	13.96
InSb	0.23	0.98	0.73	0.81	–	0.014	–	5.72	35.08	31.28	22.27
ZnS	3.8	5.3	5.2	0.07	–	0.28	–	0.14	2.54	1.50	2.75
ZnSe	2.9	4.5	4.5	0.43	–	0.14	–	0.26	3.77	2.48	3.87
ZnTe	2.56	3.64	4.26	0.92	–	0.18	–	0.26	3.74	2.14	4.30
CdTe	1.80	3.40	4.32	0.91	–	0.096	–	0.45	5.29	3.78	5.46

For many semiconductors, electrons respond to applied fields with an effective mass that depends on the crystallographic orientation of the field. In common cubic semiconductors, we find that

$$E(\mathbf{k}) = \frac{\hbar^2}{2}\left[\frac{k_\ell^2}{m_\ell^*} + \frac{k_t^2}{m_t^*}\right]. \tag{1.39}$$

Equation (1.39) describes a band with ellipsoidal constant energy surfaces. The effective mass is a tensor – the longitudinal and transverse effective masses, m_ℓ^* and m_t^*, differ. Equation (1.39) describes conduction bands at L and along Δ. Note that there are eight equivalent L points and six equivalent Δ lines in cubic crystals so there are actually many of these valleys in **k**-space.

For high applied fields, carriers may be far above the minimum, and the higher order terms in the Taylor series expansion cannot be ignored. For the conduction band, *nonparabolicity* is often described by a relation of the form

$$E(1 + \alpha E) = \frac{\hbar^2 k^2}{2m^*}, \tag{1.40}$$

where m^* is determined from eq. (1.37) at the minimum. (Equation (1.40) is obtained from approximate solutions to eq. (1.33) derived by $\mathbf{k} \cdot \mathbf{p}$ theory.) For a minimum at $\mathbf{k} = 0$,

$$\alpha_\Gamma = \frac{1}{E_{G\Gamma}}\left(1 - \frac{m_\Gamma^*}{m_0}\right)^2, \tag{1.41}$$

where $E_{G\Gamma}$ is the direct bandgap.

The simple expressions we have presented generally work well for electrons in the conduction band, but the valence bands are much more complex. In $\mathbf{k} \cdot \mathbf{p}$ theory, the shape of $E(\mathbf{k})$ is attributed to interactions between various bands. In wide bandgap semiconductors, the conduction band is well separated from the valence bands, interactions are weak, and the resulting band structure is parabolic. (But at high energies, or in narrow bandgap semiconductors, these interactions become important leading to conduction band nonparabolicity as discussed above.) The light and heavy hole valence bands, however, are degenerate at $\mathbf{k} = 0$, so the interactions are strong and the band shapes become complex. Since the split-off valence band is generally rather well-separated from the other valence bands, we expect its shape to be more nearly parabolic. Because the band structure has such a strong influence on carrier transport, it is important that we develop a descriptive understanding of the valence bands. (For a discussion of how to actually compute these band shapes, consult Datta [1.1] or Singh [1.6].)

The light and heavy hole valence bands in common semiconductors can be described by

$$E(\mathbf{k}) = ak^2[1 \mp g(\theta, \phi)]. \tag{1.42}$$

Such bands have a warped constant energy surface (the $\mp$ refers to the heavy and light hold bands respectively). The angles, θ and ϕ, are the polar and azimuthal angles of $\mathbf{k}$ with respect to the crystallographic axes. The function $g(\theta, \phi)$ is given by

$$g(\theta, \phi) = \left[b^2 + c^2(\sin^4\theta \cos^2\phi \sin^2\phi + \sin^2\theta \cos^2\theta)\right]^{1/2} \tag{1.43}$$

with

$$a = \frac{\hbar^2|A|}{2m_0}, \quad b = \frac{|B|}{|A|}, \quad c = \frac{|C|}{|A|} \tag{1.44}$$

where A, B, and C are listed in Table 1.1.

Figures 1.6–1.8 show the constant energy surfaces for the heavy, light, and split-off valence bands in Si. As shown in Fig. 1.6, the heavy hole band is warped at low energies, and the shape becomes complicated at higher hole energies. For the light-hole band displayed in Fig. 1.7, the distortion is smaller, and it is even

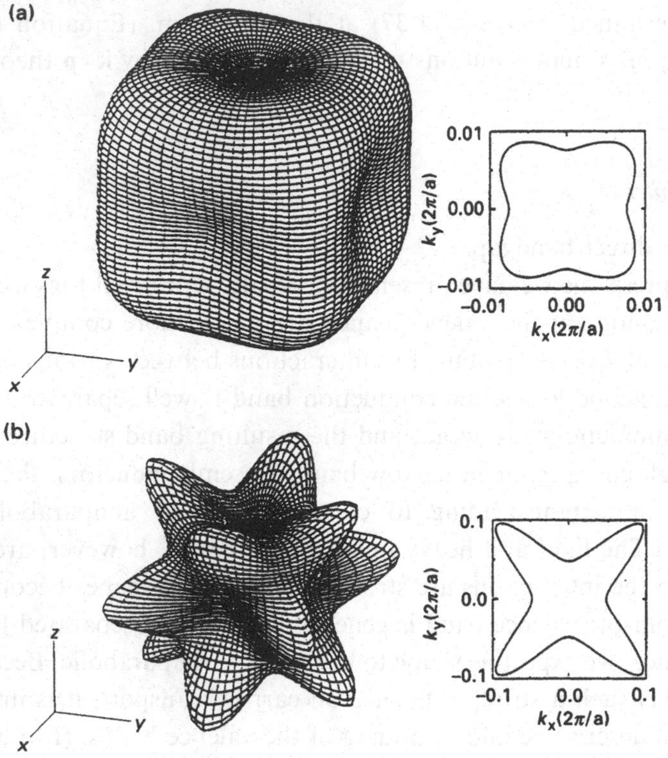

Fig. 1.6 The constant energy surfaces for the heavy hole band in Si. (a) $E = 1\,\text{meV}$, and (b) $E = 40\,\text{meV}$. From Singh [1.6]. (Reproduced with permission of The McGraw-Hill Companies.)

less for the split-off valence band, as shown in Fig. 1.8. For Si, the spin-orbit coupling is small ($\Delta_{SO} = 0.044\,\text{eV}$), so the split-off band can play a role in hole transport, but for most semiconductors, the spin-orbit coupling is much larger, and the split-off band is not typically populated by holes (e.g. in GaAs, $\Delta_{SO} = 0.35\,\text{eV}$). These examples show why hole transport is difficult to treat, even at low fields when the carriers reside near the top of the band. We will generally make use of very simple, spherical and parabolic energy band models, but it is important to recognize that a realistic description of the valence band shape is required for a quantitative treatment of hole transport.

1.2.2 Full band structure

As discussed in the previous section, the conduction band can be approximated as parabolic only near the band minima. For higher energies, we can approximate the conduction band with a nonparabolicity parameter, α, as defined in eq. (1.40). In Section 1.4.2, we will show that a spherical, nonparabolic band provides a reasonable approximation to the density of states in Si up to an energy of

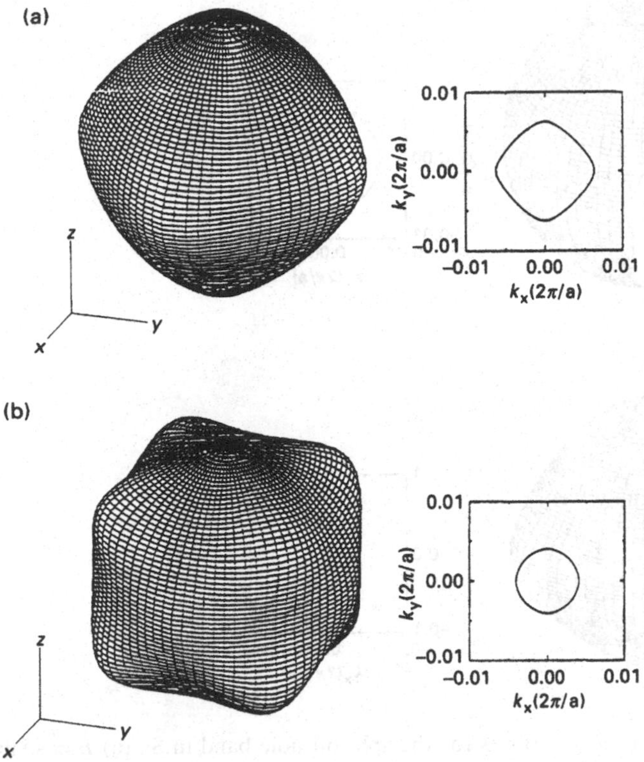

Fig. 1.7 The constant energy surfaces for the light hole band in Si. (a) $E = 1$ meV, and (b) $E = 40$ meV. From Singh [1.6]. (Reproduced with permission of The McGraw-Hill Companies.)

about $\approx$ 1–2 eV. There are important reasons, however, for examining electrons at much higher energies. Impact ionization, for example, involves electrons with a few electron volts of kinetic energy, and in Si MOSFETs an important reliability problem is the injection of electrons from the channel into the gate oxide. The energy barrier at the SiO_2 : Si interface is $\approx$ 3.1 eV. For such problems, we must abandon simple expressions for $E(\mathbf{k})$ and resort to a numerically-generated table of $E(\mathbf{k})$.

To evaluate $E(\mathbf{k})$ numerically, eq. (1.1) is solved for a bulk semiconductor ($E_{C0}(\mathbf{r}) = 0$) in the absence of scattering ($U_S(\mathbf{r}, t) = 0$). The well-developed art of such calculations is discussed by Singh [1.6]. One popular method is the pseudopotential technique which relies on the fact that the band structure is largely determined by the valence electrons. Accordingly, the effect of the core potential is subtracted out by replacing the actual potential by a *pseudopotential* which reproduces the actual potential between atoms but which is smooth through the ion core. Empirical *form factors* have been derived to fit optical bandgaps at the high symmetry locations. Using this empirical pseudopotential method, the energy band structures of most common semiconductors have been

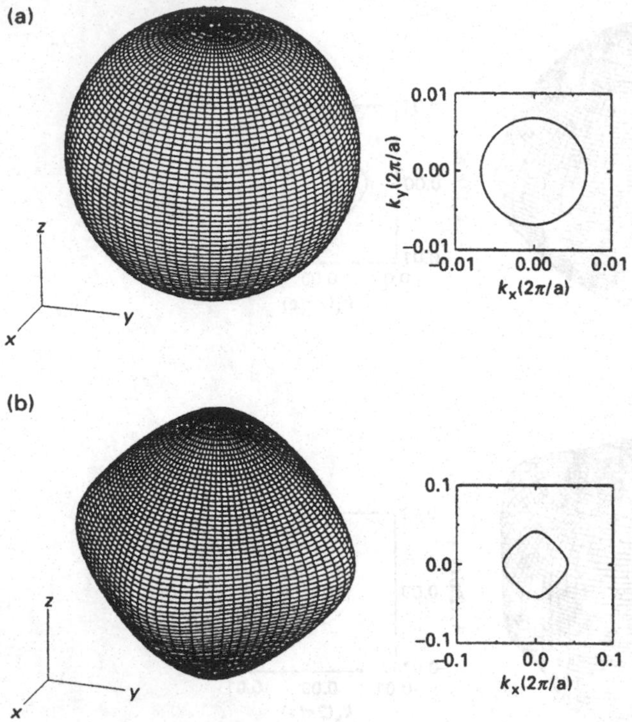

Fig. 1.8 The constant energy surfaces for the split-off hole band in Si. (a) $E = 45\,\text{meV}$, and (b) $E = 84\,\text{meV}$. From Singh [1.6]. (Reproduced with permission of The McGraw-Hill Companies.)

evaluated. The results of such calculations are generally presented as plots of $E(\mathbf{k})$ along the high symmetry lines displayed in Fig. 1.5b. Figures 1.9 and 1.10 show the results for the conduction bands of Si and GaAs [1.7].

Figures 1.9 and 1.10 show the first several conduction bands; the lowest conduction band in each case is indicated by a heavy line. For Si, we see that the lowest conduction band energy is along the x-line (a [100] direction) and occurs at about 85% of the way to the zone boundary. These are the well-known six, equivalent ellipsoidal constant energy surfaces of Si. Note that when electrons gain $\approx 0.13\,\text{eV}$ of kinetic energy, they can cross the zone boundary. More importantly, we see a second conduction band only 0.1 eV above the minimum of the first conduction band. Carriers above $\approx 0.1\,\text{eV}$ in kinetic energy may reside in either of two conduction bands. Note also that the first conduction band minimum at L lies about 1 eV above the lowest first band minimum. Under high electric fields, carriers can gain enough energy to populate these (Ge-like) valleys too.

The ellipsoidal constant energy surfaces at low energy become very complicated at high energies. In Figs. 1.11a–c we examine the constant energy contours in a cross-section of the Brillouin zone. The ellipsoidal constant energy surfaces

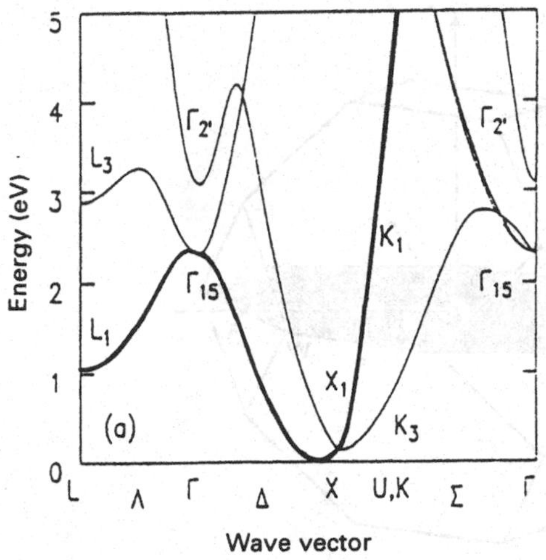

Fig. 1.9 $E(\mathbf{k})$ for the conduction bands of Si. From M. Fischetti [1.7]. (© 1991 IEEE)

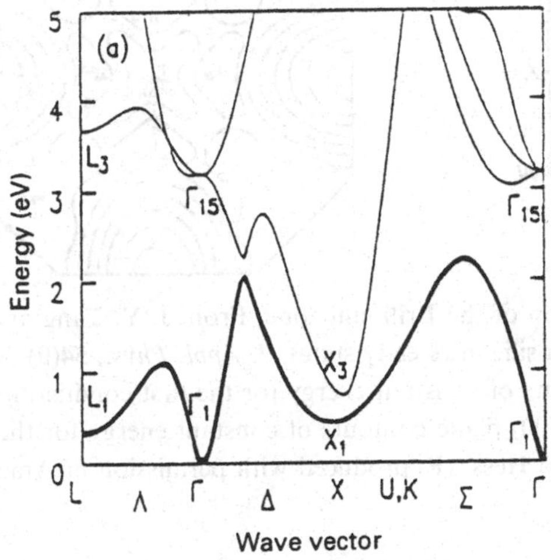

Fig. 1.10 $E(\mathbf{k})$ for the conduction bands of GaAs. From M. Fischetti [1.7]. (© 1991 IEEE)

of the first conduction band are displayed in Fig. 1.11b, and the constant energy surfaces of the second conduction band are displayed in 1.11c. The second conduction band minima lie at the zone boundary, and the constant energy surfaces are more spherical than the first. Under high electric fields, electrons populate the entire Brillouin zone, and Figs. 1.11b and c show that the constant energy surfaces between the minima cannot be described by simple analytical expressions.

The $E(\mathbf{k})$ relations for the conduction bands of GaAs are displayed in Fig. 1.10. In contrast to Si, we see that the second conduction band lies well above the

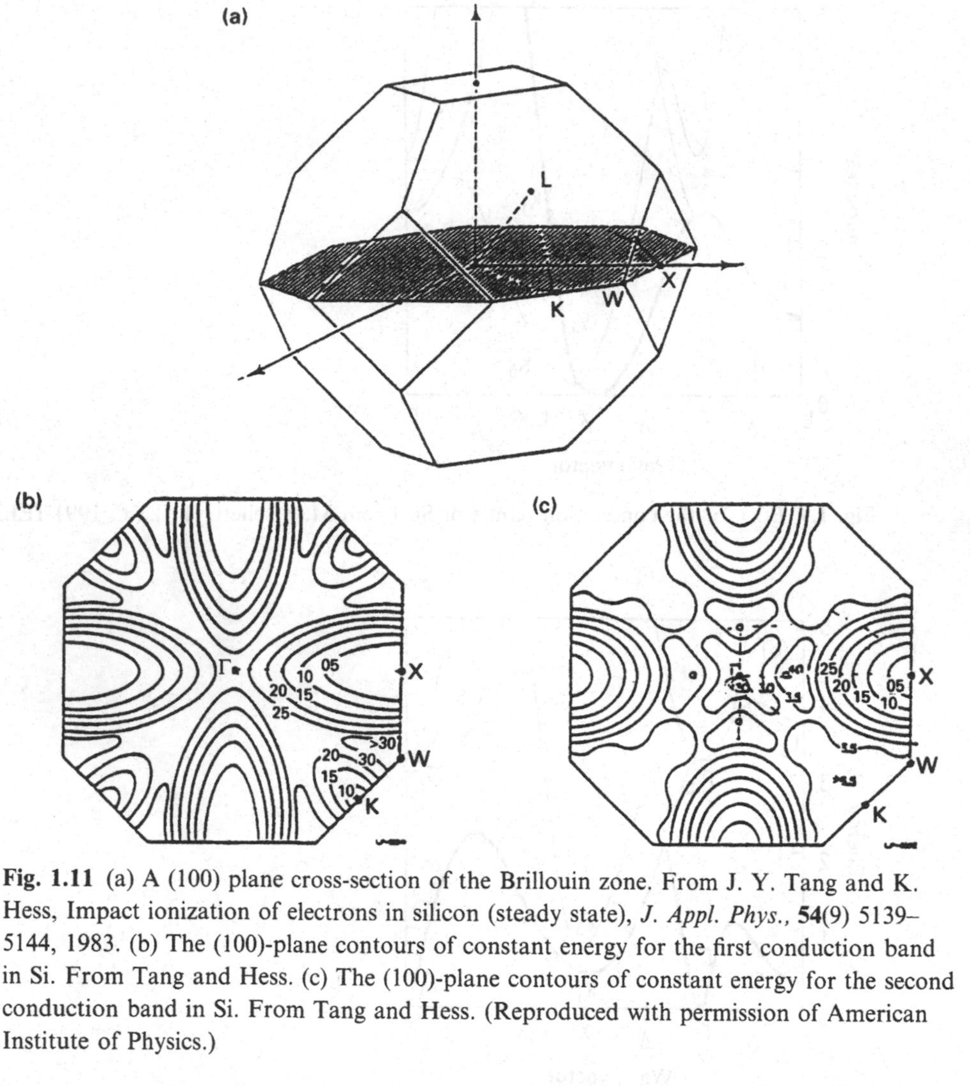

Fig. 1.11 (a) A (100) plane cross-section of the Brillouin zone. From J. Y. Tang and K. Hess, Impact ionization of electrons in silicon (steady state), *J. Appl. Phys.*, **54**(9) 5139–5144, 1983. (b) The (100)-plane contours of constant energy for the first conduction band in Si. From Tang and Hess. (c) The (100)-plane contours of constant energy for the second conduction band in Si. From Tang and Hess. (Reproduced with permission of American Institute of Physics.)

first; we can do a decent job of describing transport in GaAs with just a single conduction band. Note, however, that the first conduction band shows three minima with the lowest at the Γ point. The first conduction band minima at L (the Ge-like valleys) lie only about 0.31 eV above the Γ-valley minima, and the minima along X (the Si-like valleys) are only about 0.5 eV above the Γ-valley minimum. These valleys are easily populated under modest and high electric fields and give GaAs its distinctive transport features.

Because of the increasing importance of high energy carriers in modern devices, the use of full, numerical tables of $E(\mathbf{k})$ is common. When generating such tables, it is important to exploit symmetry to minimize the data to be stored. Consider Fig. 1.12, which shows a location in a cubic coordinate system. For a cube, one symmetry operation is reflection across the x–z plane. From repeated

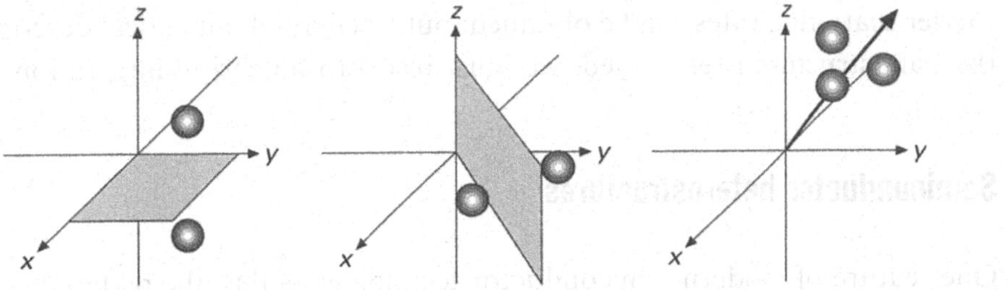

Fig. 1.12 Illustration of the symmetry operations for a cubic lattice. (a) Reflection across a (100) plane, (b) reflection across a (110) plane, and (c) rotation of $3\pi/2$ about a [111] direction.

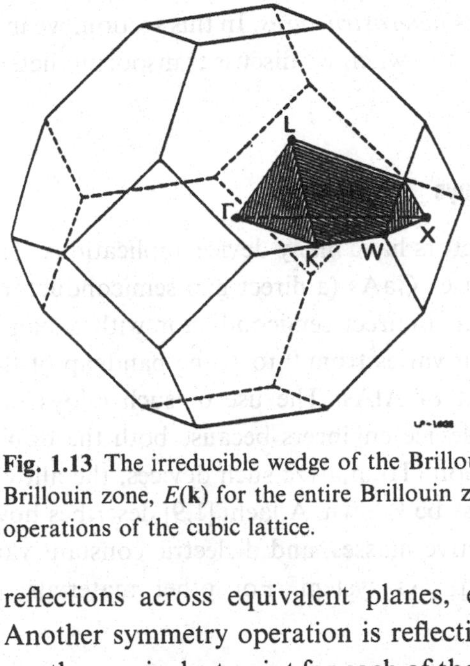

Fig. 1.13 The irreducible wedge of the Brillouin zone. Given $E(\mathbf{k})$ for this 1/48th of the Brillouin zone, $E(\mathbf{k})$ for the entire Brillouin zone is obtained by applying the symmetry operations of the cubic lattice.

reflections across equivalent planes, eight equivalent points can be formed. Another symmetry operation is reflection across a (110) plane, which generates another equivalent point for each of the first eight equivalent points. Finally, one can rotate a cube by $3\pi/2$ about a [111] direction to get three more equivalent points. The result is that for the given point, there are $8 \times 2 \times 3 = 48$ equivalent points. It is sufficient, therefore, to evaluate $E(\mathbf{k})$ in 1/48 of the Brillouin zone and to generate the other points by symmetry operations. The volume commonly used is shown in Fig. 1.13; it is known as the *irreducible wedge* of the Brillouin zone and is defined by

$$
\begin{aligned}
&0 \le k_z \le k_x \le k_y \le 2\pi/a \\
&k_x + k_y + k_z \le 3\pi/a.
\end{aligned} \tag{1.45}
$$

When simple, analytical expressions for $E(\mathbf{k})$ (e.g. eq. (1.38)) are used, we shall see that analytical expressions for quantities such as the density of states and carrier scattering rates can be obtained, but when a full, numerical description of the bandstructure is employed, we must resort to numerical integration.

1.3 Semiconductor heterostructures

One feature of modern semiconductor technology is that the material composition is readily varied as a semiconductor film is grown. This is particularly easy to accomplish in semiconductor alloys, but other combinations of different semiconductors are readily produced. Junctions between two different semiconductors are called *heterojunctions*. More complex, perhaps continuous, compositional variations are referred to as *heterostructures*. In this section, we introduce some general concepts that we will use when we discuss transport in heterostructures.

1.3.1 Band structure of semiconductor alloys

Alloys of two or more semiconductors have many device applications. The alloy $Al_xGa_{1-x}As$, for example, comprises GaAs (a direct gap semiconductor with a bandgap of 1.42 eV) and AlAs (an indirect semiconductor with a bandgap of 2.16 eV). As the AlAs mole faction varies from 0 to 1, the bandgap of the alloy varies from that of GaAs to that of AlAs. The use of such alloys offers an additional degree of freedom to device engineers because both the doping and bandgap can be varied with position. To analyze such devices, the alloy's composition-dependent properties must be known. Adachi [1.9] describes how parameters such as the bandgap, effective masses, and dielectric constant vary with alloy composition in the $Al_xGa_{1-x}As$ system. For other materials, consult Landolt-Börnstein [1.10].

1.3.2 Energy band diagrams for abrupt heterojunctions

To draw energy band diagrams for compositionally nonuniform devices, we need to know more than how the bandgap varies with composition, we must also know how the bands line up at compositional junctions. Figure 1.14 shows the experimentally observed band alignments for $Al_{0.3}Ga_{0.7}As/GaAs$ heterojunction. For $Al_xGa_{1-x}As$ heterojunctions, the offset in conduction bands is found to be about 65% of the difference in band gaps for alloy compositions below about $x = 0.5$ where the band structure is direct.

With modern epitaxial growth techniques, the alloy composition can be varied on an atomic scale to produce structures like that shown in Fig. 1.15a, which consists of a GaAs *quantum well* sandwiched between two $Al_xGa_{1-x}As$ layers.

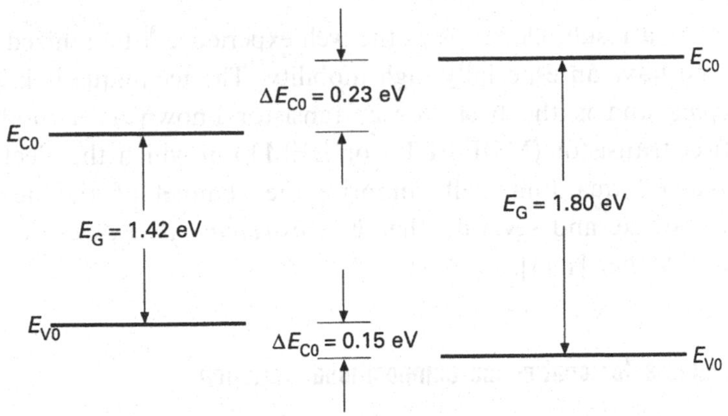

Fig. 1.14 Experimentally observed band alignments for $Al_xGa_{1-x}As$ for $x = 0$ and $x = 0.3$.

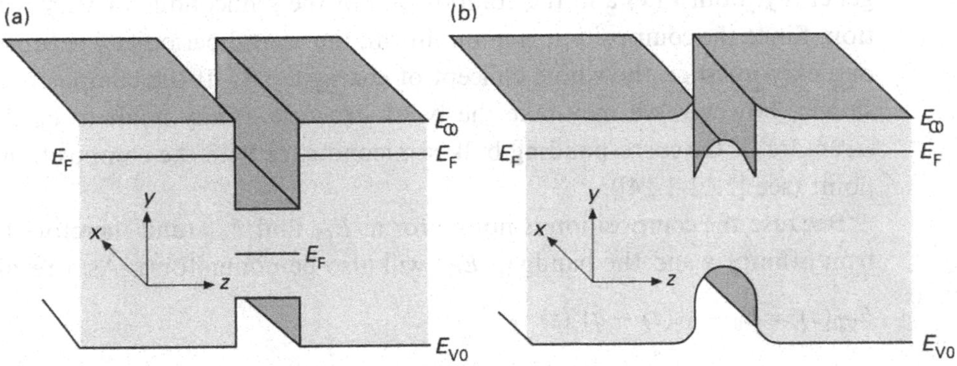

Fig. 1.15 Energy band diagram for an Al/GaAs/GaAs/AlGaAs quantum well structure. (a) For this case, we assume that the electrons are 'frozen' in place in the N-AlGaAs so that they cannot transfer to the *i*-GaAs. (b) A more realistic energy band diagram for the AlGaAs/GaAs/AlGaAs quantum well which displays the effects of mobile charge transfer to the quantum well.

Since the width of the well may be less than 100 Å, carriers within these wells are strongly influenced by quantum effects.

For this example, we assume that the $Al_xGa_{1-x}As$ layers are doped *n*-type and that the GaAs well is undoped. In Fig. 1.15a we have assumed that the electrons are frozen in place so that they cannot move down in energy from the N-AlGaAs to the *i*-GaAs. For this case, the bands are flat, and there is no electric field. A more realistic case is illustrated in Fig. 1.15b where the electrons move from the higher Fermi level to the lower one and establish equilibrium. As a consequence of the charge transfer, the AlGaAs layers are depleted and the GaAs well is accumulated. It is interesting to note that the electrons reside in the undoped quantum well – spatially separated from their parent donors in the N-AlGaAs. As a result, electrons in the well experience little ionized impurity scattering and have an especially high mobility. The technique is known as *modulation doping* and is the basis for a transistor known as a modulation doped field-effect transistor (MODFET) (or HEMT) in which the electrons in a modulation-doped quantum well comprise the channel of the field-effect transistor. This device, and several other *heterostructure devices* are described in Weisbuch and Vinter [1.11].

1.3.3 Energy band diagrams for continuous compositional variation

For a conventional, homostructure semiconductor device, the conduction and valence band edges move in response to a macroscopic potential set up by space charges. The slope of the conduction or valence band give the electric field. More generally, both $V(x)$ and the composition of the semiconductor vary with position. Since the composition is nonuniform, the crystal periodicity is broken, and one may question the whole concept of energy bands. If the composition varies slowly, however, we may take the band structure at any point to be the band structure of the corresponding bulk semiconductor with the composition at that point (see [1.12–1.14]).

Because the composition is nonuniform, E_{C0} and E_{V0} (and therefore the electron affinity χ and the bandgap E_G) will also be nonuniform. As a result,

$$E_{C0}(z) = E_0 - \chi_S(z) - qV(z) \tag{1.46a}$$

and

$$E_{V0}(z) = E_0 - \chi_S(z) - qV(z) - E_G(z). \tag{1.46b}$$

An energy band for this case might look like Fig. 1.16 which shows a semiconductor with band-bending which is due to both an electric field and to compositional variations. The slope of the conduction band gives the force of an electron, but it is impossible to deduce the electric field from the energy band diagram.

Consider the force acting on an electron in the conduction band,

$$F_e = \frac{-\mathrm{d}E_{C0}}{\mathrm{d}z} = q\frac{\mathrm{d}V(z)}{\mathrm{d}z} + \frac{\mathrm{d}\chi_S}{\mathrm{d}z} \tag{1.47a}$$

and on a hole in the valence band

$$F_h = \frac{+\mathrm{d}E_{V0}}{\mathrm{d}z} = -q\frac{\mathrm{d}V(z)}{\mathrm{d}z} - \frac{\mathrm{d}(\chi_S + E_G)}{\mathrm{d}z}. \tag{1.47b}$$

The force on electrons is not equal in magnitude and opposite in direction to the force on holes, as we would expect for forces due to electric fields. The electric field is only one component of the force on a carrier. Since we are used to

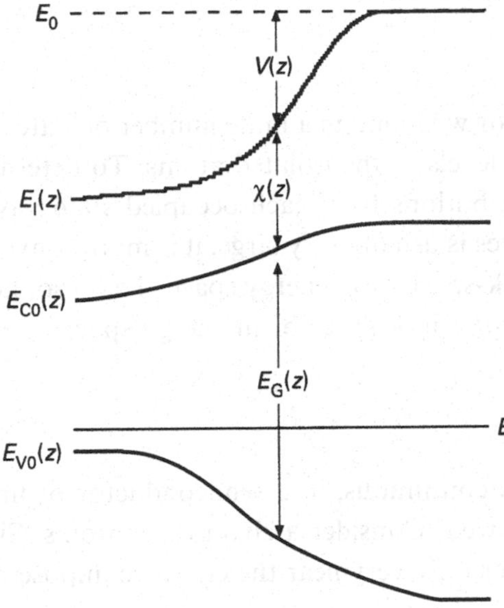

Fig. 1.16 An energy band diagram for a compositionally graded semiconductor.

thinking of electric fields producing forces on carriers, we can define *quasi-electric fields* for electrons by

$$F_e = -q\mathcal{E}(z) - q\mathcal{E}_{QN}(z) \tag{1.48a}$$

and for holes by

$$F_h = +q\mathcal{E}(z) + q\mathcal{E}_{QP}(z). \tag{1.48b}$$

With these definitions we have

$$\mathcal{E}_{QN} = -\frac{1}{q}\frac{d\chi_S}{dz} \tag{1.49a}$$

and

$$\mathcal{E}_{QP} = -\frac{1}{q}\frac{d}{dz}(\chi_S + E_G). \tag{1.49b}$$

Notice that the quasi-electric field for electrons can differ both in magnitude and direction from the quasi-electric field for holes. These quasi-electric fields give the device designer an additional degree of freedom since they can be controlled by the nonuniform composition. Notice that E_{CO} and E_{VO} are not constrained to be parallel in a heterostructure.

1.4 Counting electron states

Any finite volume of semiconductor will contain a finite number of states derived from the finite number of energy levels in the isolated atoms. To determine the macroscopic properties, the contributions from each occupied state have to be added. Because the number of states is usually very large, it is more convenient to integrate over a range of states in **k**-space or in energy space. To do so, however, we need to know the *density of states* in **k**-space or in energy space.

1.4.1 Density of states in k-space

Although we have drawn $E(k)$ as continuous, in a semiconductor of finite size only a finite number of k's is allowed. Consider a chain of N atoms. Since the precise boundary conditions matter only very near the ends, we impose periodic boundary conditions,

$$\psi(z) = \psi(z + Na), \tag{1.50}$$

for mathematical convenience. From eq. (1.32)

$$\psi(z + Na) = \mathrm{e}^{ik(z+Na)}u(z + Na) = \mathrm{e}^{ikNa}\psi(z). \tag{1.51}$$

The boundary condition eq. (1.50) then requires

$$kNa = 2\pi\ell \quad \ell = 1, 2, 3, \ldots, N$$

so only discrete values of k are given by

$$k = \frac{2\pi\ell}{Na} \quad \ell = 1, \ldots, N \tag{1.52}$$

are allowed. Since $Na = L$ the sample's length, each state occupies a space $2\pi/L$ in **k**-space. The number of states between k and $k + \mathrm{d}k$ on the curves of Fig. 1.4 is $L\mathrm{d}k/2\pi$. In three dimensions the number of states per unit volume of k-space generalizes to $L^3/8\pi^3$. We also need to multiply by two to account for the spin of the two electrons which can occupy a state. We conclude, therefore, that

$$\frac{\text{Number of electron states}}{\text{Volume of } k\text{-space}} = N_k = \frac{\Omega}{4\pi^3} \tag{1.53}$$

where $\Omega = L^3$ is the sample's volume.

We shall frequently have to evaluate sums like

$$\sum_{\mathbf{k}} g(\mathbf{k}),$$

where $g(\mathbf{k})$ is some function of **k** and the sum contains all states in the first Brillouin zone as given in eq. (1.52). It is usually convenient to think of the

$E(\mathbf{k})$ curve as continuous and to integrate rather than sum. To do so, we must properly account for the number of states between $\mathbf{k}$ and $\mathbf{k} + \mathbf{dk}$ as given by eq. (1.53). The result,

$$\boxed{\sum_{\mathbf{k}} g(\mathbf{k}) = N_k \int_{\mathbf{k}} g(\mathbf{k})d\mathbf{k}}\,, \tag{1.54}$$

is one that we shall often make use of. For device applications, we'll evaluate sums like eq. (1.54) to determine how the carrier density or current density varies with position within the device.

For carriers in a bulk semiconductor, N_k is given by eq. (1.53), and Ω is a purely conceptual box whose dimensions are large compared to an average electron's wavelength but small on the scale of the device. With the use of heterostructures, carriers can be confined in quantum wells, where they are free to move only in two dimensions, or in quantum wires where they can only move in one dimension. Equation (1.53) generalizes to

$$N_k = 2 \times \frac{L^d}{(2\pi)^d}, \tag{1.55}$$

where L is the sample size, d the dimensionality (1, 2, or 3), and the factor of two accounts for spin degeneracy. The integrals are then carried out in one, two, or three dimensions. We shall see many applications of eqs. (1.54) and (1.55).

1.4.2 Density of states in energy space

Equation (1.53) shows that the density of states in $\mathbf{k}$-space is constant. We will, however, frequently find it convenient or necessary to deal with the density of states in energy space. Figure 1.17 illustrates the relationship between $N(\mathrm{k})$ and $N(\mathrm{E})$. The states are distributed uniformly in $\mathbf{k}$-space, but not in energy space. One way to evaluate $N(E)$ is to construct a histogram. After defining bins of width ΔE, we can scan through all of the allowed $\mathbf{k}$-states, evaluate their energy, and increment the count in the appropriate energy bin. Mathematically, the number of states in a range of ΔE about E is

$$N(E)\Delta E = \sum_{\mathbf{k}} \Delta[E - E(\mathbf{k})], \tag{1.56}$$

where the sum is over all states in the Brillouin zone and $\Delta[E - E(\mathbf{k})] = 1$ if $E(\mathbf{k}) - \Delta E/2 < E < E(\mathbf{k}) + \Delta E/2$ and zero otherwise. Letting ΔE approach zero, we obtain

$$g_C(E) = \frac{N(E)}{\Omega} = \frac{1}{\Omega}\sum_{\mathbf{k}'} \delta[E - E(\mathbf{k}')]. \tag{1.57}$$

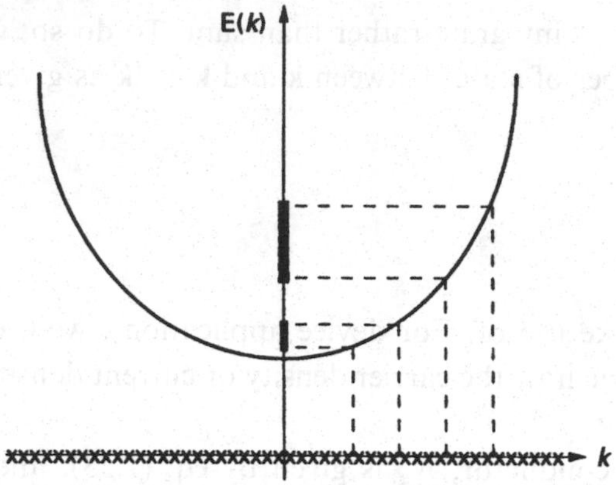

Fig. 1.17 Illustration of the density of states in **k**-space and in energy space. In **k**-space, the density of states is uniform as shown by the x's. A given number of k-states, however, occupies different ranges of energy, as shown by the shaded lines on the energy axis. In this example, $g(E)$ decreases as E increases because we are considering 1D electrons (see eq. (1.63a)).

(In this equation, $\delta(\bullet)$ is actually a Kronecker δ, which becomes a δ-function when the sum is converted to an integral.)

Equation (1.57) can be understood as a count of every state with energy E. We can prove that this is the correct result by evaluating the electron density from

$$n = \frac{1}{\Omega} \sum_{\mathbf{k}} f(\mathbf{k}), \tag{1.58}$$

where the $f(\mathbf{k})$ is the probability that the state at k is occupied. Alternatively, we can evaluate the electron density in energy space from

$$n = \int_{E_{\text{bot}}}^{E_{\text{top}}} g_C(E) f(E) \mathrm{d}E. \tag{1.59}$$

Using eq. (1.57) for the density of states, we find

$$n = \int_{E_{\text{bot}}}^{E_{\text{top}}} \frac{1}{\Omega} \sum_{\mathbf{k}'} \delta[E - E(\mathbf{k}')] f(E) \mathrm{d}E. \tag{1.60}$$

By interchanging the order of integration and summation,

$$n = \frac{1}{\Omega} \sum_{\mathbf{k}'} \int_{E_{\text{bot}}}^{E_{\text{top}}} \delta[E - E(\mathbf{k}')] f(E) \mathrm{d}E, \tag{1.61}$$

we find

$$n = \frac{1}{\Omega} \sum_{\mathbf{k}'} f[E(\mathbf{k}')]. \tag{1.62}$$

Example: Density of states calculation for 3D carriers

A simple example will illustrate the use of eq. (1.57). Recall that the density of states versus energy varies as $E^{1/2}$ for three dimensional electrons. We can obtain this result by evaluating,

$$g_C[E(\mathbf{k})] = \frac{1}{\Omega}\sum_{\mathbf{k}'} \delta[E(\mathbf{k}) - E(\mathbf{k}')].$$

To perform the sum, we convert it to an integral using the prescription given by eq. (1.54) and insert the $E(\mathbf{k})$ relation to find

$$g_C[E(\mathbf{k})] = \frac{1}{4\pi^3}\int_{k'} \delta\left(\frac{\hbar^2 k^2}{2m^*} - \frac{\hbar^2 k'^2}{2m^*}\right) 4\pi k'^2 \mathrm{d}k'.$$

Using $\delta(ax) = \delta(x)/a$, this becomes

$$g_C[E(\mathbf{k})] = \frac{2m^*}{\pi^2\hbar^2}\int_{k'} \delta(k^2 - k'^2) k'^2 \mathrm{d}k'.$$

To integrate a δ-function, we need an expression of the form, $\int \delta(x - x')f(x')\mathrm{d}x' = f(x)$, so letting $x = k^2$, we find

$$g_C(E) = \frac{2m^*}{\pi^2\hbar^2}\int_{k'} \delta(x - x')\frac{k'\mathrm{d}x'}{2} = \frac{m^*}{\pi^2\hbar^2}k = \frac{m^*\sqrt{2m^* E}}{\pi^2\hbar^3},$$

the expected result.

The fact that eq. (1.62) is identical to the correct result, eq. (1.58), verifies that eq. (1.57) is the correct expression for the density of states.

For parabolic energy bands, the density of states goes as $E^{1/2}$, as the example calculation showed. For non-parabolic energy bands, we need to repeat the calculation using eq. (1.40) for $E(k)$. (See homework problem 1.4 for the resulting expression.) As illustrated in Fig. 1.18, nonparabolicity flattens the energy bands, so there are more k-states between E and $E + \mathrm{d}E$ and the density of states increases. In general, it will be necessary to evaluate eq. (1.56) numerically using a table of $E(\mathbf{k})$.

Figure 1.19 compares the density of states in silicon assuming full, numerical energy bands, to that evaluated from the parabolic and nonparabolic expressions. The structure in the full band density of states is a result of the $E(\mathbf{k})$ relation plotted in Fig. 1.9. The sharp drop above $\approx 2\,\mathrm{eV}$ results from the fact that the first conduction band extends to only about $2\,\mathrm{eV}$. The parabolic band assumption is seen to apply only to very low energy carriers near the band minima. The non-parabolic energy band assumption provides a rough approximation to almost $2\,\mathrm{eV}$, but for very energetic carriers, the full, numerical density of states must be used.

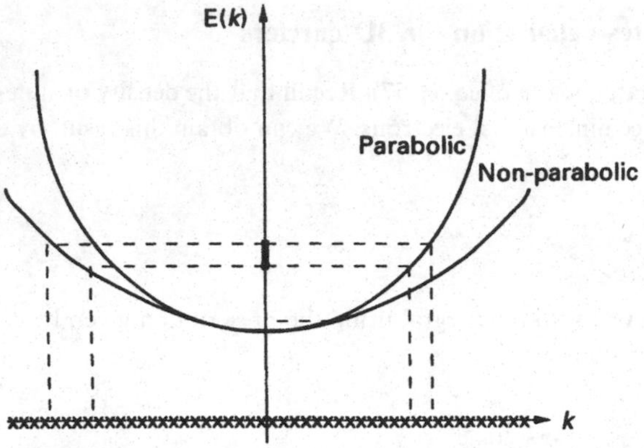

Fig. 1.18 Illustration of how conduction band nonparabolicity flattens the $E(k)$ relation and increases the density of states in energy space. For a given dE, shown in dark on the energy axis, there are more allowed k-states for the nonparabolic energy band.

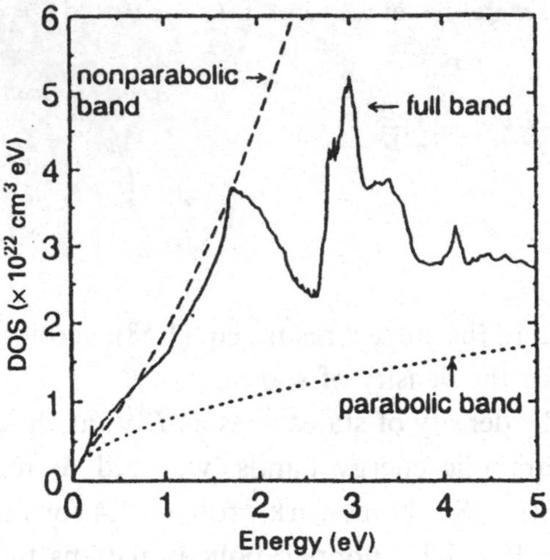

Fig. 1.19 Comparison of the density of states (DOS) for the conduction bands of silicon. The results for parabolic and non-parabolic energy bands are compared to the result using the full numerical description of the energy bands. From Kunikiyo, T. et al., *Journal of Applied Physics,* **75**(1), 297–312, 1994. (Reproduced with permission of American Institute of Physics.)

1.4.3 Density of states for confined carriers

For carriers in a bulk semiconductor, the density-of-states for parabolic energy bands goes as $E^{1/2}$, but for confined carriers, the density of states is altered. Using eqs. (1.54) and (1.55), we find the one-, two-, or three-dimensional density of states as (see homework problem 1.3)

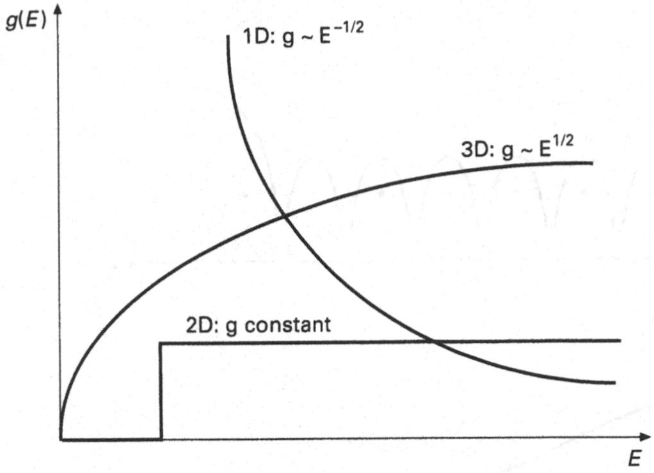

Fig. 1.20 The density of states versus energy for 1, 2, and 3-dimensional carriers with parabolic energy bands. The energy E has been referenced to E_{min}, the conduction band minima for three-dimensional carriers. For two- and one-dimensional carriers, the minimum energy is raised by quantum confinement.

$$g_{1D} = \frac{m^*}{\pi\hbar^2}\frac{2\hbar}{\sqrt{2m^*E}} \tag{1.63a}$$

for one-dimensional carriers, and

$$g_{2D} = \frac{m^*}{\pi\hbar^2} \tag{1.63b}$$

for two-dimensional carriers. These results should be compared to

$$g_{3D} = \frac{m^*}{\pi\hbar^2}\frac{\sqrt{2m^*E}}{\pi\hbar} \tag{1.63c}$$

for three-dimensional carriers. Figure 1.20 sketches the density-of-states versus energy for one-, two-, and three-dimensional carriers. Because carrier confinement is common in modern devices, we shall have to become familiar with evaluating carrier densities, average kinetic energies, scattering rates, etc. in one, two, and three dimensions.

1.5 Electron wave propagation in devices

An electron propagating within a device sees both the crystal potential and those that are applied or built-in to the device. When the applied and built-in fields are absent, only the crystal potential, $U_C(z)$, sketched in Fig. 1.21a is present. The solutions to the wave equation are well known for the crystal potentials of

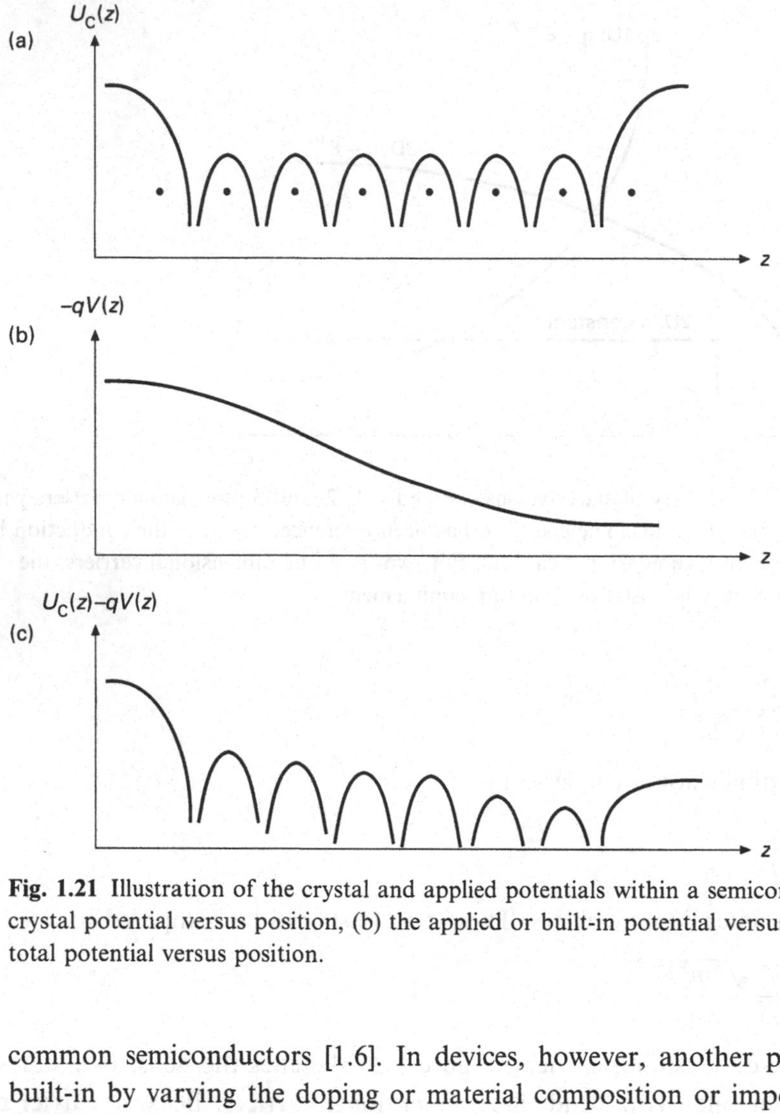

Fig. 1.21 Illustration of the crystal and applied potentials within a semiconductor. (a) The crystal potential versus position, (b) the applied or built-in potential versus position, (c) the total potential versus position.

common semiconductors [1.6]. In devices, however, another potential can be built-in by varying the doping or material composition or imposed by biasing the device. [This is the potential we refer to as $E_{C0}(\mathbf{r})$.] As sketched in Fig. 1.21b, the applied and built-in potentials often vary slowly in comparison to the crystal potential [but with modern epitaxial growth techniques, potentials that vary as rapidly as the crystal potential can also be engineered into the device (recall Fig. 1.15)]. Electrons see both the crystal and applied or built-in potentials as sketched in Fig. 1.21c.

1.5.1 The effective mass equation

When scattering can be neglected, the electron's wave function is found by solving

$$\left[-\frac{\hbar^2}{2m_0}\frac{d^2}{dz^2} + U_C(z) + E_{C0}(z)\right]\Psi(z,t) = i\hbar\frac{\partial \Psi}{\partial t}. \tag{1.64}$$

Since the Bloch function solutions to eq. (1.64) in the absence of $E_{C0}(z)$ are assumed to be known, the question of whether eq. (1.64) can be simplified by using these known solutions arises. The answer is yes, for carriers near the bottom of a simple, spherical, parabolic band, the wave equation can be written as

$$\left[-\frac{\hbar^2}{2m^*}\frac{d^2}{dz^2} + E_{C0}(z)\right]F(z,t) = i\hbar\frac{\partial F}{\partial t} \tag{1.65}$$

where $F(z, t)$ is the *envelope function*, and the actual wave function is

$$\Psi(z,t) \cong F(z,t)u_{k=0}. \tag{1.66}$$

The wave function is the product of a slowly varying envelope function and a rapidly varying Bloch function evaluated at the band minimum. Equation (1.65) is known as the single band *effective mass equation* and represents an enormous simplification of eq. (1.64) because the effects of the complicated crystal potential have been described by a single number, the effective mass. Equation (1.65) applies only when the applied or built-in potential varies slowly on the scale of the crystal potential. This certainly is not the case for the quantum well sketched in Fig. 1.15, but if the well is not too narrow, then an effective mass equation usually provides a good description of the energy levels for electrons within the well. Equation (1.65) also applies only to a parabolic band. The effective mass equation needs to be generalized when the band is non-parabolic or, in the case of valence bands, when several nearby energy bands are coupled. The derivation of the effective mass equation, and its extension to more realistic band structures, are discussed by Datta [1.1].

In devices, the contacts launch electron waves which propagate through the device according to the effective mass equation. A device can be described by specifying its energy band diagram as displayed in Fig. 1.22. The 'contacts' are heavily doped regions where $E_{C0}(z)$ is uniform; they are assumed to be near thermodynamic equilibrium so that each can be described by its own Fermi level. To compute the current through the device, we evaluate the sum

$$J_z = \frac{(-q)}{\Omega}\sum_{\mathbf{k}}\left\{f_L(\mathbf{k})\left(\frac{\hbar k_z}{m^*}\right)T_{LR}(\mathbf{k}) - f_R(\mathbf{k})\left(\frac{\hbar k_z}{m^*}\right)T_{RL}(\mathbf{k})\right\}. \tag{1.67}$$

In this equation, $\hbar k_z/m^*$ is the velocity of electrons as they are injected from the contact with wave vector, $\mathbf{k}$, $f_L(\mathbf{k})$ is the Fermi factor for the left contact, which gives the probability that such an electron is injected from the contact, and $T_{LR}(\mathbf{k})$ is the current transmission coefficient for the electron. In Section 1.1, we

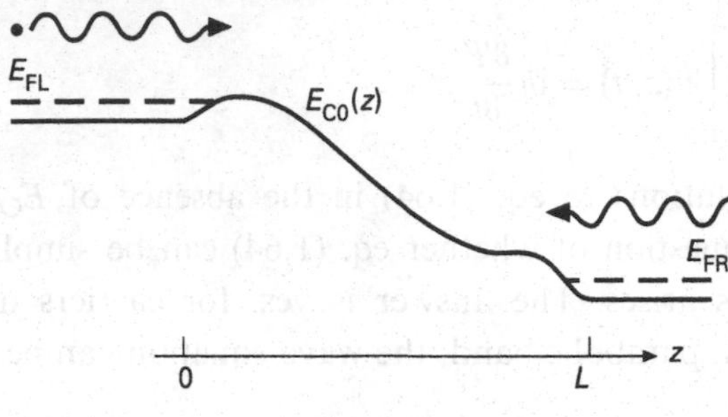

Fig. 1.22 Representation of a device by its energy band profile. Each contact, assumed to be in thermodynamic equilibrium, injects electrons into the device and absorbs electrons incident upon it.

computed $T(\mathbf{k})$ for a simple potential step. For arbitrary potential profiles, the current transmission coefficient is found by numerically solving the effective mass equation. The sum of eq. (1.67) accounts for the current due to all the electrons injected from each of the two contacts, and must, in general, be evaluated numerically.

Device analysis based on solving the effective mass equation is necessary when the potential, $E_{C0}(z)$, varies rapidly so that wave phenomena are important. Vassell et al. [1.8] describes how this technique is applied to devices.

1.5.2 Quantum confinement

Because electrons confined within a small potential well experience a rapidly varying potential, the carriers' wave nature becomes important. For the quantum well illustrated in Fig. 1.23a, electrons are confined in the $\hat{z}$-direction but are free to move in the $\hat{x}-\hat{y}$ plane. For the quantum wire illustrated in Fig. 1.23b, electrons are confined in the $\hat{x}-\hat{y}$ plane but are free to move in the $\hat{z}$-direction. Such structures can be produced with semiconductor heterojunctions, as illustrated in Fig. 1.15 for the quantum well. To describe confined carriers, the three-dimensional effective mass equation,

$$-\frac{\hbar^2}{2m^*}\nabla^2 F(\mathbf{r}) + E_{C0}(\mathbf{r})F(\mathbf{r}) = EF(\mathbf{r}) \tag{1.68}$$

must be solved. Equation (1.68) is readily solved by separating variables. For quantum wells, carriers are free to move in the $\hat{x}-\hat{y}$ plane, so we try solutions of the form

$$F(\mathbf{r}) = \phi(z)\frac{e^{ik_x x}e^{ik_y y}}{\sqrt{L_x L_y}} \tag{1.69}$$

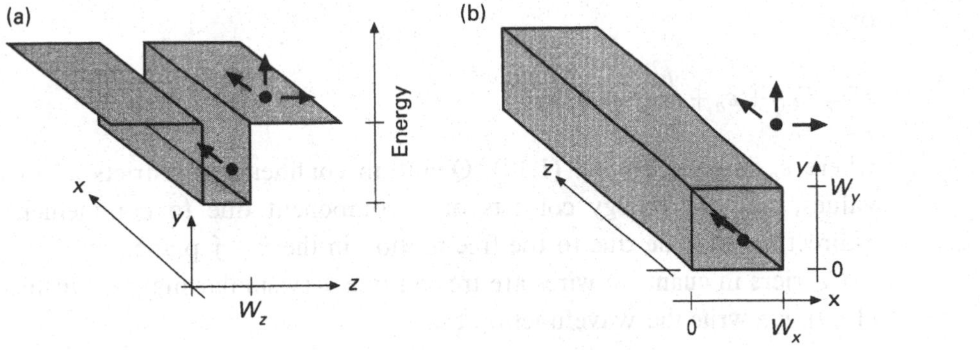

Fig. 1.23 Electrons confined in a quantum well (a) and in a quantum wire (b).

where $A = L_x L_y$ is the cross-sectional area. After substituting eq. (1.69) into eq. (1.68), we find an equation for $\phi(z)$ as

$$\frac{d^2\phi(z)}{dz^2} + k_z^2\phi(z) = 0, \tag{1.70}$$

where

$$k_z^2 = \frac{2m^*}{\hbar^2}[\varepsilon - E_{C0}(z)] \tag{1.71}$$

and

$$\varepsilon = E - \frac{\hbar^2}{2m^*}(k_x^2 + k_y^2). \tag{1.72}$$

Because $\hbar^2(k_x^2 + k_y^2)/2m^*$ is the kinetic energy we associate with motion in the $\hat{x} - \hat{y}$ plane, ε must be the energy associated with confinement in the $\hat{z}$-direction.

Equation (1.70) is identical in form to the simple, one-dimensional wave equation, eq. (1.3). The three-dimensional wave equation consists of a plane wave in the $\hat{x} - \hat{y}$ plane multiplied by a function, $\phi(z)$, which is found by solving an equation that is very similar to the one-dimensional wave equation. If the quantum well is deep, then $\phi(z)$ is given by the infinite well solutions of Section 1.1 as

$$\phi(z) = \sqrt{\frac{2}{W}}\sin k_z z, \tag{1.73}$$

where

$$k_z = n\pi/W \tag{1.74}$$

and

$$E = E_{C0} + \frac{\hbar^2 k_z^2}{2m^*} + \frac{\hbar^2}{2m^*}(k_x^2 + k_y^2)$$

or

$$E = E_{C0} + \varepsilon_n + \frac{\hbar^2}{2m^*}(k_x^2 + k_y^2), \tag{1.75}$$

where ε_n is given by eq. (1.17). Quantum confinement restricts k_z to discrete values, and the energy consists of a component due to confinement in the $\hat{z}$-direction and one due to the free motion in the $\hat{x} - \hat{y}$ plane.

Carriers in quantum wires are treated in a very similar manner. Instead of eq. (1.69), we write the wavefunction as

$$F(\mathbf{r}) = \phi(x, y)\frac{e^{ik_z z}}{\sqrt{L_z}}. \tag{1.76}$$

If the confinement potential is infinite, we find

$$\phi(x, y) = \sqrt{\frac{2}{W_x}}\sin k_x x\sqrt{\frac{2}{W_y}}\sin k_y y \tag{1.77}$$

where

$$k_x = n_x\pi/W_x \tag{1.78a}$$

and

$$k_y = n_y\pi/W_y. \tag{1.78b}$$

For quantum wires, the bottom of subbands are at the energies,

$$\varepsilon_{n_x,n_y} = \frac{\hbar^2\pi^2}{2m^*}\left(\frac{n_x^2}{W_x^2} + \frac{n_y^2}{W_y^2}\right) \tag{1.79}$$

and if the electron is moving in the $\hat{z}$-direction, its total energy is

$$E = E_{C0} + \varepsilon_{n_x,n_y} + \frac{\hbar^2 k_z^2}{2m^*}. \tag{1.80}$$

1.5.2.1 Carrier density relations for confined carriers

For three-dimensional carriers in a bulk semiconductor, there is a simple relation between the equilibrium carrier density and the location of the Fermi level. The corresponding relations for confined carriers are readily derived. Figure 1.24a shows three energy levels, or *subbands*, in a quantum well along with the position of the Fermi level. To compute the density of electrons in the well, we evaluate

$$n = \frac{1}{\Omega}\sum_{k_x,k_y,k_z} f_0[E(k_x, k_y, k_z)] \quad \text{cm}^{-3}, \tag{1.81}$$

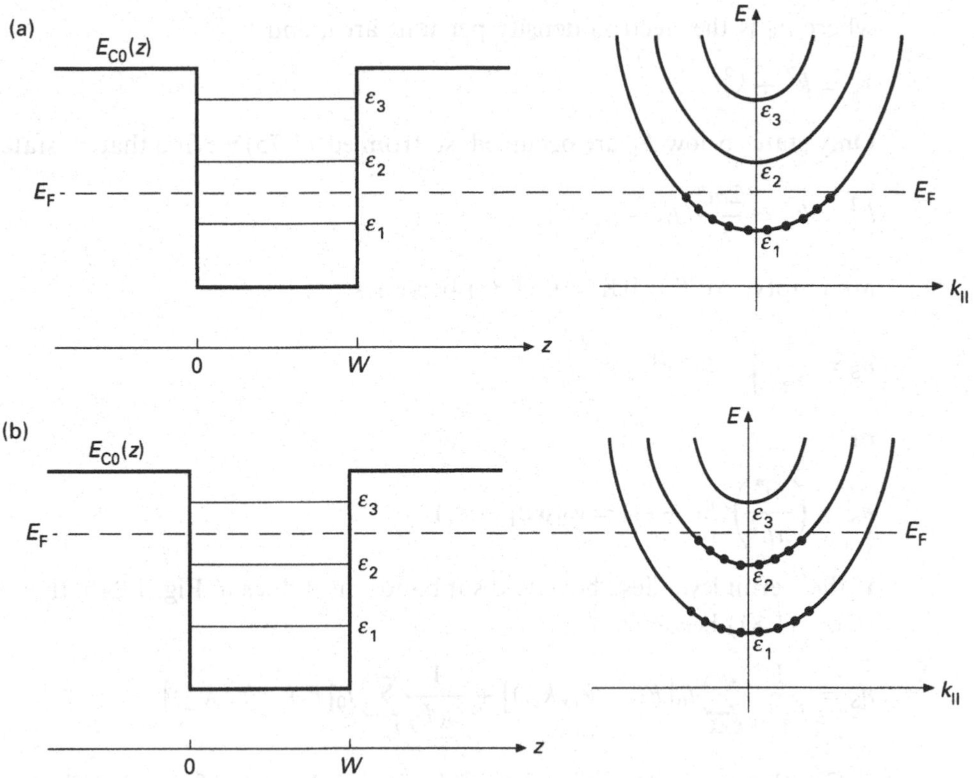

Fig. 1.24 Energy band diagram of a quantum well with the location of three subbands and the Fermi level indicated. (a) The filled circles at the right indicate occupied states and show that only the first subband is occupied. ($T = 0\,\text{K}$ is assumed.) (b) The first two subbands are occupied.

where the sum is over each state in the Brillouin zone, and f_0 gives the probability that the state at energy, E, with crystal momentum, (k_x, k_y, k_z), is occupied.

The sum, eq. (1.81), is easiest to evaluate at $T = 0\,\text{K}$ because then all states below E_F are occupied, and all those above E_F are empty. For the example illustrated in Fig. 1.24a, only states with $k_z = k_{z1}$ are occupied, so eq. (1.81) becomes

$$n = \frac{1}{W}\frac{1}{L_x L_y}\sum_{k_x, k_y} f_0[E(k_x, k_y, k_{z1})]. \tag{1.82}$$

Using eqs. (1.54) and (1.55), we convert the sum over wave vectors in the $\hat{x} - \hat{y}$ plane to an integral as

$$n_S = nW = \frac{1}{2\pi^2}\int_0^\infty f_0(E) 2\pi k_{\|} \mathrm{d}k_{\|} \ \text{cm}^{-2}, \tag{1.83}$$

where n_S is the electron density per unit area, and

$$k_{\parallel}^2 = k_x^2 + k_y^2. \tag{1.84}$$

Only states below E_F are occupied, so from eq. (1.75) we find that all states above

$$k_{\parallel}^2 = k_F^2 = \frac{2m^*}{\hbar^2}(E_F - \varepsilon_1) \tag{1.85}$$

are empty. At $T = 0\,\text{K}$, eq. (1.83) becomes

$$n_S = \frac{1}{2\pi^2}\int_0^{k_F} 2\pi k_{\parallel} \mathrm{d}k_{\parallel} = \frac{k_F^2}{2\pi}$$

or

$$n_S = \left(\frac{m^*}{\pi\hbar^2}\right)(E_F - \varepsilon_1) = g_{2D}(E_F - \varepsilon_1). \tag{1.86}$$

If the Fermi level lies above two subbands, as it does in Fig. 1.24b, then the sum in eq. (1.81) becomes

$$n_S = \frac{1}{L_x L_y}\sum_{k_x,k_y} f_0[E(k_x, k_y, k_{z1})] + \frac{1}{L_x L_y}\sum_{k_x,k_y} f_0[E(k_x, k_y, k_{z2})]. \tag{1.87}$$

For this case, the carrier density is simply the sum of the contributions from the two subbands,

$$n_S = g_{2D}(E_F - \varepsilon_1) + g_{2D}(E_F - \varepsilon_2). \tag{1.88}$$

The corresponding results for finite temperatures are also readily derived. We begin with eq. (1.81) but use the Fermi function,

$$f_0 = \frac{1}{1 + \mathrm{e}^{(\varepsilon_1 + \hbar^2 k_{\parallel}^2/2m^* - E_F)/k_B T}}. \tag{1.89}$$

Alternatively, we can work in energy space and evaluate

$$n_S = \int_{\varepsilon_1}^{\infty} \frac{g_{2D}\mathrm{d}E}{1 + \mathrm{e}^{(E - E_F)/k_B T}}. \tag{1.90}$$

In either case, for one occupied subband, we find

$$n_S = g_{2D} k_B T \ln(1 + \mathrm{e}^{(E_F - \varepsilon_1)/k_B T})\ \text{cm}^{-2}, \tag{1.91}$$

which is analogous to

$$n = N_C \mathcal{F}_{1/2}[(E_F - E_C)/k_B T]\ \text{cm}^{-3} \tag{1.92}$$

for three-dimensional electrons. Here, $\mathcal{F}_{1/2}$ is the Fermi–Dirac integral of order 1/2, and

$$N_C = 2\left(\frac{2\pi m^* k_B T}{h^2}\right)^{3/2} \tag{1.93}$$

is the effective density of states. When additional subbands are occupied, eq. (1.91) is easily generalized by adding the contributions from the additional subbands.

Quantum confinement often occurs in modern devices such as heterostructure field effect transistors in which the channel comprises carriers confined in a quantum well [1.10]. Even in the conventional, silicon MOSFET, carriers are confined within a nearly triangular potential well at the oxide–silicon interface. The wave functions of confined electrons are qualitatively different from the plane waves that describe three-dimensional, bulk electrons. There is a close analogy between these confined electrons and electromagnetic waves in a waveguide. The various subbands are analogous to the waveguide modes; occupied subbands correspond to propagating modes, unoccupied subbands to evanescent modes. This analogy can even be exploited to build electron devices analogous to optical or microwave devices.

1.6 Semiclassical electron dynamics

For conventional devices, the applied or built-in potentials vary slowly in comparison to the crystal potential, so that wave phenomena such as reflections and tunneling are absent and electron motion can be described by classical physics. When the potential is nearly constant, the bottom of the band is simply shifted,

$$E(k, z) = E_{C0}(z) + E(k). \tag{1.94}$$

$E_{C0}(z)$ is interpreted as the bottom of the conduction band, and $E(k)$ represents only the kinetic energy. Since $E_{C0} = \text{constant} - qV(z)$, where $V(z)$ is the electrostatic potential, it varies with position only when applied or built-in fields are present.

Equation (1.94) is illustrated in Fig. 1.25. As the electron wave packet (centered at k_0) moves without scattering, its total energy remains constant. From eq. (1.94) we have

$$\begin{aligned}\frac{\mathrm{d}E(k_0, z)}{\mathrm{d}t} &= \frac{\partial E}{\partial k_0}\cdot\frac{\mathrm{d}k_0}{\mathrm{d}t} + \frac{\partial E}{\partial z}\cdot\frac{\mathrm{d}z}{\mathrm{d}t}\\ &= v_g\cdot\hbar\frac{\mathrm{d}k_0}{\mathrm{d}t} + \frac{\partial E_{C0}(z)}{\partial z}\cdot v_g.\end{aligned}$$

Because energy must be conserved, $\mathrm{d}E/\mathrm{d}t = 0$ and we conclude that

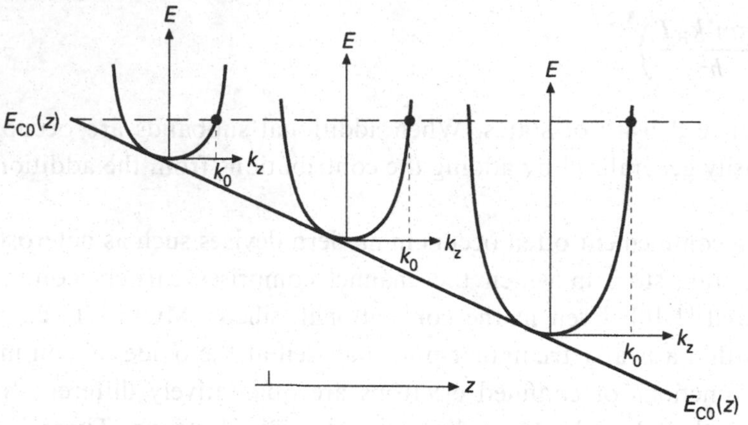

Fig. 1.25 Motion of an electron wave packet centered at $k_z = k_0$ across a region of slowly varying potential. After Datta [1.1].

$$\frac{\mathrm{d}(\hbar \mathbf{k}_0)}{\mathrm{d}t} = -\nabla E_{C0}(z) = \mathbf{F}_e \tag{1.95a}$$

where $\mathbf{k}_0$ is the wave vector at the center of the wave packet. Because eq. (1.95a) is so similar to the equation of motion for classical particles, with $\hbar\mathbf{k}_0$ playing the role of momentum, $\hbar\mathbf{k}_0$ is termed the *crystal momentum.*

For heterostructures, the effective mass may vary with position, so

$$E_C(z, k) = E_{C0} + \frac{\hbar^2 k^2}{2m^*(z)}.$$

For heterostructures, the equation of motion generalizes to (see homework problem 1.13)

$$\frac{\mathrm{d}(\hbar \mathbf{k}_0)}{\mathrm{d}t} = -\nabla E_{C0}(z) - \nabla\left(\frac{\hbar^2 k^2}{2m^*(z)}\right) \tag{1.95b}$$

In this case, only the first term represents a real, physical force on carriers.

The analogy of $\hbar\mathbf{k}_0$ to momentum is also apparent from the velocity of a Bloch electron as given by eq. (1.14). For spherical, parabolic bands described by eq. (1.38), we obtain

$$\upsilon_g = \frac{\hbar \mathbf{k}_0}{m^*}, \tag{1.96}$$

which looks like momentum divided by mass. But for nonparabolic bands described by eq. (1.40), the group velocity is

$$v_g = \frac{\hbar \mathbf{k}_0}{m^*[1 + 2\alpha E(k)]}. \tag{1.97}$$

In the semiclassical view of electron transport, the electron wave packet is treated as a particle; the uncertainty in the momentum is assumed to be small so that the electron's energy is sharply defined, the uncertainty in the electron's position is assumed to be small in comparison to the distance over which applied and built-in potentials vary significantly. The motion of the center of this wave packet is described by eq. (1.95), which looks like the classical relation between force and momentum. The velocity of the electron, eq. (1.14), corresponds to the velocity of a classical particle only for spherical, parabolic bands. This semiclassical treatment of carrier dynamics is the basis for each of the following chapters, but collisions involve rapidly varying potentials and must be treated quantum mechanically.

1.7 Scattering of electrons by the random potential, $U_S(\mathbf{r}, t)$

Bloch waves move through the lattice unimpeded by the crystal potential. Occasionally, however, the electron encounters a perturbation caused when a lattice vibration moves an atom or by impurities or defects which may be present. When an electron encounters such a perturbation it scatters – scattering 'knocks' an electron wave packet centered at k_0 to k_0'. Frequent scattering tends to 'wash out' the interference effects due to the carrier's wave nature. Scattering plays a dominant role in transport, and it is important that we know $S(k_0, k_0')$, the *transition rate* from k_0 to k_0'. We now present a brief derivation of the expression for $S(k_0, k_0')$ in terms of $U_S(z, t)$, the perturbing potential. For a proper derivation, consult a quantum mechanics textbook such as Datta [1.1]. The intent here is to indicate how the result is derived and some of its limitations. We will make extensive use of the result, known as *Fermi's Golden Rule*, to calculate scattering rates for electrons in semiconductors, so the reader should develop a familiarity with its use.

1.7.1 Fermi's Golden Rule

The wave equation, [eq. (1.1)] is written as

$$[H_0 + U_S(z, t)]\Psi(z, t) = i\hbar \frac{\partial \Psi(z, t)}{\partial t}, \tag{1.98}$$

where H_0 is the Hamiltonian operator for the unperturbed problem (the problem without the scattering potential). We assume that the unperturbed problem:

$$H_0\psi_k = E(k)\psi_k \tag{1.99}$$

$$\Psi_k^0(z, t) = \psi_k(z)e^{-iE(k)t/\hbar} \tag{1.100}$$

has been solved. These solutions form a complete, orthonormal set, so we can express the solution to the perturbed problem as linear combinations of them:

$$\Psi(z, t) = \sum_k c_k(t)\Psi_k^0(z, t) = \sum_k c_k(t)\psi_k(z)e^{-iE(k)t/\hbar}. \tag{1.101}$$

Now consider the situation sketched in Fig. 1.26 – an electron wave packet centered at $k = k_0$ enters, interacts with $U_S(z, t)$, and emerges centered at k_0'. At $t = 0$ we have

$$\begin{aligned} c_{k_0}(t = 0) &= 1 \\ c_k(t = 0) &= 0 \qquad (k \neq k_0). \end{aligned} \tag{1.102}$$

After the scattering event, the probability of finding the electron with wave vector, k_0', is

$$P(k = k_0') = \lim_{t\to\infty} \left|c_{k_0'}(t)\right|^2 \tag{1.103}$$

so the scattering rate from k_0 to k_0' is

$$S(k_0, k_0') = \lim_{t\to\infty} \frac{|c_{k_0'}(t)|^2}{t}. \tag{1.104}$$

(To allow $t \to \infty$ in these expressions, without another collision occurring, collisions must be infrequent.)

To find c_k, we insert eq. (1.101) in eq. (1.98) and obtain

$$U_S(z, t)\sum_k c_k(t)\psi_k e^{-iE(k)t/\hbar} = i\hbar \sum_k \frac{\partial c_k}{\partial t}\psi_k e^{-iE(k)t/\hbar}. \tag{1.105}$$

Next, we multiply both sides by $\psi_{k_0'}^* e^{iE(k_0')t/\hbar}$, integrate over position, and make use of the orthogonality of eigenfunctions, to find

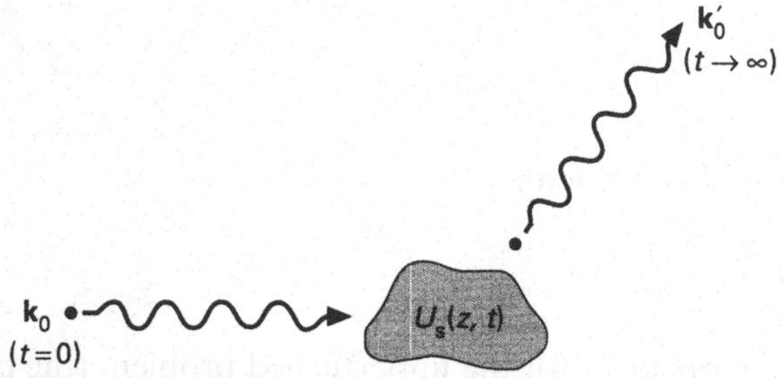

Fig. 1.26 Scattering of a wave packet centered at $k = k_0$ to one centered at $k = k_0'$.

$$i\hbar \frac{\partial c_{k_0'}}{\partial t} = \sum_k H_{k_0' k} c_k(t) e^{i(E(k_0')-E(k))t/\hbar}, \tag{1.106}$$

where

$$H_{k_0', k_0}(t) \equiv \int_{-L/2}^{+L/2} \psi_{k_0'}(z) U_S(z, t) \psi_{k_0}(z) dz \tag{1.107}$$

is the *matrix element* of the scattering potential between states k_0' and k. We have normalized the wavefunctions over a length L, which becomes a volume, Ω, in three dimensions.

Since we assume that the scattering is weak, $c_{k_0} \simeq 1$ for all time, and the other c_k's are always small. With this approximation (the so-called *Born approximation*) the sum in eq. (1.106) can be approximated by one term as

$$i\hbar \frac{\partial c_{k_0'}}{\partial t} = H_{k_0' k_0}(1) e^{i[E(k_0')-E(k_0)]t/\hbar},$$

which can be integrated to find

$$c_{k_0'}(t) = \frac{1}{i\hbar} \int_0^t H_{k_0' k_0} e^{i[E(k_0')-E(k_0)]t/\hbar} dt + c_{k_0'}(0). \tag{1.108}$$

Because the final state k_0' was empty at $t = 0$, $c_{k_0'}(0) = 0$.

Let's specify the time-dependent matrix element as

$$H_{k_0', k_0}(t) = H^{a,e}_{k_0' k_0} e^{\mp i\omega t}. \tag{1.109}$$

(The significance of the a and e superscripts which apply the minus and plus signs respectively will be explained shortly.) With eq. (1.109) for the matrix element, eq. (1.108) can be integrated as

$$c_{k_0'} = \frac{1}{i\hbar} H^{a,e}_{k_0' k_0} \frac{e^{i(E(k_0')-E(k_0)\mp\hbar\omega)t/\hbar} - 1}{i[E(k_0') - E(k_0) \mp \hbar\omega]/\hbar}. \tag{1.110}$$

When we define

$$\Lambda = \left[E(k_0') - E(k_0) \mp \hbar\omega\right]/\hbar, \tag{1.111}$$

then eq. (1.110) can be written as

$$c_{k_0'}(t) = \frac{1}{i\hbar} H^{a,e}_{k_0' k_0} e^{i\Lambda t/2} \frac{\sin(\Lambda t/2)}{\Lambda t/2} t. \tag{1.112}$$

Now, according to eq. (1.104), we find the transition rate as

$$S(k_0, k_0') = \lim_{t\to\infty} \frac{\left|H^{a,e}_{k_0'k_0}\right|^2}{t\hbar^2} \left[\frac{\sin(\Lambda t/2)}{\Lambda t/2}\right]^2 t^2. \tag{1.113}$$

For large t, the function in brackets is very sharply peaked near the origin and looks like a δ-function. The strength of the δ-function is determined from the area under the curve. Recall that

$$\int_{-\infty}^{\infty} \frac{\sin^2 x}{x^2} \mathrm{d}x = \pi,$$

so $\sin^2 x/x^2$ can be replaced by $\delta(x)$. Using $x = \Lambda t/2$, we find the replacement

$$\lim_{t\to\infty} \frac{\sin^2(\Lambda t/2)}{(\Lambda t/2)^2} = \frac{2\pi}{t}\delta(\Lambda) \tag{1.114}$$

which can be inserted in eq. (1.113) to find

$$\boxed{\begin{aligned} S(k_0, k_0') = {} & \frac{2\pi}{\hbar}\left|H^{a}_{k_0'k_0}\right|^2 \delta(E(k_0') - E(k_0) - \hbar\omega) \\ & + \frac{2\pi}{\hbar}\left|H^{e}_{k_0'k_0}\right|^2 \delta(E(k_0') - E(k_0) + \hbar\omega) \end{aligned}} \tag{1.115}$$

The δ-function in eq. (1.115) simply expresses conservation of energy and applies when scattering is weak, so that time can approach infinity in eq. (1.113). For frequent scattering, there is an uncertainty in the final energy, given by eq. (1.12), which is known as *collisional broadening*. The first term in eq. (1.115) contributes when $E(k_0') = E(k_0) + \hbar\omega$; an energy of $\hbar\omega$ has been absorbed. The second contributes when $E(k_0') = E(k_0) - \hbar\omega$; an energy of $\hbar\omega$ has been emitted.

Equation (1.115) is the basic result of scattering theory that we will apply to carriers in semiconductors. The result is known as *Fermi's Golden Rule*. To apply the Golden Rule, the scattering potential must be identified so that the matrix element can be evaluated. For electrons in semiconductors, the wave functions for the unperturbed problem are Bloch waves. When the matrix element, eq. (1.107) is evaluated for Bloch waves, one finds [1.5]:

$$H_{k'k} = I(k, k')U_S(k - k') \tag{1.116}$$

where

$$I(k, k') \equiv \int_{\text{cell}} u^*_{k'}(z)u_k(z)\mathrm{d}z \tag{1.117}$$

is called the *overlap integral* (the integral is over a unit cell), and

$$U_S(k-k') = \int_{-L/2}^{+L/2} \frac{e^{-ik'z}}{\sqrt{L}} U_S(z) \frac{e^{+ikz}}{\sqrt{L}} dz. \tag{1.118}$$

For a parabolic band, $I(k, k') \simeq 1$ [1.5] and

$$\boxed{H_{k'k} \simeq \frac{1}{L}\int_{-L/2}^{+L/2} e^{-ik'z} U_S(z) e^{ikz} dz}\ , \tag{1.119}$$

which is just what we would have obtained from eq. (1.107) using plane waves rather than Bloch waves. When we evaluate scattering rates in Chapter 2, we'll keep the algebra to a minimum by assuming that the energy bands are parabolic and employ eq. (1.119), but for quantitative work, overlap integrals should be considered.

1.7.2 Examples

To illustrate how the Golden Rule is applied to scattering problems, we consider two simple, but illustrative, examples. First, we consider scattering from a δ-function potential, which might approximate a short range scattering potential. Second, we consider a periodic perturbing potential, which might represent, for example, a lattice vibration.

Example: Scattering from a δ-function potential

Consider a perturbing potential of the form

$$U_S(z) = A_0 \delta(z). \tag{1.120}$$

From eq. (1.119) we find

$$H_{k'k} = \frac{A_0}{L}. \tag{1.121}$$

and from eq. (1.115), the transition rate becomes

$$S(k, k') = \frac{2\pi}{\hbar}\frac{A_0^2}{L^2}\delta[E(k') - E(k)]. \tag{1.122}$$

This time-independent scattering potential elastically scatters electrons with a transition rate that is proportional to the squared magnitude of the scattering potential. The δ-function potential is an approximate description of ionized impurity scattering when it is strongly screened by free carriers.

Example: Scattering from a periodic potential

As a second example, consider the scattering potential,

$$U_S(z,t) = A_\beta^{a,e} e^{\pm i(\beta z - \omega t)}. \tag{1.123}$$

For plane waves confined to a normalization length, $-L/2 \le z \le L/2$,

$$\psi(z) = \frac{1}{\sqrt{L}} e^{ikz}, \tag{1.124}$$

and the matrix element becomes

$$H_{k'k} = \int_{-L/2}^{L/2} \frac{A_\beta^{ae}}{L} e^{i(k-k'\pm\beta)z} dz. \tag{1.125}$$

If the normalization length is long, the oscillating, exponential factor ensures that no net contribution to the integral will result unless $k' = k \pm \beta$, so

$$H_{k'k} = A_\beta^{a,e} \delta_{k',k\pm\beta}, \tag{1.126}$$

where the Kronecker delta, δ_{ij}, is defined to be one if $i = j$ and zero for $i \ne j$. For this scattering potential, the transition rate is

$$S(k,k') = \frac{2\pi}{\hbar} \left| A_\beta^{a,e} \right|^2 \delta[E(k') - E(k) \mp \hbar\omega] \delta_{k',k\pm\beta}, \tag{1.127}$$

which is, again, proportional to the squared magnitude of the perturbing potential. The δ-function in eq. (1.127) states that

$$E(k') = E(k) \pm \hbar\omega, \tag{1.128}$$

which is a statement of conservation of energy. For time-dependent scattering potentials like eq. (1.123), carriers either absorb or emit energy. To satisfy the Kronecker δ,

$$\hbar k' = \hbar k \pm \hbar\beta, \tag{1.129}$$

which we interpret as a statement of conservation of momentum. The scattered momentum has either absorbed or emitted momentum. (The momentum-conserving Kronecker δ was absent in eq. (1.122) because the δ-function scattering potential contained Fourier components with all momenta.) This scattering potential is a good description of the perturbing potential due to lattice vibrations (phonons).

1.8 Lattice vibrations (phonons)

Because much of the scattering in semiconductors is due to lattice vibrations, it is important that we understand their basic properties. If an atom is displaced from its equilibrium position, the bonding forces tend to push it back, so it oscillates about its equilibrium site. Since lattice waves propagate in a periodic medium, they have properties much like those of Bloch waves. Figure 1.27a shows a typical dispersion relation, ω versus β, observed for elastic waves in cubic semiconductors like silicon and gallium arsenide. (We label the wave vector by $\boldsymbol{\beta}$

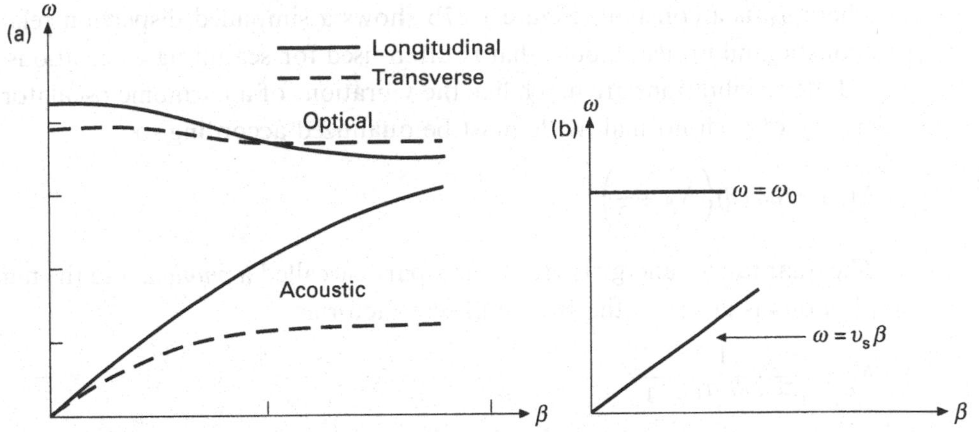

Fig. 1.27 (a) Typical dispersion relation for elastic waves propagating along a high-symmetry direction in cubic semiconductors. (b) Simplified dispersion relation useful when only longitudinal lattice vibrations near the center of the Brillouin zone are considered. After Datta [1.1]. (Reproduced with permission from Addison-Wesley)

rather than **k** to distinguish elastic waves from electron waves.) Six types of elastic wave exist – three *acoustic* modes, and three *optical* modes. Acoustic modes are like sound waves in that adjacent atoms are displaced in the same direction – only the magnitude of the displacement varies from atom to atom. Of the three acoustic modes, one is longitudinal (LA) and two are transverse (TA). For longitudinal waves, atoms are displaced in the direction of propagation; the two transverse modes, in which atoms are displaced in a transverse direction, are degenerate in cubic silicon and GaAs.

In Chapter 2 we shall establish that the scattering of electrons within a valley is due to lattice vibrations with wave vectors very near the origin of the Brillouin zone. For small β, the dispersion relation for acoustic modes can be approximated by

$$\omega(\beta) = \upsilon_s \beta, \tag{1.130}$$

where υ_s is the sound velocity.

Optical modes differ from acoustic modes in that adjacent atoms are displaced out of phase. (The term arises because such vibrations can interact strongly with light.) As shown in Fig. 1.27a, the dispersion relation for optical modes displays relatively little variation with wave vector. When electrons are scattered by optical phonons and remain within the same valley, only small wave vectors are involved and the dispersion relation can be approximated as

$$\omega(\beta) = \omega_0, \tag{1.131}$$

where ω_0 is a constant. Figure 1.27b shows a simplified dispersion relation for acoustic and optical modes that is often used for scattering calculations.

Lattice vibrations are much like the vibrations of a harmonic oscillator, so the energy of each normal mode must be quantized according to

$$E(\beta) = \hbar\omega(\beta)\left(N_\beta + \frac{1}{2}\right). \tag{1.132}$$

The quantum of energy is viewed as a particle called a *phonon*, and the number of phonons is given by the Bose–Einstein factor as

$$N_\beta = \frac{1}{e^{\hbar\omega(\beta)/k_B T_L} - 1}. \tag{1.133}$$

For

$\hbar\omega(\beta) \ll k_B T_L$

eq. (1.133) reduces to

$$N_\beta \simeq \frac{k_B T_L}{\hbar\omega(\beta)}, \tag{1.134}$$

which is known as *equipartition* and is usually valid for acoustic phonons – except at very low temperatures. Equation (1.134) is easy to understand; $k_B T_L$ is the thermal energy and $\hbar\omega_\beta$ is the energy of the phonon at β, so eq. (1.134) just tells us how many phonons are needed to account for the thermal energy. In Chapter 2, we shall describe how phonons, both acoustic and optical, scatter carriers.

1.9 Summary

A simple approach for treating carrier motion within conventional devices has been outlined. This semiclassical approach treats carriers as particles whose dynamics, between collisions, are governed by eq. (1.14) and eq. (1.95), which are analogous to Newton's Laws. Carrier scattering, however, is treated by quantum mechanics using Fermi's Golden Rule. The semiclassical approach is applicable when the applied and built-in potentials vary slowly on the scale of an electron's wavelength. Room-temperature, thermal average electrons in silicon have a wavelength of about 120 Å and about 240 Å in GaAs, so the semiclassical approach may be questioned in ultra-small devices. Many devices contain quantum wells, and the carriers within such wells clearly display their wave nature. Quantum confinement alters the wavefunctions of electrons confined in potential wells, but transport within the confined region can often be described semiclassically. Our focus in this text is on the semiclassical transport of three dimensional carriers, but we shall also from time to time consider the transport of carriers confined in quantum wells and wires. An introduction to quantum transport, in which the electron's wave nature is essential, is contained in Chapter 9.

References and further reading

The quantum mechanical foundations which underlie device analysis are thoroughly discussed by Datta in a volume of the Modular Series on Solid State Devices.

1.1 Datta, S. *Quantum Phenomena*, Vol. VIII of the Modular Series on Solid State Devices. Addison-Wesley, Reading, Mass., 1989.

An introductory quantum mechanics text discusses the principles of wave mechanics; my own favorite is

1.2 Bohm, D. *Quantum Theory*. Prentice-Hall, Englewood Cliffs, NJ, 1951.

The basics of energy band theory and phonons are treated in introductory solid-state physics texts such as

1.3 Ashcroft, N. W. and Mermin, N. D. *Solid-State Physics*. Saunders College, Philadelphia, PA, 1976.

1.4 Harrison, W. *Solid State Theory*. Dover, New York, 1980.

Nag derives eq. (1.116), the matrix element for Bloch waves, and discusses the evaluation of overlap integrals.

1.5 Nag, B. *Electron Transport in Compound Semiconductors*. Springer-Verlag, New York, 1980.

For descriptions of the energy band structure of important semiconductors, and of the $\mathbf{k} \cdot \mathbf{p}$ method for approximating $E(k)$ near a band minima, consult

1.6 Singh, J. *Physics of Semiconductors and their Heterostructures*. McGraw-Hill, New York, 1993.

The complete, numerically evaluated bandstructures of several common semiconductors are presented in

1.7 Fischetti, M. V. Monte Carlo simulation of transport in technologically significant semiconductors of the diamond and zinc-blende structures – Part I: homogeneous transport. *IEEE Transactions in Electron Devices*, **38**, 634–49, 1991.

Applications of the effective mass equation to 'quantum devices' are described in

1.8 Vassell, M., Lee, J. and Lockwood, H. Multibarrier tunneling in $Ga_{1-x}Al_x/GaAs$ heterostructures, *Journal of Applied Physics*, **54**, 5206–13, 1983.

Adachi describes the properties of the alloy, $Al_xGa_{1-x}As$

1.9 Adachi, S. GaAs, AlAs, and $Al_xGa_{1-x}As$: material parameters for use in research and device applications. *Journal of Applied Physics*, **58**, R1–R29, 1985.

For information on materials parameters for other semiconductors, consult

1.10 Landolt-Börnstein, *Numerical Data and Functional Relationships in Science and Technology*. New Series, Vol 17a, Physics of Group IV Elements and III-V Compounds, ed. by O. Madelung. Springer, Berlin, 1982.

Many modern devices make use of heterostructures. For an introduction to these structures and devices, good starting points are

1.11 Weisbuch, C. and Vinter, B. *Quantum Semiconductor Structures*. Academic Press, Inc., Boston, 1991.

1.12 Marshak, A. H. and van Vliet, K. M. Electrical currents in solids with position-dependent band structure. *Solid-State Electronics*, **21**, 417–28, 1978.

1.13 Marshak, A. H. and van Vliet, K. M. Carrier densities and emitter efficiency in degenerate materials with position-dependent band structure. *Solid State Electronics*, **21**, 429–34, 1978.

1.14 Van Vliet, C. M., Nainaparampil, J. J. and Marshak, A. H. Ehrenfest derivation of the mean forces acting in materials with non-uniform band structure: a canonical approach. *Solid-State Electronics*, **38**, 217–23, 1995.

Problems

1.1 Assume the scattering potential shown below and assume that electrons are free to move in the z direction only.

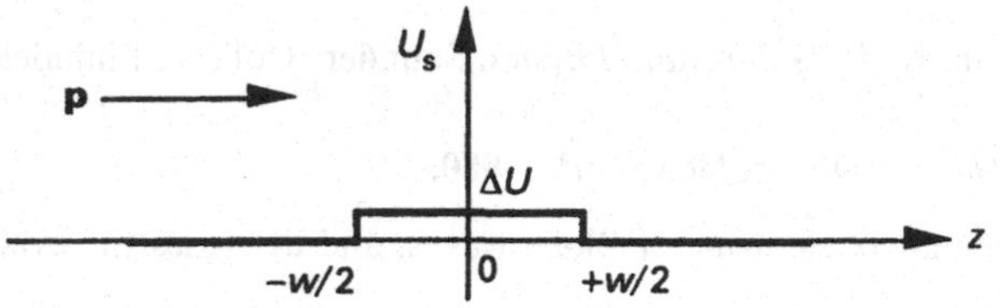

(a) Work out an expression for the transition rate, $S(\mathbf{p}, \mathbf{p}')$ for one-dimensional electrons. Be sure to normalize the wavefunction over a length L.

(b) An incident electron with crystal momentum $\mathbf{p}$ can only make a transition to one different state, $\mathbf{p}'$. What is that state?

(c) Explain what would happen if the sign of ΔU were to change.

1.2 Consider the effect of a perturbing potential that is constant in both space and time,

$$U_S(z, t) = U_0$$

and answer the following questions.

(a) Obtain an expression for the transition rate, $S(k, k')$.

(b) Interpret your answer to part (a). What does your result imply about the motion of electrons through regions of uniform potential?

1.3 The densities of states in one, two, and three dimensions can each be expressed as the sum in **k**-space as given by eq. (1.57).

(a) Evaluate the two-dimensional density of states, and show that the result is eq. (1.63b).

(b) Evaluate the density of states for one-dimensional electrons and show that the result is eq. (1.63a).

1.4 When evaluating the density of states in energy space, we have assumed parabolic energy bands, but energy bands are typically nonparabolic.

(a) For three-dimensional carriers with parabolic energy bands, the density of states goes as $E^{1/2}$, as given by eq. (1.63c). Work out the corresponding result for 3D carriers with nonparabolic energy bands as given by eq. (1.40). Show that the density of states for nonparabolic energy bands is the parabolic band result multiplied by $\sqrt{1+\alpha E}(1+2\alpha E)$.

(b) For two-dimensional carriers with parabolic energy bands, the density of states is constant, as given by eq. (1.63b). Work out the corresponding result for 2D carriers with nonparabolic energy bands as given by eq. (1.40). Show that the density of states for nonparabolic energy bands is the parabolic band result multiplied by $(1+2\alpha E)$.

1.5 The following problem concerns electrons in a quantum well of width, W, with one subband at $E=\varepsilon_1$. Assume equilibrium conditions and that the Fermi level is located above ε_1. Answer the following questions assuming parabolic energy bands.

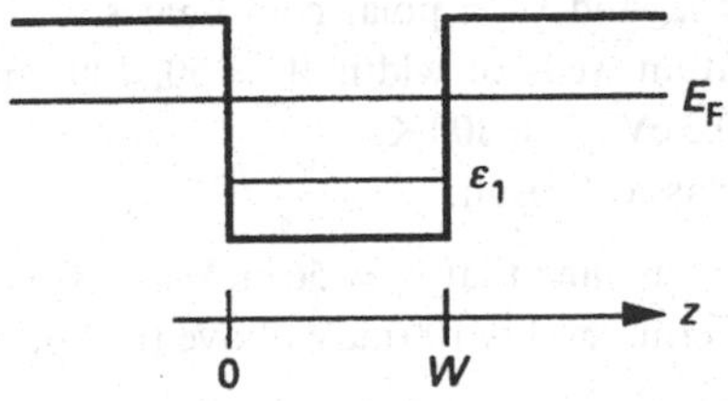

(a) Write an expression, involving sums over momentum space, which gives the average kinetic energy (due to its motion in the x–y plane) per electron.

(b) Convert the sum to an integral over momentum or **k**-space.

(c) Write an expression, involving an integral over energy space, which gives the average kinetic energy per electron.

(d) Assume $T=0\mathrm{K}$ and evaluate the average kinetic energy per electron. You may work in either energy or momentum space.

1.6 Obtain an expression for the concentration per unit area of electrons in a quantum well as a function of the Fermi level position, $n_S(E_F)$. You should assume that the temperature is finite, and do not assume that the semiconductor is nondegenerate.

(a) Find $n_S(E_F)$ when one subband is occupied. For this part, you should work in energy-space using eq. (1.90).

(b) Repeat part (a) but do the work in **k**-space using eq. (1.83). For both parts (a) and (b), show that the answer is eq. (1.91).

(c) What is the result when two subbands are occupied?

(d) Explain how the results depend on the shape of the quantum well (i.e., does eq. (1.91) hold for parabolic or triangular quantum wells? What changes, if anything?).

1.7 For an infinite depth GaAs quantum well of width $W=200\,\text{Å}$ at $T=0\,\mathrm{K}$,

(a) How many subbands are occupied if $n_S=5\times10^{11}\,\mathrm{cm}^{-2}$?

(b) How many if $n_S=5\times10^{12}\,\mathrm{cm}^{-2}$?

1.8 For the quantum well problems, we have been asking about the total number of electrons per unit area within the well – not how the electrons are distributed within the well. The carrier density as a function of position within the well is found from

$$n(z)=\sum_{k_x,k_y,k_z} f(E)\phi^*(z)\phi(z)$$

where

$$\phi(z) = \sqrt{\frac{2}{W}} \sin \frac{n\pi z}{W}.$$

(a) Show that the carrier concentration within the well is

$$n(z) = \frac{2m^* k_B T}{\pi \hbar^2 W} \sum_{n=1}^{\infty} \sin^2 \frac{n\pi z}{W} \ln(1 + e^{-\eta_n})$$

where

$$\eta_n = \frac{\left(\frac{\hbar^2 n^2 \pi^2}{2m^* W^2} - E_F\right)}{k_B T}.$$

Hint: It's easiest to perform the integral over k_x and k_y in polar coordinates.

(b) Compute and plot $n(z)$ versus z for quantum wells of width $W = 50$, 100, and 500 Å. Assume that $m^* = 0.067 m_0$, $E_{C0} - E_F = 0.3$ eV, $T = 300$ K.

(c) Compare the results of part (b) with the classical result.

1.9 Consider a 100 Å wide GaAs quantum well, and assume that $\varepsilon_1 = 56$ meV and $\varepsilon_2 = 225$ meV above the bottom of the quantum well. If the Fermi level is 100 meV above the bottom of the well, then

(a) What is n_S at $T = 300$ K?

(b) If the electrons were considered to be three-dimensional, what n_S would be computed?

1.10 The quantum well that confines carriers at the AlGaAs/GaAs interface in a MODFET is approximately triangular. The first two energy levels are given by

$$E_1 = \gamma_1 n_S^{2/3}$$

and

$$E_2 = \gamma_2 n_S^{2/3}$$

where

$$\gamma_1 = 2.5 \times 10^{-12}\ \text{eV} - m^{4/3}$$

and

$$\gamma_2 = 3.2 \times 10^{-12}\ \text{eV} - m^{4/3}.$$

(See M. Shur, *GaAs Devices and Circuits*, p. 519, Plenum Press, New York, 1987, for a derivation of this result.) If we require that only one subband be occupied, what is the maximum number of electrons per square centimeter that can be accommodated in the well at $T = 0$ K?

1.11 Verify the results, eqs. (1.77)–(1.80) for electrons in a quantum wire.

1.12 Assume $E_C(z)$ is as follows:

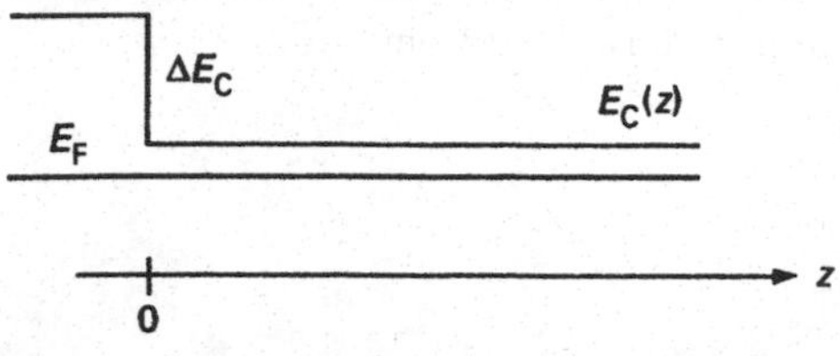

(a) Compute $\phi^*_{P_z}(z)\phi_{P_z}(z)$ assuming $\Delta E_C = \infty$.
(b) Compute $n(z)$ for $x > 0$, approximate $f_R(k)$ using Boltzmann statistics.
(c) Sketch $n(z)$ and compare it with the classical value. Show that differences occur when z is with Λ of $z = 0$.

$$\left(\Lambda = \sqrt{\frac{2\hbar^2}{m^* k_B T}}\right)$$

1.13 Prove that the equation of motion for a semiconductor heterostructure in which m^* varies slowly with position is given by eq. (1.95b).

1.14 In the so-called tight binding method for computing bandstructures, the $E(k)$ relation for a one-dimensional lattice is given by

$$E(k) = A - B\cos(ka),$$

where A and B are constants, and a is the lattice spacing. Using this band structure for one-dimensional electrons, answer the following questions.
(a) Plot the $E(k)$ relation for $-\pi/a \leq k \leq \pi/a$.
(b) Plot the velocity, $\upsilon(k)$, for $-\pi/a \leq k \leq \pi/a$.
(c) Assuming that the electric field is $-\mathcal{E}_0$, how long would it take an electron to travel from $k = 0$ to $k = \pi/2a$?
(d) Given the answer, T_0, from part (c), how far would the electron go in this time?
(e) Compute the density of states, $g(E)$, for this one-dimensional semiconductor.

1.15 Consider a metal-semiconductor barrier as shown below:

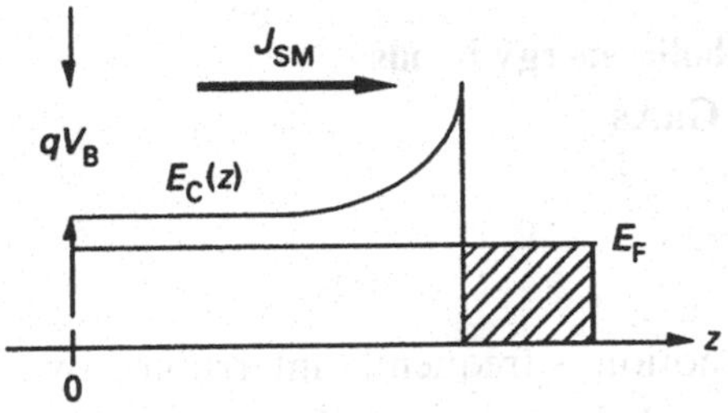

(a) Write an expression, involving a sum, for the electron current injected from the semiconductor to the metal, J_{SM}.
(b) Convert the sum to an integral. Be sure to show the limits of integration.
(c) Sketch the transmission coefficient versus k_z expected from (1) quantum mechanical and (2) classical considerations.
(d) Set up the problem for computing the classical (thermionic emission) current J_{SM}. Show the formula that has to be integrated, but do not integrate it.
(e) Evaluate the integral and show that the result is the expected thermionic emission relation.

2 Carrier scattering

As carriers traverse a device, their motion is frequently interrupted by collisions with impurity atoms, phonons, crystal defects, or with other carriers. In this chapter we examine carrier scattering and evaluate the transition rate, $S(\mathbf{p}, \mathbf{p}')$, which is the probability per unit time that a carrier with crystal momentum $\mathbf{p}$ scatters to a state with crystal momentum $\mathbf{p}'$. Our approach is based on Fermi's Golden Rule as described in Section 1.7. The first step is to identify the scattering potential then to evaluate the matrix element,

$$H_{\mathbf{p}'\mathbf{p}} = \frac{1}{\Omega}\int_{-\infty}^{\infty} e^{-i\mathbf{p}'\cdot\mathbf{r}/\hbar}\, U_S(r) e^{i\mathbf{p}\cdot\mathbf{r}/\hbar} \mathrm{d}^3 r. \tag{2.1}$$

In this chapter, we keep the mathematics to a minimum by evaluating matrix elements using plane wave electron wave functions rather than the actual Bloch functions. The overlap integral due to the cell periodic part of the Bloch function [recall eq. (1.116)] is unity when the constant energy surfaces are spherical and the energy varies parabolically with momentum (which we will simply call spherical, parabolic bands). In practice, however, the energy bands are rarely parabolic, and overlap integrals have to be evaluated [1.5].

Having evaluated the matrix element, we find the transition rate as

$$S(\mathbf{p}, \mathbf{p}') = \frac{2\pi}{\hbar} |H_{\mathbf{p}'\mathbf{p}}|^2 \delta(E(\mathbf{p}') - E(\mathbf{p}) - \Delta E), \tag{2.2}$$

where ΔE is the change in energy (if any) caused by the scattering event. (Note that the order of $\mathbf{p}$ and $\mathbf{p}'$ in $H_{p'p}$ is, by definition, opposite to their order in $S(\mathbf{p}, \mathbf{p}')$.)

The effects of scattering are conveniently summarized by evaluating characteristic times from the transition rate, $S(\mathbf{p}, \mathbf{p}')$. We discuss these *relaxation times* in the following section. The perturbing potentials for several common scattering mechanisms are then identified in Section 2.2. Next, some general features of carrier scattering are described in Section 2.3 followed by a step by step examination of various scattering mechanisms. We will not attempt to catalog $S(\mathbf{p}, \mathbf{p}')$ for all possible scattering mechanisms but will focus, instead, on scattering by ionized impurities, phonons, and by other carriers because they tend to be the most important. Finally, in Section 2.14, we summarize the general features of scattering in common semiconductors. For a first reading, Sections 2.4–2.13 can be bypassed.

2.1 Relaxation times

Consider a beam of energetic carriers injected into a semiconductor at time $t = 0$, with their momenta, $\mathbf{p}_0$, aligned along the $\hat{z}$-axis. Figure 2.1 illustrates how collisions affect such carriers. The transition rate is the rate at which carriers out-scatter from a specific initial state to a specific final state. The scattering rate,

$$\frac{1}{\tau(\mathbf{p}_0)} = \sum_{\mathbf{p}',\uparrow} S(\mathbf{p}_0, \mathbf{p}')[1 - f(\mathbf{p}')], \tag{2.3a}$$

is the rate at which carriers with a specific momentum $\mathbf{p}_0$ scatter to any other state. Alternatively, $\tau(\mathbf{p}_0)$ is the average time between collisions (also known as the lifetime of the state, $\mathbf{p}_0$). The vertical arrow below the sum is to indicate that the sum over final states includes only those whose spin is parallel to that of the incident carrier (the scattering mechanisms we consider do not flip the carrier's spin). The factor of $[1 - f(\mathbf{p}')]$, where $f(\mathbf{p}')$ is the probability that the state at $\mathbf{p}'$ is occupied, gives the probability of finding an empty final state. For a non-degenerate semiconductor, there is a high probability that the state at $\mathbf{p}'$ is empty, so eq. (2.3a) becomes

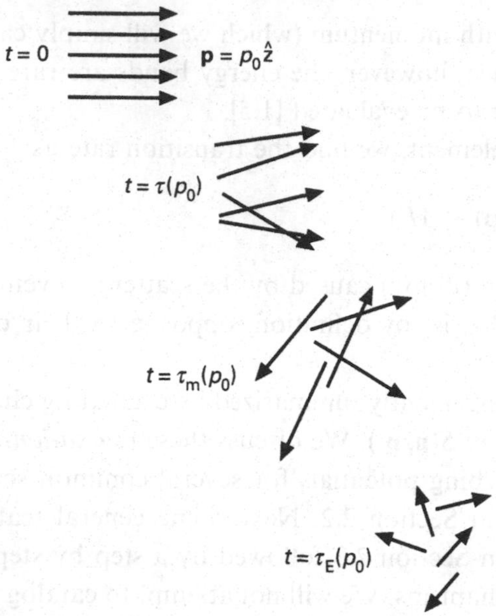

Fig. 2.1 Illustration of how collisions affect a group of electrons injected at $t = 0$ with momentum $\mathbf{p}_0$.

$$\frac{1}{\tau(\mathbf{p}_0)} = \sum_{\mathbf{p}',\uparrow} S(\mathbf{p}_0, \mathbf{p}'). \tag{2.3b}$$

This sum is much easier to evaluate because it does not depend on knowing how the states are occupied.

Some important scattering mechanisms are not isotropic. Instead, they tend to deflect carriers by small angles. Such a case is illustrated in Fig. 2.1, which shows that even after $\tau(\mathbf{p}_0)$ seconds, the carriers can retain a memory of their incident momentum. To evaluate the rate at which the $\hat{z}$-directed momentum is relaxed, we need to weight each collision by the fractional change in the $\hat{z}$-directed momentum. The result is

$$\frac{1}{\tau_m(\mathbf{p}_0)} = \sum_{\mathbf{p}',\uparrow} S(\mathbf{p}_0, \mathbf{p}')(1 - p_z'/p_{z0}) = \sum_{\mathbf{p}',\uparrow} S(\mathbf{p}_0, \mathbf{p}')[1 - (p'/p_0)\cos\alpha], \tag{2.4}$$

where α is the polar angle between the incident and scattered momenta as illustrated in Fig. 2.2a. $\tau_m(\mathbf{p}_0)$ is known as the *momentum relaxation time* and, as Fig. 2.1 indicates, is the time required to randomize the momentum.

We may also be interested in the time required to dissipate carrier energy which is denoted by the length of the vectors in Fig. 2.1. As Fig. 2.1 indicates, it is quite possible to relax the injected momentum by elastic scattering without

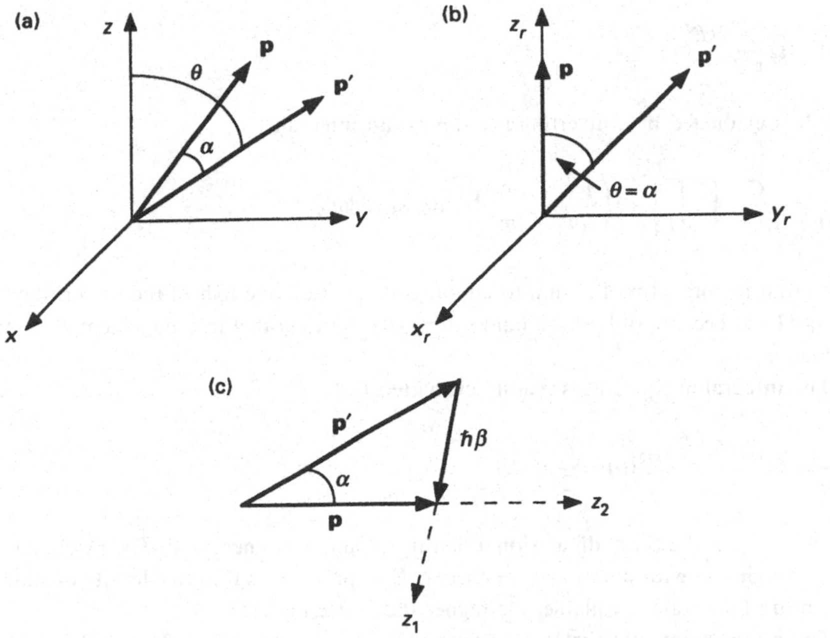

Fig. 2.2 (a) Illustration of a scattering event showing the incident and scattered momenta and the polar angle, α. (b) The same scattering event with the coordinate axes oriented so that $\theta = \alpha$. (c) Definition of $\hbar\boldsymbol{\beta}$, the momentum change resulting from scattering. We also show two choices for orienting the z-axis. The first choice, denoted by z_1, is used to evaluate the matrix element in eq. (2.34), and the second choice, denoted by z_2, is used for evaluating the momentum relaxation time in eq. (2.38).

affecting the energy of the carriers. To find the energy relaxation rate, we weight each collision by the fractional change in energy and find

$$\frac{1}{\tau_E(\mathbf{p}_0)} = \sum_{\mathbf{p}',\uparrow} S(\mathbf{p}_0, \mathbf{p}')\left[1 - \frac{E(\mathbf{p}')}{E(\mathbf{p}_0)}\right], \tag{2.5}$$

where $\tau_E(\mathbf{p}_0)$ is the *energy relaxation time*. We routinely evaluate $\tau(\mathbf{p}_0)$, $\tau_m(\mathbf{p}_0)$, and $\tau_E(\mathbf{p}_0)$ from $S(\mathbf{p}_0, \mathbf{p}')$ because knowledge of these three characteristic times concisely summarizes how collisions affect carriers.

Example: Scattering by a δ-function potential

In Section 1.7, we evaluated $S(\mathbf{p}, \mathbf{p}')$ for two simple perturbing potentials. For the δ-function perturbing potential, we found

$$S(\mathbf{p}, \mathbf{p}') = \frac{C}{\Omega}\delta(E' - E), \tag{2.6}$$

where C is a constant and Ω a normalization volume. For this potential, the scattering rate is

$$\frac{1}{\tau(p)} = \frac{C}{\Omega}\sum_{\mathbf{p}',\uparrow} \delta(E' - E) \tag{2.7}$$

which is evaluated by converting the sum to an integral,

$$\frac{1}{\tau(\mathrm{p})} = \frac{C}{8\pi^3\hbar^3}\int_0^\infty \int_0^\pi \int_0^{2\pi} \delta\left(\frac{p'^2}{2m^*} - \frac{p^2}{2m^*}\right) p'^2 \mathrm{d}p' \sin\theta \mathrm{d}\theta \mathrm{d}\phi. \tag{2.8}$$

Note that in converting the sum to an integral, we used one-half of the prescription stated in eq. (1.53) because only those final states with spin parallel to the incident electron's are available.

The integral in eq. (2.8) is readily evaluated to find

$$\frac{1}{\tau(\mathrm{p})} = C\frac{(2m^*)^{3/2}}{4\pi^2\hbar^3} E^{1/2}(p) = \frac{C}{2} g_C(E), \tag{2.9}$$

where $g_C(E)$ is the three-dimensional density of states at energy, E. The result states that the scattering rate for a carrier with energy, E, is proportional to the density of states at E. The more final states available, the higher the scattering rate.

To find the momentum relaxation time, we make use of eq. (2.4). The resulting integral is much like eq. (2.8) but with a second term as well. If we align our coordinate axes so that $\hat{z}$ is directed along the incident momentum, then $\alpha = \theta$, as illustrated in Fig. 2.2b. The second contribution to the integral then involves integration over $\cos\theta \sin\theta \, d\theta$ which integrates to zero. We conclude that $\tau_m = \tau$ for this scattering potential. More generally, the momentum relaxation time is always equal to the scattering time when the transition rate is isotropic [that is, $S(\mathbf{p}, \mathbf{p}')$ contains no dependence on θ or ϕ].

Finally, we consider the energy relaxation time. It is apparent that this scattering mechanism is elastic, so $E' - E$, and from eq. (2.5) we conclude that $\tau_E = \infty$. The result simply states that elastic scattering cannot relax energy.

2.2 The perturbing potential

The first step is to identify the perturbing potential responsible for scattering, so that the matrix element can be evaluated. In this section, we identify the perturbing potentials for the most common scattering mechanisms, ionized impurity and phonon scattering. In following sections, the transition and scattering rates will be evaluated, and some additional mechanisms will be considered.

2.2.1 Ionized impurity scattering

Carriers are scattered when they encounter the electric field of an ionized impurity. We might assume that the scattering potential is Coulombic,

$$U_S(\mathbf{r}) = \frac{q^2}{4\pi\kappa_S\varepsilon_0 r}, \tag{2.10}$$

but the ionized impurity attracts mobile carriers which *screen* the potential. The electrostatic potential due to both the ionized impurity and mobile carriers is found by solving Poisson's equation in spherical coordinates,

$$\frac{1}{r^2}\frac{\mathrm{d}}{\mathrm{d}r}\left(r^2\frac{\mathrm{d}V}{\mathrm{d}r}\right) = \frac{q}{\kappa_S\varepsilon_0}(n - N_D^+), \tag{2.11}$$

where an n-type semiconductor has been assumed. Space charge neutrality dictates that $n = n_0 = N_D^+$ (with a corresponding electrostatic potential of $V = V_0$) on a macroscopic scale, but on a microscopic scale, perturbations in the potential and carrier density exist. By writing the potential and carrier density as $V = V_0 + \delta V$ and $n = n_0 + \delta n$ and substituting in eq. (2.11), we find

$$\frac{1}{r^2}\frac{\mathrm{d}}{\mathrm{d}r}\left(r^2\frac{\mathrm{d}\delta V}{\mathrm{d}r}\right) = \frac{q\delta n}{\kappa_S\varepsilon_0}. \tag{2.12}$$

To find δn, recall that

$$n = N_C e^{[E_F - E_C(\mathbf{r})]/k_B T}$$

for a non-degenerate semiconductor and that

$$E_C(\mathbf{r}) = \text{constant} - qV(\mathbf{r}),$$

so small perturbations in carrier density can be related to perturbations in the potential by

$$\delta n = \frac{\partial n}{\partial V}\delta V = \frac{q n_0}{k_B T}\delta V.$$

After inserting this result in Poisson's equation, we find

$$\frac{1}{r^2}\frac{\mathrm{d}}{\mathrm{d}r}\left(r^2\frac{\mathrm{d}\delta V}{\mathrm{d}r}\right) = \frac{q^2 n_0}{\kappa_S\varepsilon_0 k_B T}\delta V = \delta V / L_D^2, \tag{2.13}$$

where

$$L_D \equiv \sqrt{\frac{\kappa_S\varepsilon_0 k_B T}{q^2 n_0}} \tag{2.14}$$

is known as the *Debye length*. [For degenerate semiconductors, Fermi–Dirac statistics apply, and the screening length, eq. (2.14), must be generalized as discussed in homework problem 2.3.]

The solution to eq. (2.13),

$$\delta V = \frac{A}{r}e^{-r/L_D}, \tag{2.15}$$

shows that perturbations in potential decay exponentially with distance from the perturbation. If the perturbation is due to an ionized impurity, we should recover the Coulombic potential as $r \to 0$, so

$$A = \frac{q^2}{4\pi\kappa_S\varepsilon_0}.$$

For an *n*-type semiconductor, therefore, the appropriate perturbing potential is the *screened Coulomb potential,*

$$\boxed{U_S(r) = \frac{q^2}{4\pi\kappa_S\varepsilon_0 r}e^{-r/L_D}} \quad . \tag{2.16}$$

On the other hand, when the mobile carrier density is low, as it is in the depletion region of a *p–n* junction, the unscreened Coulomb potential [eq. (2.10)] is used.

2.2.2 Phonon scattering

Because a semiconductor's band structure is determined by the crystal potential, it is influenced by changes in lattice spacing. A semiconductor under pressure has a perturbed lattice constant and band structure as sketched in Fig. 2.3a and 2.3b. For a small change in lattice constant, we expect that

$$\delta E_C = D_C \frac{\delta a}{a} \tag{2.17a}$$

and that

$$\delta E_V = D_V \frac{\delta a}{a}, \tag{2.17b}$$

where D_C and D_V, the *deformation potentials*, can be deduced from experiments and have been characterized for common semiconductors (refer to Table 2.1 at the end of this chapter). The change in effective mass with lattice spacing is small and is neglected. Lattice vibrations deform the lattice producing a 'grating' in the band edges as sketched in Fig. 2.3b. Electrons and phonons interact when carrier waves scatter off of this grating.

Consider an elastic wave,

$$u(x, t) = A_\beta e^{+i(\beta x - \omega t)} + A_\beta^* e^{-i(\beta x - \omega t)}, \tag{2.18}$$

propagating in a one-dimensional lattice. (Writing it in this form ensures that the wave is real, see Datta [1.1].) The use of a continuum description in eq. (2.18) is justified because only long wavelength (small β) phonons are effective in intra-valley scattering. Four acoustic phonons, which displace neighboring atoms in

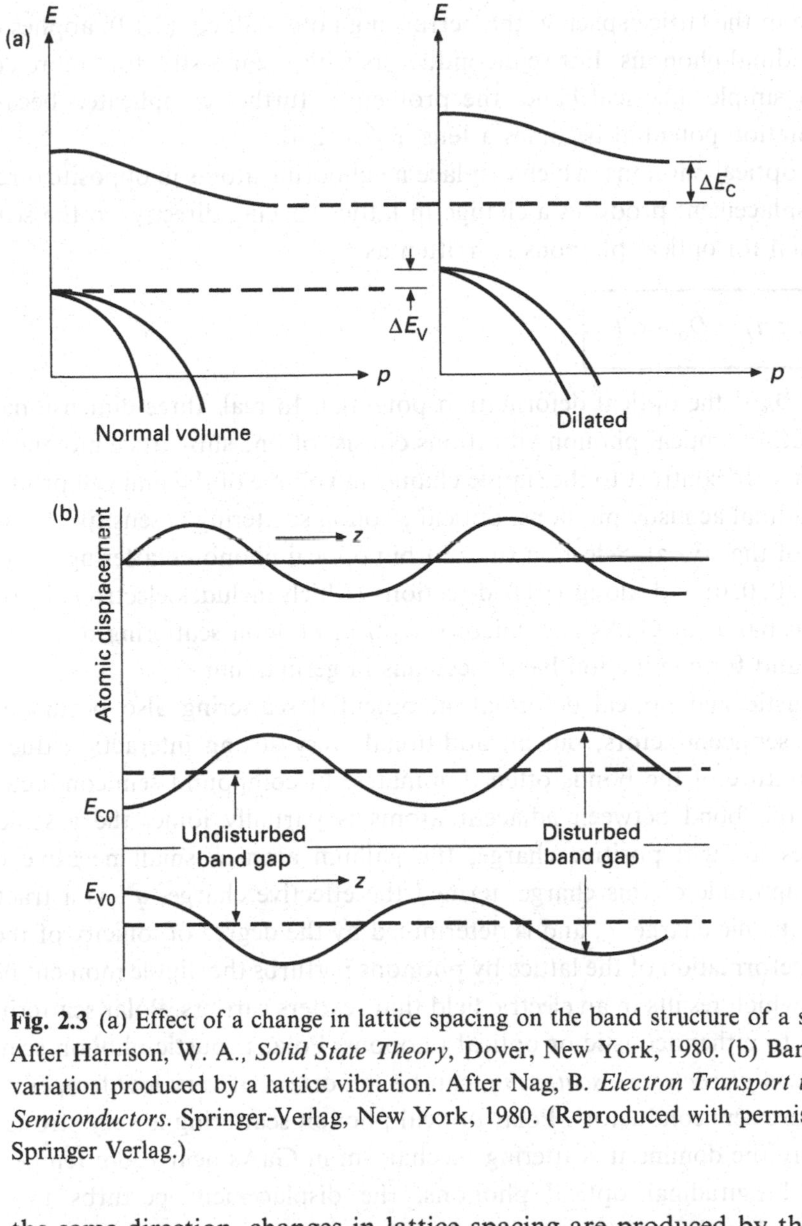

Fig. 2.3 (a) Effect of a change in lattice spacing on the band structure of a semiconductor. After Harrison, W. A., *Solid State Theory*, Dover, New York, 1980. (b) Band edge variation produced by a lattice vibration. After Nag, B. *Electron Transport in Compound Semiconductors*. Springer-Verlag, New York, 1980. (Reproduced with permission from Springer Verlag.)

the same direction, changes in lattice spacing are produced by the strain $\partial u/\partial x$ not by the displacement, $u(x, t)$. Motivated by eq. (2.17a), we write the interaction potential for acoustic phonons as

$$U_{AP}(x, t) = D_A \frac{\partial u}{\partial x} \quad . \tag{2.19}$$

To extend these arguments to three-dimensional crystals, the phonon's polarization must be considered. Since transverse elastic waves produce no first-order change in the lattice spacing, the perturbing potential, eq. (2.19), applies only to longitudinal phonons. For semiconductors with a band structure more complex than a simple spherical band, the problem is further complicated because the deformation potential becomes a tensor [2.1, 2.4].

For optical phonons, which displace neighboring atoms in opposite directions, the displacement produces a change in lattice spacing directly, so the scattering potential for optical phonons is written as

$$\boxed{U_{OP}(x,t) = D_o u(x,t)}\,, \tag{2.20}$$

where D_o is the optical deformation potential. In real, three-dimensional semiconductors, optical phonon vibrations consist of one sub-lattice moving against the other. In contrast to the simple change in volume of the unit cell produced by longitudinal acoustic phonons, optical phonon scattering is sensitive to the symmetry of the crystal. Selection rules forbid optical phonon scattering of electrons at $\mathbf{p} = (0, 0, 0)$ and along $\langle 100 \rangle$ directions (which includes electrons in the conduction bands of GaAs and silicon). Optical phonon scattering does occur for holes, and for conduction band electrons in germanium.

Acoustic and optical deformation potential scattering also occurs in compound semiconductors, but an additional, very strong interaction due to the polar nature of the bonds often dominates. In compound semiconductors like GaAs the bond between adjacent atoms is partially ionic; the arsenic atom acquires a slight positive charge, the gallium atom a small negative charge. The magnitude of this charge, termed the effective charge, q^*, is a fraction of the electronic charge, q, and is determined by the degree of ionicity of the bond [2.1]. Deformation of the lattice by phonons perturbs the dipole moment between atoms which results in an electric field that scatters carriers. Polar scattering may be due to either acoustic or optical phonons. Polar acoustic phonon scattering, known as *piezoelectric scattering*, can be important at very low temperatures in very pure semiconductors. Polar optical phonon scattering is very strong and is typically the dominant scattering mechanism in GaAs near room temperature.

For longitudinal optical phonons, the displacement perturbs the dipole moment directly according to

$$\delta p = q^* u. \tag{2.21}$$

The electric field due to the perturbed dipole moment is obtained from the relation,

$$D = \varepsilon_0 \mathcal{E} + \delta P, \tag{2.22}$$

and from the assumption of zero macroscopic, free charge (that is, $\nabla \cdot D = 0$). (In eq. (2.22), δP is the change in the dipole moment per unit volume produced by the elastic wave.) Since the fields and polarization are due to elastic waves, they vary as

$$D = D_x e^{i\beta x} + D_x^* e^{-i\beta x} = |D| \cos(\beta x + \phi).$$

Taking the divergence,

$$\nabla \cdot D = \frac{\partial D}{\partial x} i\beta |D| \sin(\beta x + \phi) = 0,$$

which implies that $D = 0$. From eq. (2.22) with $D = 0$, we find

$$\mathcal{E} = -\frac{\delta P}{\varepsilon_0}, \tag{2.23}$$

where

$$\delta P = \frac{\delta p}{V_u}, \tag{2.24}$$

with V_u being the volume of a unit cell. From eq. (2.23), we find the electric field as

$$\mathcal{E} = -\frac{q^* u}{\varepsilon_0 V_u}, \tag{2.25}$$

which is integrated to find the interaction potential,

$$U_S(x, t) = -q \int \mathcal{E} dx. \tag{2.26}$$

For polar optical phonon scattering, the resulting perturbing potential is

$$\boxed{U_{POP} = \frac{qq^* u}{i\beta V_u \varepsilon_0}} \quad . \tag{2.27}$$

It is common to measure the strength of the polar interaction by the low and high frequency dielectric constants, which are easily measured, rather than by the effective charge, q^*, on the dipole. The two approaches are related by [2.1]

$$\left(\frac{q^*}{V_u}\right)^2 = \frac{\varepsilon_0 \rho \omega_0^2}{\kappa_0} \left(\frac{\kappa_0}{\kappa_\infty} - 1\right), \tag{2.28}$$

which is understood as follows. At low frequencies, the dielectric constant includes a contribution due to the dipole between atoms, but at high frequencies this dipole cannot respond to the signal so that $\kappa_\infty < \kappa_0$. The factor, $\kappa_0/\kappa_\infty - 1$, is a measure of the strength of the dipole.

In polar semiconductors, acoustic phonons also produce an electrostatic perturbation known as piezoelectric scattering. This effect is much weaker than polar optical phonon scattering, but it can be important at very low temperatures when the number of optical phonons is small and carriers don't have sufficient thermal energy to emit them. As shown in homework problem 2.4, the perturbing potential for piezoelectric scattering is

$$\boxed{U_{PZ} = \frac{q e_{PZ}}{\kappa_S \varepsilon_0} u}\ , \tag{2.29}$$

where e_{PZ} is the *piezoelectric constant.*

2.2.3 Discussion

In the following several sections, the transition rates, $S(\mathbf{p}, \mathbf{p}')$, and the various relaxation times of interest are computed from the interaction potentials identified here. We will also consider some additional scattering mechanisms, such as electron–electron scattering. The mathematical procedure is straightforward, but it can get tedious, so before we do the calculations, we discuss some general features of carrier scattering which will come out of these calculations.

2.3 General features of phonon scattering

The general features of phonon scattering are easy to describe and understand. In this section, we describe the general characteristics of the most common types of scattering.

2.3.1 Scattering from a static potential

The simplest model for a static perturbing potential is a δ-function

$$U_S(\mathbf{r}) = A_0 \delta(\mathbf{r}). \tag{2.30a}$$

As shown in Section 1.7, this perturbing potential leads to a transition rate of the form

$$S(\mathbf{p}, \mathbf{p}') = C\delta(E - E'), \tag{2.30b}$$

which, as shown in Section 2.1, produces a scattering rate that is proportional to the density of states,

$$\frac{1}{\tau} \propto g_C(E). \tag{2.30c}$$

This basic result, that the scattering rate is proportional to the density of states, is a common feature of carrier scattering. The more states at a given energy, the more ways a carrier can scatter to it. The δ-function approximation is a useful approximation for highly localized, static scattering potentials.

2.3.2 Nonpolar phonon scattering

A second common scattering potential is the oscillating potential produced by lattice vibrations. This potential can be written as

$$U_S = K_\beta u_\beta = K_\beta A_\beta e^{\pm i(\beta\cdot \mathbf{r} - \omega t)}, \tag{2.31a}$$

where A_β is related to the amplitude of the oscillation and K_β to the deformation potentials discussed in Section 2.2. As shown in Section 1.7, this type of perturbing potential leads to a transition rate of the form

$$S(\mathbf{p}, \mathbf{p}') = C_\beta \delta\left[E(\mathbf{p}') - E(\mathbf{p}) \mp \hbar\omega(\boldsymbol{\beta})\right]\delta_{\mathbf{p}',\mathbf{p}\pm\hbar\boldsymbol{\beta}}, \tag{2.31b}$$

where the δ-function expresses conservation of energy and the Kronecker δ conservation of momentum.

In Section 2.5 we will demonstrate that the requirement that both energy and momentum be conserved restricts the phonons that can be involved in intravalley scattering processes to those near the center of the Brillouin zone. As a result, we can use the simplified phonon dispersion relation shown in Fig. 1.27b. For acoustic phonons, the result is that the phonon energy is small and can be neglected near room temperature where the average kinetic energy of a carrier is much larger, so we can simplify eq. (2.31b) for elastic, acoustic deformation potential (ADP) scattering as

$$S(\mathbf{p}, \mathbf{p}') = C\delta\left[E(\mathbf{p}') - E(\mathbf{p})\right]\delta_{\mathbf{p}',\mathbf{p}\pm\hbar\boldsymbol{\beta}}. \tag{2.31c}$$

(it turns out that the strength of the transition as measured by C is independent of β in this case). Equation (2.31c) is much like eq. (2.30b), so we find that the scattering rate for this case is also proportional to the density of states.

For optical phonon scattering, Fig. 1.27b shows that the phonon energy is large, but that it is independent of β (this also applies to scattering of carriers between valleys by optical or acoustic phonons). For optical phonon scattering, eq. (2.31b) simplifies to

$$S(\mathbf{p}, \mathbf{p}') = C_\beta \delta\left[E(\mathbf{p}') - E(\mathbf{p}) \mp \hbar\omega_0\right]\delta_{\mathbf{p}',\mathbf{p}\pm\hbar\boldsymbol{\beta}}. \tag{2.31d}$$

As shown in Section 2.6, this expression can be integrated over the final states to find the transition rate as

$$\frac{1}{\tau} \propto g_C(E \pm \hbar\omega_0), \tag{2.31e}$$

where the top sign applies to scattering by absorbing an optical phonon and the bottom to scattering by emitting an optical phonon (phonon emission cannot occur unless the carrier's kinetic energy exceeds the optical phonon energy). Equation (2.31e) simply states that the scattering rate is proportional to the density of states at the final energy. Figure 2.4a compares the scattering rates for elastic ADP scattering to that for inelastic, optical deformation potential (ODP) scattering. Note the presence of a threshold energy for ODP scattering. Below the threshold, carriers can scatter only by absorbing optical phonons.

2.3.3 Scattering by electrostatic interactions

The δ-function potential is not a particularly good approximation for ionized impurity scattering because carriers feel the potential only when they encounter it directly. For more realistic potentials, we expect that the amount that a carrier is deflected will be determined by the strength of the electrostatic interaction and by the length of time that the carrier feels the potential. The result is that slower moving carriers scatter more than faster moving carriers. The effect is particularly strong for the momentum relaxation rate as illustrated in Fig. 2.4b. These considerations apply not only to ionized impurity scattering, but to any electrostatic interaction. Consider polar optical phonon (POP) scattering, the strongest mechanism in polar semiconductors like GaAs. As shown in Fig. 2.4b, POP scattering displays the general features of optical phonon scattering with a clear threshold energy for phonon emission, but the scattering rate decreases

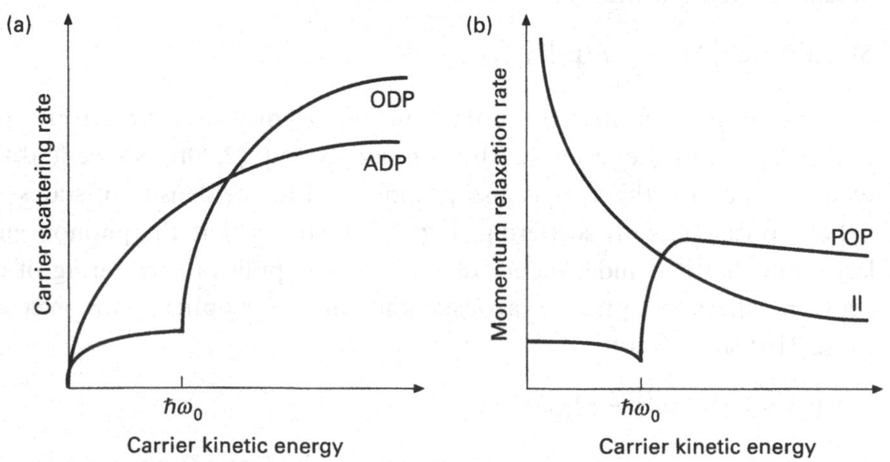

Fig. 2.4 General features of carrier scattering in semiconductors. (a) Nonpolar phonon scattering, acoustic deformation potential (ADP) in the elastic limit and optical deformation potential (ODP). (b) Scattering by electrostatic interactions, ionized impurity (II) and polar optical phonon (POP).

at high energy (as for ionized impurity scattering) because the faster moving electrons spend less time in the vicinity of the perturbing potential.

2.3.4 Discussion

Figure 2.4 summarizes the general features of scattering in common semiconductors. We expect the scattering rate to be proportional to the density of final states, except for electrostatic interactions which become weaker for high energy carriers. The sketch in Fig. 2.4 assumes carriers are near the bottom of a band where the density of state goes as $\sqrt{E}$, but for higher energy carriers, we expect the scattering rate to mimic the density of states as displayed in Fig. 1.19. For analytical calculations of scattering of low energy carriers, it is often convenient to express the scattering rate in power law form,

$$\tau(E) = \text{constant} \times E^s, \tag{2.32}$$

where s is a characteristic exponent. When the bands are parabolic and the scattering goes as the density of states, then $s = -1/2$. For ionized impurity scattering, on the other hand, $s = +3/2$.

In the next several sections, we will work out the scattering rate expressions for various mechanism; readers not interested in these details may proceed to Section 2.14.

2.4 Scattering by ionized impurities

Having identified the perturbing potential for ionized impurity scattering in Section 2.2.1, we evaluate the matrix element from eq. (2.1) as

$$H_{\mathbf{p}'\mathbf{p}} = \frac{1}{\Omega}\left(\frac{q^2}{4\pi\kappa_S\varepsilon_0}\right)\int \mathrm{e}^{-\mathrm{i}\mathbf{p}'\cdot\mathbf{r}/\hbar}\left(\frac{\mathrm{e}^{-r/L_D}}{r}\right)\mathrm{e}^{\mathrm{i}\mathbf{p}\cdot\mathbf{r}/\hbar}\mathrm{d}^3r$$

or

$$H_{\mathbf{p}'\mathbf{p}} = \frac{1}{\Omega}\left(\frac{q^2}{4\pi\kappa_S\varepsilon_0}\right)\int_0^{2\pi}\mathrm{d}\phi\int_0^{\pi}\int_0^{\infty}\mathrm{e}^{-r/L_D}\mathrm{e}^{\mathrm{i}(\mathbf{p}-\mathbf{p}')\cdot\mathbf{r}/\hbar}r\mathrm{d}r\sin\theta\mathrm{d}\theta.$$

According to the geometry of the scattering event (displayed in Fig. 2.2c),

$$\hbar\boldsymbol{\beta} = \mathbf{p}' - \mathbf{p} \tag{2.33a}$$

and

$$\hbar\beta = 2p\sin\left(\frac{\alpha}{2}\right), \tag{2.33b}$$

the latter follows from Fig. 2.2c because $p = p'$ for elastic scattering. With the aid of eq. (2.33a), the matrix element becomes

$$H_{\mathbf{p}'\mathbf{p}} = \frac{1}{\Omega}\left(\frac{q^2}{4\pi\kappa_S\varepsilon_0}\right)\int_0^\infty\int_0^{2\pi}\int_{-1}^{+1} re^{r/L_D}e^{-i\beta r\cos\theta}\mathrm{d}(\cos\theta)\mathrm{d}\phi\mathrm{d}r, \tag{2.34}$$

which can be integrated to find

$$H_{\mathbf{p}'\mathbf{p}} = \frac{q^2}{\Omega\kappa_S\varepsilon_0}\frac{1}{\beta^2 + 1/L_D^2}. \tag{2.35}$$

According to eq. (2.2), the transition rate due to scattering from a single ionized impurity is

$$S(\mathbf{p}, \mathbf{p}') = \frac{2\pi}{\hbar}\left(\frac{q^2}{\Omega\kappa_S\varepsilon_0}\right)^2\frac{\delta(E' - E)}{[\beta^2 + 1/L_D^2]^2}.$$

Finally, we multiply by $N_I\Omega$, the number of impurities in the normalization volume, and make use of eq. (2.33b) to write

$$\boxed{S(\mathbf{p}, \mathbf{p}') = \frac{2\pi N_I q^4}{\hbar\kappa_S^2\varepsilon_0^2\Omega}\frac{\delta(E' - E)}{\left[4\left(\frac{p}{\hbar}\right)^2\sin^2\left(\frac{\alpha}{2}\right) + \frac{1}{L_D^2}\right]^2}} \tag{2.36a}$$

Equation (2.36a) describes scattering from an ionized impurity whose potential is screened by mobile carriers. When mobile carriers are absent, $L_D \to \infty$ and eq. (2.36a) becomes

$$S(\mathbf{p}, \mathbf{p}') = \frac{2\pi N_I q^4}{\hbar\kappa_S^2\varepsilon_0^2\Omega}\frac{\delta(E' - E)}{16\left(\frac{p}{\hbar}\right)^4\sin^4\left(\frac{\alpha}{2}\right)}. \tag{2.36b}$$

Equation (2.36b) applies when mobile carriers are absent which occurs, for example, in the depletion region of a *p–n* junction. When the carrier density is very high, L_D is small so that $1/L_D^2$ in the denominator of eq. (2.36a) dominates. For this strongly screened case,

$$S(\mathbf{p}, \mathbf{p}') = \frac{2\pi N_I q^4 L_D^4}{\hbar\kappa_S^2\varepsilon_0^2\Omega}\delta(E' - E), \tag{2.36c}$$

and the transition rate has the form of eq. (2.6) which describes scattering from a δ-function potential.

A plot of $S(\mathbf{p}, \mathbf{p}')$ versus α (the polar angle between the incident and scattered momenta) is displayed in Fig. 2.5. The important point to note is that ionized

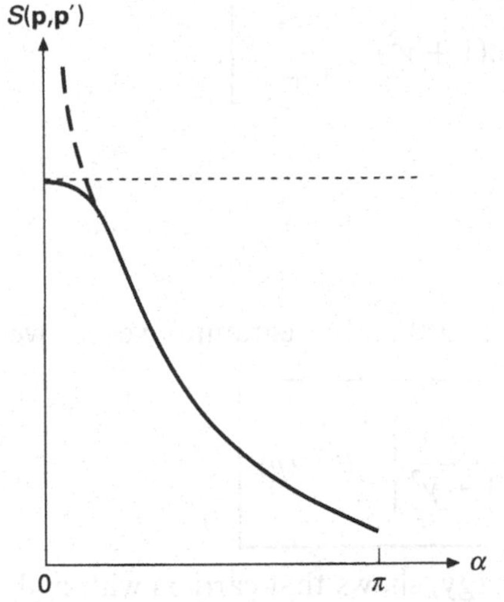

Fig. 2.5 $S(\mathbf{p}, \mathbf{p}')$ versus α for ionized impurity scattering; the solid line is for a screened Coulomb potential, the dashed line for the unscreened ($L_D \to \infty$) Coulomb potential, and the dotted line for the strongly screened ($L_D \to 0$) Coulomb potential.

impurity scattering tends to deflect carriers by small angles – except when it is strongly screened in which case ionized impurity scattering is isotropic.

From the transition rate, the various relaxation times can also be evaluated. Evaluation of the scattering rate is the subject of homework problem 2.1. To find the momentum relaxation time, we evaluate

$$\begin{aligned}\frac{1}{\tau_m(p)} &= \sum_{\mathbf{p}',\uparrow} S(\mathbf{p}, \mathbf{p}')\left(1 - \frac{p'}{p}\cos\alpha\right) \\ &= \frac{\Omega}{8\pi^3\hbar^3}\int_0^{2\pi}\int_{-1}^{1}\int_0^{\infty} S(\mathbf{p}, \mathbf{p}')\left(1 - \frac{p'}{p}\cos\alpha\right)p'^2 \mathrm{d}p' \mathrm{d}(\cos\theta)\mathrm{d}\phi.\end{aligned} \tag{2.37}$$

If we orient the $\hat{z}$ axis so that it points along the initial momentum (see Fig. 2.2c), then $\alpha = \theta$, $\mathbf{p} = p\hat{z}$, $p'_z = p'\cos\theta$, and eq. (2.37) becomes

$$\frac{1}{\tau_m(p)} = \frac{N_1 q^4}{2\pi\kappa_S^2\varepsilon_0^2\hbar^4}\int_0^{\infty}\int_{-1}^{1}\frac{(1-\cos\theta)\mathrm{d}(\cos\theta)}{\left[4(p/\hbar)^2\sin^2\left(\frac{\theta}{2}\right) + \frac{1}{L_D^2}\right]}\delta(E' - E)p'^2\mathrm{d}p' \tag{2.38}$$

(we also made use of the fact that $p = p'$ because ionized impurity scattering is elastic). With the substitution, $x = (1 - \cos\theta)$, the first integral in eq. (2.38) becomes

$$I_\theta = \int_0^2 \frac{xdx}{\left[2(p/\hbar)^2 x + \frac{1}{L_D^2}\right]^2} = \frac{1}{4(p/\hbar)^4}\left[\ln(1+\gamma^2) - \frac{\gamma^2}{1+\gamma^2}\right],$$

where

$$\gamma^2 \equiv 4L_D^2(p/\hbar)^2 = 8m^* E(p) L_D^2/\hbar^2. \tag{2.39}$$

After inserting this result for I_θ in eq. (2.38) and integrating over p', we find

$$\boxed{\tau_m(p) = \frac{16\sqrt{2m^*}\pi\kappa_S^2\varepsilon_0^2}{N_I q^4}\left[\ln(1+\gamma^2) - \frac{\gamma^2}{1+\gamma^2}\right]^{-1} E^{3/2}(p)}. \tag{2.40}$$

Figure 2.6, a plot of $1/\tau_m(p)$ versus energy, shows that carriers with high kinetic energy have long momentum relaxation times. The increase in τ_m with energy occurs because rapidly moving carriers are deflected less by ionized impurities. As a consequence, the influence of ionized impurity scattering decreases at high temperatures or at high electric fields because both increase the carriers' kinetic energy. Equation (2.40), which assumes a screened Coulomb potential, is a simple version of a theory due to Brooks and Herring.

2.4.1 Unscreened Coulomb scattering

When the free carrier density is low, screening is minimal, and the transition rate is given by eq. (2.36b). When we attempt to use eq. (2.36b), however, we are

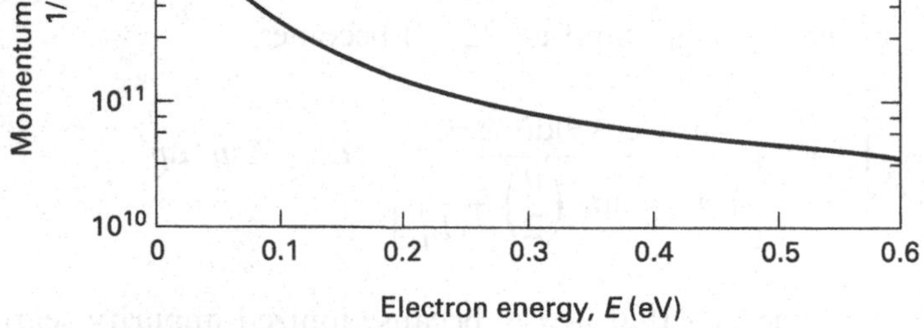

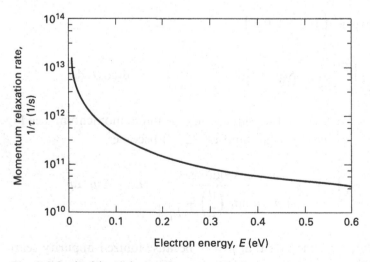

Fig. 2.6 Ionized impurity momentum scattering rate versus energy for electrons in GaAs ($N_i = 10^{17}$ cm^{-3}, $T = 300$ K).

confronted with the unpleasant fact that it predicts that the probability of scattering approaches infinity as the angle of deflection approaches zero. When the carrier is far from an ionized impurity, as measured by the *impact parameter* defined in Fig. 2.7, it is deflected very little. But if the carrier's impact parameter is greater than

$$b_{\max} = 1/2N_I^{-1/3}, \tag{2.41}$$

which is half spacing between impurities, then the scattering simply occurs from a neighboring impurity. The result is that a minimum scattering angle exists, so the infinity in $S(\mathbf{p}, \mathbf{p}')$ is never approached. Since the impact parameter and the angle of deflection are related by Rutherford's theory as

$$b = \frac{q^2}{8\pi\kappa_S\varepsilon_0 E(p)}\cot(\alpha/2), \tag{2.42}$$

the minimum deflection angle is determined from

$$\frac{1}{\sin^2(\theta_{\min}/2)} = 1 + \gamma_{\mathrm{CW}}^2, \tag{2.43}$$

where

$$\gamma_{\mathrm{CW}} = \frac{4\pi\kappa_S\varepsilon_0 E(p)}{q^2 N_{\mathrm{I}}^{1/3}} = \frac{b_{\max}}{[q^2/8\pi\kappa_S\varepsilon_0 E(p)]}. \tag{2.44}$$

The subscript CW is for Conwell and Weisskopf who first performed the calculation about to be described.

The momentum scattering rate in the Conwell–Weisskopf approach is obtained from eq. (2.38) by letting $L_D \to \infty$. If we orient the coordinate system so that $\alpha = \theta$ and integrate from $\theta_{\min}$ to π, we find

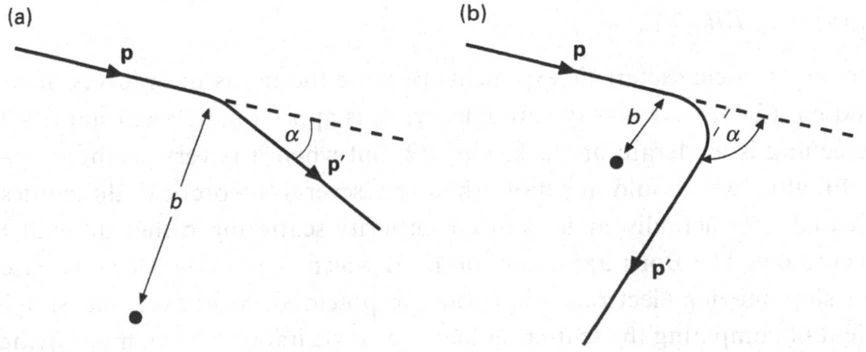

Fig. 2.7 Effect of impact parameter on ionized impurity scattering. (a) large impact parameter, (b) small impact parameter.

$$\frac{1}{\tau_m(p)} = \frac{N_1 q^4}{2\pi\kappa_S^2\varepsilon_0^2\hbar^4}\int_0^\infty\int_{-1}^{\cos\theta_{\min}}\frac{(1-\cos\theta)\mathrm{d}(\cos\theta)}{16(p/\hbar)^4\sin^4(\theta/2)}\delta(E'-E)p'^2\mathrm{d}p'. \tag{2.45}$$

Equation (2.45) works out much like the integrals in the Brooks–Herring approach with the result that

$$\boxed{\tau_m(p) = \frac{16\pi\sqrt{2m^*}\kappa_S^2\varepsilon_0^2}{N_1 q^4}\left[\frac{1}{\ln(1+\gamma_{\mathrm{CW}}^2)}\right]E^{3/2}(p)} . \tag{2.46}$$

Except for the term in brackets, eq. (2.46) is much like the Brooks–Herring result.

2.4.2 Strongly screened Coulomb scattering

When the mobile carrier density is very high, the Coulomb potential is strongly screened, and the transition rate is given by eq. (2.36c). Because the result has the simple form due to a δ-function scattering potential, the results of the example considered in Section 2.1 apply, and we find

$$\frac{1}{\tau_m(\mathrm{p})} = \frac{1}{\tau(\mathrm{p})} = \frac{\pi N_1}{\hbar}\left(\frac{q^2 L_D^2}{\kappa_S\varepsilon_0}\right)^2 g_C(E). \tag{2.47}$$

2.4.3 Discussion

We have outlined three simple treatments of ionized impurity scattering; the first assumed that free carriers were present to screen the impurity potential, the second assumed that free carrier screening was very weak, and the third assumed that free carrier screening was very strong. The final results [eqs. (2.40), (2.46), and (2.47)] can all be written in the form,

$$\tau_m(p) = \tau_0(E/k_B T)^s, \tag{2.48}$$

where s is a characteristic exponent. Because the terms in brackets in eq. (2.40) and eq. (2.46) vary slowly with energy, τ_0 is approximately constant. When the screening is moderate or weak, $s = 3/2$, but when it is very strong, $s = -1/2$.

Finally, we should mention that the several theoretical difficulties we've glossed over actually make ionized impurity scattering rather difficult to treat accurately. The Born approximation, on which the Golden Rule is based, fails for slow moving electrons in a Coulomb potential. Moreover, our simple expedient of computing the scattering due to a single impurity then multiplying by the number of impurities does not account for the interference effects that occur as the electron wave propagates through the array of randomly placed scatterers.

The expressions developed in this section comprise the standard approach for treating ionized impurity scattering in semiconductors; they serve reasonably well for many applications. The limitations of this approach are described in Anderson et al. [2.10].

2.5 Energy-momentum conservation in phonon scattering

When a carrier collides with a phonon, both energy and momentum are conserved. Energy and momentum conservation impose constraints on the maximum wave vector change and, therefore, on which phonons may participate in scattering events. Our purpose in this section is to demonstrate that for *intravalley* scattering, in which the carrier resides in the same valley before and after scattering, the phonons involved are those with wave vectors near the center of the Brillouin zone. This result is important because it means that we can employ the simple dispersion relations sketched in Fig. 1.27b.

Conservation of energy states that

$$E(\mathbf{p}') = E(\mathbf{p}) \pm \hbar\omega(\boldsymbol{\beta}), \tag{2.49}$$

where $\hbar\omega(\boldsymbol{\beta})$ is the energy of the phonon and $\pm$ denotes scattering by phonon absorption or by emission. For spherical, parabolic energy bands, eq. (2.49) becomes

$$\frac{p'^2}{2m^*} = \frac{p^2}{2m^*} \pm \hbar\omega(\beta). \tag{2.50}$$

Momentum conservation also applies and states that

$$\mathbf{p}' = \mathbf{p} \pm \hbar\boldsymbol{\beta}, \tag{2.51}$$

where $\hbar\boldsymbol{\beta}$ is the momentum of the phonon. The dot product of eq. (2.51) with itself can be inserted in eq. (2.50) and re-arranged to write

$$\hbar\beta = 2p\left[\mp\cos\theta \pm \frac{\omega}{\upsilon(p)\beta}\right], \tag{2.52}$$

where θ is the polar angle between $\mathbf{p}$ and $\boldsymbol{\beta}$. Equation (2.52) simultaneously states energy and momentum conservation. Since $\cos\theta$ is restricted to range between -1 and $+1$, eq. (2.52) determines the minimum and maximum values for the wave vectors of the phonons involved in intravalley carrier scattering.

2.5.1 Intravalley acoustic phonon scattering

For acoustic phonons with wave vectors near the zone center, ω is proportional to β (recall Fig. 1.27). If we call the slope the sound velocity, υ_S, then $\omega/\beta = \upsilon_S$ and eq. (2.52) becomes

$$\hbar\beta = 2p\left[\mp\cos\theta \pm \frac{\upsilon_S}{\upsilon(p)}\right]. \tag{2.53}$$

For phonon absorption, the maximum wave vector occurs when $\theta = \pi$ and when $\theta = 0$ for phonon emission, so we conclude that

$$\hbar\beta_{\mathbf{max}}(\mathrm{AP}) = 2\mathrm{p}\left[1 \pm \frac{\upsilon_S}{\upsilon(p)}\right]. \tag{2.54}$$

Since $\upsilon_s \simeq 10^5$ cm/sec and $\upsilon(p) \simeq 10^7$ cm/sec for a thermal average carrier,

$$\hbar\beta_{\mathrm{max}}(AP) \simeq 2p \tag{2.55}$$

for a typical acoustic phonon scattering event. As shown in Fig. 2.8, eq. (2.55) is what we expect for elastic scattering, which suggests that acoustic phonon scattering is nearly elastic. Note also that eq. (2.54) and the requirement that the wave vector be positive dictate that a carrier can't scatter by emitting an acoustic phonon unless its velocity exceeds the sound velocity.

To gauge the magnitude of β_{max}, it should be compared to π/a, the approximate half-width of the Brillouin zone. From eq. (2.55) we find that

$$\frac{\beta_{\mathrm{max}}}{\pi/a} = \frac{2m^*\upsilon(p)}{\hbar\pi/a} \simeq 1/4$$

[for $m^* \simeq m_0$, $\upsilon(p) \cong 10^7$ cm/sec, and $a = 5$ Å]. We conclude that acoustic phonon scattering involves phonons with wave vectors near the center of the Brillouin zone.

The maximum change in carrier energy resulting from acoustic phonon scattering is readily estimated from

$$\Delta E_{\mathrm{max}} = \hbar\omega_{\mathrm{max}} = \hbar\beta_{\mathrm{max}}\upsilon_S \simeq 10^{-3}\ \mathrm{eV}.$$

Acoustic phonon scattering is often taken to be elastic because ΔE is small compared to $k_B T_L$, except at very low temperatures.

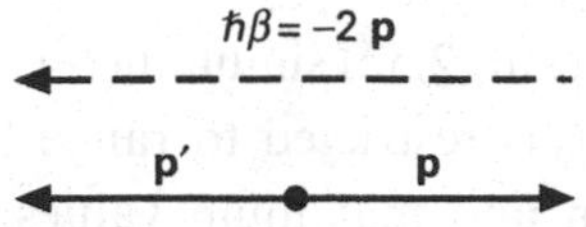

Fig. 2.8 Illustration of maximum phonon wave vector for elastic scattering.

2.5.2 Intravalley optical phonon scattering

For optical phonons, ω is nearly constant at $\omega(\boldsymbol{\beta}) \simeq \omega_0$ (recall again Fig. 1.27). According to Table 2.1 at the end of this chapter, the optical phonon energy, $\hbar\omega_0$, is typically tens of milli-electron volts, which is comparable to the thermal energy of an average carrier at room temperature. As a consequence, optical phonon scattering can rarely be considered elastic. With a constant $\omega(\boldsymbol{\beta})$, eq. (2.52) produces a quadratic equation for β that can be solved to obtain

$$\hbar\beta = p\left[\mp\cos\theta \pm \sqrt{\cos^2\theta \pm \frac{\hbar\omega_0}{E(p)}}\right], \tag{2.56}$$

from which we obtain the maximum wave vector of the optical phonon as

$$\hbar\beta_{\max}(OP) = p\left[1 + \sqrt{1 \pm \frac{\hbar\omega_0}{E(p)}}\right]. \tag{2.57}$$

Because the maximum wave vector must be a positive number, eq. (2.57) states that carriers cannot scatter by emitting optical phonons unless their energy exceeds the optical phonon energy. When typical numbers are inserted in eq. (2.57), we find that only optical phonons near the center of the Brillouin zone participate in intravalley scattering.

Application of energy and momentum conservation has identified the phonons involved in intravalley scattering as those whose wave vectors are near the center of the Brillouin zone. Acoustic phonon scattering is usually considered to be elastic, but optical phonon scattering can be considered elastic only for high energy carriers. Figures 2.9 and 2.10 illustrate typical carrier wave vectors before and after scattering from acoustic and optical phonons.

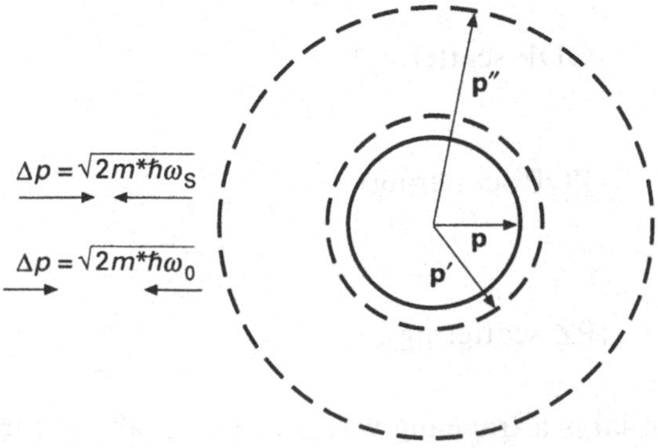

Fig. 2.9 Illustration of carrier scattering by absorbing an acoustic or optical phonon. The incident momentum is **p**. The momentum after scattering by absorbing an acoustic phonon is **p**′. The momentum after scattering by absorbing an optical phonon is **p**″. [2.1].

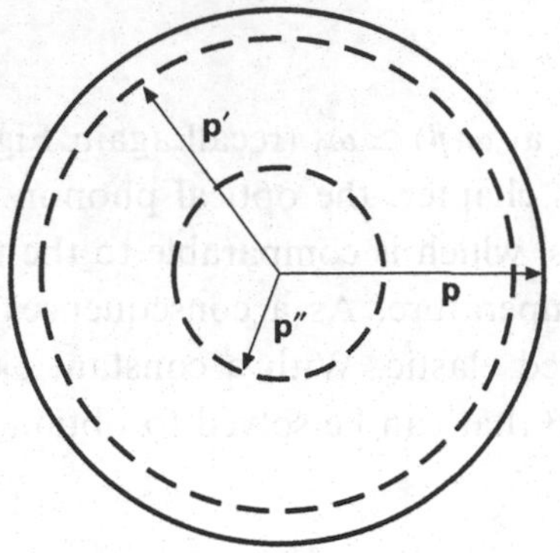

Fig. 2.10 Illustration of carrier scattering by emitting an acoustic or optical phonon. The incident momentum is $\mathbf{p}$. The momentum after scattering by emitting an acoustic phonon is $\mathbf{p}'$, and the momentum after scattering by emitting an optical phonon is $\mathbf{p}''$. [2.1].

2.6 Procedure for evaluating phonon scattering rates

In the next two sections, we evaluate the scattering and momentum relaxation rates for phonon scattering. Because the same general procedure is used for all types of phonon scattering, we present the general procedure first, then examine the various types of phonon scattering separately. Recall first from Section 2.2, that we can express the perturbing potential for phonon scattering as

$$U_S = K_\beta u_\beta, \tag{2.58}$$

where u_β is a Fourier component of the lattice vibration as given by eq. (2.18). From the results of Section 2.2.2, we have

$$|K_\beta|^2 = \beta^2 D_A^2 \qquad \text{(ADP scattering)} \tag{2.59a}$$

$$|K_\beta|^2 = D_o^2 \qquad \text{(ODP scattering)} \tag{2.59b}$$

$$|K_\beta|^2 = \left(\frac{\rho q^2 \omega_0^2}{\beta^2 \kappa_0 \varepsilon_0}\right)\left(\frac{\kappa_0}{\kappa_\infty} - 1\right) \qquad \text{(POP scattering)} \tag{2.59c}$$

$$|K_\beta|^2 = \left(\frac{q e_{PZ}}{\kappa_S \varepsilon_0}\right)^2 \qquad \text{(PZ scattering).} \tag{2.59d}$$

Because the perturbing potential is a traveling wave, we can evaluate the matrix element by analogy with eq. (1.127) in Section 1.7 to find

$$|H_{\mathbf{p}'\mathbf{p}}|^2 = |K_\beta|^2 |A_\beta|^2 \delta_{\mathbf{p}',\mathbf{p}\pm\hbar\boldsymbol{\beta}}. \tag{2.60}$$

To evaluate the transition rate, we insert eq. (2.60) in eq. (2.2) to find

$$S(\mathbf{p}, \mathbf{p}') = \frac{2\pi}{\hbar}|K_\beta|^2|A_\beta|^2\delta_{\mathbf{p},\mathbf{p}\pm\hbar\boldsymbol{\beta}}\delta(E' - E \mp \hbar\omega_\beta). \tag{2.61}$$

The two δ-functions in this equation are simply expressions of conservation of momentum and energy. To deal with the product of two δ-functions, we re-express them as a single δ-function expressing conservation of both momentum and energy. Beginning with

$$\mathbf{p}' = \mathbf{p} \pm \hbar\boldsymbol{\beta} \tag{2.62}$$

and taking its dot product with itself, we find

$$\mathbf{p}' \cdot \mathbf{p}' = p^2 \pm 2\hbar\mathbf{p} \cdot \boldsymbol{\beta} + \hbar^2\beta^2 \tag{2.63}$$

which, assuming parabolic energy bands ($E = p^2/2m^*$), we use to write

$$E' - E \mp \hbar\omega_\beta = \pm\frac{\hbar p\beta\cos\theta}{m^*} + \frac{\hbar^2\beta^2}{2m^*} \mp \hbar\omega_\beta \tag{2.64}$$

or

$$E' - E \mp \hbar\omega_\beta = \hbar\upsilon\beta\left(\pm\cos\theta + \frac{\hbar\beta}{2p} \mp \frac{\omega_\beta}{\upsilon\beta}\right). \tag{2.65}$$

By inserting this expression inside the δ-function in eq. (2.61), we obtain

$$\delta_{\mathbf{p}',\mathbf{p}\pm\hbar\boldsymbol{\beta}}\delta(E' - E \mp \hbar\omega_\beta) \rightarrow \frac{1}{\hbar\upsilon\beta}\delta\left(\pm\cos\theta + \frac{\hbar\beta}{2p} \mp \frac{\omega_\beta}{\upsilon\beta}\right) \tag{2.66}$$

(here, we made use of the fact that $\delta(ax) = \delta(x)/a$). Equation (2.66) replaces the product of two δ-functions, which express momentum and energy conservation, with a single δ-function that expresses both momentum and energy conservation. Finally, inserting eq. (2.66) in eq. (2.61), we find the transition rate for carriers scattering from $\mathbf{p}$ to $\mathbf{p}'$ as

$$S(\mathbf{p}, \mathbf{p}') = \frac{2\pi}{\hbar^2\upsilon\beta}|K_\beta|^2|A_\beta|^2\delta\left(\pm\cos\theta + \frac{\hbar\beta}{2p} \mp \frac{\omega_\beta}{\upsilon\beta}\right). \tag{2.67}$$

Before we proceed to evaluate scattering rates from eq. (2.2), however, we should recognize that $|A_\beta|^2$, the magnitude squared of the lattice vibration, is a classical concept. We need to deduce the proper quantum mechanical expression to replace $|A_\beta|^2$.

Recall that, classically, the energy of a vibration is proportional to the square of its magnitude. Quantum mechanically, however, the energy is quantized according to $E = (N + 1/2)\hbar\omega$, where $N = 0, 1, 2, \ldots$ is an integer. To relate $|A_\beta|^2$ to the classical energy, recall that the kinetic energy of a vibration is

$$KE = \frac{1}{2} M \left| \frac{du}{dt} \right|^2, \tag{2.68}$$

where M is the mass of the quantity oscillating. The lattice displacement with wavevector β is a real quantity given by

$$u(x, t) = A_\beta e^{i(\beta x - \omega t)} + A_\beta^* e^{-i(\beta x - \omega t)} \tag{2.69a}$$

or

$$u(x, t) = 2|A_\beta| \cos(\beta x - \omega_\beta t + \phi_\beta). \tag{2.69b}$$

By differentiating (2.69b) and inserting the result in eq. (2.68), we find the maximum kinetic energy as

$$KE|_{\max} = \frac{1}{2} \rho \Omega \omega_\beta^2 4 |A_\beta|^2 = E. \tag{2.70}$$

The energy oscillates between kinetic and potential energy, but their sum is constant, so the maximum kinetic energy is also the total, time-independent energy. By equating eq. (2.70) to the quantum mechanical result, we find

$$|A_\beta|^2 = \frac{\left(N_{\omega_\beta} + \frac{1}{2}\right)\hbar}{2\rho\Omega\omega_\beta}. \qquad \text{(wrong!)}$$

The result is correct on average, but more careful considerations show that when the lattice scattering event occurs by absorbing a quantized lattice vibration (i.e. a phonon), then we should make the replacement,

$$|A_\beta|^2 \rightarrow \frac{N_{\omega_\beta}\hbar}{2\rho\Omega\omega_\beta} \quad \text{(ABS)} \tag{2.71a}$$

and when it occurs by emitting a phonon,

$$|A_\beta|^2 \rightarrow \frac{(N_{\omega_\beta} + 1)\hbar}{2\rho\Omega\omega_\beta} \quad \text{(EMS).} \tag{2.71b}$$

We can write the general replacement more compactly as

$$\boxed{|A_\beta|^2 \rightarrow \frac{\hbar}{2\rho\Omega\omega_\beta}\left(N_{\omega_\beta} + \frac{1}{2} \mp \frac{1}{2}\right)}, \tag{2.71c}$$

where the top sign applies for phonon absorption and the bottom for phonon emission. This discussion should be considered as a simple plausibility argument for the prescription for the conversion to quantum mechanics as given by eq. (2.71c). For a more thorough treatment of the problem, consult Datta ([1.1], pp. 122–130).

After inserting the prescription, (2.71c), into the transition rate, eq. (2.67), we obtain

$$\boxed{S(\mathbf{p},\mathbf{p}') = C_\beta\left(N_\beta + \frac{1}{2} \mp \frac{1}{2}\right)\delta\left(\pm\cos\theta + \frac{\hbar\beta}{2p} \mp \frac{\omega}{\upsilon\beta}\right)}, \tag{2.72}$$

where

$$C_\beta = \frac{\pi m^* D_A^2}{\hbar\rho \upsilon_S p\Omega} \quad \text{(ADP)} \tag{2.73a}$$

$$C_\beta = \frac{\pi m^* D_o^2}{\hbar\rho\omega_0 \beta p\Omega} \quad \text{(ODP)} \tag{2.73b}$$

$$C_\beta = \frac{\pi m^* q^2\omega_0}{\hbar\kappa_0\varepsilon_0\beta^3 p\Omega}\left(\frac{\kappa_0}{\kappa_\infty} - 1\right) \quad \text{(POP)} \tag{2.73c}$$

$$C_\beta = \left(\frac{q e_{PZ}}{\kappa_S \varepsilon_0}\right)^2 \frac{\pi m^*}{\hbar\rho \upsilon_S \beta^2 p\Omega} \quad \text{(PZ).} \tag{2.73d}$$

2.6.1 Evaluation of the scattering rate

To evaluate the scattering rate, we sum the transition rate over all final states,

$$\frac{1}{\tau(p)} = \sum_{\mathbf{p}',\uparrow} S(\mathbf{p},\mathbf{p}') = \sum_{\boldsymbol{\beta}} S(\mathbf{p},\mathbf{p}'). \tag{2.74}$$

The second expression occurs because the mapping from $\mathbf{p}'$ to $\boldsymbol{\beta}$ is unique, so it doesn't matter if we take the sum over final states, $\mathbf{p}'$, or over $\boldsymbol{\beta}$ (recall Fig. 2.2). Converting the sum to an integral by the usual procedure,

$$\frac{1}{\tau} = \frac{\Omega}{8\pi^3}\int_0^{2\pi} \mathrm{d}\phi \int_0^\infty \left(N_\beta + \frac{1}{2} \mp \frac{1}{2}\right) C_\beta \beta^2 \mathrm{d}\beta \int_{-1}^{1} \delta\left(\pm\cos\theta + \frac{\hbar\beta}{2p} \mp \frac{\omega}{\upsilon\beta}\right)\mathrm{d}(\cos\theta). \tag{2.75}$$

Consider the integral over $\mathrm{d}(\cos\theta)$ first (recall that θ is the angle between $\mathbf{p}$ and $\boldsymbol{\beta}$). This integral is of the form

$$\int_{x_1}^{x_2} \delta(x - x_0)\mathrm{d}x,$$

where $x = \cos\theta$ and $x_0 = \mp(\hbar\beta/2p) + (\omega/\upsilon\beta)$. This integral is unity, as long as $x_1 < x_0 < x_2$, and it is zero otherwise. If β is either too large or too small, then the argument of the δ-function will never go to zero, so the effect of the δ-

function in $\cos\theta$ is simply to restrict the integration over β. Equation (2.75) becomes

$$\frac{1}{\tau} = \frac{\Omega}{4\pi^2}\int_{\beta_{min}}^{\beta_{max}}\left(N_\beta + \frac{1}{2} \mp \frac{1}{2}\right)C_\beta \beta^2 d\beta \,. \tag{2.76}$$

The minimum and maximum phonon wavevectors, β_{min} and β_{max}, are the minimum and maximum values of β for which the argument of the δ-function goes to zero. These are simply the minimum and maximum phonon wavevectors for which both energy and momentum are conserved, and the results are just those stated in Section 2.5.

2.6.2 Evaluation of the momentum relaxation rate

For the momentum relaxation rate, we evaluate a sum of the form

$$\frac{1}{\tau_m(p)} = \sum_\beta S(\mathbf{p}, \mathbf{p}')\left(1 - \frac{p'}{p}\cos\alpha\right), \tag{2.77}$$

where α is the angle between $\mathbf{p}$ and $\mathbf{p}'$. The geometry of the scattering event was illustrated in Fig. 2.2; recall again that we have identified θ as the angle between $\mathbf{p}$ and β'. Note that

$$\left(1 - \frac{p'}{p}\cos\alpha\right) = 1 - \frac{\mathbf{p}\cdot(\mathbf{p} \pm \hbar\boldsymbol{\beta})}{p^2} = \mp\frac{\hbar\beta\cos\theta}{p}. \tag{2.78}$$

With eq. (2.75), we can write the momentum relaxation rate as

$$\frac{1}{\tau_m} = \frac{\Omega}{8\pi^3}\int_0^{2\pi} d\phi \int_0^{\infty}\left(N_\beta + \frac{1}{2} \mp \frac{1}{2}\right)C_\beta\beta^2 d\beta \int_{-1}^{1}\delta\left(\pm\cos\theta + \frac{\hbar\beta}{2p} \mp \frac{\omega}{\upsilon\beta}\right) \times \left(\mp\frac{\hbar\beta\cos\theta}{p}\right)d(\cos\theta). \tag{2.79}$$

In this case, the integral over θ is of the form

$$\int_{x_1}^{x_2}\delta(x - x_0)f(x)dx,$$

which is $f(x_0)$ for $\beta_{min} < \beta < \beta_{max}$. The result is that the momentum relaxation rate becomes

$$\boxed{\frac{1}{\tau_{\mathrm{m}}} = \frac{\Omega}{4\pi^2}\int_{\beta_{\min}}^{\beta_{\max}}\left(N_\beta + \frac{1}{2} \mp \frac{1}{2}\right) C_\beta \left(\frac{\hbar\beta}{2p} \mp \frac{\omega}{\upsilon\beta}\right)\frac{\hbar\beta^3}{p}\,\mathrm{d}\beta} \quad (2.80)$$

Equations (2.76) and (2.80) are the main results of this section. They give the scattering and momentum relaxation rates for phonon scattering in general. To evaluate these expressions, we identify the appropriate C_β from eqs. (2.73), the appropriate minimum and maximum phonon wavevectors from Sec. 2.5, then evaluate the integrals. For acoustic phonons, $\omega = \omega_{\mathrm{S}} = \beta\upsilon_{\mathrm{S}}$, and for optical phonons $\omega = \omega_0 \approx$ constant.

2.7 Deformation potential scattering

As discussed in Section 2.2, the perturbing potential for phonon scattering can be the deformation potential, or in polar materials there is also an electrostatic interaction. In this section we treat deformation potential scattering by nonpolar acoustic or optical phonons.

2.7.1 Acoustic deformation potential (ADP) scattering

To evaluate scattering rates for acoustic phonon scattering via the deformation potential scattering, we begin with eq. (2.76) and insert eq. (2.73a) to find

$$\frac{1}{\tau} = \frac{m^* D_{\mathrm{A}}^2}{4\pi\hbar\rho\upsilon_{\mathrm{s}}p}\int_{\beta_{\min}}^{\beta_{\max}}\left(N_{\omega_s} + \frac{1}{2} \mp \frac{1}{2}\right)\beta^2\,\mathrm{d}\beta. \quad (2.81)$$

At room temperature, the number of acoustic phonons is large, so $N_{\omega_s} \simeq N_{\omega_s} + 1$. Recalling eq. (1.134), we can invoke equipartition, $N_{\omega_s} \simeq k_{\mathrm{B}}T_{\mathrm{L}}/\hbar\omega_s$, because $\hbar\omega_s \ll k_{\mathrm{B}}T_{\mathrm{L}}$. With this approximation, we simplify eq. (2.81) for room temperature applications as

$$\frac{1}{\tau} = \frac{m^* D_{\mathrm{A}}^2 k_{\mathrm{B}}T_{\mathrm{L}}}{2\pi\hbar^2 c_l p}\int_{\beta_{\min}}^{\beta_{\max}}\beta\,\mathrm{d}\beta = \frac{m^* D_{\mathrm{A}}^2 k_{\mathrm{B}}T_{\mathrm{L}}}{4\pi\hbar^2 c_l p}(\beta_{\max}^2 - \beta_{\min}^2), \quad (2.82)$$

where we have used

$$\frac{\omega_s}{\beta} = \upsilon_s = \sqrt{\frac{c_l}{\rho}}. \quad (2.83)$$

to relate the sound velocity, υ_{s}, elastic constant, c_l, and mass density, ρ. Equation (2.82) is the sum of transitions due to acoustic phonon absorption and emission.

To evaluate eq. (2.82), we need to specify the minimum and maximum phonon wavevectors involved. From eq. (2.53), and the reasonable assumption that ADP scattering is approximately elastic near room temperature, we find that $\hbar\beta_{\max} = 2m^* \upsilon(p)$ and $\hbar\beta_{\min} = 0$ which can be inserted in eq. (2.82) to find

$$\boxed{\frac{1}{\tau(p)} = \frac{1}{\tau_m(p)} = \frac{\pi D_A^2 k_B T_L}{\hbar c_l} g_C(E)}, \tag{2.84}$$

where $g_C(E)$ is the density of states defined in eq. (2.9). The result, which shows that the scattering rate is proportional to the number of final states available, could have been obtained by simpler methods, but the general procedure employed here will prove useful for more complex scattering potentials. We have indicated in eq. (2.84) that the scattering and momentum relaxation rates are equal; acoustic phonon scattering is isotropic. This can be verified by evaluating $1/\tau_m$ from eq. (2.80), but the two rates are equal only when acoustic phonon scattering can be regarded as elastic (at low temperatures, ADP scattering is anisotropic). Finally, it is worth noting and remembering that $\tau_m(p)$ for ADP scattering has the power law form of (2.32) with $s = -1/2$. Figure 2.11a is a plot of the ADP scattering rate versus energy for electrons in silicon.

2.7.2 Optical deformation potential (ODP) scattering

The treatment of optical deformation potential (ODP) scattering proceeds much like ADP scattering, except that it can't be considered as elastic unless the carrier energy is very high. Beginning with eq. (2.76) and using eq. (2.73b), we find

$$\frac{1}{\tau} = \frac{m^* D_o^2}{4\pi\hbar\rho\omega_o}\left(N_o + \frac{1}{2} \mp \frac{1}{2}\right)\int_{\beta_{\min}}^{\beta_{\max}} \beta d\beta, \tag{2.85}$$

where $\hbar\omega_o$ is the optical phonon energy. After integrating this expression and obtaining $\beta_{\min}$ and $\beta_{\max}$ from eq. (2.56), we find

$$\boxed{\frac{1}{\tau(p)} = \frac{1}{\tau_m(p)} = \frac{\pi D_o^2}{2\rho\omega_o}(N_o + 1/2 \mp 1/2) g_C(E \pm \hbar\omega_o)}. \tag{2.86}$$

Note that the density of states is nonzero for positive arguments only and that optical phonon scattering is also isotropic. Equation (2.86) states that a carrier with any energy can scatter by absorbing optical phonons, but only those whose energy exceeds $\hbar\omega_o$ can emit optical phonons. The result is that a plot of the scattering rate versus carrier energy displays a threshold at $\hbar\omega_o$ above which the scattering rate greatly increases. Figure 2.11b is a sketch illustrating how the ODP scattering rate varies with energy.

High energy carriers shed their energy by emitting optical phonons, and the energy loss can be characterized by the energy relaxation time,

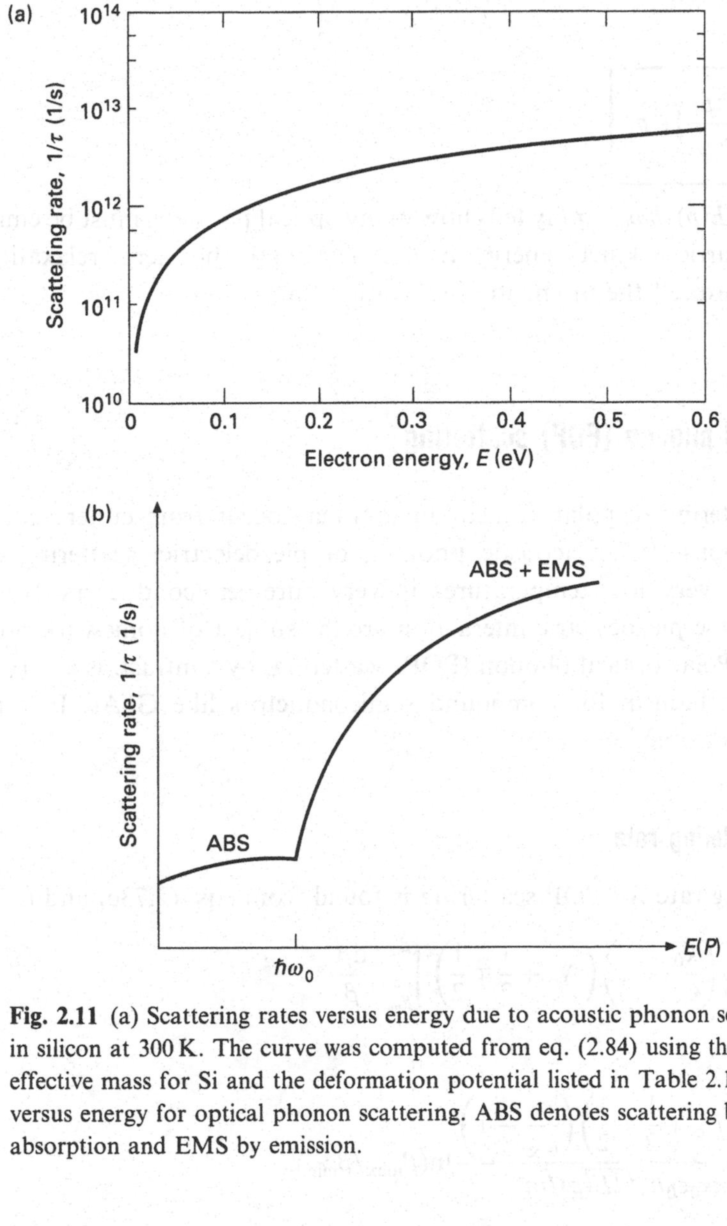

Fig. 2.11 (a) Scattering rates versus energy due to acoustic phonon scattering of electrons in silicon at 300 K. The curve was computed from eq. (2.84) using the density of states effective mass for Si and the deformation potential listed in Table 2.1. (b) Scattering rate versus energy for optical phonon scattering. ABS denotes scattering by optical phonon absorption and EMS by emission.

$$\frac{1}{\tau_E(p)} = \sum_{\mathbf{p}',\uparrow} S(\mathbf{p}, \mathbf{p}')(1 - E'/E). \tag{2.87}$$

For very high energy carriers, phonon emission greatly exceeds absorption, so $E(\mathbf{p}') = E(\mathbf{p}) - \hbar\omega_o$, and we find the energy relaxation rate as

$$\frac{1}{\tau_E(p)} = \sum_{\mathbf{p}',\uparrow} S(\mathbf{p}, \mathbf{p}')\frac{\hbar\omega_o}{E(p)} = \frac{\hbar\omega_o}{E(p)}\frac{1}{\tau(p)}$$

or

$$\boxed{\tau_E(p) = \left(\frac{E(p)}{\hbar\omega_o}\right)\tau(p)} \quad . \tag{2.88}$$

The fraction $E(p)/\hbar\omega_o$ simply tells how many optical phonons must be emitted to remove the carrier's kinetic energy, $E(p)$. Accordingly, the energy relaxation time may greatly exceed the momentum relaxation time.

2.8 Polar optical phonon (POP) scattering

Phonon scattering in polar semiconductors may occur from either acoustic or optical phonons. Polar acoustic phonon, or piezoelectric, scattering can be important at very low temperatures in very pure semiconductors. Scattering rates due to the piezoelectric interaction are the subject of homework problems 2.4 and 2.5. Polar optical phonon (POP) scattering, by contrast, is a very strong scattering mechanism for compound semiconductors like GaAs. It is neither elastic nor isotropic.

2.8.1 The POP scattering rate

The scattering rate for POP scattering is found from eqs. (2.73c) and (2.76), as

$$\frac{1}{\tau} = \frac{m^* q^2 \omega_o}{4\pi\hbar\kappa_0\varepsilon_0 p}\left(\frac{\kappa_0}{\kappa_\infty} - 1\right)\left(N_o + \frac{1}{2} \mp \frac{1}{2}\right)\int_{\beta_{min}}^{\beta_{max}} \frac{d\beta}{\beta} \tag{2.89}$$

or

$$\frac{1}{\tau(p)} = \frac{q^2\omega_o\left(N_o + \frac{1}{2} \mp \frac{1}{2}\right)\left(\frac{\kappa_0}{\kappa_\infty} - 1\right)}{4\pi\kappa_0\varepsilon_0\hbar\sqrt{2E(p)/m^*}} \ln(\beta_{max}/\beta_{min}), \tag{2.90}$$

and all that remains is to specify β_{max} and β_{min}.

To find the maximum and minimum phonon wave vectors, we set the argument of the δ-function in eq. (2.72) to zero and find a quadratic equation,

$$\beta^2 \pm \frac{(2p\cos\theta)}{\hbar}\beta \mp \frac{2p\omega_o}{\hbar\upsilon} = 0, \tag{2.91}$$

whose solutions give those values of β which satisfy energy and momentum conservation for a given scattering angle θ. Solving for β_{max} and β_{min} (from $-1 \le \cos\theta \le +1$), we find

$$\beta_{\max} = \frac{p}{\hbar}\left(1 \pm \sqrt{1 \pm \frac{\hbar\omega_o}{E(p)}}\right) \tag{2.92}$$

and

$$\beta_{\min} = \frac{p}{\hbar}\left(\mp 1 \pm \sqrt{1 \pm \frac{\hbar\omega_o}{E(p)}}\right), \tag{2.93}$$

which are very similar to the results of Section 2.5. With these results, the POP scattering rate becomes

$$\frac{1}{\tau(p)} = \frac{q^2\omega_o\left(\frac{\kappa_0}{\kappa_\infty} - 1\right)}{2\pi\kappa_0\varepsilon_0\hbar\sqrt{2E(p)/m^*}}\left[N_o \sinh^{-1}\left(\frac{E(p)}{\hbar\omega_o}\right)^{1/2} + (N_o + 1)\sinh^{-1}\left(\frac{E(p)}{\hbar\omega_o} - 1\right)^{1/2}\right], \tag{2.94}$$

where the first term represents POP absorption, the second POP emission. It is understood that the second term applies only when $E(p) > \hbar\omega_o$, so that emission can occur. In going from eq. (2.90) to eq. (2.94), we made use of the identity

$$\sinh^{-1}(x) = \ln\left[x + \sqrt{1 + x^2}\right].$$

Figure 2.12 is a plot of the POP scattering rate versus carrier energy for electrons in GaAs at room temperature. The onset of phonon emission at $E(p) \simeq \hbar\omega_o \simeq 35\,\text{meV}$ is readily apparent. Notice that in contrast to ADP and ODP scattering, the POP scattering rate is roughly constant at high energies.

2.8.2 The POP energy relaxation time

The energy relaxation rate due to POP scattering is found by weighting each transition by the fractional change in energy. For high energy electrons, POP emission dominates and from eq. (2.5) and eq. (2.94) we find

$$\tau_E(p) = \left(\frac{E(p)}{\hbar\omega_o}\right)\tau(p) = \left(\frac{2\pi\kappa\varepsilon_0}{q^2\omega_o^2\left(\frac{\kappa_0}{\kappa_\infty} - 1\right)}\right)\frac{E(p)\sqrt{2E(p)/m^*}}{(N_o + 1)\sinh^{-1}\left(\frac{E(p)}{\hbar\omega_o} - 1\right)^{1/2}}. \tag{2.95}$$

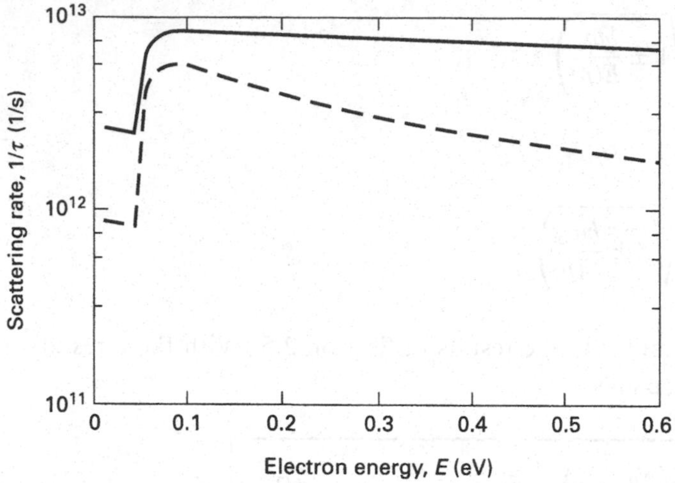

Fig. 2.12 Polar optical phonon scattering rates versus energy for electrons in GaAs at room temperature. Parabolic bands are assumed. The solid line is $1/\tau(p)$, the dashed line $1/\tau_m(p)$.

2.8.3 The POP momentum relaxation time

The moment relaxation time is found from eq. (2.80), which works out much like the sum in eq. (2.89) with the result

$$\frac{1}{\tau_m(p)} = \frac{q^2\omega_o\left(\frac{\kappa_0}{\kappa_\infty} - 1\right)}{4\pi\kappa_0\varepsilon_0\hbar\sqrt{2[E(p)/m^*]}}\left[N_o\sqrt{1+\frac{\hbar\omega_o}{E(p)}} + (N_o+1)\sqrt{1-\frac{\hbar\omega_o}{E(p)}}\right.$$
$$\left. -\frac{\hbar\omega_o N_o}{E(p)}\sinh^{-1}\left(\frac{E(p)}{\hbar\omega_o}\right)^{1/2} + \frac{\hbar\omega_o(N_o+1)}{E(p)}\sinh^{-1}\left(\frac{E(p)}{\hbar\omega_o} - 1\right)^{1/2}\right]. \quad (2.96)$$

The momentum relaxation rate versus energy is also plotted in Fig. 2.12. Notice that τ_m exceeds τ, which is a consequence of the fact that POP scattering favors small angle scattering events which have little effect on momentum relaxation.

2.9 Intervalley scattering

As displayed in Fig. 2.13, the constant energy surfaces for electrons in Si and GaAs consist of several valleys. For Si, the valleys are energetically equivalent and lie along [100] directions near the zone boundary. Two types of intervalley scattering are possible in Si; 'g-type' processes move a carrier from a given valley to one on the opposite side of the same axis (e.g., from a valley along $\langle 100\rangle$ to one along $\langle\bar{1}00\rangle$). The 'f-type' processes move a carrier to one of the remaining valleys. Both g- and f-type scattering produce very large changes in momentum,

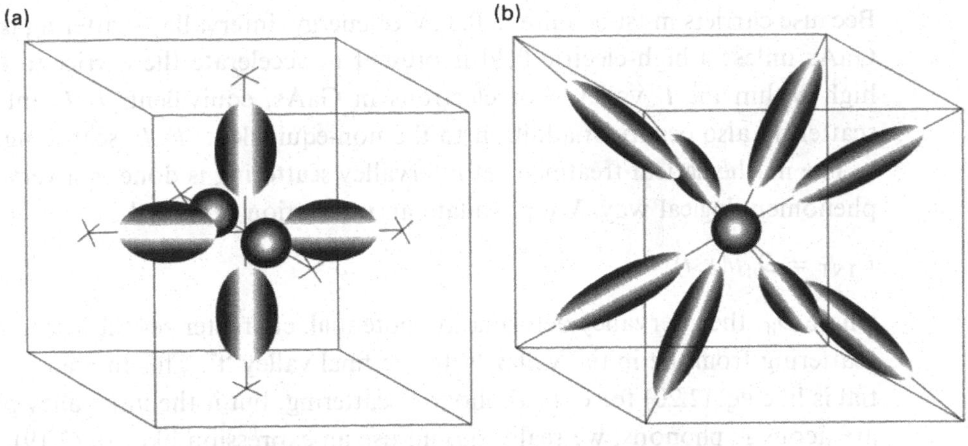

Fig. 2.13 (a) constant energy surfaces for silicon, and (b) constant energy surfaces for gallium arsenide.

so they require phonons with wave vectors near the zone boundary. Such phonons are termed *intervalley phonons* and may be either acoustic or optical phonons. Note from Fig. 1.27a that near the zone boundary the energies of both acoustic and optical phonons are comparable and are somewhat less than the longitudinal optical phonon energy, $\hbar\omega_0$. The specific phonons involved in g- and f-type scattering are listed in Table 2.1 at the end of this chapter.

Intervalley scattering in GaAs is somewhat different because the valleys shown in Fig. 2.13b are not energetically equivalent. The central, Γ, valley lies about 0.3 eV below the ellipsoidal, L, valleys located along $\langle 111 \rangle$ directions. An illustration in energy–momentum space of a Γ to L transition is shown in Fig. 2.14.

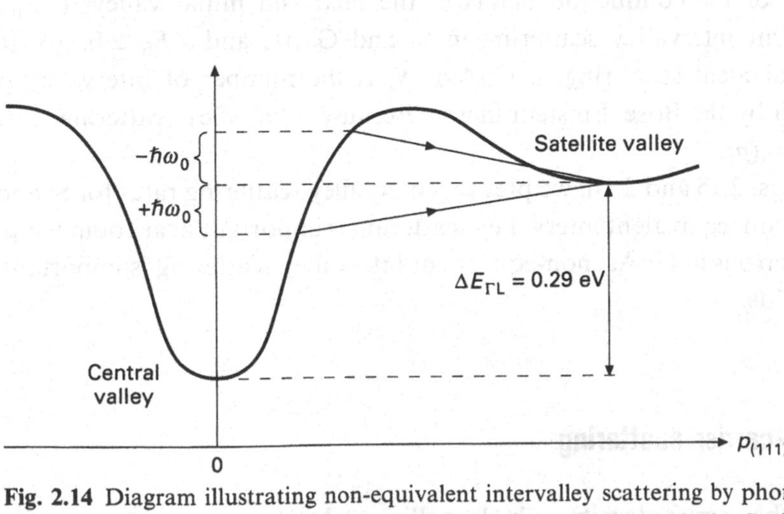

Fig. 2.14 Diagram illustrating non-equivalent intervalley scattering by phonon absorption and emission in GaAs. (From Seeger, *Semiconductor Physics*. 3rd edn, Springer-Verlag, New York, 1985.)

Because carriers must acquire $\simeq 0.3$ eV of energy, intervalley scattering is rare in GaAs unless a high-electric field is present to accelerate the carrier to energies high within the Γ valley. For electrons in GaAs, equivalent, L–L, intervalley scattering also occurs in addition to the non-equivalent, Γ–L, scattering.

The mathematical treatment of intervalley scattering is done in a very simple, phenomenological way. We postulate an interaction potential,

$$U_{\text{INT}} = D_{\text{if}} u(x, t), \tag{2.97}$$

where D_{if}, the intervalley deformation potential, characterizes the strength of the scattering from the initial valley 'i' to the final valley 'f'. This interaction potential is like eq. (2.20) for optical phonon scattering, but if the intervalley phonons are acoustic phonons, we really should use an expression like eq. (2.19). Such a choice would simply change our final result by a constant, which could be absorbed in the definition of D_{if}.

When we evaluated the scattering rates for optical phonon scattering, we assumed that the phonon energy was constant. For intervalley scattering, the phonon momentum is large and nearly constant, so $\hbar\omega_{\text{if}}$, the intervalley phonon energy, can be assumed to be constant. As a consequence, the intervalley scattering rate works out just like that for nonpolar optical phonon scattering. By analogy with eq. (2.86) we find

$$\frac{1}{\tau(p)} = \frac{1}{\tau_m(p)} = \frac{\pi D_{\text{if}}^2 Z_f}{2\rho\omega_{\text{if}}}\left(N_i + \tfrac{1}{2} \mp \tfrac{1}{2}\right) g_{\text{Cf}}(E \pm \hbar\omega_{\text{if}} - \Delta E_{\text{fi}}), \tag{2.98}$$

where Z_f is the number of final valleys available for scattering, and g_{Cf} is the density of states in the final valley. The term ΔE_{fi} is the difference between the bottom of the conduction bands in the final and initial valleys ($\Delta E_{\text{fi}} = 0$ for equivalent intervalley scattering in Si and GaAs, and $\Delta E_{\text{fi}} \simeq 0.3$ eV for Γ–L, non-equivalent scattering in GaAs). N_i is the number of intervalley phonons as given by the Bose–Einstein factor. Because intervalley scattering is isotropic, $\tau(p) = \tau_m(p)$.

In Figs. 2.15 and 2.16, we plot the intervalley scattering rates for Si and GaAs. For silicon, equivalent intervalley scattering is important near room temperature. For electrons in GaAs, non-equivalent intervalley scattering is important only at high fields.

2.10 Carrier–carrier scattering

When the carrier density is high, collisions between carriers are an important scattering mechanism. Two types of processes must be distinguished – a binary

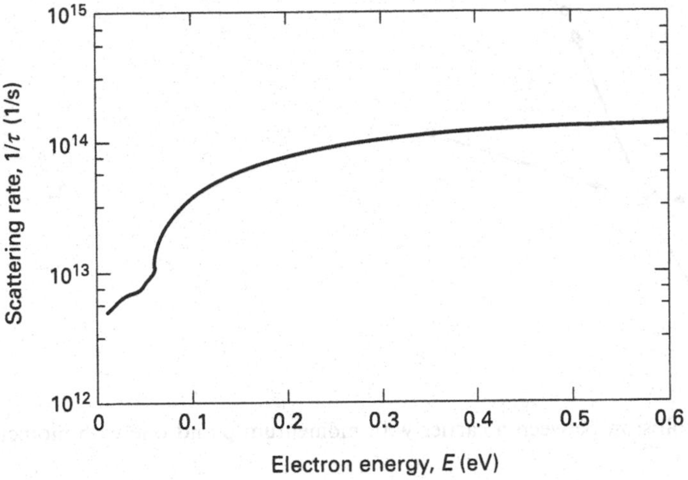

Fig. 2.15 Equivalent intervalley scattering rates in silicon due to f-type and g-type scattering processes. The intervalley deformation potentials and phonon energies are listed in Table 2.1.

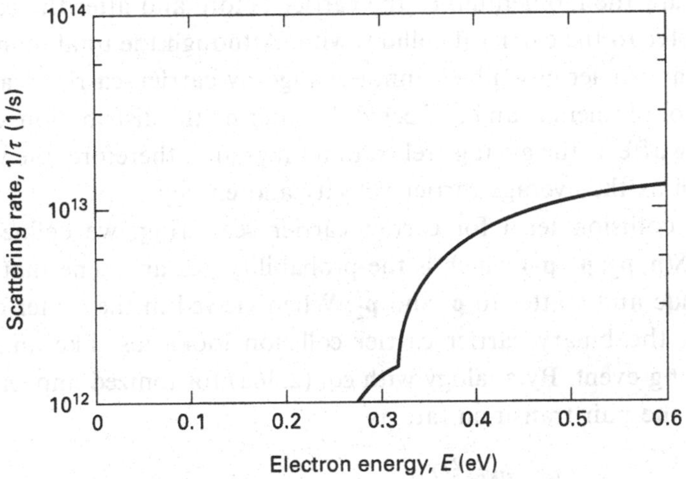

Fig. 2.16 Non-equivalent, Γ–L, scattering rate in GaAs at room temperature.

process in which one carrier collides with another and a collective process in which a carrier interacts with the plasma comprised by the carriers.

2.10.1 Binary carrier-carrier scattering

For the binary process, which is depicted in Fig. 2.17, momentum and energy conservation dictate that

$$\mathbf{p} + \mathbf{p}_2 = \mathbf{p}' + \mathbf{p}_2' \tag{2.99a}$$

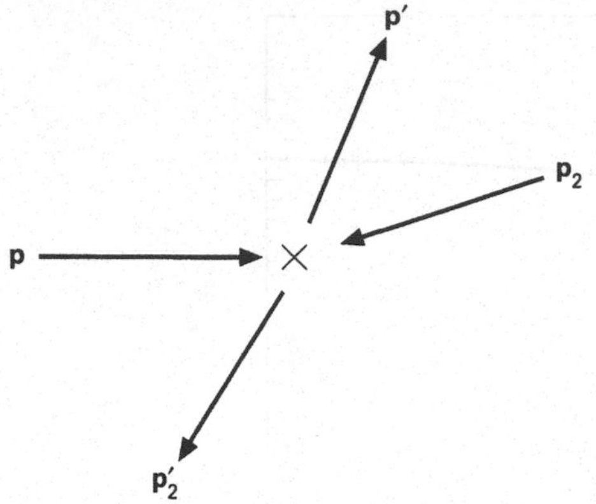

Fig. 2.17 Binary collision between a carrier with momentum **p** and one with momentum $\mathbf{p}_2$.

and

$$E(\mathbf{p}) + E(\mathbf{p}_2) = E(\mathbf{p}') + E(\mathbf{p}_2'), \tag{2.99b}$$

where **p** and $\mathbf{p}'$ are the momentum of the carrier before and after the collision, and $\mathbf{p}_2$ and $\mathbf{p}_2'$ refer to the carrier it collides with. Although the total momentum and energy of the carrier ensemble cannot change by carrier–carrier scattering, the distribution of momenta can be affected. By altering the distribution, carrier–carrier scattering affects the average relaxation times and, therefore, the value of observables such as the average carrier velocity and energy.

To write the collision term for carrier–carrier scattering, we define a *pair transition rate*, $S(\mathbf{p}, \mathbf{p}_2; \mathbf{p}', \mathbf{p}_2')$ which is the probability per unit time that carrier at **p** and $\mathbf{p}_2$ collide and scatter to $\mathbf{p}'$ and $\mathbf{p}_2'$. When viewed in the center-of-mass reference frame, the binary carrier–carrier collision looks just like an ionized impurity scattering event. By analogy with eq. (2.36a) for ionized impurity scattering, we write the pair transition rate as

$$S(\mathbf{p}, \mathbf{p}_2; \mathbf{p}', \mathbf{p}_2') = \frac{2\pi q^4/\hbar\kappa_S^2\varepsilon_0^2\Omega}{[4(p/\hbar)^2\sin^2(\alpha/2) + 1/L_D^2]^2} \times \delta_{\mathbf{p}+\mathbf{p}_2,\mathbf{p}'+\mathbf{p}_2'} \times \delta[E(\mathbf{p}) + E(\mathbf{p}_2) - E(\mathbf{p}') - E(\mathbf{p}_2')]. \tag{2.100}$$

(For details of this derivation, consult Ridley [2.1].)

To evaluate the scattering rate due to binary carrier–carrier scattering, we weight the pair transition rate by the probability that a target carrier is present and by the probability that the final states at $\mathbf{p}'$ and $\mathbf{p}_2'$ are empty. The result is summed over the final states, $\mathbf{p}'$ and over the target states, $\mathbf{p}_2$, to obtain

$$\frac{1}{\tau(p)} = \sum_{\mathbf{p}_2}\sum_{\mathbf{p}'} S(\mathbf{p}, \mathbf{p}_2; \mathbf{p}', \mathbf{p}_2') f(\mathbf{p}_2)[1 - f(\mathbf{p}')][1 - f(\mathbf{p}_2')]. \tag{2.101}$$

[A separate sum over $\mathbf{p}_2'$ is not performed because it is uniquely determined from $\mathbf{p}$, $\mathbf{p}_2$, and $\mathbf{p}'$ according to eq. (2.99a).] In eq. (2.101), $f(\mathbf{p}')$ is the probability that the state at $\mathbf{p}'$ is occupied and is known as the distribution function. To find $f(\mathbf{p}_2)$, a complicated integrodifferential equation known as the Boltzmann Transport Equation (BTE) must be solved. The requirement that we know the distribution function makes carrier–carrier scattering extremely difficult to treat. The BTE, and techniques to solve it for the distribution function, are the subjects of Chapter 3.

2.10.2 Collective carrier-carrier scattering

In addition to binary collisions, carriers also interact with oscillations in the carrier density. Such fluctuations are accompanied by electric fields which oppose the fluctuation and produce oscillations at the *plasma frequency*,

$$\omega_{\mathrm{p}} = \left(\frac{q^2 n}{\kappa_S \varepsilon_0 m^*}\right)^{1/2}, \tag{2.102}$$

which can be sustained if

$$\omega_{\mathrm{p}} \tau \geq 1, \tag{2.103}$$

(to ensure that collisions don't damp out the oscillations). Because τ is on the order of one picosecond, eq. (2.103) implies that plasma oscillations can be sustained if the carrier density exceeds about 10^{17} cm^{-3}. Carriers which scatter from plasma oscillations interact with the ensemble of carriers rather than with a single carrier as in the binary process.

The charge density oscillation of the plasma can be written as

$$\rho_\beta = A_\beta \mathrm{e}^{\pm i(\beta x - \omega_p t)} \tag{2.104}$$

and the electric field it produces as

$$\mathcal{E}_\beta = \int \frac{\rho_\beta \mathrm{d}x}{\kappa_S \varepsilon_0} = \frac{\rho_\beta}{i\beta \kappa_S \varepsilon_0}. \tag{2.105}$$

The interaction potential is obtained by integrating again to find

$$U_{\mathrm{PL}} = -q \int \mathcal{E}_\beta \mathrm{d}x = \frac{q\rho_\beta}{\kappa_S \varepsilon_0 \beta^2} \tag{2.106}$$

from which the matrix element for carrier–plasma scattering is obtained as

$$H_{\mathbf{p}',\mathbf{p}} = \frac{qA_\beta}{\kappa_S \varepsilon_0 \beta^2} \delta_{\mathbf{p}',\mathbf{p}\pm\hbar\boldsymbol{\beta}}. \tag{2.107}$$

So far we have treated the plasma oscillation classically, but its energy should be quantized in units of $\hbar\omega_p$. By equating the classical, electrostatic energy to its quantum mechanical counterpart, we find

$$A_\beta^2 \to \frac{\hbar\omega_p \kappa_S \varepsilon_0 \beta^2}{2\Omega}(N_p + 1/2 \mp 1/2), \tag{2.108}$$

N_p represents the number of *plasmons* (quantized plasma oscillations) as given by the Bose–Einstein factor. When eq. (2.108) is inserted in eq. (2.107) and used in the Golden Rule, the transition rate is found to be

$$S(\mathbf{p}, \mathbf{p}') = \frac{\pi q^2 \omega_p}{\kappa_S \varepsilon_0 \beta^2 \Omega}\left(N_p + \frac{1}{2} \mp \frac{1}{2}\right)\delta_{\mathbf{p}',\mathbf{p}\pm\hbar\boldsymbol{\beta}}\delta(E - E \mp \hbar\omega_p). \tag{2.109}$$

Using eq. (2.66) for the δ-functions, this becomes,

$$S(\mathbf{p}, \mathbf{p}') = C_\beta\left(N_p + \frac{1}{2} \mp \frac{1}{2}\right)\delta\left(\pm\cos\theta + \frac{\hbar\beta}{2p} \mp \frac{\omega}{\upsilon\beta}\right), \tag{2.110a}$$

where

$$C_\beta = \frac{\pi m^* q^2 \omega_p}{\hbar \kappa_S \varepsilon_0 \beta^3 p \Omega}. \tag{2.110b}$$

Since the transition rate for plasmon scattering is so similar to that for POP scattering [compare eqs. (2.110) with eqs. (2.72) and (2.73c)], the results of Section 2.8 may be used directly. By analogy with eq. (2.90), we find the scattering rate for plasmon scattering as

$$\frac{1}{\tau(p)} = \frac{q^2 \omega_p (N_p + 1/2 \mp 1/2)}{4\pi\kappa_S \varepsilon_0 \hbar\sqrt{2E(p)/m^*}} \ln(\beta_{max}/\beta_{min}), \tag{2.111}$$

where β_{max} and β_{min} are determined by energy and momentum conservation as in eqs. (2.92) and (2.93), but some care is required when specifying β_{max}. A large β_{max} refers to short wavelength oscillations, but about one Debye length is required to screen out the charge of a carrier. When β_{max} exceeds about $1/L_D$, the scattering should be treated as a binary collision. Since eq. (2.111) does not apply to binary collisions, β_{max} is replaced by β_{co}, which is equal to β_{max} from eq. (2.92) or $1/L_D$, whichever is smaller. The scattering rate for plasmon scattering becomes

$$\frac{1}{\tau(p)} = \frac{q^2\omega_p\left(N_p + \frac{1}{2} \mp \frac{1}{2}\right)}{4\pi\kappa_S\varepsilon_0\hbar\sqrt{2E(p)/m^*}} \ln\left[\frac{\hbar\beta_{co}}{p(\mp 1 \pm \sqrt{1 \pm \hbar\omega_p/E(p)})}\right]. \tag{2.112}$$

In Fig. 2.18 we plot the scattering rate versus energy for electron–plasmon scattering in GaAs at room temperature. For high carrier densities, plasmon scattering is an important component of the total scattering rate, but when the electron density exceeds about 10^{18} cm^{-3}, the computations become more involved because the plasma oscillations couple with the LO phonons [2.9], and the scattering rate from these *coupled modes* must be evaluated. Finally, we should note that although the electron–plasmon scattering rate may be high, the effects of electron–plasmon scattering can be subtle. If the plasma oscillations are not heavily damped by phonon scattering, then the total momentum of the electron ensemble is conserved. One electron may lose momentum to the plasma, but another will gain momentum. In this case, electron–plasmon scattering is similar to binary electron–electron scattering, and the biggest effect may be simply to change the shape of the electron distribution function.

2.11 Phonon scattering of confined carriers

In a bulk semiconductor, carriers are free to move in three dimensions, but in modern semiconductor devices, carriers are often confined in quantum wells where they can move in only two dimensions. Important examples are silicon

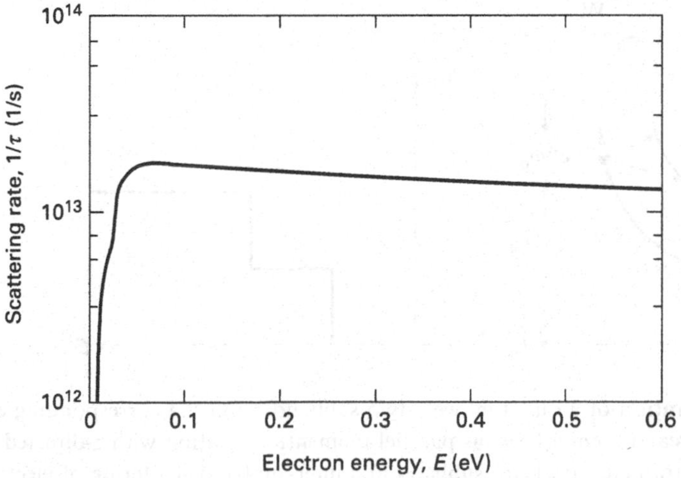

Fig. 2.18 Electron–plasmon scattering rate versus energy for electrons in GaAs at room temperature. The electron density is $n_0 = 1.0 \times 10^{17}$ cm^{-3}.

MOSFETs and modulation-doped III-V FETs where the carriers in the channel are confined near the surface in a potential well. The scattering rates for these two-dimensional carriers are different than for three-dimensional carriers. This section is an introduction to the treatment of carrier scattering in reduced dimensional structures.

2.11.1 Brief review of quantum confinement

Quantum wells

Figure 2.19 reviews some fundamentals of quantum wells. In Fig. 2.19a we show a GaAs quantum well; the confining barriers are a wider bandgap semiconductor such as $Al_xGa_{1-x}As$. Because of the confinement in the z-direction, k_z is quantized. For an infinitely deep well, momentum conservation does not apply. Since the carriers are unconfined in the x–y plane, however, momentum conservation continues to hold in the plane.

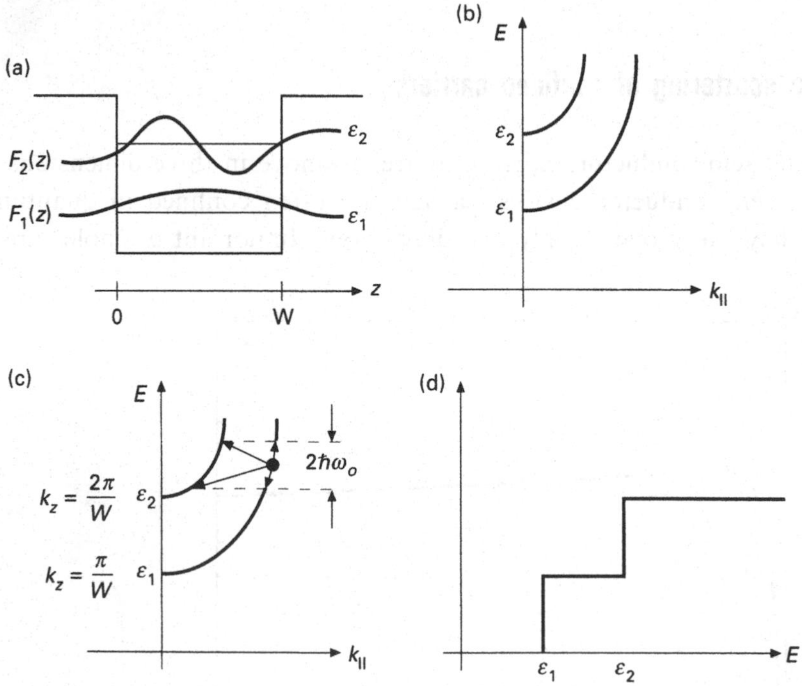

Fig. 2.19 (a) Illustration of a quantum well, two subbands, and the corresponding envelope functions, (b) The carrier energy versus parallel momentum relation with z-directed momentum as a parameter, (c) Intra-subband and inter-subband scattering of electrons in a quantum well. Scattering by optical phonons is assumed. (d) Expected shape of the scattering rate versus energy (mimics the density of states).

The calculation of two-dimensional (2D) scattering rates proceeds much like it does for three-dimensional (3D) electrons, but the proper wave function for confined carriers must be used. We should also consider the possibility of 2D phonons as well as 2D electrons, but for many quantum wells, the elastic constants of the well are similar to those of the surrounding media, so 3D phonons can be assumed. While the calculations proceed differently in two dimensions, we still expects that the overall scattering rate will be proportional to the density-of-states, except for electrostatic interactions which favor small angle deflections. Instead of the $1/\tau \propto E^{1/2}$ result for simple energy bands in 3D, we recall that the density of states for carriers in a quantum well is piecewise constant at

$$k_z = \frac{n\pi}{W} \qquad n = 1, 2, 3, \ldots n_{\max} \tag{2.113}$$

where W is the width of the quantum well. Also sketched in Fig. 2.19a are the electron envelope functions for the first two levels. Again, for an infinitely deep well, the envelope functions are given by

$$F_n(z) = \sqrt{\frac{2}{W}} \sin k_z z. \tag{2.114}$$

Electrons are confined in the z-direction, but they are free to move in the x–y plane. The total wavefunction for the electron is

$$\varphi(x, y, z) = F_n(z) \frac{e^{i\mathbf{k}_\parallel \cdot \boldsymbol{\rho}}}{\sqrt{A}}, \tag{2.115}$$

where A is the cross-sectional area. These specific results apply only to infinitely deep wells, but the underlying concepts for finite depth wells are the same.

Because the momentum in the z-direction is quantized, the electron's energy is also quantized.

$$E = \varepsilon_n + \frac{\hbar^2 k_\parallel^2}{2m^*} = \frac{n^2 \hbar^2 \pi^2}{2m^* W^2} + \frac{\hbar^2 k_\parallel^2}{2m^*}. \tag{2.116}$$

The $E(k_\parallel)$ relation is sketched in Fig. 2.19b.

Figure 2.19c is a simple picture of carrier scattering in a quantum well. In *intra-subband* scattering (analogous to intraband scattering in a bulk semiconductor), k_z does not change. In *inter-subband* scattering, however, k_z does change. This picture is flawed, however, because momentum conservation does not strictly apply. Recall the uncertainty relation,

$$\Delta z \Delta p_z \geq \hbar. \tag{2.117}$$

Because the uncertainty in the carrier's position has been reduced by confining it in a well of width, W, the uncertainty in the z-directed momentum can be large.

For wide wells, momentum conservation is approximately valid and sometimes assumed, but for thin wells,

$$g_{2D}(E) = \frac{m^*}{\pi\hbar^2} \tag{2.118}$$

for each subband, so the scattering rate would reflect the piecewise constant density of states as sketched in Fig. 2.19d. In this section, we'll discuss ADP scattering of 2D electrons to illustrate how such calculations proceed and to establish the important features of phonon scattering of confined carriers.

Confinement in an inversion layer

Confinement potentials can also be produced by electrostatic effects and heterojunctions. A common example, the metal–oxide–silicon (MOS) system, is illustrated in Fig. 2.20. The gate voltage and the energy barrier at the $Si:SiO_2$ interface confines carriers near the oxide–silicon interface. If the number of electrons in the inversion layer is small, then the confining potential can be approximated by a triangular well as shown in Fig. 2.20b. In this case, the problem can be solved analytically and the wave functions are Airy functions (recall homework problem 1.10). Note that there are two series of subbands, ladder 1 and ladder 2 (also called the unprimed and primed series). Ladder 1 is associated with the two valleys whose ellipsoids have their long axes normal to the surface, and ladder 2 with the remaining four valleys whose short axes are normal to the surface. For ladder 1, the confinement energy is determined by the longitudinal effective mass, m_l^* and for ladder 2 by the transverse effective mass, m_t^*.

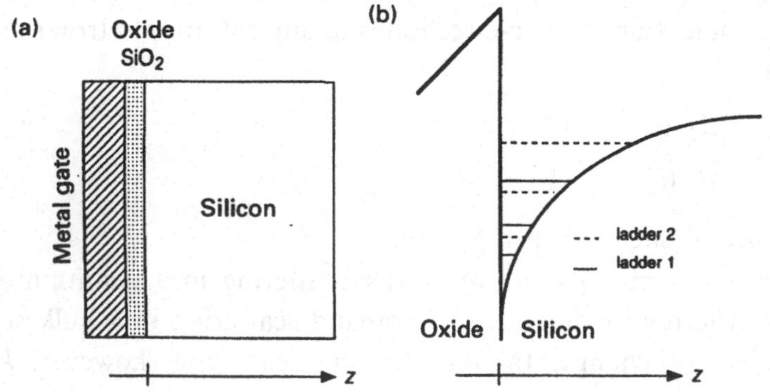

Fig. 2.20 (a) Illustration of a metal–oxide–silicon (MOS) structure used to confine electrons near the oxide–silicon interface. A voltage $V_G > 0$ produces a confining potential for minority carrier electrons. (b) The conduction band energy versus position versus distance from the oxide–silicon interface. The energy levels for confined electrons are also sketched. Ladder 1 is associated with electrons that respond with the longitudinal effective mass and ladder 2 with those that respond with the transverse effective mass.

Because the triangular well approximation is crude, numerical solutions are often used. It is usually necessary to solve the effective mass equation and Poisson's equation simultaneously to determine the energy levels and the wavefunctions of inversion layer electrons. We begin by solving an effective mass equation like eq. (1.65),

$$-\frac{\hbar^2}{2m_z^*}\frac{d^2F_i(z)}{dz^2} - qV(z)F_i(z) = \varepsilon_i F_i, \tag{2.119}$$

where the confining potential is the electrostatic potential determined by the gate voltage and the band bending in the semiconductor, and $m_z^* = m_l^*$ for ladder 1 and $m_z^* = m_t^*$ for ladder 2. The numerical solution to the eigenvalue problem in eq. (2.119) with an assumed potential profile $V(z)$ gives the energy levels for ladders 1 and 2 along with the wavefunctions.

After solving eq. (2.119) for the energy levels, the next step is to solve Poisson's equation to correct the assumed electrostatic potential. We solve

$$\frac{d^2V(z)}{dz^2} = -\frac{q}{\kappa_S\varepsilon_0}\left[(N_D^+(z) - N_A^-(z)) - \sum_i \frac{N_i}{n_S}|F_i(z)|^2\right]. \tag{2.120}$$

The first term in brackets is the ionized dopant charge, and the second term is the contribution from the confined electrons. The sum is over all of the occupied levels in ladders 1 and 2, n_S is the total concentration per cm^2 of confined electrons and N_i is the contribution of each level as given by

$$N_i = n_v\left(\frac{m_d^*}{\pi\hbar^2}\right)k_BT\ln\left[1 + \exp\left(\frac{E_F - \varepsilon_i}{k_BT}\right)\right], \tag{2.121}$$

which is similar to eq. (1.91). Here n_v is the valley degeneracy (two for ladder 1 and four for ladder 2), and m_d^* is the density-of-states effective mass per valley. (For ellipsoidal bands, $m_d^* = \sqrt{m_l^* m_t^*}$.)

The numerical solution to eq. (2.120) corrects the electrostatic potential, which can then be inserted in eq. (2.119) to obtain an improved solution to the effective mass equation. The process proceeds until the solution, $V(z)$, converges. Figure 2.21 shows the results of typical calculations. For low carrier densities (Fig. 2.21a), the potential well is triangular, but for the higher densities (Fig. 2.21b), the potential is highly nonlinear. Figure 2.21 also shows the two ladders of energy levels, the ground state wavefunctions for both ladders and the first excited state for the first ladder. When the energy levels and wavefunctions are only known as numerical tables, we should not expect to obtain closed-form expressions for the scattering rates.

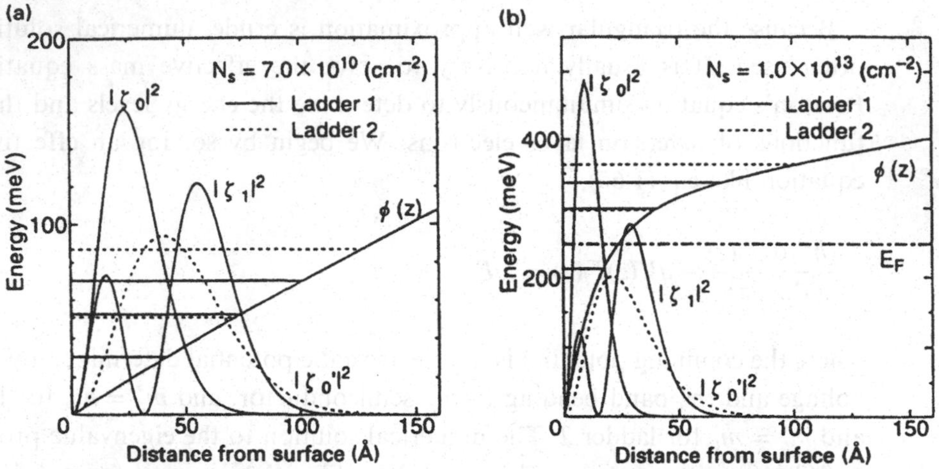

Fig. 2.21 Results of a self-consistent solution of the effective mass and Poisson equations for an MOS structure. (a) Low density of confined carriers (b) High density of confined carriers. (From Yamakawa, S. et al. *Journal of Applied Physics,* **79**, 911, 1996). Reproduced with permission of American Institute of Physics.)

2.11.2 Phonon scattering rate for 2D carriers

The transition rate for carrier scattering continues to be given by eq. (2.2), but the matrix element,

$$H_{\mathbf{p'p}} = \int_{-\infty}^{+\infty} \varphi_{\mathbf{p'}}^* U_S \varphi_{\mathbf{p}} \mathrm{d}^3\mathbf{r}, \tag{2.122}$$

works out differently. The perturbing potential for electron–phonon scattering, U_S, is still given by eqs. (2.58) and (2.59) if we assume three-dimensional phonons, but the initial and final wavefunctions are different. Writing

$$\mathbf{p} = \mathbf{p}_{\|} + p_z\hat{\mathbf{z}} \tag{2.123a}$$

$$\mathbf{p}' = \mathbf{p}_{\|}' + p_z\hat{\mathbf{z}} \tag{2.123b}$$

and

$$\boldsymbol{\beta} = \boldsymbol{\beta}_{\|} + \beta_z\hat{\mathbf{z}} \tag{2.123c}$$

eq. (2.122) becomes

$$H_{\mathbf{p'p}} = \int_{-\infty}^{+\infty} F_{\mathrm{f}}^*(z)\frac{\mathrm{e}^{-\mathrm{i}\mathbf{p}_{\|}'\cdot\boldsymbol{\rho}/h}}{\sqrt{A}}\left(A_\beta K_\beta \mathrm{e}^{\pm\mathrm{i}\boldsymbol{\beta}_{\|}\cdot\boldsymbol{\rho}}\mathrm{e}^{\pm\mathrm{i}\beta_z z}\right)F_{\mathrm{i}}(z)\frac{\mathrm{e}^{\mathrm{i}\mathbf{p}_{\|}\cdot\boldsymbol{\rho}/\hbar}}{\sqrt{A}}\,\mathrm{d}z\mathrm{d}\rho. \tag{2.124}$$

A subscript, i, refers to an initial subband and f to the final subband. The integral over the transverse plane gives a δ-function expressing momentum conservation in the plane, so eq. (2.124) becomes

$$H_{\mathbf{p}'\mathbf{p}} = I_{\text{fi}}(\beta_z) A_\beta K_\beta \delta_{\mathbf{p}'_\parallel, \mathbf{p}_\parallel \pm \hbar \beta_\parallel}, \tag{2.125}$$

where

$$I_{\text{fi}} = \int_{-\infty}^{+\infty} F_{\text{f}}^*(z) F_{\text{i}}(z) e^{\pm i\beta_z z} dz. \tag{2.126}$$

Finally, we write the transition rate for 2D carriers from eq. (2.2) as

$$S(\mathbf{p}, \mathbf{p}') = \frac{2\pi}{\hbar} |I_{fi}(\beta_z)|^2 |K_\beta|^2 |A_\beta|^2 \delta_{\mathbf{p}'_\parallel, \mathbf{p}_\parallel \pm \hbar \beta_\parallel} \delta(E' - E \mp \hbar\omega), \tag{2.127}$$

which is exactly like the corresponding result for 3D carriers [eq. (2.61)] except for two things: (1) momentum conservation only applies in the plane, and (2) the appearance of the *form factor*, $I_{fi}(\beta_z)$. The 2D transition rate is identical to the 3D expression, except for the replacement,

$$\boxed{\delta_{p', p \pm \hbar \beta} \rightarrow \delta_{\mathbf{p}'_\parallel, \mathbf{p}_\parallel \pm \hbar \beta_\parallel} |I_{\text{fi}}(\beta_z)|^2}, \tag{2.128}$$

which results from the fact that z-directed momentum is not strictly conserved.

To evaluate the scattering rate, we begin with eq. (2.3) and write

$$\frac{1}{\tau_{\text{fi}}} = \sum_{\mathbf{p}'} \frac{2\pi}{\hbar} |I_{\text{fi}}(\beta_z)|^2 |K_\beta|^2 |A_\beta|^2 \delta_{\mathbf{p}'_\parallel, \mathbf{p}_\parallel \pm \hbar \beta_\parallel} \delta(E' - E \mp \hbar\omega) \tag{2.129}$$

or

$$\frac{1}{\tau_{\text{fi}}} = \sum_{\mathbf{p}'_\parallel} \frac{2\pi}{\hbar} \left\{ \sum_{\mathbf{p}'_z} |I_{\text{fi}}(\beta_z)|^2 |K_\beta|^2 |A_\beta|^2 \right\} \delta_{\mathbf{p}'_\parallel, \mathbf{p}_\parallel \pm \hbar \beta_\parallel} \delta(E' - E \mp \hbar\omega). \tag{2.130}$$

Defining a quantity,

$$|D_{\text{2D}}(\beta)|^2 = \sum_{\mathbf{p}'_z} |I_{\text{fi}}(\beta_z)|^2 |K_\beta|^2 |A_\beta|^2, \tag{2.131}$$

we can express the scattering rate as

$$\frac{1}{\tau_{\text{fi}}} = \sum_{\mathbf{p}'_\parallel} \frac{2\pi}{\hbar} |D_{\text{2D}}(\beta)|^2 \delta_{\mathbf{p}'_\parallel, \mathbf{p}_\parallel \pm \hbar \beta_\parallel} \delta(E' - E \mp \hbar\omega). \tag{2.132}$$

The corresponding expression for 3D carriers is

$$\frac{1}{\tau} = \sum_{\mathbf{p}'} \frac{2\pi}{\hbar} |D_{\text{3D}}(\beta)|^2 \delta_{\mathbf{p}', \mathbf{p} \pm \hbar \beta} \delta(E' - E \mp \hbar\omega), \tag{2.133}$$

where

$$|D_{\text{3D}}(\beta)|^2 = |K_\beta|^2 |A_\beta|^2. \tag{2.134}$$

If the δ-functions in eq. (2.133) are converted to a single δ-function expressing momentum and energy conservation, and the sum is converted to an integral, then the 3D result is simply eq. (2.75). Comparing eqs. (2.132) and (2.133), we see that the 2D and 3D scattering rate expressions are identical except for the difference between D_{2D} and D_{3D} and the two-dimensional sum in eq. (2.132) and the three-dimensional sum in eq. (2.133).

2.11.3 ADP scattering rate of 2D carriers

Using eq. (2.59a) for $|K_\beta|^2$ and eq. (2.71c) in the equipartition approximation [eq. (1.134)], we find

$$|K_\beta|^2|A_\beta|^2 = \frac{D_A^2 k_B T_L}{2c_l \Omega}, \tag{2.135}$$

which allows us to write

$$|D_{2D}(\beta)|^2 = \frac{D_A^2 k_B T_L}{2c_l \Omega} \frac{W}{2\pi} \int_0^\infty |I_{fi}(\beta_z)|^2 d\beta_z. \tag{2.136}$$

Using $\Omega/W = A$, the cross-sectional area and inserting eq. (2.126) for $|I_{fi}(\beta_z)|^2$, we have

$$|D_{2D}(\beta)|^2 = \frac{D_A^2 k_B T_L}{4\pi c_l A} \int_{-\infty}^{+\infty} F_f^*(z) F_i(z) dz \int_{-\infty}^{+\infty} F_f(z') F_i^*(z') dz' \int_0^\infty e^{\pm i\beta_z(z-z')} d\beta_z \tag{2.137}$$

where we have replaced the dummy variable, z, with z' in the last integral. Doing the integral over β_z first and using

$$\frac{1}{2\pi} \int_{-\infty}^{+\infty} e^{\pm i\beta_z(z-z')} d\beta_z = \delta(z - z'), \tag{2.138}$$

we obtain

$$|D_{2D}(\beta)|^2 = \frac{D_A^2 k_B T_L}{2c_l A} \int_{-\infty}^{+\infty} |F_f(z)|^2 |F_i(z)|^2 dz = \frac{D_A^2 k_B T_L}{2c_l A} \frac{1}{W_{fi}}. \tag{2.139}$$

The quantity

$$\frac{1}{W_{fi}} \equiv \int_{-\infty}^{+\infty} |F_f(z)|^2 |F_i(z)|^2 dz, \tag{2.140}$$

describes the effective extent of the interaction in the z-direction. It is the principal difference between the 2D and 3D results. If the well has a simple shape, this quantity can be evaluated analytically. More generally, the integral has to be performed numerically.

For an infinite, square well, the envelope functions are sine and cosine functions and the integral in eq. (2.140) works out to be

$$\frac{1}{W_{ii}} = \int_{-\infty}^{+\infty} |F_j(z)|^2 |F_i(z)|^2 \mathrm{d}z = \frac{3}{2W} \quad i = j \tag{2.141a}$$

$$\frac{1}{W_{ji}} = \int_{-\infty}^{+\infty} |F_j(z)|^2 |F_i(z)|^2 \mathrm{d}z = \frac{1}{W} \quad i \neq j. \tag{2.141b}$$

The general result for both intra- and inter-subband scattering in an infinite, square well is

$$\frac{1}{W_{\mathrm{fi}}} = \frac{\{2 + \delta_{\mathrm{fi}}\}}{2W}. \tag{2.142}$$

Finally using eq. (2.139) in eq. (2.132), we have

$$\frac{1}{\tau_{\mathrm{fi}}} = \left(\frac{\pi D_{\mathrm{A}}^2 k_{\mathrm{B}} T_{\mathrm{L}}}{\hbar c_{\mathrm{l}}} \frac{1}{W_{\mathrm{fi}}} \right) \frac{1}{A} \sum_{\mathbf{p}'_{\parallel}} \delta_{\mathbf{p}'_{\parallel}, \mathbf{p}_{\parallel} \pm \hbar \beta_{\parallel}} \delta(E' - E \mp \hbar\omega). \tag{2.143}$$

The sum in eq. (2.143) is similar to one we've worked out before. In three dimensions, the result was one-half of the three-dimensional density of states. In two dimensions it is just one-half of the two-dimensional density of states as given by eq. (2.118). After multiplying by two to account for ABS and EMS, the final scattering rate for ADP scattering of 2D carriers is

$$\boxed{\frac{1}{\tau_{\mathrm{fi}}} = \frac{\pi D_{\mathrm{A}}^2 k_{\mathrm{B}} T_{\mathrm{L}}}{\hbar c_{\mathrm{l}}} \frac{1}{W_{\mathrm{fi}}} g_{\mathrm{2Df}}(E)}. \tag{2.144a}$$

Equation (2.144a) should be compared to eq. (2.84), the corresponding result for 3D carriers. The two differences are the replacement of the 3D density of states by the 2D of states and the appearance of the $1/W_{\mathrm{fi}}$ factor.

Equation (2.144a) describes ADP scattering from subband 'i' to subband 'f'; g_{2Df} is the two-dimensional density of states for the subband, 'f.' For intra-subband scattering in an infinite, square well, eq. (2.144a) reduces to

$$\frac{1}{\tau_{\mathrm{ii}}} = \frac{3\pi D_{\mathrm{A}}^2 k_{\mathrm{B}} T_{\mathrm{L}}}{2\hbar c_{\mathrm{l}} W} g_{\mathrm{2Di}}(E) \tag{2.144b}$$

and for inter-subband scattering to

$$\frac{1}{\tau_{\mathrm{fi}}} = \frac{\pi D_{\mathrm{A}}^2 k_{\mathrm{B}} T_{\mathrm{L}}}{\hbar c_{\mathrm{l}} W} g_{\mathrm{2Df}}(E). \tag{2.144c}$$

The density of states for 2D carriers is constant with energy for each subband. Adding the density of states for a quantum well with three subbands, we get the results in Fig. 2.22a. The 2D ADP scattering rate is proportional to the 2D density of states, so for intra-subband scattering, we get the results sketched in Fig. 2.22b. Inter-subband scattering can also be induced by ADP scattering; the acoustic phonons carry little energy, so an electric field must accelerate carriers to energies exceeding the bottom of the next subband before scattering to a

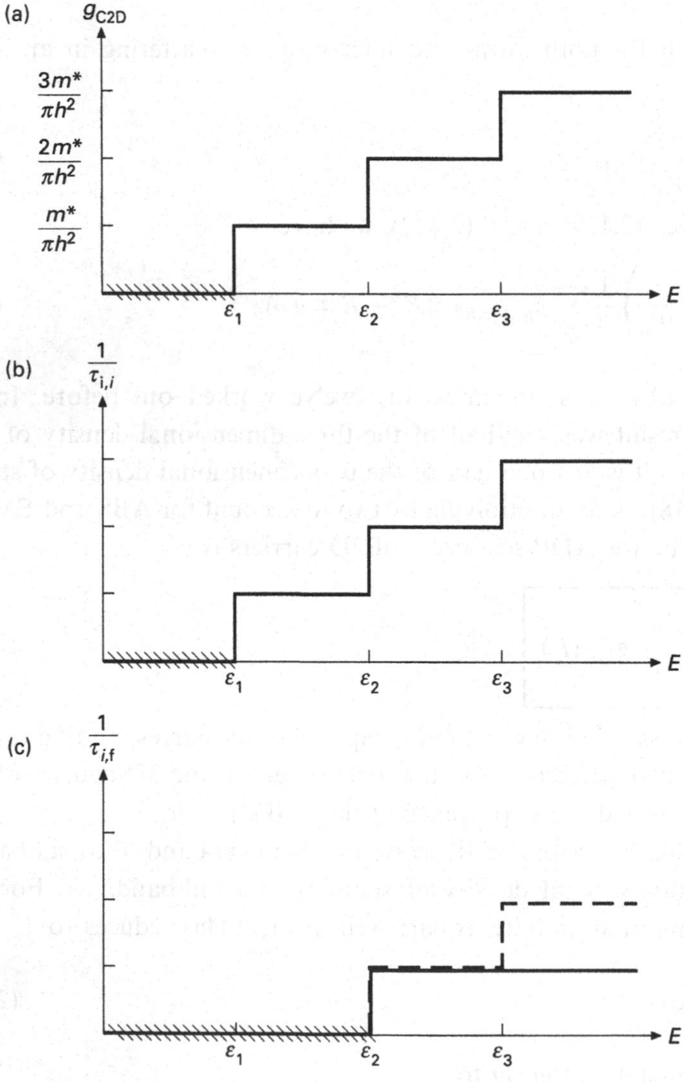

Fig. 2.22 (a) The two-dimensional density of states versus energy for electrons in a quantum well. The subband energies are labeled, E_1, E_2, and E_3. (b) Acoustic deformation potential scattering rate versus energy for 2D electrons. Intra-subband scattering is assumed. (c) Inter-subband scattering by acoustic phonons. The solid line is for scattering from subband 1 to subband 2 and the dashed line is the total scattering rate for transitions from subband 1 to 2 and from subband 1 to 3.

higher subband can occur. Figure 2.22c illustrates inter-subband scattering of electrons by acoustic phonons.

2.11.4 Intervalley scattering of 2D carriers

Very similar procedures can be used to evaluate the scattering rates of confined carriers due to other mechanisms. For example, in addition to inter-subband scattering, intervalley subband scattering also occurs in semiconductor like silicon. The results for 3D carriers, eq. (2.98) becomes

$$\frac{1}{\tau(p)} = \frac{1}{\tau_m(p)} = \left(\frac{\pi D_{\mathrm{if}}^2 Z_f}{2\rho\omega_{\mathrm{if}}}\frac{1}{W_{\mathrm{if}}}\right)(N_{\mathrm{i}} + \tfrac{1}{2} \mp \tfrac{1}{2})g_{2\mathrm{Df}}(E \pm \hbar\omega_{\mathrm{if}} - \Delta E_{\mathrm{fi}}). \tag{2.145}$$

Again, the only difference is the replacement of the 3D density of states by the 2D density of states and the appearance of the effective interaction length, W_{fi} which depends on the envelope functions of the confined carriers.

2.11.5 Discussion

In this section, we have discussed one example calculation of scattering rates for carriers confined in a quantum well and have quoted the result for one more. For confined carriers, the matrix element needs to be evaluated with the proper wave function for confined carriers – not with the plane wavefunctions used for bulk electrons. We found that momentum conservation applies in the plane of the quantum well but that it is 'fuzzy' in the $\hat{z}$-direction. Both 2D and 3D scattering rates generally vary as the density of states, but the staircase variation of g_{2D} with energy leads to very abrupt changes in the 2D scattering rate versus energy. Similar procedures can be used to evaluate the scattering rates due to other mechanisms, such as POP scattering. Such calculations are discussed in the references listed at the end of this chapter.

2.12 Scattering at a surface

Confined electrons are subject to all the scattering mechanisms that affect 3D carriers, but some additional mechanisms also occur. One is inter-subband scattering, but another is the possibility that electrons can scatter off the boundaries of the confining potential. If the confining potential is smooth, there is no effect, but it is usually atomically rough, so electrons may scatter. Consider the $Si : SiO_2$ interface in an MOS structure. As sketched in Fig. 2.23, the oxide growth process leaves a rough surface, so the thickness of the oxide fluctuates. Since the gate is

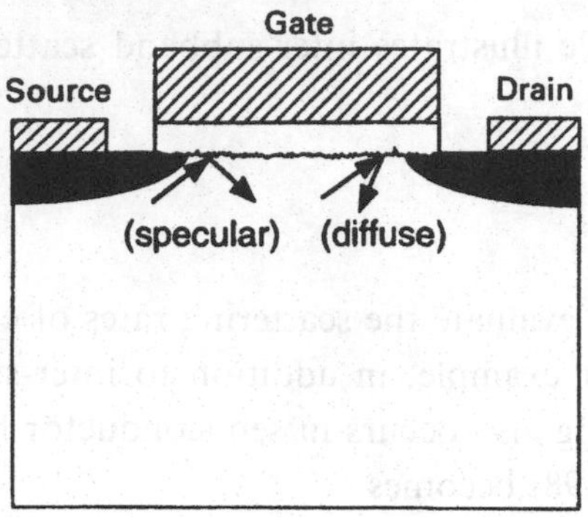

Fig. 2.23 Illustration of surface roughness scattering of electrons at the Si : SiO_2 interface in an MOS structure. Scattering is specular from a smooth interface but diffuse from a rough interface.

an equipotential, this produces fluctuations in the confining potential along the channel. The result is that the subband minimum will fluctuate which gives rise to scattering. It should be expected that this effect will increase as the gate voltage increases, which confines carriers more closely to the surface. For high gate biases, surface roughness scattering can be the dominant mechanism at the Si : SiO_2 interface. Other effects occur too. For example, charges at the Si : SiO_2 interface will produce ionized impurity scattering. At low inversion layer densities, charged impurity scattering will dominate, but at high inversion layer densities, screening will reduce charge impurity scattering, and interface roughness scattering dominates.

To evaluate the scattering rate due to surface roughness, we first need to identify the perturbing potential. A fluctuation in the oxide thickness will produce a fluctuation in the confining potential,

$$V[z + \delta(\boldsymbol{\rho})] = V(z) + \delta(\boldsymbol{\rho})\frac{\mathrm{d}V(z)}{\mathrm{d}z}, \tag{2.146}$$

where $\boldsymbol{\rho}$ is a vector in the plane of the interface and $\delta(\boldsymbol{\rho})$ is a random function that measures the surface roughness. The perturbing potential is simply

$$U_{\mathrm{SR}} = q\mathcal{E}_z\delta(\boldsymbol{\rho}), \tag{2.147}$$

where $\mathcal{E}_z$ is the electric field. Evaluating the matrix element in the usual way, we find

$$H_{\mathbf{p}'\mathbf{p}} = \int_{-\infty}^{+\infty} F_{\mathrm{f}}^*(z)(q\mathcal{E}_z)F_{\mathrm{i}}(z)\mathrm{d}z \times \frac{1}{A}\int_{-\infty}^{+\infty} \delta(\boldsymbol{\rho})\mathrm{e}^{\mathrm{i}(\mathbf{p}_\parallel - \mathbf{p}'_\parallel)\cdot\boldsymbol{\rho}/\hbar}\mathrm{d}\boldsymbol{\rho}. \tag{2.148a}$$

The first integral is seen to be a weighted average of the electric field; we will term this the effective normal electric field, $\mathcal{E}_{\mathrm{eff}}$. The second term is the Fourier transform of $\delta(\rho)$, which is the power spectrum, $S(\boldsymbol{\beta}_{\|})$. Using these definitions, we write the matrix element as

$$H_{\mathbf{p}',\mathbf{p}} = -q\mathcal{E}_{\text{eff}} S(\boldsymbol{\beta}_{\|}) \tag{2.148b}$$

and the scattering rate as

$$S(\mathbf{p}_{\|}, \mathbf{p}'_{\|}) = \frac{2\pi}{\hbar}(q\mathcal{E}_{\text{eff}})^2 \left|S(\boldsymbol{\beta}_{\|})\right|^2 \delta(E' - E). \tag{2.149}$$

The thickness fluctuations can be described by an exponential autocorrelation function

$$\left\langle \delta(\rho)\delta(\rho' - \rho)\right\rangle = \Delta^2 \mathrm{e}^{-\rho/L}, \tag{2.150}$$

where Δ is the rms amplitude of the fluctuations and L is their correlation length (roughly the distance between fluctuations). Surface roughness is controlled by processing conditions and can be measured by scanning probe microscopy. Typical values are $\Delta \approx 2 - 4\,\text{Å}$ and $L \approx 10 - 30\,\text{Å}$. [Some authors use a Gaussian autocorrelation function instead of the exponential in eq. (2.150).] The power spectrum of eq. (2.150) is

$$\left|S(\boldsymbol{\beta}_{\|})\right|^2 = \pi\Delta^2 L^2 \frac{1}{\left[1 + (L^2\beta_{\|}^2/2)\right]}. \tag{2.151}$$

When eq. (2.151) is inserted in (2.149), the scattering rate can be evaluated. Note the strong dependence on the strength of the electric field and on the amplitude of the interface roughness.

2.13 Scattering rates for nonparabolic energy bands

When working out expressions for scattering rates, we have been assuming spherical, parabolic energy bands. For silicon and germanium, the constant energy surfaces are ellipsoids, but appropriate averages of band structure dependent parameters can often be used to reduce the problem to an 'equivalent', spherical band problem [2.5, 2.7]. Under high applied fields, however, carriers may be accelerated to energies far above the band minima. At such energies, the bands are definitely not parabolic. To evaluate scattering rates for high energy carriers, the non-parabolicity of the energy bands must be considered. In this section, we outline briefly how scattering rates are evaluated when the energy bands are spherical but nonparabolic.

Any spherical energy band can be described as

$$\frac{p^2}{2m^*} = \gamma[E(p)], \tag{2.152a}$$

where m^* is evaluated from the curvature of $E(p)$ at $p = 0$, and γ is some function of energy. For spherical, parabolic bands

$$\gamma(E) = E(p), \tag{2.152b}$$

but if nonparabolicity is described by eq. (1.40), then

$$\gamma(E) = E(p)[1 + \alpha E(p)]. \tag{2.152c}$$

To compute scattering rates, we first need to find $S(\mathbf{p}, \mathbf{p}')$ from Fermi's Golden Rule. When the bands are spherical and parabolic, the overlap integral $I(\mathbf{p}, \mathbf{p}')$ is unity, but for a nonparabolic band it can be substantially less than one.

The occurrence of non-unity overlap integrals is only one consequence of nonparabolicity. The various sums also work out differently. Consider the sum,

$$\frac{1}{\Omega}\sum_{\mathbf{p}',\uparrow} \delta[E(\mathbf{p}') - E(\mathbf{p})] = \frac{1}{2\pi^2\hbar^3}\int_0^\infty \delta[E(\mathbf{p}') - E(\mathbf{p})]p'^2 \mathrm{d}p' \tag{2.153}$$

which we worked out in Section 2.1 and found the result to be one-half of the density of states. For nonparabolic bands

$$p'^2 = 2m^*\gamma(E')$$

so

$$p'^2\mathrm{d}p' = \sqrt{2}\left[\gamma(E')\right]^{1/2} m^{*3/2}\frac{\mathrm{d}\gamma}{\mathrm{d}E'}\mathrm{d}E'. \tag{2.154}$$

After substituting eq. (2.154) into eq. (2.153), we find

$$\frac{1}{\Omega}\sum_{\mathbf{p}',\uparrow} \delta[E(p') - E(p)] = \frac{(2m^*)^{3/2}}{4\pi^2\hbar^3}\sqrt{\gamma(E')}\frac{\mathrm{d}\gamma}{\mathrm{d}E'}\bigg)_{E'=E} \tag{2.155}$$

which is the density of states (for one of the two spins) for a nonparabolic band.

With the technique outlined above, band nonparabolicity can be accounted for in the various scattering rates. It is essential to do so when high-field transport is analyzed. For example, when nonparabolicity is included, we find that the POP scattering rate is nearly constant with energy (for energies considerably above the optical phonon energy) in contrast to the decreasing behavior displayed in Fig. 2.12, which was based on parabolic energy bands.

2.14 Electron scattering in Si and GaAs

The expressions we've developed in this chapter describe electron scattering in common semiconductors. We haven't treated hole scattering because it is complicated by the degenerate heavy and light hole bands with their warped constant energy surfaces [2.6]. For energetic carriers, overlap integrals need to be considered and the non-parabolicity accounted for as discussed in Section 2.13. For very energetic carriers, even this is not adequate and a detailed, numerical description of $E(\mathbf{p})$ is required.

2.14.1 Common scattering mechanisms in semiconductors

Common scattering mechanisms can be classified as shown in Fig. 2.24. The total scattering rate is the sum of the rates for each of the individual processes,

$$\Gamma(\mathbf{p}) = \sum_i \frac{1}{\tau_i(\mathbf{p})}, \tag{2.156}$$

where the index, i, labels the various scattering mechanisms listed in Fig. 2.24. To evaluate the scattering rate versus energy, the important scattering mechanisms need to be identified for the particular semiconductor and conditions under consideration. Scattering occurs by defects, phonons, and by other carriers. Defect scattering includes scattering by both ionized and neutral impurities

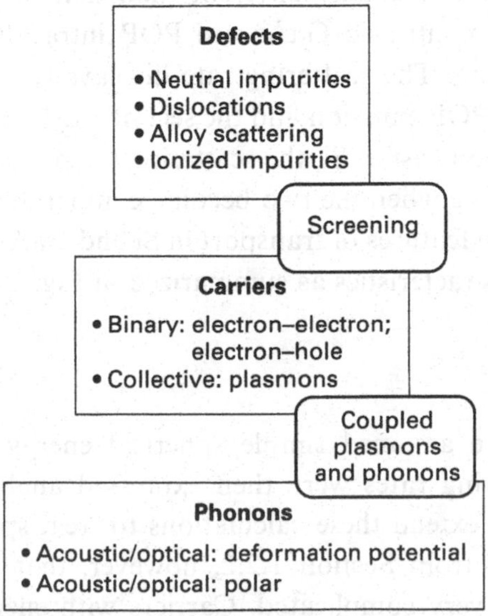

Fig. 2.24 Classification of carrier scattering mechanisms.

and by crystal defects such as dislocations. For semiconductor alloys, variations in the alloy composition also produce scattering (see homework problem 2.6). Phonon scattering occurs by the deformation potential in covalent semiconductors and by both the deformation potential and by polar interactions in ionic semiconductors. Carrier–carrier scattering includes both binary collisions and interactions with the carrier plasma. Free carriers can also influence the other scattering process by screening the perturbing potential (see Section 2.15). In polar semiconductors, free carrier plasma oscillations can also couple with the longitudinal optical phonons. For high-quality, intrinsic, crystalline semiconductors, defect scattering is minimal, as is carrier–carrier scattering if the carrier density is low. Figure 2.25 plots electron scattering rates versus energy for intrinsic Si and GaAs. For these conditions, scattering is dominated by intra- and intervalley phonon scattering.

2.14.2 Electron scattering in Si and GaAs

For pure silicon, acoustic deformation potential and equivalent intervalley scattering are the dominant mechanisms. Near room temperature, the acoustic deformation and equivalent intervalley scattering rates for thermal average electrons are comparable. Because of the several phonons involved, intervalley scattering rises faster than $\sqrt{E}$ as the onset for various emission processes is met. As a consequence, the high energy scattering rate is dominated by equivalent intervalley scattering.

The scattering rate versus energy for electrons in intrinsic GaAs is displayed in Fig. 2.25b and shows a characteristic distinctly different than that of Si. The important scattering mechanisms in intrinsic GaAs are POP intravalley and $\Gamma - L$ and $L - L$ intervalley scattering. The scattering rate displays two thresholds; the first (at $E \cong 0.03$ eV) is for POP emission and the second (at $E \cong 0.3$ eV) for $\Gamma - L$ intervalley scattering. In contrast to Si, the scattering rate remains low until the onset of intervalley scattering when the two become comparable. We'll see in later chapters that many of the features of transport in Si and GaAs can be understood from their scattering characteristics as summarized in Fig. 2.25.

2.14.3 Full band treatment of scattering

For most of this chapter, we have assumed simple spherical energy bands. Expressions for the various scattering rates were then expressed analytically. In Section 2.13, we showed how to extend these calculations to treat spherical, nonparabolic energy bands. Recall from Section 1.2.2, however, that at high energies, the bandstructure can be very complicated. Carriers with energies of a few electron volts can cause reliability problems by being injected into the gate

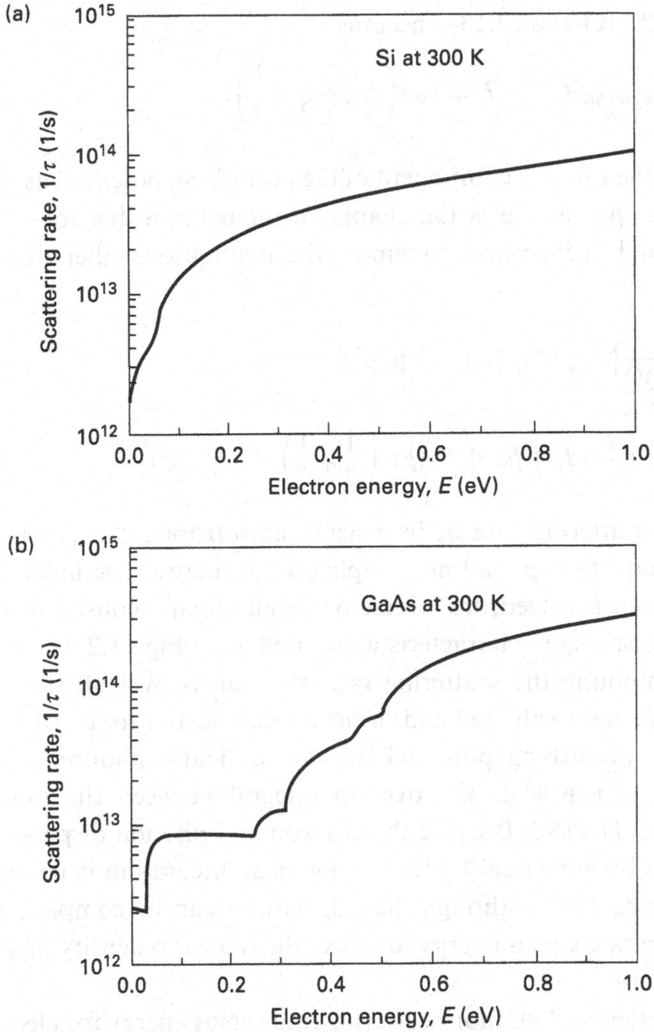

Fig. 2.25 (a) Electron scattering rate versus energy for intrinsic Si at room temperature. This curve was computed by assuming ellipsoidal, nonparabolic energy bands. (b) Electron scattering rate versus energy for intrinsic GaAs at room temperature. This curve was computed by assuming spherical, non-parabolic energy bands. (Courtesy of Amitava Das.)

oxide of a MOSFET, or they can lead to breakdown by impact ionization. When evaluating the scattering rates of such carriers, the bandstructure will generally be available only as a numerical table of values. Scattering rates are still evaluated from eqs. (2.1)–(2.3), but the integrals must be performed numerically.

A general form for phonon scattering was given in eq. (2.133),

$$\frac{1}{\tau} = \sum_{\mathbf{p}'} \frac{2\pi}{\hbar} \left| D_{3D}(\beta) \right|^2 \delta_{\mathbf{p}',\mathbf{p}\pm\hbar\beta} \delta(E' - E \mp \hbar\omega),$$

which, using eqs. (2.71c) and (2.134) becomes

$$\frac{1}{\tau} = \sum_{\mathbf{p}'} \frac{\pi}{\rho\omega\Omega} |K_\beta|^2 \delta_{\mathbf{p}',\mathbf{p}\pm\hbar\boldsymbol{\beta}} \delta(E' - E \mp \hbar\omega)\left(N_\omega + \frac{1}{2} \mp \frac{1}{2}\right). \tag{2.157}$$

Recall that $|K_\beta|^2$ is the Fourier component of the perturbing potential as given by eq. (2.59) and that $\hbar\boldsymbol{\beta} = \mathbf{p}' - \mathbf{p}$ is the change in momentum due to scattering. Equation (2.157) can be generalized to remove the assumptions inherent in it. We write the result as

$$\frac{1}{\tau_{\eta,v}(\mathbf{p})} = \sum_{\nu',\boldsymbol{\beta}} \frac{\pi}{\rho\omega_\eta(\boldsymbol{\beta})\Omega} \left|\Delta_{\eta,\nu'}(\boldsymbol{\beta})\right|^2 \left|I(\nu,\nu';\mathbf{p},\mathbf{p}')\right|^2$$
$$\times\, \delta_{\mathbf{p}',\mathbf{p}\pm\hbar\boldsymbol{\beta}} \delta(E' - E \mp \hbar\omega)\left(N_\eta(\boldsymbol{\beta}) + \frac{1}{2} \mp \frac{1}{2}\right). \tag{2.158}$$

Note first that the scattering rate is, in general, anisotropic; it depends on the particular momentum state, $\mathbf{p}$, and not simply on the energy. The index η refers to the type of phonon involved, transverse or longitudinal, acoustic or optical. The full phonon dispersion characteristic is used [i.e. Fig. 1.27(a) not Fig. 1.27(b)]. When computing the scattering rate, the sum is over all the possible momentum states, $\boldsymbol{\beta}$, and to all the bands a carrier may scatter to, ν'. $|\Delta_{\eta,\nu'}(\boldsymbol{\beta})|^2$ is the strength of the perturbing potential for the particular phonon and band involved, and $|I(\nu,\nu';\mathbf{p},\mathbf{p}')|^2$ is the overlap integral between the initial and final states [recall eq. (1.116)]. Because the electron and phonon dispersion relations are only given by numerical tables, a numerical integration is necessary to evaluate the scattering rate. Although the calculations can be complex, we still expect the scattering rate versus energy to reflect the full band density of states as illustrated in Fig. 1.19.

Figure 2.26 plots the total phonon scattering rate versus energy for electrons in silicon. In general, the scattering rates are anisotropic, depending on the location of the carrier in the Brillouin zone. The rates plotted in Fig. 2.26 were averaged over constant energy surfaces. Also shown in Fig. 2.26 is the scattering rate calculated assuming spherical, nonparabolic energy bands. We note first, that the scattering rate versus energy does have the expected shape of the density of states. Spherical nonparabolic energy bands are a good assumption up to about 2 eV. For higher energies, detailed full band calculations are essential.

2.15 Screening

When a charge is placed in an electron gas, mobile electrons move to cancel out the charge, and the net result is a screened potential which alters the Coulomb

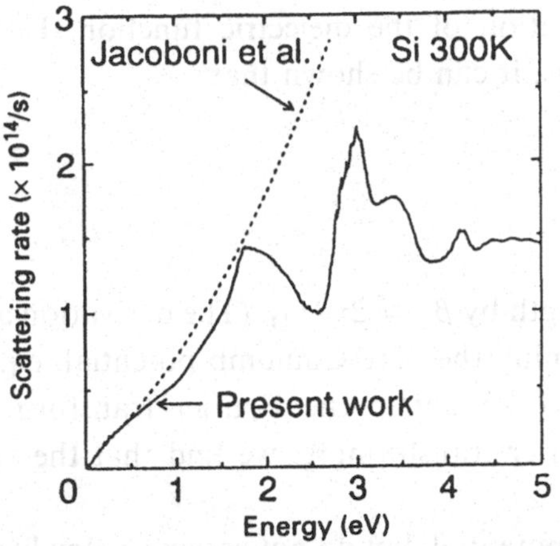

Fig. 2.26 (a) Total electron–phonon scattering rate versus energy for intrinsic silicon. Also shown, as a dotted line, is the scattering rate evaluated from spherical nonparabolic energy bands. (From T. Kunikiyo, et al., *Journal of Applied Physics,* **75**(1), 299, 1994.) (Reproduced with permission of American Institute of Physics.)

potential of the point charge. For ionized impurity and electron–electron scattering, we treated screening by replacing the bare Coulomb potential, eq. (2.10), with the screened Coulomb potential, eq. (2.16). The Debye length, eq. (2.14), is the distance over which the perturbation is screened out. To derive the screening length in Section 2.2.1, we used a simple procedure whose underlying approximations are not at all clear. The response of the total charge distribution to the charge perturbation is actually a complicated, many body problem that must be solved self-consistently to treat screening. A good introduction can be found in Ferry [2.15]; what follows is a brief synopsis of some important points.

A review of the screening derivation in Section 2.2.1 reveals that we assumed an equilibrium relation between the carrier density and the electrostatic potential. In effect, we assumed that carriers are distributed in momentum space in a thermal equilibrium Maxwellian or Fermi–Dirac distribution. As we'll discuss in Chapter 3, the actual distribution function can be much different. In addition, when the perturbing potential varies rapidly in time or space, electrons are not able to fully respond, so the screening length should depend on the frequency and spatial variation of the perturbing potential. Screening should become weaker at high frequencies and for rapid spatial variations.

Screening is most conveniently treated in Fourier space. If a perturbing potential is applied, the true potential that results from screening is given by

$$V_{\text{true}}(\boldsymbol{\beta}) = \frac{V_{\text{bare}}(\boldsymbol{\beta})}{\varepsilon(\boldsymbol{\beta}, \omega)}, \tag{2.159}$$

where $V_{\text{bare}}(\boldsymbol{\beta})$ is the Fourier component of the bare perturbing potential, and $\varepsilon(\boldsymbol{\beta}, \omega)$ is the wave-vector and frequency-dependent dielectric function.

Ferry [2.15] discusses the calculation of the dielectric function. For static, slowly varying perturbing potentials, it can be shown that

$$\varepsilon(\boldsymbol{\beta} \to 0, 0) = \varepsilon_0 \kappa_S \left(1 + \frac{\beta_D^2}{\beta^2}\right), \tag{2.160}$$

where β_D is related to the Debye length by $\beta_D = 2\pi / L_D$. (The derivation assumes that $\beta \ll \beta_D$.) By Fourier transforming the bare Coulomb potential, eq. (2.10), and using the result in eq. (2.159) we obtain the Fourier transform of the screened potential. By inverse Fourier transforming, we find that the screened potential is just eq. (2.16).

If we continue to assume a static potential, but do not assume a slowly varying potential, eq. (2.160) can be generalized as [2.15]

$$\varepsilon(\boldsymbol{\beta}, 0) = \varepsilon_0 \kappa_S \left(1 + \frac{\beta_D^2 F(\beta)}{\beta^2}\right), \tag{2.161}$$

where $F(\beta)$ is a function that approaches unity as $\beta \to 0$ and zero as $\beta \to \infty$, which means there is full screening for small β (slowly varying perturbing potential) and no screening for large β (rapidly varying perturbing potential).

Figure 2.27 illustrates how momentum-dependent screening affects ionized impurity scattering. Recall that $\hbar\boldsymbol{\beta} = \mathbf{p}' - \mathbf{p}$, which is small when the carrier is scattered while far away from the perturbing potential as illustrated in Fig. 2.27a. In this case, Debye screening applies $[F(\beta) = 1]$. A large change in momentum, however, implies that the carrier made a close approach to the impurity as

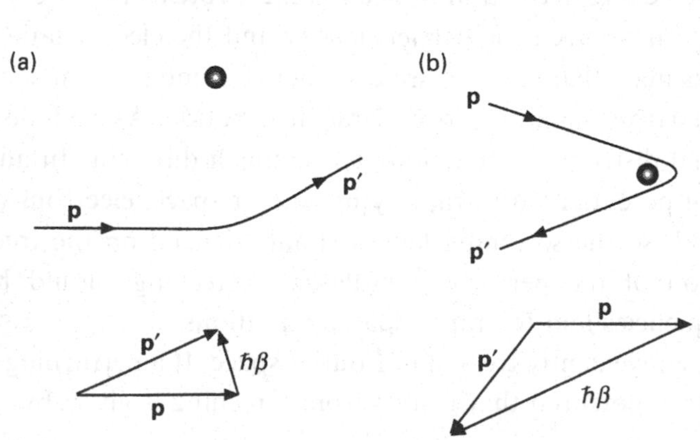

Fig. 2.27 Illustration of how momentum-dependent screening influences ionized impurity scattering. (a) An electron far away from the impurity is deflected by a small amount, small β, (b) an electron that makes a close approach is deflected by a large amount, large β. The first electron sees the Debye length screened potential, and the second sees the bare ionized impurity potential.

sketched in Fig. 2.27b. If the electron approaches closer than a Debye length, the Coulomb potential is not screened at all [$F(\beta) = 0$].

Finally, we consider time-dependent perturbing potentials. It should be apparent that if the perturbing potential varies too rapidly in time, the electrons will not have time to respond and screen the potential. If the electron response is out of phase with the perturbing potential, anti-screening can even result. The natural response of the electron gas is determined by the plasma frequency, ω_p as given by eq. (2.102). The plasma frequency increases as the square root of the carrier density, so for low and moderate electron densities, the optical phonon energy is much higher. This is the reason that we did not screen polar optical phonon scattering in Section 2.8. At higher electron densities, however, ω_p can approach or even exceed ω_0, and the role of screening has to be considered. We have treated phonon and plasmon scattering as two separate mechanisms, but in reality screening couples them. Scattering from these coupled plasmon-phonon modes can be treated by evaluating a dielectric response function for the coupled phonon-plasmon system [2.15].

2.16 Summary

The purpose of this chapter was to illustrate the procedures and a few of the tricks commonly employed to evaluate scattering rates for carriers in semiconductors. Important scattering mechanisms for common semiconductors were also identified, and the scattering rate versus energy characteristics of pure Si and GaAs were described. The method employed is based on Fermi's Golden Rule and proceeds in a straightforward manner once the interaction potential is identified. The scattering rate, and the momentum and energy relaxation rates are evaluated directly from the transition rate, $S(\mathbf{p}, \mathbf{p}')$. Various material parameters needed for scattering rate calculations in silicon and gallium arsenide are listed in Table 2.1. Although our discussion centered on electron scattering in Si and GaAs, the general features of the results are typical of those observed in other covalent and polar semiconductors.

Since our motives were to introduce the basic procedure and to establish the approximate magnitude of the various scattering rates and their functional dependence on carrier energy, a simple approach was adopted. In practice, however, overlap integrals must be treated and the energy bands cannot be described as spherical and parabolic. For high energy electrons in GaAs, the conduction band non-parabolicity is important, and for electrons in silicon, the ellipsoidal nature of the conduction band minima must be included. For holes, scattering between light and heavy hole bands must be included along with the warped constant energy surfaces described in Chapter 1. For the very high energies

Table 2.1. *Transport parameters for silicon and gallium arsenide*

Parameter	Symbol	Value in Si	Value in GaAs
Mass density (g/cm^3)	ρ	2.329	5.36
Lattice constant (Angstroms)	a_0	5.43	5.462
Low frequency dielectric constant	κ_0	11.7	12.90
High frequency dielectric constant	κ_∞	–	10.92
Piezoelectric constant (C/m^2)	e_{pz}	–	0.160
Longitudinal acoustic velocity ($\times 10^5$ cm/s)	υ_l or υ_s	9.04	5.24
Transverse acoustic velocity ($\times 10^5$ cm/s)	υ_t	5.34	3.0
Longitudinal optical phonon energy (eV)	$\hbar\omega_o$	0.063	0.03536
Electron effective mass ratio (lowest valley)	m^* $(m_l m_t^*)$	– (0.916, 0.19)	0.067
Electron effective mass ratio (upper valley)	m^* α $(m_l m_t)$	– – 1.59, 0.12	0.22 (L) 0.58 (X)
Non-parabolicity parameter parameter (eV^{-1})	α	0.5	0.610 (Γ) 0.461 (L) 0.204 (X)
Electron acoustic deformation potential (eV)	D_A	9.5	7.01 (Γ) 9.2 (L) 9.0 (X)
Electron optical deformation potential ($\times 10^8$ eV/cm)	D_o	–	3.0 (L)
Optical phonon energy (eV)	ω_o		0.0343 (L)
Hole acoustic deformation potential (eV)	D_A	5.0	3.5
Hole optical deformation potential (eV/cm)	D_o	6.00	6.48
Intervalley parameters *g*-type (X–X)	D_{if}, E_{in} ($\times 10^8$ eV/cm)(eV)	0.5, 0.012 (TA) 0.8, 0.019 (LA) 11.0, 0.062 (LO)	–
Intervalley parameters *f*-type (X–X)	D_{if}, E_{in} ($\times 10^8$ eV/cm)(eV)	0.3 0.019 (TA) 2.0 0.047 (LA) 2.0 0.059 (TO)	–
Intervalley parameters (X–L)	D_{if}, E_{in} ($\times 10^8$ eV/cm)(eV)	2.0 0.058 2.0 0.055 2.0 0.041 2.0 0.017	–

Table 2.1. *(continued)*

Parameter	Symbol	Value in Si	Value in GaAs
Intervalley deformation potential ($\times 10^8$ eV/cm)	$D_{\Gamma L}, D_{\Gamma X}$	–	10, 10.
	D_{LL}, D_{LX}		10, 5.0
	D_{XX}		7.0
Intervalley phonon energy (eV)	$E_{\Gamma L}, E_{\Gamma X}$	–	0.0278, 0.0299
	E_{LL}, E_{LX}		0.0290, 0.0293
	E_{XX}		0.0299
Energy separation between valleys (eV)	$\Delta E_{L\Gamma}$	–	0.29
	$\Delta E_{X\Gamma}$		

typical of carriers in modern devices, a full, numerical treatment of the energy bands is essential. Finally, there is the critical issue of screening to consider. Many of these refinements to the basic procedure are discussed in the chapter references.

References and further reading

For a clear discussion of the physics of carrier scattering in semiconductors, consult

2.1 Ridley, B. K. *Quantum Process in Semiconductors.* Clarendon Press, Oxford, 1982.

For comprehensive treatments of electron scattering in semiconductors, refer to

2.2 Zawadzki, W. Mechanisms of electron scattering in semiconductors. In *Handbook of Semiconductors*, Vol. 1, ed. by Moss, T. S. Chapter 12. North-Holland Publishing Co., New York, 1982.

2.3 Gantmakher, V. F. and Levinson, Y. B. *Carrier Scattering in Metals and Semiconductors.* North Holland Publishing Co., New York, 1987.

Electron scattering, with emphasis on high-field effects, is discussed in Chapter 2 of

2.4 Reggiani, L. *Hot Electron Transport in Semiconductors.* Topics in Applied Physics, Vol. 58, ed. by Reggiani, L., Springer-Verlag, New York, 1985.

Jacoboni and Reggiani describe the treatment of ellipsoidal energy bands and the scattering of holes in

2.5 Jacoboni, C. and Reggiani, L. The Monte Carlo method for the solution of charge transport in semiconductors with applications to covalent materials. *Review of Modern Physics,* **55**, 645, 1983.

Scattering rates for holes in p-type GaAs are discussed by Wiley in

2.6 Wiley, J. D. Mobility of holes in III-V compounds. In *Semiconductors and Semimetals.* Vol. 10, ed. by Willardson, R. K. and Beer, A. C. Academic Press, New York, 1975.

Rode reviews the important scattering mechanisms under low applied fields in group IV, III-V, and II-VI semiconductors.

2.7 Rode, D. L. Low-field electron transport. In *Semiconductors and Semimetals*. Vol. 10, ed. by Willardson, R. K. and Beer, A. C. Academic Press, New York, 1975.

A classic and very readable account of electrons, phonons, scattering, and transport is

2.8 Ziman, J. *Electrons and Phonons*. Clarendon Press, Oxford, UK, 1960.

An introductory discussion of phonons in polar solids is presented in Chapter 27 of

2.9 Ashcroft, N. W. and Mermin, N. D. *Solid State Physics*. Saunder College/Holt, Rinehart and Winston, Philadelphia, PA, 1976.

A critical comparison of the standard treatment of ionized impurity scattering with experimental results is the subject of

2.10 Anderson, D. A., Apsley, N., Davies, P. and Giles, P. L. Compensation in heavily doped n-type InP and GaAs. *Journal of Applied Physics*, **58**, 3059–67, 1985.

For the theory of phonon scattering of two-dimensional electrons, consult

2.11 Goodnick, S. M. and Lugli, P. Hot carrier relaxation in quasi-2D systems. In *Hot Carriers in Semiconductor Nanostructures: Physics and Applications*. Ed. by Shah, J. Chapter III.I, pp. 191–233, 1992.

2.12 Ridley, B. K. The electron–phonon interaction in quasi-two-dimensional semiconductor quantum well structures, *Journal of Physics C: Solid State Physics*, **15**, 5899–917, 1982.

The interface roughness scattering of electrons at the Si : SiO_2 interface is discussed in

2.13 Jungemann, Chr., Emunds, A. and Engl, W. L. Simulation of linear and nonlinear electron transport in homogeneous silicon inversion layers. *Solid-State Electronics*, **36**, 1529–40, 1993.

2.14 Yamakawa, S., Ueno, H., Taniguchi, K., Hamaguchi, C., Miyatsuji, K. and Ravaioli, U. Study of interface roughness dependence of electron mobility in Si inversion layers using the Monte Carlo method. *Journal of Applied Physics*, **79**, 911–16, 1996.

Ferry presents a good discussion of screening in Chapter 12 of

2.15 Ferry, D. K. *Semiconductors*. Macmillan, New York, 1991.

Problems

2.1 (a) Show that the Brooks–Herring scattering rate,

$$\frac{1}{\tau(p)} = \sum_{\mathbf{p}',\uparrow} S(\mathbf{p}, \mathbf{p}')$$

is

$$\tau(p) = \frac{32\sqrt{2m^*}\pi\kappa_S^2\varepsilon_0^2}{N_i q^4}\frac{(1+\gamma^2)}{\gamma^4}E^{3/2}(p)$$

(b) Evaluate the scattering rate for thermal average electrons in room-temperature GaAs doped at $N_D = 10^{17}$ cm^{-3}. Explain why the scattering rate is so much higher than the momentum relaxation rate plotted in Fig. 2.6.

2.2 (a) Show that the Conwell–Weisskopf scattering rate is

$$\frac{1}{\tau(p)} = N_I \pi b_{\max}^2 \frac{\sqrt{2m^* E(p)}}{m^*}.$$

(b) Provide a simple, physical explanation of the Conwell–Weisskopf expression for $1/\tau(p)$ terms of the cross-section for scattering, $\pi b_{\max}^2$.

(c) Evaluate and plot the scattering rate for thermal average electrons in GaAs at room-temperature and compare it with the momentum relaxation rate. You should plot τ_m/τ versus N_I for $10^{14} < N_I < 10^{18}$ cm^{-3}. Explain in physical terms why $\tau(p)$ and $\tau_m(p)$ differ, and explain why $\tau_m(p)/\tau(p)$ decreases with N_I.

2.3 Equation (2.14), the Debye length, is the screening length for a nondegenerate semiconductor. Derive a more general expression for the screening length by removing the assumption that the semiconductor is nondegenerate. You should express the result as the nondegenerate result times a correction factor.

2.4 Using arguments similar to those in Section 2.2, derive an expression for the interaction potential for piezoelectric scattering, and show that the result is eq. (2.29). Begin with,

$$D = \kappa_S \varepsilon_0 \mathcal{E} + e_{PZ} \frac{\partial u}{\partial x}$$

where e_{PZ} is the *piezoelectric constant*. Show that the piezoelectric scattering potential is

$$U_{PZ}(x, t) = \frac{q e_{PZ}}{\kappa_S \varepsilon_0} u(x, t).$$

This result is sometimes stated in terms of the *electromechanical coupling coefficient*, K^2, where

$$\frac{K^2}{(1 - K^2)} \equiv \frac{e_{PZ}^2}{\kappa_S \varepsilon_0 v_S}$$

and v_S is the longitudinal sound velocity.

2.5 Use the scattering potential for piezoelectric scattering derived in problem 2.4 to answer the following:

(a) Assume equipartition and show that the matrix element for piezoelectric scattering is

$$|H_{p'p}|^2 = \left(\frac{q e_{PZ}}{\kappa_S \varepsilon_0}\right)^2 \frac{k_B T_L}{2 c_l \beta^2 \Omega} \delta_{p', p \pm \hbar \beta} = |K_\beta|^2 |A_\beta|^2 \delta_{p', p \pm \hbar \beta}.$$

What is $|K_\beta|^2$ for piezoelectric scattering?

(b) Write an expression for $S(\mathbf{p}, \mathbf{p}')$ for piezoelectric scattering. Your expression should be in the form of eq. (2.72). What is C_β for piezoelectric scattering?

(c) Evaluate $1/\tau_m(p)$ assuming that piezoelectric scattering is elastic.

2.6 For alloys of compound semiconductors such as $Al_xGa_{1-x}As$ microscopic fluctuations in the alloy composition, x, produce perturbations in the conduction and valence band edges. The transition rate for alloy scattering is

$$S(\mathbf{p}, \mathbf{p}') = \frac{2\pi}{\hbar} \left(\frac{3\pi^2}{16}\right) \frac{|\Delta U|^2}{N\Omega} \delta(E' - E)$$

where N is the concentration of atoms and

$$\Delta U = x(1 - x)(\chi_{GaAs} - \chi_{AlAs})$$

where χ = electron affinity.

(a) Explain why the alloy scattering rate vanishes at $x = 0$ and at $x = 1$.
(b) Derive an expression for $\tau_{\mathrm{m}}(p)$ for alloy scattering.

2.7 Acoustic phonon scattering was assumed to be elastic when evaluating the momentum relaxation rate (2.84). Work out an expression for $1/\tau_{\mathrm{m}}(p)$ due to ADP scattering *without* assuming that the scattering is elastic, and show that the result is nearly equal to equation (2.84) near room temperature.

2.8 Compute the energy relaxation rate due to ADP scattering. Assume energetic carriers so that phonon emission dominates, and assume that spontaneous emission dominates so that $N_{\omega_s} + 1 \approx 1$.
(a) Show that

$$\tau_{\mathrm{E}} = \frac{k_{\mathrm{B}} T_{\mathrm{L}}}{2m^* \upsilon_{\mathrm{S}}^2} \tau_m.$$

(b) Assuming GaAs, evaluate the ratio for thermal average electrons at $T_{\mathrm{L}} = 300\,\mathrm{K}$ and at $T_{\mathrm{L}} = 77\,\mathrm{K}$.

2.9 Compute and compare the momentum relaxation times due to ionized impurity scattering under the following circumstances (you may use either the Brooks–Herring or Conwell–Weiskopf approach):
(a) Find $1/\tau_m$ for electrons with the thermal average energy $E = 3k_{\mathrm{B}}T_{\mathrm{L}}/2$, in GaAs doped at $N_{\mathrm{D}} = 10^{18}\ \mathrm{cm}^{-3}$. Assume a lattice temperature of $T_{\mathrm{L}} = 300\,\mathrm{K}$.
(b) Find $1/\tau_m$ for electrons with $E = 0.3\,\mathrm{eV}$ in GaAs doped at $N_{\mathrm{D}} = 10^{18}\ \mathrm{cm}^{-3}$. Such electrons can be produced by a heterojunction launching ramp as displayed in Fig. 3.2.

2.10 Consider the optical deformation potential scattering of confined electrons. Assume a rectangular well with infinitely high barriers and answer the following.
(a) Derive an expression for the transition rate $S(\mathbf{p}, \mathbf{p}')$.
(b) Obtain an expression for the scattering rate.
(c) Plot the total scattering rate versus energy for ODP scattering. Assume that $\hbar\omega_0 = E_1$, $N(\omega_0) = 1/4$, and that $E_2 = 4E_1$ and $E_3 = 9E_1$. Compare your answer to Figure 8 of Ridley's paper [2.12].

3 The Boltzmann transport equation

To completely specify the operation of a device, we should know the state of each carrier within the device. If the carriers behave as classical particles, we should know each carrier's position and momentum as a function of time. A direct approach would consist of solving Newton's equations,

$$\frac{d\mathbf{p}_i}{dt} = (-q)\boldsymbol{\mathcal{E}} + \mathbf{R}(\mathbf{r}, \mathbf{p}, t) \tag{3.1}$$

and

$$\frac{d\mathbf{r}_i}{dt} = v_i(t), \tag{3.2}$$

for each of the $i = 1, \ldots, N$ carriers in the device. In these equations $\mathbf{p}_i(t)$ is the momentum of carrier i, $\mathbf{r}_i(t)$ its position, and $\mathbf{R}$ is the random force due to impurities or lattice vibrations. Alternatively, we could ask: what is the probability of finding a carrier with crystal momentum $\mathbf{p}$, at location $\mathbf{r}$, at time t? The answer is $f(\mathbf{r}, \mathbf{p}, t)$ where $f(\mathbf{r}, \mathbf{p}, t)$, the *distribution function*, is a number between zero and one. To find $f(\mathbf{r}, \mathbf{p}, t)$ we solve the *Boltzmann Transport Equation*. The distribution function describes the average distribution of carriers in both position and momentum and can be used to obtain various quantities of interest such as the carrier, current, and kinetic energy densities. Our purpose in this chapter is to derive and discuss the Boltzmann Transport Equation (BTE) and to show how it is solved to obtain the distribution function.

3.1 The distribution function, $f(\mathbf{r}, \mathbf{p}, t)$

Before formulating an equation for $f(\mathbf{r}, \mathbf{p}, t)$, let's discuss what it is and how it is used. The equilibrium distribution function is simply the Fermi–Dirac function,

$$f_0(\mathbf{p}) = \frac{1}{1 + e^{[E_C(\mathbf{r},\mathbf{p}) - E_F]/k_B T_L}}, \tag{3.3}$$

where E_F is the Fermi level, T_L the lattice temperature, and

$$E_C(\mathbf{r}, \mathbf{p}) = E_{C0}(\mathbf{r}, t) + E(\mathbf{p}) \tag{3.4}$$

is the sum of the carrier's potential, $E_{C0}(\mathbf{r}, t)$, and kinetic, $E(\mathbf{p})$, energies. If the energy band is assumed to be spherical and parabolic, then $E(\mathbf{p}) = p^2/2m^*$. For such cases, the equilibrium distribution function depends only on the magnitude of $\mathbf{p}$, so, for a nondegenerate semiconductor, eq. (3.3) becomes:

$$f_0(\mathbf{r}, \mathbf{p}) = e^{[E_F - E_{C0}(\mathbf{r})]/k_B T_L} \times e^{-p^2/2m^* k_B T_L}, \tag{3.5}$$

which is plotted in Fig. 3.1a. The distribution function shows that the zero velocity state has the highest probability of being occupied, but that a significant number of high velocity states are also occupied.

To find the total number of carriers we simply add up the carriers in each momentum state. The average carrier density is

$$n(\mathbf{r}, t) = \frac{1}{\Omega} \sum_{\mathbf{p}} f(\mathbf{r}, \mathbf{p}, t), \tag{3.6}$$

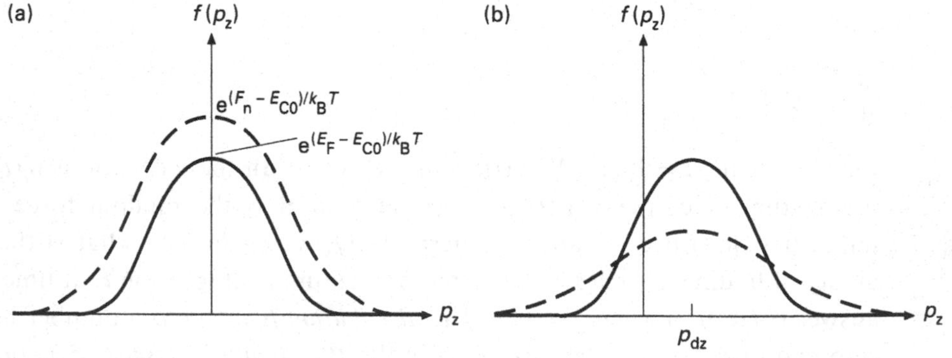

Fig. 3.1 (a) Illustration of Maxwellian distribution functions. The solid line is the equilibrium, non-degenerate, distribution function, and the dashed line is an equilibrium Maxwellian with the Fermi level replaced by the quasi-Fermi level. (b) Illustration of displaced Maxwellian distribution functions. The solid line assumes that the carrier temperature is equal to the lattice temperature, and the dashed line assumes a carrier temperature that exceeds that of the lattice.

where the sum is over all states in the first Brillouin zone. Equation (3.6) is the average carrier density in a small normalization volume, Ω, centered about $\mathbf{r}$. Similarly, we find the electron current density by weighting the sum by velocity,

$$\mathbf{J}(\mathbf{r}, t) = \frac{(-q)}{\Omega} \sum_{\mathbf{p}} \upsilon(\mathbf{p}) f(\mathbf{r}, \mathbf{p}, t) = \frac{(-q)}{\Omega} \sum_{\mathbf{p}} \frac{\mathbf{p}}{m^*} f(\mathbf{r}, \mathbf{p}, t), \tag{3.7}$$

and the average kinetic energy density from

$$W(\mathbf{r}, t) = \frac{1}{\Omega} \sum E(\mathbf{p}) f(\mathbf{r}, \mathbf{p}, t) = \frac{1}{\Omega} \sum_{\mathbf{p}} \frac{p^2}{2m^*} f(\mathbf{r}, \mathbf{p}, t). \tag{3.8}$$

In each of the above two equations, the second expression on the right-hand side follows for spherical, parabolic, energy bands.

Because it is easier to integrate over $\mathbf{p}$ than to sum over a very large number of states, we usually convert the sums to integrals as prescribed by eq. (1.54). After transforming from wave vector to momentum space, the prescription becomes

$$\sum_{\mathbf{p}} g(\mathbf{p}) = \frac{\Omega}{4\pi^3\hbar^3} \int_{\mathbf{p}} g(\mathbf{p}) d\mathbf{p} \tag{3.9}$$

for three-dimensional electrons. Although the integration is to be performed only over the states in one Brillouin zone, the integration limit is extended to infinity in practice because the probability that states are occupied falls off very rapidly with energy. Because we shall make frequent use of eq. (3.9), it is worthwhile to consider a few examples of its use (a nondegenerate semiconductor in equilibrium is assumed).

Example: Evaluation of macroscopic quantities from the equilibrium distribution function

Carrier density:
The equilibrium carrier density is evaluated from

$$n_0(\mathbf{r}) = \frac{1}{\Omega} \sum_{\mathbf{p}} f_0(\mathbf{r}, \mathbf{p}) = \frac{1}{4\pi^3\hbar^3} \int_{\mathbf{p}} e^{[E_F - E_{C0}(\mathbf{r}) - p^2/2m^*]/k_B T_L} d\mathbf{p}.$$

After performing the integral we find

$$n_0(\mathbf{r}) = \frac{1}{4} \left(\frac{2m^* k_B T_L}{\pi\hbar^2} \right)^{3/2} e^{(E_F - E_{C0})/k_B T_L} \equiv N_C e^{(E_F - E_{C0})/k_B T_L}, \tag{3.10}$$

a familiar result which relates the equilibrium carrier density to the Fermi level.

Average kinetic energy per carrier:
The average kinetic energy density in equilibrium is found from

$$W_0(\mathbf{r}) = \frac{1}{\Omega} \sum_{\mathbf{p}} \frac{p^2}{2m^*} f_0(\mathbf{r}, \mathbf{p}) = \frac{1}{8\pi^3\hbar^3 m^*} \int_{\mathbf{p}} p^2 e^{[E_F - E_{C0}(\mathbf{r}) - p^2/2m^*]/k_B T_L} d\mathbf{p},$$

which gives

$$\frac{W_0}{n} \equiv u_0 = \frac{3}{2} k_B T_L \tag{3.11}$$

for u_0, the average kinetic energy per carrier in equilibrium. The kinetic energy component associated with the degree of freedom along the $\hat{z}$-axis can also be evaluated from

$$W_{zz}(\mathbf{r}) = \frac{1}{\Omega} \sum_{\mathbf{p}} \frac{p_z^2}{2m^*} f_0(\mathbf{r}, \mathbf{p}) = \frac{1}{8\pi^3 \hbar^3 m^*} \int_{\mathbf{p}} p_z^2 e^{[E_F - E_{C0}(\mathbf{r}) - p^2/2m^*]/k_B T_L} \, d\mathbf{p}$$

which evaluates to

$$\frac{W_{zz}}{n} \equiv u_{zz} = \frac{1}{2} k_B T_L. \tag{3.12}$$

It is also easy to show that $u_{xx} = u_{yy} = u_{zz}$, which demonstrates that the kinetic energy is equally distributed among the three degrees of freedom; the result is known as *equipartition of energy*. The total kinetic energy per carrier is simply $u = u_{xx} + u_{yy} + u_{zz}$.

Energy-related tensor:
A tensor, W_{ij}, closely related to the scalar W, can also be defined. This tensor has nine components defined by

$$W_{ij}(\mathbf{r}) \equiv \frac{1}{\Omega} \sum \frac{p_i \upsilon_j}{2} f(\mathbf{r}, \mathbf{p}), \tag{3.13}$$

where i and j range over the three coordinate axes. After evaluating eq. (3.13) component by component in equilibrium, we find that the tensor is diagonal and that

$$\frac{W_{ij}}{n} \equiv u_{ij} = \frac{k_B T_L}{2} \delta_{ij} \tag{3.14}$$

(recall that δ_{ij}, the Kronecker delta, is zero unless $i = j$). The kinetic energy density, W, is simply the trace of W_{ij}, that is

$$W = \sum_i W_{ii} \equiv W_{ii}. \tag{3.15}$$

The second expression on the right-hand side defines the *summation convention*; when repeated subscripts occur, it is understood that a sum over the three coordinate axes is to be performed.

Current density:
Finally, we note that the equilibrium electron current density

$$\mathbf{J}_{n0}(\mathbf{r}) = (-q) \frac{1}{\Omega} \sum_{\mathbf{p}} \frac{\mathbf{p}}{m^*} f_0(\mathbf{r}, \mathbf{p})$$

is zero because the velocity, $\boldsymbol{\upsilon}(\mathbf{p} = \mathbf{p}/m^*$, is odd in momentum while f_0 is even [that is, $\boldsymbol{\upsilon}(\mathbf{p}) = -\boldsymbol{\upsilon}(-\mathbf{p})$ and $f_0(-\mathbf{p} = f_0(\mathbf{p})$].

The fundamental problem in device analysis is to find $f(\mathbf{r}, \mathbf{p}, t)$ because it defines the average state of the carriers in the device. We shall find that it is exceedingly difficult to deduce $f(\mathbf{r}, \mathbf{p}, t)$ in realistic devices, so a reasonable guess for the distribution function is often used for device analysis. For example, the *quasi-Fermi level* is introduced so that the nonequilibrium distribution function can be written as

$$f(\mathbf{r}, \mathbf{p}, t) = \frac{1}{1 + e^{[E_C(\mathbf{r},\mathbf{p}) - F_n(\mathbf{r},p)]/k_B T_L}}, \tag{3.16}$$

where $F_n(\mathbf{r}, t)$, the quasi Fermi level, plays the role that E_F did in equilibrium. That this cannot be the correct distribution function is apparent by observing that it is even in momentum and therefore predicts that current can never flow. Nevertheless, it is not an unreasonable guess considering that average carrier velocities are frequently small. If we measure the spread in velocities by the velocity at which $f(\mathbf{r}, \mathbf{p})$ drops to $1/e$ of its peak, we find a spread of $\sqrt{2k_B T_L/m^*}$ or about 10^7 cm/sec for a semiconductor with $m^* \simeq m_0$. Since average velocities in a device are often much smaller than this, the assumption that the average velocity is zero may not be so bad. The non-degenerate approximation to eq. (3.16) is also plotted in Fig. 3.1a.

A better guess for $f(\mathbf{r}, \mathbf{p}, t)$ might be to assume that it retains its equilibrium shape but that its peak, or average, momentum is displaced from the origin. If the average carrier velocity is υ_d, then we would write $f(\mathbf{r}, \mathbf{p}, t)$ in this *displaced Maxwellian* approximation as

$$f(\mathbf{r}, \mathbf{p}, t) = e^{[F_n(\mathbf{r},t) - E_{C0}(\mathbf{r},t)]/k_B T_L} \times e^{-|\mathbf{p}-\mathbf{p}_d|^2/2m^* k_B T_L} \tag{3.17}$$

where

$$\mathbf{p}_d = m^* \boldsymbol{\upsilon}_d, \tag{3.18}$$

and $\boldsymbol{\upsilon}_d$ is the average carrier velocity. By computing the carrier density, we find

$$n(\mathbf{r}, t) = N_C e^{[F_n(\mathbf{r},t) - E_{C0}(\mathbf{r},t)]/k_B T_L}$$

which is much like eq. (3.10), but the kinetic energy density per carrier is

$$u(\mathbf{r}) = \frac{W(\mathbf{r})}{n} = \frac{1}{2} m^* \upsilon_d^2 + \frac{3}{2} k_B T_L \tag{3.19}$$

instead of eq. (3.11). We interpret this result as the sum of *drift energy*, due to the average carrier velocity, and *thermal energy* due to the collisions of carriers with random lattice vibrations. The fact that the temperature is the lattice temperature implies that phonon scattering is strong enough to ensure that the carriers and lattice are in equilibrium.

We could further improve this approximation to $f(\mathbf{r}, \mathbf{p}, t)$ by replacing the lattice temperature, T_L, with T_C, the carrier temperature. The lattice and carriers

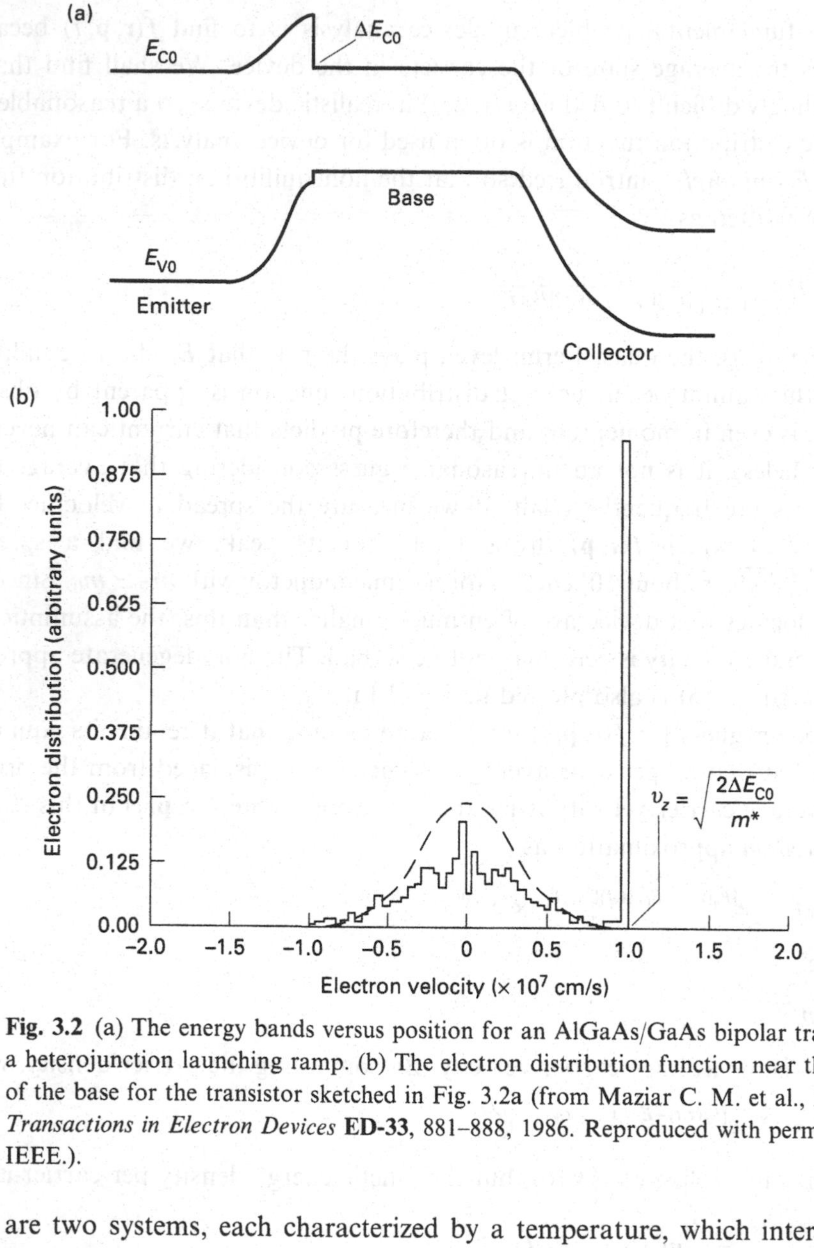

Fig. 3.2 (a) The energy bands versus position for an AlGaAs/GaAs bipolar transistor with a heterojunction launching ramp. (b) The electron distribution function near the beginning of the base for the transistor sketched in Fig. 3.2a (from Maziar C. M. et al., *IEEE Transactions in Electron Devices* **ED-33**, 881–888, 1986. Reproduced with permission from IEEE.).

are two systems, each characterized by a temperature, which interact through collisions. The carriers and the lattice exchange energy through collisions, but the rate of energy transfer is frequently too slow to make $T_C = T_L$. Under high fields, $T_C > T_L$ high-field transport is sometimes referred to as *hot-carrier transport*. Figure 3.1b displays two displaced Maxwellian distribution functions, one with the carrier temperature equal to that of the lattice and another with a higher carrier temperature.

Our guesses for nonequilibrium distribution functions have been guided by knowledge of the equilibrium distribution function. Since each guess resembles

Table 3.1. *Various quantities of interest for a Maxwellian and displaced Maxwellian distribution function**

Quantity	Notation	Maxwellian	Displaced Maxwellian
Distribution function	$f(\mathbf{r}, \mathbf{p}, t)$	$e^{\eta_C(\mathbf{r},t)}e^{-p^2/2m^*k_BT_C}$	$e^{\eta_C(\mathbf{r},t)}e^{-\lvert\mathbf{p}-\mathbf{p}_d\rvert^2/2m^*k_BT_C}$
Carrier density	$n(\mathbf{r}, t)$	$N_C e^{\eta_C(\mathbf{r},t)/k_BT_C}$	$N_C e^{\eta_C(\mathbf{r},t)/k_BT_C}$
Average velocity	υ_d	0	$\mathbf{p}_d/m^*$
Average kinetic energy per carrier	u	$\frac{3}{2}k_BT_C$	$\frac{3}{2}k_BT_C + p_d^2/2m^*$
Kinetic energy component	u_{zz}	$\frac{1}{2}k_BT_C$	$\frac{1}{2}k_BT_C + p_{dz}^2/2m^*$
Energy-related tensor	W_{ij}/n	$\frac{1}{2}k_BT_C\delta_{ij}$	$\frac{1}{2}k_BT_C\delta_{ij} + p_{di}p_{dj}/2m^*$

* The quantity η_C is defined by $\eta_C = (F_n - E_{C0}/k_BT_C)$ where T_C is the carrier temperature.

$f_0(\mathbf{r}, \mathbf{p})$, we expect them to be valid in benign situations – when the device is near equilibrium. But in devices the carriers may be very far from equilibrium so that we have little guidance as to what to expect for $f(\mathbf{r}, \mathbf{p}, t)$. For example, consider the energy band diagram for a heterojunction bipolar transistor as sketched in Fig. 3.2. The heterojunction 'launching ramp' at the emitter acts like an electron gun shooting electrons into the base with their initial momentum plus $\delta p_z = \sqrt{2m^*\Delta E_{C0}}$. (This device is made with a wide band gap $Al_xGa_{1-x}As$ emitter and a GaAs base for example.) If the base is so thin that little scattering takes place, and if δp_z is much larger than the thermal spread of velocities entering from the emitter, then $f(z, p_z) \simeq \delta(p_z - \sqrt{2m^*\Delta E_{C0}})$ as shown in Fig. 3.2b (which was obtained by Monte Carlo simulation as described in Chapter 6).

It is apparent from this example that we need an equation which we can solve $f(\mathbf{r}, \mathbf{p}, t)$ since it may be difficult to guess its form. The equation is called the Boltzmann Transport Equation and is derived next. Nevertheless, a Maxwellian or displaced Maxwellian is a frequently-used approximation. Table 3.1 summarizes the important specific results for Maxwellian distribution functions.

3.2 The Boltzmann transport equation

The distribution function gives the probability of finding carriers at time, t, located at position $\mathbf{r}$, with momentum $\mathbf{p}$. The Boltzmann transport equation (BTE) is just a bookkeeping equation for $f(\mathbf{r}, \mathbf{p}, t)$ (much like the continuity equation for carriers) which accounts for all possible mechanisms by which f may change. The BTE can be obtained in several different ways. Because of its importance, we present two derivations in this section. The first derives the BTE in terms of carrier in- and out-flows, and the second is based on carrier trajectories in the six-dimensional position–momentum space.

3.2.1 Derivation of the BTE by carrier conservation in position–momentum space

The BTE is much like the familiar particle continuity equation; the difference is that it describes particle flow in a six-dimensional position–momentum, or *phase*, space. The BTE can be derived by careful bookkeeping. Consider a region in two-dimensional position–momentum space as shown in Fig. 3.3. In a time δt, f may increase within the region shown if the in-flow exceeds the out-flow (in both position *and* momentum space) or if there is a net generation of carriers or if collisions send carriers from other cells to the one shown (collisions are assumed to instantaneously change the carriers' momentum but not their position). Conservation of carriers requires that

$$(\delta f \delta r \delta p) = [f(r) - f(r + \delta r)]\upsilon \delta t \delta p + [f(p) - f(p + \delta p)]F \delta t \delta r \\ + [s(r, p, t) + \partial f / \partial t|_{\text{coll}}] \delta t \delta r \delta p. \tag{3.20}$$

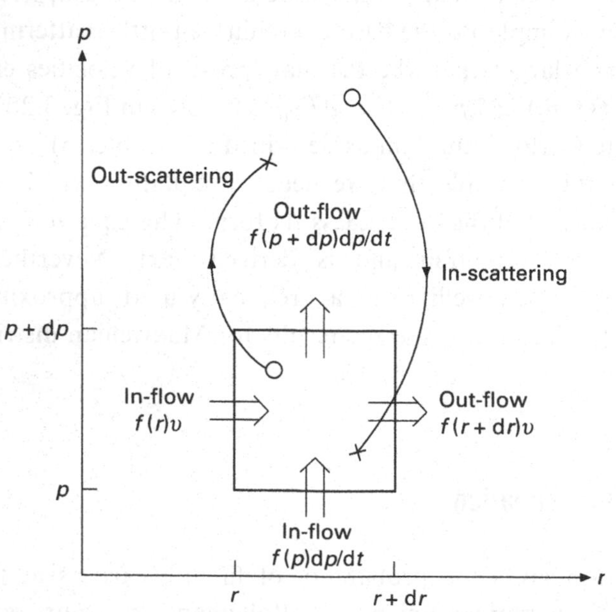

Fig. 3.3 Cell in two-dimensional position–momentum space.

After rearranging terms and letting δt, δr and δp approach zero, we find

$$\frac{\partial f}{\partial t} = -\upsilon\frac{\partial f}{\partial r} - F\frac{\partial f}{\partial p} + \left.\frac{\partial f}{\partial t}\right|_{coll} + s(r, p, t), \tag{3.21}$$

which is the BTE for one space and one momentum dimension. Generalizing this result to a six-dimensional position–momentum space, we find

$$\boxed{\frac{\partial f}{\partial t} + \boldsymbol{\upsilon}\cdot\nabla_r f + \mathbf{F}\cdot\nabla_p f = \left.\frac{\partial f}{\partial t}\right|_{coll} + s(\mathbf{r}, \mathbf{p}, t)}\ , \tag{3.22}$$

where

$$\nabla_r f \equiv \frac{\partial f_x}{\partial x}\hat{x} + \frac{\partial f_y}{\partial y}\hat{y} + \frac{\partial f_z}{\partial z}\hat{z} \tag{3.23a}$$

and

$$\nabla_p f \equiv \frac{\partial f_x}{\partial p_x}\hat{x} + \frac{\partial f_y}{\partial p_y}\hat{y} + \frac{\partial f_z}{\partial p_z}\hat{z}. \tag{3.23b}$$

Equation (3.22) is the Boltzmann transport equation; its solution provides the distribution function from which macroscopic quantities of interest are readily evaluated. For classical particles, $\boldsymbol{\upsilon} = \mathbf{p}/m^*$, but for electrons in semiconductors, $\boldsymbol{\upsilon} = \nabla_p E(\mathbf{p})$. The two expressions are equal only when $E(\mathbf{p}) = p^2/2m^*$.

One can understand the BTE by analogy with the familiar hole continuity equation,

$$\frac{\partial \rho}{\partial t} = -\nabla\cdot(\mathbf{J}_\rho/q) + \left.\frac{\partial \rho}{\partial t}\right|_{G-R}, \tag{3.24}$$

which simply states that the increase in carrier density, ρ, within a small volume centered at $\mathbf{r}$ is due to the net in-flow of carriers plus the net increase due to generation-recombination processes. (Note that we use ρ in eq. (3.24) to denote the hole density rather than the usual p to avoid confusion with momentum.) In Eq. (3.24), $-\nabla\cdot(\mathbf{J}_\rho/q)$ represent the net in-flow of particles to an elementary volume centered at $\mathbf{r}$. Similarly, $-(\boldsymbol{\upsilon}\cdot\nabla_r f + \mathbf{F}\cdot\nabla_p f)$ in the BTE represents net in-flows. The first term is an in-flow in position space and the second is an in-flow in momentum space. The 'generation-recombination' term in the BTE consists of two components. The first describes actual carrier generation recombination processes such as photogeneration or recombination through defects by the function $s(\mathbf{r}, \mathbf{p}, t)$. Collisions send carriers from one momentum state to another and, therefore, also produce sources and sinks in momentum space.

3.2.2 Derivation of the BTE from trajectories in position-momentum space

The BTE can also be derived from a much different viewpoint, one that is useful not only for deriving the BTE, but also, as we shall see in Section 3.6, for solving the BTE. Carriers in a semiconductor move and change momentum in response to an electric field. We call the path a carrier takes in the six-dimensional phase space, its trajectory $[\mathbf{r}(t), \mathbf{p}(t)]$. Figure 3.4 shows several trajectories in a simple, two-dimensional phase space. Consider a carrier at position A on trajectory 2 at time t. Carriers at position A were at position A′ at time $t - \mathrm{d}t$. The probability that a state at A is occupied, therefore, is simply the probability that the state A′ was occupied at a time $\mathrm{d}t$ earlier. That is,

$$f(r, p, t) = f(r - \upsilon \mathrm{d}t, p - F\mathrm{d}t, t - \mathrm{d}t), \tag{3.25}$$

or

$$\frac{\mathrm{d}f}{\mathrm{d}t} = 0 \quad \text{(no scattering)}. \tag{3.26}$$

Equation (3.26) states that if we follow a carrier as it moves in phase space, the probability of occupation does not change. If the state was occupied at $t = 0$, then it is always occupied; if it was empty at $t = 0$, then it is always empty. In deriving eq. (3.26), however, we made no allowance for scattering or carrier generation. As shown in Fig. 3.4, the probability that the state at A is occupied can change if carriers out-scatter to other trajectories or if they in-scatter from other trajectories. The probability may also change if there is a source, s, of carriers present. Consequently, eq. (3.26) must be modified to

$$\frac{df}{dt} = \left.\frac{\partial f}{\partial t}\right|_{coll} + s \tag{3.27}$$

in order to account for scattering. To obtain the BTE explicitly, we expand the total derivative on the left-hand side of eq. (3.27) to find

$$\frac{\mathrm{d}f}{\mathrm{d}t} = \frac{\partial f}{\partial t} + \frac{\partial f}{\partial r}\frac{dr}{dt} + \frac{\partial f}{\partial p}\frac{dp}{dt} = \frac{\partial f}{\partial t} + \frac{\partial f}{\partial r}\upsilon + \frac{\partial f}{\partial p}F = \left.\frac{\partial f}{\partial t}\right|_{\mathrm{coll}} + s \tag{3.28}$$

which is the one-dimensional analog of eq. (3.22). We will find this viewpoint useful when we discuss path integral solutions to the BTE in Section 3.6.

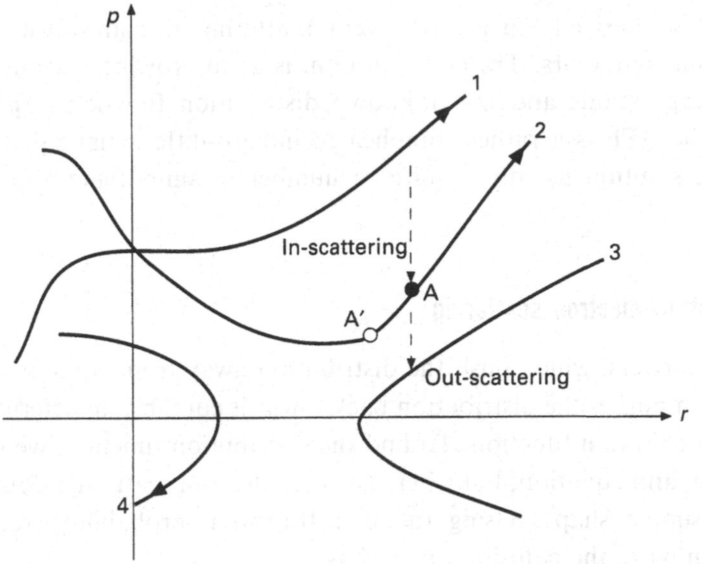

Fig. 3.4 Illustration of trajectories in position–momentum space. Carrier's move along a trajectory according to Newton's Laws. Occasionally they scatter to another trajectory. Scattering instantly changes the carrier's momentum, but does not affect its position.

3.3 The collision integral and the relaxation time approximation

Scattering may alter $f(\mathbf{r}, \mathbf{p}, t)$ by two processes – carriers at $\mathbf{p}'$ could be scattered to $\mathbf{p}$ thereby increasing f (the in-scattering process) or carriers at $\mathbf{p}$ could scatter out decreasing f (the out-scattering process). The net rate of increase of $f(\mathbf{r}, \mathbf{p}, t)$ due to collisions is a result of the competition between in-scattering and out-scattering and is given by

$$\left.\frac{\partial f}{\partial t}\right|_{coll} = \sum_{\mathbf{p}'} f(\mathbf{p}')[1 - f(\mathbf{p})]S(\mathbf{p}', \mathbf{p}) - \sum_{\mathbf{p}'} f(\mathbf{p})[1 - f(\mathbf{p}')]S(\mathbf{p}, \mathbf{p}') = \hat{C}f, \qquad (3.29a)$$

where $\hat{C}$ is the *collision operator*. For the in-scattering process, $f(\mathbf{p}')$ gives the probability that a carrier is at $\mathbf{p}'$, and $[1 - f(\mathbf{p})]$ is the probability that the state at $\mathbf{p}$ is empty. The transition rate $S(\mathbf{p}', \mathbf{p})$ is the probability per second that a carrier at $\mathbf{p}'$ will scatter to $\mathbf{p}$ (assuming that state $\mathbf{p}'$ is occupied and that state $\mathbf{p}$ is empty). The sum is over $\mathbf{p}'$ – all of the possible states from or to which carriers may scatter. In non-degenerate semiconductors $f(\mathbf{r}, \mathbf{p}, t) \ll 1$ and the $[1 - f(\mathbf{p})]$ terms in eq. (3.29a) can be set to unity, so

$$\left.\frac{\partial f}{\partial t}\right|_{coll} = \sum_{\mathbf{p}'} f(\mathbf{p}')S(\mathbf{p}', \mathbf{p}) - \sum_{\mathbf{p}'} f(\mathbf{p})S(\mathbf{p}, \mathbf{p}'). \qquad (3.29b)$$

In Chapter 2 we derived $S(\mathbf{p}, \mathbf{p}')$ for several scattering mechanisms important in common semiconductors. The collision term is a sum (or integral) involving these known expressions and the unknown distribution function, $f(\mathbf{p})$. As a consequence, the BTE is a rather complicated integro-differential equation for $f(\mathbf{r}, \mathbf{p}, t)$ whose solution usually requires a number of simplifying approximations.

3.3.1 The role of electron–electron scattering

Forces on the carriers, which push the distribution away from equilibrium and scattering, which pushes the distribution back towards equilibrium determine the shape of the distribution function. To find the distribution function, we need to solve the Boltzmann equation, but when electron–electron scattering dominates, it assumes a simple shape. Using the pair transition probability [recall eq. (2.100)], we can write the collision integral as

$$\left.\frac{\partial f}{\partial t}\right|_{coll} = -\sum_{\mathbf{p}', \mathbf{p}_2} S(\mathbf{p}, \mathbf{p}_2; \mathbf{p}', \mathbf{p}_2')f(\mathbf{p})f(\mathbf{p}_2) + \sum_{\mathbf{p}', \mathbf{p}_2'} S(\mathbf{p}', \mathbf{p}_2'; \mathbf{p}, \mathbf{p}_2)f(\mathbf{p}')f(\mathbf{p}_2'). \tag{3.30}$$

Equation (3.30) describes processes in which carriers at $\mathbf{p}$ out-scatter from a target carrier at $\mathbf{p}_2$ (first term) and in which carriers at $\mathbf{p}'$ in-scatter from a target carrier at $\mathbf{p}_2'$ (second-term). (Note that eq. (3.30) assumes nondegenerate conditions because there are no $(1-f)$ state-filling terms present.) Figure 2.17 illustrated the electron–electron scattering process. Notice from eq. (2.100) that the probability for a forward transition is equal to the probability for a reverse transition, that is

$$S(\mathbf{p}, \mathbf{p}_2; \mathbf{p}', \mathbf{p}_2') = S(\mathbf{p}', \mathbf{p}_2'; \mathbf{p}, \mathbf{p}_2). \tag{3.31}$$

Using eq. (3.31) in eq. (3.30) and performing the second sum over $\mathbf{p}_2$ instead of $\mathbf{p}_2'$ (which is permissible because the two are related by momentum conservation), we find

$$\left.\frac{\partial f}{\partial t}\right|_{coll} = -\sum_{\mathbf{p}', \mathbf{p}_2} S(\mathbf{p}, \mathbf{p}_2; \mathbf{p}', \mathbf{p}_2')\left[f(\mathbf{p})f(\mathbf{p}_2) - f(\mathbf{p}')f(\mathbf{p}_2')\right]. \tag{3.32}$$

Eventually, in-scattering and out-scattering balance and steady-state conditions are achieved. Then $\partial f/\partial t|_{coll} = 0$, which occurs when

$$f(\mathbf{p})f(\mathbf{p}_2) = f(\mathbf{p}')f(\mathbf{p}_2'). \tag{3.33}$$

To solve eq. (3.33), let's try a Maxwellian distribution function,

$$f(p) = e^{Kp^2}, \tag{3.34}$$

where K is a constant. After inserting eq. (3.34) into eq. (3.33), we find

$$e^{K(p^2+p_2^2)} = e^{K(p'^2+p_2'^2)}. \tag{3.35}$$

Equation (3.35) is indeed satisfied because electron–electron scattering conserves energy, so we conclude that in steady state, electron–electron scattering forces the carrier distribution functions to a Maxwellian shape. The constant, K, is related to the average carrier energy and, therefore, to the carrier temperature ($\langle E\rangle = 3/4m^*K = 3/2k_BT_C$). So when the carrier concentration is high (typically $\approx 10^{17}-10^{18}\ \text{cm}^{-3}$), electron–electron scattering dominates and the distribution function can safely be assumed to be Maxwellian with the spread of the distribution, or carrier temperature, being determined by the average kinetic energy of the carriers. For such cases, the problem is greatly simplified because there is no need to solve for the shape of the distribution function. Finally note that our argument assumed simple, parabolic energy bands and nondegenerate conditions. For degenerate conditions, electron–electron scattering forces the carrier distribution to a Fermi–Dirac shape rather than to a Maxwellian.

3.3.2 The relaxation time approximation

We conclude this section by introducing a commonly-used simplification for the collision integral. In the following section, we explore solutions to the BTE using this simplification, and in Section 3.5 we examine its validity. We begin by writing the non-equilibrium distribution function as

$$f(\mathbf{r}, \mathbf{p}, t) = f_S(\mathbf{r}, \mathbf{p}, t) + f_A(\mathbf{r}, \mathbf{p}, t), \tag{3.36}$$

where the first term is symmetric in momentum and is assumed to be large while the second term is a small, anti-symmetric component. Note that although the symmetric component is large, it can carry no current. It is apparent that in equilibrium, $f_S = f_0$ and $f_A = 0$. Out of equilibrium, we assume that f_S has the same form but with E_F replaced with F_n as in eq. (3.16). Under high applied fields, we would also need to replace the lattice temperature with the carrier temperature.

Using eq. (3.36) in the non-degenerate collision integral, we find

$$\left.\frac{\partial f}{\partial t}\right|_{\text{coll}} = \left.\frac{\partial f_S}{\partial t}\right|_{\text{coll}} + \left.\frac{\partial f_A}{\partial t}\right|_{\text{coll}} \tag{3.37}$$

where

$$\left.\frac{\partial f_S}{\partial t}\right|_{\text{coll}} \equiv \sum_{\mathbf{p}'} f_S(\mathbf{p}')S(\mathbf{p}', \mathbf{p}) - f_S(\mathbf{p})S(\mathbf{p}, \mathbf{p}') \tag{3.38a}$$

and

$$\left.\frac{\partial f_A}{\partial t}\right|_{\text{coll}} \equiv \sum_{\mathbf{p}'} f_A(\mathbf{p}')S(\mathbf{p}', \mathbf{p}) - f_A(\mathbf{p})S(\mathbf{p}, \mathbf{p}'). \tag{3.38b}$$

In equilibrium, $f_S = f_0$ and $\partial f_S/\partial t|_{\text{coll}}$ vanishes. When the non-equilibrium f_S retains its equilibrium form with the carrier temperature equal to that of the lattice, $\partial f_S/\partial t|_{\text{coll}}$ also vanishes [see homework problem (3.12)]. For high electric fields, however, the carriers are heated, so $T_C > T_L$, and $\partial f_S/\partial t|_{\text{coll}}$ is non-zero. We conclude that for low-field transport when the carriers are in equilibrium with the lattice, $\partial f_S/\partial t|_{\text{coll}}$ may be set to zero.

To continue, we need to approximate $\partial f_A/\partial t|_{\text{coll}}$. A plausible form is

$$\left.\frac{\partial f_A}{\partial t}\right|_{\text{coll}} = -\frac{f_A(\mathbf{r}, \mathbf{p}, t)}{\tau_f(\mathbf{r}, \mathbf{p}, t)}, \tag{3.39}$$

where τ_f is a characteristic time which describes how the distribution function relaxes. Putting these considerations together, we obtain

$$\left.\frac{\partial f}{\partial t}\right|_{coll} = \left.\frac{\partial f_S}{\partial t}\right|_{coll} - \frac{f_A}{\tau_f} \tag{3.40a}$$

for high applied fields and

$$\boxed{\left.\frac{\partial f}{\partial t}\right|_{\text{coll}} = -\frac{f_A}{\tau_f}} \tag{3.40b}$$

for low applied fields. Both results are known as the relaxation time approximation (RTA), but in this text the term RTA will refer only to eq. (3.40b) which is valid for low applied fields.

To illustrate the meaning of eq. (3.40b), consider the BTE under spatially uniform conditions with no applied force. From eq. (3.22) and eq. (3.40b), we find

$$\frac{\partial f}{\partial t} = -\frac{(f - f_0)}{\tau_f} \tag{3.41}$$

(if we assume near-equilibrium conditions so that $f_S \simeq f_0$). According to eq. (3.41), the semiconductor responds to small perturbations in the distribution function by trying to restore equilibrium $\partial f/\partial t < 0$ if $f > f_0$ and $\partial f/\partial t > 0$ if $f < f_0$. The solution to eq. (3.41) is

$$f(t) = f_0 + [f(0) - f_0]e^{-t/\tau_f}, \tag{3.42}$$

which states that perturbations decay exponentially with the characteristic time τ_f.

It is far from obvious that eq. (3.38b) can be written in the simple form, eq. (3.40b). Of course, the collision term may always be written in this form if we simply equate the two expressions and view the result as a defining relation for τ_f. But the approximation is useful only if τ_f is independent of the distribution function. We expect that this can be true only under low-field conditions where the perturbations are small, but as we show in Section 3.5, even under low-field conditions, the RTA is valid only when the scattering mechanism is either elastic, isotropic or both.

3.4 Solving the BTE in the relaxation time approximation

In this section we examine solutions to the BTE for a few simple cases. A systematic treatment of low-field transport in the relaxation time approximation is deferred until Chapter 4.

3.4.1 Equilibrium

We consider equilibrium first. Recall that the equilibrium distribution function is

$$f_0 = \frac{1}{e^{\Theta} + 1}, \tag{3.43}$$

where

$$\Theta = \left(E_{C0} + p^2/2m^* - E_F\right)/k_B T_L \tag{3.44}$$

(spherical, parabolic energy bands are again assumed). The BTE, eq. (3.22), becomes

$$\upsilon \cdot \nabla_r f_0 + \mathbf{F} \cdot \nabla_p f_0 = 0, \tag{3.45}$$

because nothing changes with time in equilibrium. Equation (3.45) can be written as

$$\upsilon \cdot \frac{\partial f_0}{\partial \Theta} \nabla_r \Theta + \mathbf{F} \cdot \frac{\partial f_0}{\partial \Theta} \frac{\upsilon}{k_B T_L} = 0 \tag{3.46}$$

since $\nabla_p \Theta = \upsilon / k_B T_L$. Now if we permit E_{C0}, E_F, and T_L to vary with position, then eq. (3.46) becomes

$$\nabla_r \left(\frac{E_{C0}(\mathbf{r}) + p^2/2m^* - E_F}{T_L} \right) + \frac{\mathbf{F}}{T_L} = 0. \tag{3.47}$$

Using $F = -\nabla_r E_{C0}(\mathbf{r})$, this expression can be expanded to find

$$-\frac{1}{T_L}\nabla_r E_F + \left[E_{C0}(\mathbf{r}) + p^2/2m^* - E_F\right]\nabla_r\left(\frac{1}{T_L}\right) = 0, \tag{3.48}$$

which must hold for every p so each of the two terms must independently be zero. We conclude that in equilibrium – even when built-in fields make E_{C0} position-dependent, that

$$\nabla_r E_F = 0 \tag{3.49a}$$

$$\nabla_r T_L = 0 \quad . \tag{3.49b}$$

This solution to the BTE demonstrates that both the Fermi-level and temperature are constant in equilibrium.

3.4.2 Uniform electric field with a constant relaxation time

Next, consider an n-type semiconductor under bias and assume that it is long, so that we need not be concerned with boundary conditions, that it is uniform, so $\nabla_r f = 0$, and that it is in steady-state, so $\partial f/\partial t = 0$. If a small electric field is applied, we expect that $f(\mathbf{r}, \mathbf{p})$ will be close to $f_0(\mathbf{p})$ and may hope that the RTA is valid. To simplify the calculations, we'll assume for now that $\tau_f = \tau_0$, a constant. Under these assumptions, eq. (3.22) becomes

$$(-q)\boldsymbol{\mathcal{E}} \cdot \nabla_p f = -\frac{f_A}{\tau_0}. \tag{3.50}$$

If we assume that $f \simeq f_0$, then $\nabla_p f$ may be approximated by $\nabla_p f_0$ so

$$f = f_0 + f_A = f_0 - \tau_0(-q)\boldsymbol{\mathcal{E}} \cdot \nabla_p f_0 = f_0 - \frac{q\tau_0 f_0}{k_B T_L}\boldsymbol{\mathcal{E}} \cdot \upsilon \tag{3.51}$$

is the solution to the BTE with the force on electrons arising from an applied electric field $\boldsymbol{\mathcal{E}}$.

To examine the shape of the non-equilibrium f, assume $\boldsymbol{\mathcal{E}}$ is directed along $\hat{z}$. Then eq. (3.51) becomes

$$f = f_0 + f_A = f_0 + q\tau_0\mathcal{E}_z\frac{\partial f_0}{\partial p_z}, \tag{3.52}$$

which may be interpreted as a Taylor series expansion for $f_0(p_x, p_y, p_z + q\tau_0\mathcal{E}_z)$ if $\mathcal{E}_z$ is small. We conclude that the non-equilibrium distribution function is simply

$$f(p_x, p_y, p_z) = f_0(p_x, p_y, p_z + q\tau_0\mathcal{E}_z), \tag{3.53}$$

which is the equilibrium distribution function shifted by $q\tau_0\mathcal{E}_z$ in a direction opposite to the applied field. In this simple case, the non-equilibrium distribution function is a displaced Maxwellian. As shown in Fig. 3.5, the distribution has a net negative velocity and current will flow. To gauge the magnitude of this displacement let $\tau_0 = 0.1$ picosecond and $\mathcal{E}_z = 1\,\text{kV/cm}$. Then the displacement is $q\tau_0\mathcal{E}_z = 1.6 \times 10^{-27}$ kgm/sec. A Brillouin zone half width is roughly $\hbar\pi/a$. Assuming a lattice spacing of $a \simeq 5\,\text{Å}$, we find that the distribution function is displaced from origin by only $\simeq 10^{-7}$ of the Brillouin zone width, so the displacement is typically quite small.

From the low-field solution, eq. (3.51), the electron current,

$$J_{nz} = (-q)n\langle\upsilon_z\rangle, \tag{3.54}$$

can also be evaluated. The average electron velocity is evaluated from the sum,

$$\langle\upsilon_z\rangle = \frac{\sum_{\mathbf{p}} \upsilon_z(f_0 + f_A)}{\sum_{\mathbf{p}}(f_0 + f_A)}, \tag{3.55}$$

where f_A is the deviation from equilibrium as given by eq. (3.51). Notice that f_0 is even in momentum whereas f_A is odd. Under low fields, $f_A \ll f_0$ so it can be ignored when evaluating the sum in the denominator, but the product, $\upsilon_z f_0$, is odd so only the $\upsilon_z f_A$ term contributes to the sum in the numerator. Evaluating $\langle\upsilon_z\rangle$ from this sum is a good exercise, but we can find the answer very easily because τ_0 is assumed to be constant. According to eq. (3.53), $f(\mathbf{p})$ is simply f_0 shifted in momentum by $(-q)\tau_0\mathcal{E}_z$ in the p_z direction. Because the distribution function is symmetrical,

$$\langle\upsilon_z\rangle = \frac{\langle p_z\rangle}{m^*} = (-q)\tau_0\frac{\mathcal{E}_z}{m^*}, \tag{3.56}$$

and the electron current becomes

$$J_{nz} = nq\mu_n\mathcal{E}_z, \tag{3.57}$$

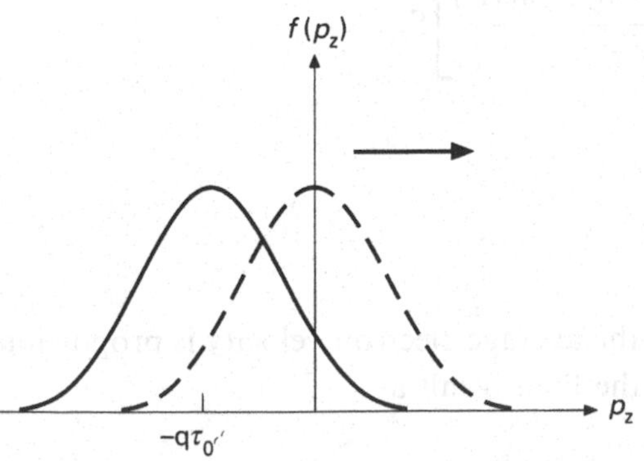

Fig. 3.5 The nonequilibrium distribution function for electrons in the presence of a small applied electric field. The dashed line is the equilibrium distribution function, $f_0(p)$.

where

$$\mu_n \equiv \frac{q\tau_0}{m^*} \tag{3.58}$$

is the electron mobility. Under low fields, the current is directly proportional to the electric field; we have derived Ohm's law. High carrier mobilities are associated with long relaxation times and small effective masses.

3.4.3 Uniform electric field with energy-dependent relaxation time

For most common scattering mechanisms, τ_f is a function of the carrier's kinetic energy, so the nonequilibrium distribution function is not a simple displacement of f_0. Equation (3.51) is still a valid solution where τ_0 is replaced by $\tau_f(E)$, but the average velocity is more difficult to evaluate.

To evaluate the average velocity, we begin with eq. (3.55) and insert eq. (3.51) to find

$$\langle \upsilon_z \rangle = \frac{\sum_{\mathbf{p}} \upsilon_z f_A}{\sum_{\mathbf{p}} f_0} = (-q)\mathcal{E}_z \left[\frac{\sum_{\mathbf{p}} \upsilon_z^2 \tau_f(E) f_0(E)}{nk_B T_L} \right], \tag{3.59}$$

which can be written as

$$\langle \upsilon_z \rangle = \frac{(-q)}{m^*} \left[\frac{\sum_{\mathbf{p}} (1/2m^* \upsilon_z^2) \tau_f(E) f_0(E)}{(1/2nk_B T_L)} \right] \mathcal{E}_z.$$

According to equipartition, the denominator of the term in brackets is simply one-third of the average equilibrium kinetic density. Using this, and the spherical symmetry of the problem, we find

$$\langle \upsilon_z \rangle = \frac{(-q)}{m^*} \left[\frac{(1/3) \sum_{\mathbf{p}} (1/2m^* \upsilon^2) \tau_f(E) f_0(E)}{(1/3)\langle E \rangle} \right] \mathcal{E}_z,$$

which we can write as

$$\langle \upsilon_z \rangle = \frac{(-q)}{m^*} \frac{\langle E\tau_f(E) \rangle}{\langle E \rangle} \mathcal{E}_z. \tag{3.60}$$

Equation (3.60) shows that the average electron velocity is proportional to the electric field. We can express the final result as

$$\langle \upsilon_z \rangle = -\mu_n \mathcal{E}_z,$$

where the mobility is

$$\boxed{\mu_n \equiv \frac{q\langle\langle\tau_f\rangle\rangle}{m^*}} \tag{3.61}$$

where

$$\boxed{\langle\langle\tau_f\rangle\rangle \equiv \frac{\langle E\tau_f(E)\rangle}{\langle E\rangle}} \tag{3.62}$$

The quantity $\langle\langle\tau_f\rangle\rangle$ was introduced to make the result for an energy-dependent relaxation time look like eq. (3.58), the corresponding result for a constant relaxation time. The double brackets are to indicate that $\langle\langle\tau\rangle\rangle$ is not a simple average of $\tau_f(E)$ over the symmetric component of the distribution function; it is a specially defined 'average' that arises in transport calculations.

In the following section we'll specify $\tau_f(E)$ and show that for several common scattering mechanisms, it can be expressed in *power-law* form as

$$\tau_f(E) = \tau_0[E(p)/k_B T_L]^s, \tag{3.63}$$

where τ_0 is a constant and 's' a characteristic exponent. When this form is inserted in eq. (3.62), we find

$$\langle\langle\tau_f\rangle\rangle = \tau_0 \frac{\sum_{\mathbf{p}}(p^2/2m^*)(p^2/2m^* k_B T_L)^s e^{-p^2/2m^* k_B T_L}}{\sum_{\mathbf{p}}(p^2/2m^*)e^{-p^2/2m^* k_B T_L}},$$

which can be converted to an integral and evaluated. Since the integrations over the angular coordinates cancel from the numerator and denominator, we find

$$\langle\langle\tau_f\rangle\rangle = \tau_0 \frac{\int_0^\infty (p^2/2m^* k_B T_L)^s e^{-p^2/2m^* k_B T_L} p^4 dp}{\int_0^\infty e^{-p^2/2m^* k_B T_L} p^4 dp}.$$

With the substitution, $y = p^2/2m^* k_B T_L$,

$$\langle\langle\tau_f\rangle\rangle = \tau_0 \frac{\int_0^\infty y^{s+3/2} e^{-y} dy}{\int_0^\infty y^{3/2} e^{-y} dy}.$$

After recalling the definition of the Γ-function,

$$\Gamma(p) \equiv \int_0^\infty y^{p-1} e^{-y} dy, \tag{3.64}$$

Table 3.2. *Characteristic exponents for common power law scattering mechanisms**

Scattering mechanisms	Exponent s	Hall factor
Acoustic phonon	−1/2	1.18
Ionized impurity (weakly screened)	+3/2	1.93
Ionized impurity (strongly screened)	−1/2	1.18
Neutral impurity	0	1.00
Piezoelectric	+1/2	1.10

* The meaning of the term Hall factor is discussed in Chapter 4.

we can write the result as

$$\boxed{\langle\langle \tau_f \rangle\rangle = \tau_0 \frac{\Gamma(s+5/2)}{\Gamma(5/2)}}. \tag{3.65}$$

The average relaxation time is seen to be a function of the exponent, s, which characterizes the energy-dependence of $\tau_f(E)$. The following three properties of the Γ function enable us to evaluate $\langle\langle \tau \rangle\rangle$ for any exponent, s,

$$\boxed{\begin{aligned} &\Gamma(n) = (n-1)! \qquad \text{(for integer } n) \\ &\Gamma\left(\frac{1}{2}\right) = \sqrt{\pi} \\ &\Gamma(p+1) = p\Gamma(p) \end{aligned}}. \tag{3.66}$$

For example, if $s = -1/2$, then

$$\langle\langle \tau_f \rangle\rangle = \tau_0 \frac{\Gamma(2)}{\Gamma(5/2)} = \tau_0 \frac{1}{3/2 \times 1/2 \times \Gamma(1/2)} = \frac{4}{3\sqrt{\pi}} \tau_0.$$

The results of Chapter 2 show that the power law form describes scattering by acoustic phonons. It also describes ionized impurity scattering, but then τ_0 varies slowly with energy. Because the variation with energy is slow, eq. (3.65) is still a good approximation. In this case, τ_0 is evaluated at the energy which maximizes the integrand. In Table 3.2, the characteristic exponents for several common scattering mechanisms are listed.

3.5 Validity of the relaxation time approximation

The relaxation time approximation is widely used because it makes solving the BTE so easy. Chapter 4 presents a formal theory for low-field, or linear, transport theory based on the RTA. Our objective for now is to identify the conditions under which the RTA is valid and then to specify its form. As demonstrated in the previous section, the RTA [as we defined it in eq. (3.40b)] is valid only under low applied fields so that the carriers and lattice are in equilibrium. Even for such cases, however, stringent conditions must be met in order to apply the RTA.

Our goal is to demonstrate that the collision integral can be written as

$$\left.\frac{\partial f_A}{\partial t}\right|_{\text{coll}} = \sum_{\mathbf{p}'} S' f_A' - S f_A = -\frac{f_A}{\tau_f} \tag{3.67}$$

where $S' = S(\mathbf{p}', \mathbf{p})$, $f_A' = f_A(\mathbf{p}')$, $S = S(\mathbf{p}, \mathbf{p}')$ and $f_A = f_A(\mathbf{p})$. In equilibrium, $\partial f/\partial t|_{\text{coll}} = 0$, so $S_0' = S_0 f_0/f_0'$. If we assume that this relation applies out of equilibrium as well, we can use it to eliminate S' in eq. (3.67) and find

$$\frac{1}{\tau_f} = \sum_{\mathbf{p}'} S\left[1 - \frac{f_0 f_A'}{f_0' f_A}\right]. \tag{3.68}$$

Use of the equilibrium relation between S' and S is justified because the transition rate is often determined by strong, short-range potentials that do not change out of equilibrium. Equation (3.68) *could* be regarded as a definition of τ_f, but when the term, relaxation time approximation, is used, we mean something much more specific. To be useful, τ_f should not depend on the distribution function nor on the driving force which determines $f(\mathbf{r}, \mathbf{p}, t)$.

Notice that if the second term in eq. (3.68) sums to zero, then τ_f will depend only on the scattering processes as described by $S(\mathbf{p}, \mathbf{p}')$ and the relaxation time approximation will be valid. Notice also that f_0 and f_A do not depend on $\mathbf{p}'$, that f_0' is an even function of $\mathbf{p}'$, and that f_A' is odd [recall eq. (3.51)]. As a consequence, $f_0 f_A'/f_0' f_A$ is odd in $\mathbf{p}'$, so when $S(\mathbf{p}, \mathbf{p}')$ is even in $\mathbf{p}'$, the second term on the right-hand of eq. (3.68) sums to zero. The transition rate $S(\mathbf{p}, \mathbf{p}')$, is even for isotropic scattering, so we find

$$\boxed{\frac{1}{\tau_f(\mathbf{p})} = \sum_{\mathbf{p}'} S(\mathbf{p}, \mathbf{p}') = \frac{1}{\tau(\mathbf{p})} \quad \text{(isotropic scattering)}} \quad . \tag{3.69}$$

We conclude that the relaxation time approximation is valid when the scattering is isotropic and that the characteristic time, $\tau_f(\mathbf{p})$, is just the average time between collisions, $\tau(\mathbf{p})$.

From eq. (3.68), we can also demonstrate that elastic scattering will produce a τ_f that is independent of the distribution function. To demonstrate so, first recall that the form for f_A obtained in Section 3.4 involved a dot product between the applied force and the carrier's velocity [see eq. (3.51)]. In Chapter 4 we'll show more generally that for spherical bands under low fields, the general form for f_A is

$$f_A = g(p^2)\cos\theta, \tag{3.70}$$

where θ is the angle between **p** and the generalized force (which may include both electric fields and temperature and concentration gradients). The function g depends on the strength of the driving force and on the carrier energy. Using eq. (3.70) for f_A in eq. (3.59), we find

$$\frac{1}{\tau_f} = \sum_{\mathbf{p}'} S\left[1 - \frac{f_0 g' \cos\theta'}{f_0' g \cos\theta}\right]. \tag{3.71}$$

Since both f_0 and g depend on the carrier energy, $f_0 = f_0'$ and $g = g'$ when the scattering is elastic and only $\cos\theta'/\cos\theta$ need be evaluated.

Figure 3.6 shows the geometry for a particular scattering event with the initial momentum along $\hat{z}$ and the applied force in the $\hat{y} - \hat{z}$ plane at an angle θ to **p**. Since

$$\cos\theta = \frac{\mathbf{F}\cdot\mathbf{p}}{Fp} \tag{3.72}$$

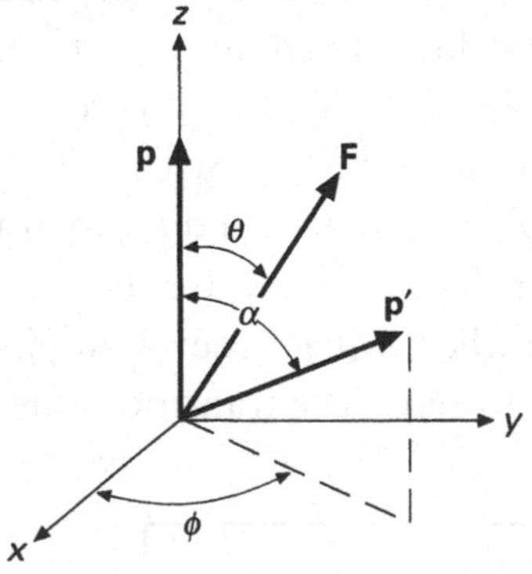

p: $p_x = 0$ $p_y = 0$ $p_z = p$

F: $F_x = 0$ $F_y = F\sin\theta$ $F_z = F\cos\theta$

p′: $p_x' = p'\sin\alpha\cos\phi$ $p_y' = p'\sin\alpha\sin\phi$ $p_z' = p'\cos\alpha$

Fig. 3.6 Coordinate system illustrating a scattering event. The incident carrier has momentum **p**, the scattered carrier, **p**′, and the applied force is **F**.

we find

$$\cos\theta' = \sin\theta\sin\alpha\sin\phi + \cos\theta\cos\alpha, \tag{3.73}$$

or

$$\frac{\cos\theta'}{\cos\theta} = \tan\theta\sin\alpha\sin\phi + \cos\alpha. \tag{3.74}$$

When eq. (3.74) is inserted in eq. (3.71), the term involving ϕ will integrate to zero because $S(\mathbf{p}, \mathbf{p}')$ is independent of ϕ for spherical bands. For elastic scattering, we conclude that eq. (3.68) becomes

$$\boxed{\frac{1}{\tau_f(\mathbf{p})} = \sum_{\mathbf{p}'} S(\mathbf{p}, \mathbf{p}')(1 - \cos\alpha) \quad \text{(elastic scattering)}}, \tag{3.75}$$

which again depends only on scattering processes and not on the distribution function. Since α is the polar angle between the incident and scattered carrier, eq. (3.75) says that collisions which don't change the direction of $\mathbf{p}$ very much, don't count very much.

We conclude that *under low fields when the scattering is elastic or isotropic* the collision term in the BTE can be approximated by eq. (3.40b) with a relaxation time, τ_f, that depends only on the nature of the scattering process. For isotropic scattering $\tau_f = \tau$, the average time between collisions. For elastic scattering a weighting factor of $(1 - \cos\alpha)$ appears.

3.6 Numerical solution to the BTE

Although the relaxation time approximation makes it easy to solve the BTE, it cannot be applied to semiconductors such as GaAs for which the dominant scattering mechanisms are neither elastic nor isotropic, and it does not work well under high fields in any semiconductor. For such cases, numerical techniques are necessary. In this section, we describe a numerical technique that is widely-used for low-field transport, another one that is often used for high-field transport, and briefly examine techniques for solving the Boltzmann equation for devices.

3.6.1 Rode's iterative method

An iterative technique for solving the BTE under steady-state, spatially homogeneous, low field conditions is introduced in this section. The technique is due to

Rode and is described in [3.6]. (Homework problem 3.13 asks you to fill in the details of the derivation sketched below.)

We begin with the steady-state, spatially homogeneous BTE for a nondegenerate semiconductor,

$$(-q)\boldsymbol{\mathcal{E}} \cdot \nabla_p f = \sum_{\mathbf{p}'} S(\mathbf{p}', \mathbf{p}) f(\mathbf{p}') - \sum_{\mathbf{p}'} S(\mathbf{p}, \mathbf{p}') f(\mathbf{p}) = I(\mathbf{p}) - \frac{f(\mathbf{p})}{\tau(\mathbf{p})}, \tag{3.76}$$

where

$$I(\mathbf{p}) \equiv \sum_{\mathbf{p}'} S(\mathbf{p}', \mathbf{p}) f(\mathbf{p}') \tag{3.77}$$

is a 'source' term which describes the in-scattering of carriers, and

$$\frac{1}{\tau(\mathbf{p})} \equiv \sum_{i=1}^{N} \sum_{\mathbf{p}'} S_i(\mathbf{p}, \mathbf{p}') \tag{3.78}$$

is the out-scattering rate. The sum over i is to include the N different scattering mechanisms described by S_i. Note that the in-scattering contribution depends on the unknown distribution function, but the out-scattering part, $1/\tau(\mathbf{p})$, is known.

To proceed, we write the unknown distribution function as $f = f_0 + f_A$ and decompose the scattering processes into elastic and inelastic components, S^{el} $(\mathbf{p}, \mathbf{p}')$ and $S^{in}(\mathbf{p}, \mathbf{p}')$. With these assumptions, eq. (3.76) becomes

$$f_A = \frac{\hat{I}(\mathbf{p}) - (-q)\boldsymbol{\mathcal{E}} \cdot \nabla_p f_0}{\nu_{el} + 1/\tau_{in}}, \tag{3.79}$$

where

$$\frac{1}{\tau_{in}} = \sum_{\mathbf{p}} S^{in}(\mathbf{p}, \mathbf{p}'). \tag{3.80}$$

$$\nu_{el} = \sum_{\mathbf{p}} S^{el}(\mathbf{p}, \mathbf{p}')(1 - \cos\alpha). \tag{3.81}$$

and

$$\hat{I}(\mathbf{p}) = \sum_{\mathbf{p}'} S^{in}(\mathbf{p}', \mathbf{p}) f(\mathbf{p}') [f_A(\mathbf{p}') f_0(\mathbf{p}) / f_0(\mathbf{p}')]. \tag{3.82}$$

Equation (3.81) arises because elastic scattering satisfies the relaxation approximation. Inelastic scattering is treated differently because we don't assume that it is isotropic.

Using eq. (3.70), $f_A = g(p)\cos\theta$, where θ is the angle between $\mathbf{p}$ and $\boldsymbol{\mathcal{E}}$, and assuming that $\boldsymbol{\mathcal{E}}$ is oriented along the z-axis, we can express eq. (3.79) as

$$g(p)\cos\theta = \frac{\hat{I}(p) - (-q)\mathcal{E}\dfrac{\partial f_0}{\partial p}\cos\theta}{\nu_{\text{el}} + 1/\tau_{\text{in}}}. \tag{3.83}$$

The unknown in eq. (3.82) is the function $g(p)$. Equation (3.83) can be solved for $g(p)$ to obtain (see homework problem 3.13)

$$g(p) = \frac{\tilde{I}(p) - (-q)\mathcal{E}\dfrac{\partial f_0}{\partial p}}{\nu_{\text{el}} + 1/\tau_{\text{in}}} \tag{3.84}$$

where

$$\tilde{I}(p) \equiv \sum_{\mathbf{p}'} S^{\text{in}}(\mathbf{p}', \mathbf{p})[f_0(p)/f_0(p')]g(p')\cos\alpha. \tag{3.85}$$

Note that $g(p)$ appears on both sides of eq. (3.84).

Equation (3.84) can be solved by approximating the right-hand side using a guess for $g(p)$ [we denote the kth guess by g^k]. This guess is then iteratively refined according to the prescription:

$$g^{k+1}(p) = \frac{\tilde{I}(p) - (-q)\mathcal{E}\dfrac{\partial f_0}{\partial p}}{\nu_{\text{el}} + 1/\tau_{\text{in}}}. \tag{3.86}$$

Equation (3.86) describes a convergent, iterative process which begins with a guess for the distribution function [3.6]. Using the guess, the right-hand side of eq. (3.86) is evaluated, a better guess obtained, and the process repeated until convergence is achieved.

To implement the iterative solution technique, a grid in energy space, like that shown in Fig. 3.7, is defined. A guess, $g^0(p)$, for the value of g at each node in the grid is then made. With this guess, the in-scattering term, $\tilde{I}(p)$ can be computed by numerical integration. A new guess, $g^1(p)$, for the value of g at each node in the grid is then obtained from eq. (3.86). The process continues until the distribution function, or some quantity such as the average velocity, does not change appreciably from one iteration to the next.

An easy guess with which to begin the iteration is the equilibrium solution, where $g = 0$. In this case, $\tilde{I}(p) = 0$, and eq. (3.86) gives the first approximation as

$$g^1(p) = q\left[\frac{1}{(\nu_{\text{el}} + 1/\tau_{\text{in}})}\right]\mathcal{E}\frac{\partial f_0}{\partial p}. \tag{3.87}$$

Equation (3.87) is exactly eq. (3.52), with a relaxation time, $\tau_f = 1/(\nu_{\text{el}} + 1/\tau_{\text{in}})$, which demonstrates that the first step of this iterative process is identical to the relaxation time approximation. The iterative process, however, must continue until it converges. For some scattering processes, the RTA is valid and one

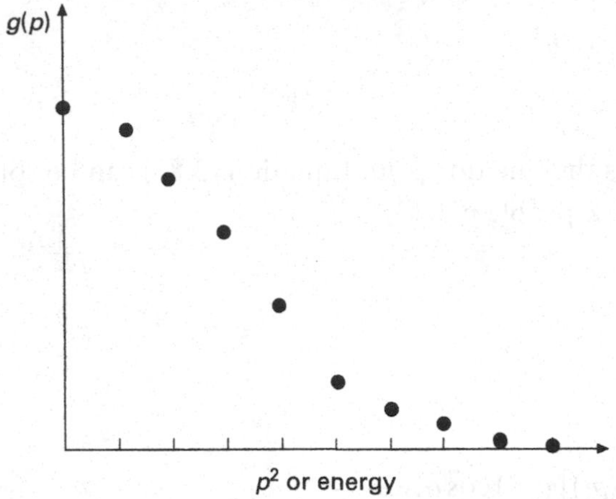

Fig. 3.7 Discretization of momentum space for numerical solution by the iterative technique.

iteration suffices, but for others the RTA produces poor results and additional iterations are required.

3.6.2 Path integral solution to the BTE

Rode's technique is limited to low-field, or near-equilibrium conditions, where the sequence converges rapidly. A more general solution can also be obtained [3.8]. We begin with the BTE for a nondegenerate semiconductor and express it in a form similar to eq. (3.76),

$$\frac{\partial f}{\partial t} + (-q)\boldsymbol{\mathcal{E}} \cdot \nabla_p f + \upsilon \cdot \nabla_r f + \frac{f(\mathbf{p})}{\tau(\mathbf{p})} = I(\mathbf{p}), \tag{3.88}$$

where $I(\mathbf{p})$, given by eq. (3.77) describes in-scattering and involves the unknown distribution function and $\tau(\mathbf{p})$, given by eq. (3.78) describes out-scattering. Next, we define a fictitious scattering process known as *self-scattering* by

$$S_{\text{eff}}(\mathbf{p}, \mathbf{p}') \equiv \sum_{i=1}^{N} S(\mathbf{p}, \mathbf{p}') + \Omega(p)\delta_{\mathbf{p},\mathbf{p}'}, \tag{3.89}$$

where the sum is over the N physical scattering processes present and the second term is an additional, fictitious, scattering mechanism. Note that the added term has no physical effect because it is only non-zero when the scattered momentum is equal to the incident momentum. The self-scattering term is not essential, it just simplifies the derivation (and in Chapter 6 we'll see that it is also useful in Monte Carlo simulation). We select the magnitude of the fictitious scattering so that the total out-scattering rate is constant,

$$\sum_{\mathbf{p}'} S_{\text{eff}}(\mathbf{p}, \mathbf{p}') = \frac{1}{\tau(p)} + \Omega(p) = \Gamma, \tag{3.90}$$

where Γ is a constant. The meaning of eq. (3.90) is illustrated in Fig. 3.8; an additional, fictitious scattering rate with no physical effect, $\Omega(p)$, is added to the real out-scattering rate, $1/\tau(p)$, so that the total scattering rate becomes a constant. Using eqs. (3.89) and (3.90), the BTE, eq. (3.88) becomes

$$\frac{\partial f}{\partial t} + (-q)\boldsymbol{\mathcal{E}} \cdot \nabla_p f + \upsilon \cdot \nabla_r f + \Gamma f(\mathbf{p}) = I'(\mathbf{p}), \tag{3.91}$$

where

$$I'(p) = \sum_{i=1}^{N} \sum_{\mathbf{p}'} \left[S_i(\mathbf{p}', \mathbf{p}) + \left(\Gamma - \frac{1}{\tau(p)} \right) \delta_{\mathbf{p},\mathbf{p}'} \right] f(\mathbf{p}'). \tag{3.92}$$

Equation (3.91) can be solved for two cases: (i) spatially uniform conditions and (ii) time-independent conditions.

Spatially homogeneous path integral solution:
Setting $\nabla_r = 0$ in eq. (3.88), we find

$$\frac{\partial f}{\partial t} + (-q)\boldsymbol{\mathcal{E}} \cdot \nabla_p f + \Gamma f(\mathbf{p}) = I'(\mathbf{p}), \tag{3.93}$$

which can be solved by introducing the *path variables*,

$$\tilde{t} = t \tag{3.94a}$$

$$\tilde{\mathbf{p}} = \mathbf{p} + q\boldsymbol{\mathcal{E}}t. \tag{3.94b}$$

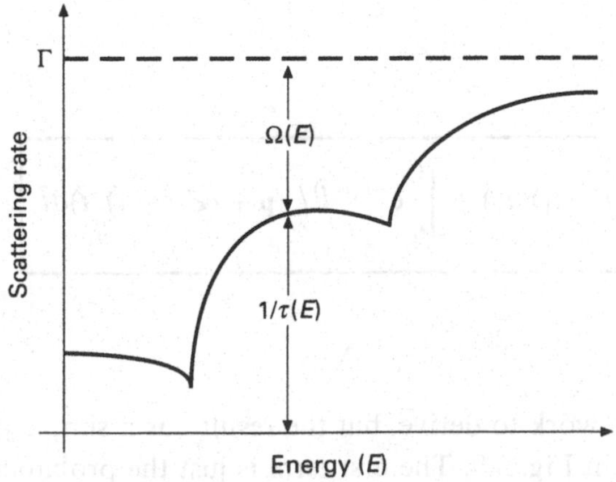

Fig. 3.8 Illustration of how an energy-dependent self-scattering rate is used to make the total scattering rate a constant.

In response to an electric field, electrons move along a trajectory in momentum space (recall Fig. 3.4). The path variables define a moving coordinate system in which the electrons are stationary. With this change of variables,

$$\frac{\partial f}{\partial \mathbf{p}} = \frac{\partial f}{\partial \tilde{\mathbf{p}}} \tag{3.95a}$$

and

$$\frac{\partial f}{\partial t} = \frac{\partial f}{\partial \tilde{t}} + \frac{\partial f}{\partial \tilde{\mathbf{p}}} \cdot (q\boldsymbol{\mathcal{E}}). \tag{3.95b}$$

After inserting eqs. (3.95) in the BTE, eq. (3.93), we find

$$\partial f(\tilde{\mathbf{p}} - q\boldsymbol{\mathcal{E}}\tilde{t}, \tilde{t}) + \Gamma f(\tilde{\mathbf{p}} - q\boldsymbol{\mathcal{E}}\tilde{t}, \tilde{t}) = I'(\tilde{\mathbf{p}} - q\boldsymbol{\mathcal{E}}\tilde{t}, \tilde{t}). \tag{3.96}$$

To solve eq. (3.96), note that

$$\mathrm{e}^{-\Gamma\tilde{t}} \frac{\partial (f\mathrm{e}^{\Gamma\tilde{t}})}{\partial \tilde{t}} = \frac{\partial f}{\partial \tilde{t}} + \Gamma f. \tag{3.97}$$

Using eq. (3.97) in eq. (3.96), we find

$$\frac{\partial (f\mathrm{e}^{\Gamma\tilde{t}})}{\partial \tilde{t}} = \mathrm{e}^{\Gamma\tilde{t}} I'(\tilde{\mathbf{p}} - q\boldsymbol{\mathcal{E}}\tilde{t}, \tilde{t}), \tag{3.98}$$

which can be integrated between $\tilde{t}_1$ and $\tilde{t}_2$ to find

$$\mathrm{e}^{\Gamma\tilde{t}_2} f(\tilde{\mathbf{p}} - q\boldsymbol{\mathcal{E}}\tilde{t}_2, \tilde{t}_2) - \mathrm{e}^{\Gamma\tilde{t}_1} f(\tilde{\mathbf{p}} - q\boldsymbol{\mathcal{E}}\tilde{t}_1, \tilde{t}_1) = \int_{\tilde{t}_1}^{\tilde{t}_2} \mathrm{e}^{\Gamma\tilde{t}} I'(\tilde{\mathbf{p}} - q\boldsymbol{\mathcal{E}}\tilde{t}, \tilde{t})\mathrm{d}\tilde{t}. \tag{3.99}$$

Finally, we return to the original variables, inserting

$$t = \tilde{t}_2 \tag{3.100a}$$

$$\mathbf{p} = \tilde{\mathbf{p}} - q\boldsymbol{\mathcal{E}}\tilde{t}_2 \tag{3.100b}$$

in eq. (3.99) to find

$$\boxed{f(\mathbf{p}, t) = \mathrm{e}^{-\Gamma(t-t_1)} f(\mathbf{p} + q\boldsymbol{\mathcal{E}}(t - t_1), t_1) + \int_{t_1}^{t} \mathrm{e}^{-\Gamma(t-\tilde{t})} I'(\mathbf{p} + q\boldsymbol{\mathcal{E}}(t - \tilde{t}), \tilde{t})\mathrm{d}\tilde{t}}\,. \tag{3.101}$$

Equation (3.101) took some work to derive, but the result has a simple physical interpretation as illustrated in Fig. 3.9. The first term is just the probability that an electron upwind on the trajectory can move from the state $(\mathbf{p} + q\boldsymbol{\mathcal{E}}(t - t_1), t_1)$ to the point $(\mathbf{p}, t)$ without scattering.

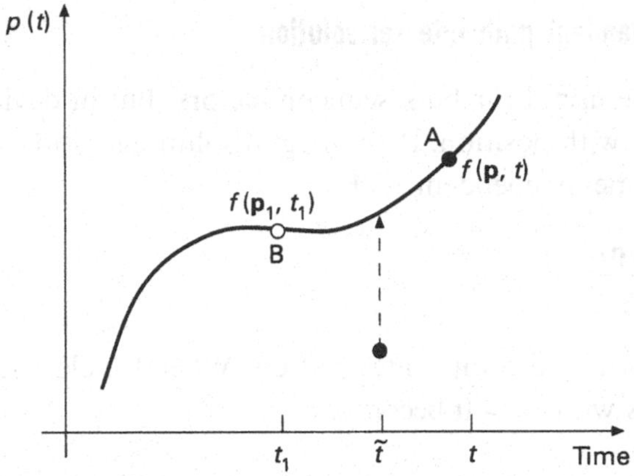

Fig. 3.9 Illustration of the path integral solution to the Boltzmann Transport Equation. A collisionless trajectory is shown. Electrons at the state $A(\mathbf{p}, t)$ at time t were at state B $[\mathbf{p} + q\boldsymbol{\mathcal{E}}(t - t_1)]$ at an earlier time, t_1. The probability that electrons at state B at time t_1 arrive at state A at time t, without out-scattering is $e^{-\Gamma(t-t_1)}f(p + q\mathcal{E}(t - t_1))$. Similarly, the contribution from electrons that in-scatter to the trajectory at time $t = \tilde{t}$ then travel without scattering to state A is also shown.

The second term is the probability that an electron in-scatters at time $\tilde{t}$ (where $t_1 < \tilde{t} < t$) to a point $(\mathbf{p} + q\boldsymbol{\mathcal{E}}(t - \tilde{t}), \tilde{t})$ on the trajectory, then travels without scattering to the point $(\mathbf{p}, t)$. The integral sums the contributions for in-scattering at all times from t_1 to t.

From eq. (3.101), we can also obtain the steady-state solution by letting $t_1 \to -\infty$. As shown in homework problem 3.15, the result can be expressed as

$$\boxed{f(\mathbf{p}) = \int_0^{\infty} e^{-\Gamma t} I'(\mathbf{p} + q\boldsymbol{\mathcal{E}}t, t)\mathrm{d}t}. \tag{3.102}$$

Equation (3.102) is a Fredholm integral equation of the second kind; it contains the unknown distribution function on both sides. This equation can be solved by Kellog iteration as

$$\boxed{f^{k+1}(\mathbf{p}) = \int_0^{\infty} e^{-\Gamma t} \sum_{\mathbf{p}'} \left[\sum_i S_i(\mathbf{p} + q\boldsymbol{\mathcal{E}}t, \mathbf{p}') + \left(\Gamma - \frac{1}{\tau(\mathbf{p} + q\boldsymbol{\mathcal{E}}t)} \right) \delta_{\mathbf{p},\mathbf{p}'} \right] f^k(\mathbf{p}')\mathrm{d}t}, \tag{3.103}$$

which shows how to obtain an improved guess from the kth guess for $f(\mathbf{p})$.

Space-dependent, time-independent path integral solution

Homogeneous solutions are useful for bulk semiconductors, but in devices, the distribution function varies with position. Path integral solutions can be derived for the space-dependent, time-independent BTE

$$\frac{\partial f}{\partial z}+\frac{(-q)\mathcal{E}_z}{\upsilon_z}\frac{\partial f}{\partial p_z}+\frac{\Gamma}{\upsilon_z}f=\frac{I'(\mathbf{p})}{\upsilon_z}, \tag{3.104}$$

but the presence of the electric field complicates matters. When the electric field is zero eq. (3.104) for carriers with $\upsilon_z > 0$ becomes

$$\frac{\partial f^+}{\partial z}+\frac{\Gamma}{\upsilon_z}f^+=\frac{I'(\mathbf{p})}{\upsilon_z}, \tag{3.105}$$

which can be integrated to obtain

$$f^+(\mathbf{p},z)=f^+(\mathbf{p},0)e^{-\Gamma z/\upsilon_z}+\int_0^z I'(\mathbf{p},z')\mathrm{e}^{-\Gamma(z-z')/\upsilon_z}\,\mathrm{d}z' \tag{3.106}$$

(see homework problem 3.16). The first term on the right is $f(\mathbf{p})$ at the contact multiplied by the probability that the electron out-scatters on it's way to position, z. The second term is the probability that an electron in-scatters to $\mathbf{p}$ at position, z, multiplied by the probability of out-scattering as it travels to location, z.

The presence of an electric field complicates matters because: (i) backward-travelling carriers can be converted to forward-travelling carriers by the electric field, (ii) some carriers injected from a contact may not reach location, z, and (iii) carriers at location, z, may have a different kinetic energy or $\mathbf{p}$ than they were injected with. A solution for the field-dependent case can be constructed by evaluating the different ways that carriers can populate the distribution at location, z (i.e. by travelling ballistically from a contact or by scattering at an intermediate location).

3.6.3 Solution by orthogonal polynomial expansion

Another useful technique for solving the BTE is by expanding the solution in a set of orthogonal polynomials [3.9, 3.10]. To solve the Boltzmann equation generally requires at least a three-dimensional solution because momentum space is three-dimensional. When the spatial variation is along a single direction and the energy bands are isotropic, however, then the symmetry simplifies the solution. Recall that for low fields, the solution has the form [see eq. (3.51)]

$$f(z,\mathbf{p})=f_0(z,p)+g(p)\cos\theta \tag{3.107}$$

and, therefore, depends only on the magnitude of the momentum (or the energy) and on the angle, θ, between the momentum vector and the electric field which is assumed to point along the z-axis (in this case, there is azimuthal symmetry about the direction of the electric field). Rode's iterative technique, therefore, only needed to work in a one-dimensional energy space and provides a numerical solution for $g(p)$.

Recall that the Legendre polynomials are defined as

$$P_0 = 1 \tag{3.108a}$$

$$P_1 = \cos\theta \tag{3.108b}$$

$$P_2 = \frac{3}{2}\cos^2\theta - \frac{1}{2} \tag{3.108c}$$

etc.

and that they are orthogonal when integrated against $\mathrm{d}(\cos\theta)$,

$$\int_{-1}^{1} P_l P_{l'} \mathrm{d}(\cos\theta) = \frac{2}{2l+1}\delta_{ll'}. \tag{3.108d}$$

For spatial variation in one dimension with azimuthal symmetry about the z-axis, the distribution function can be written as an infinite sum of Legendre polynomials,

$$f(z, \mathbf{p}, t) = \sum_{l=0}^{\infty} f_l(z, p, t) P_l(\cos\theta), \tag{3.109}$$

where $f_l(z, p, t)$ is the weighting coefficient for the Legendre polynomial, P_l. The first two terms of eq. (3.109) are

$$f(z, \mathbf{p}, t) = f_0(z, p, t) + f_1(z, p, t)\cos\theta, \tag{3.110}$$

which shows that the low-field solution [eq. (3.107)] consists of the first two terms of a Legendre polynomial expansion. Under low-field conditions, we assume that f_0 is the equilibrium distribution function, but in general we should solve for f_0, and under high-field conditions, we expect to find a heated Maxwellian for f_0.

To indicate how an orthogonal polynomial solution proceeds, consider the one-dimensional BTE,

$$\frac{\partial f}{\partial t} + \upsilon_z \frac{\partial f}{\partial z} - q\mathcal{E}\frac{\partial f}{\partial p_z} = \left.\frac{\partial f}{\partial t}\right|_{\text{coll}} \tag{3.111a}$$

or

$$\frac{\partial f}{\partial t} + \upsilon \cos\theta \frac{\partial f}{\partial z} - q\mathcal{E}\cos\theta \frac{\partial f}{\partial p} = \left.\frac{\partial f}{\partial t}\right|_{\text{coll}} \tag{3.111b}$$

where we have assumed that the electric field points along the z-axis. Equation (3.109) can now be inserted into eq. (3.111b). If the result is multiplied by P_l and integrated against d(cos θ), we obtain an infinite set of equations for the Legendre polynomial coefficients, P_l.

To keep the mathematics simple, let's work with two terms of the expansion and insert eq. (3.110) in eq. (3.111b) to find

$$\begin{aligned}\frac{\partial f_0}{\partial t} + \frac{\partial f_1}{\partial t} + \upsilon\frac{\partial f_0}{\partial z}\cos\theta + \upsilon\frac{\partial f_1}{\partial z}\cos^2\theta - q\mathcal{E}\upsilon\frac{\partial f_0}{\partial E}\cos\theta - q\mathcal{E}\upsilon\frac{\partial f_1}{\partial E}\cos^2\theta \\ = \left.\frac{\partial f_0}{\partial t}\right|_{\text{coll}} + \left.\frac{\partial f_1}{\partial t}\right|_{\text{coll}} \cos\theta,\end{aligned} \tag{3.112}$$

where the collision terms are obtained by inserting eq. (3.110) into the collision integral, eq. (3.29b). This two-term expansion is actually quite good for semiconductors like Si where the dominant scattering mechanisms are isotropic. Now if we integrate eq. (3.112) against d(cos θ), we find

$$\frac{\partial f_0}{\partial t} + \frac{\upsilon}{3}\left[\frac{\partial f_1}{\partial z} - q\mathcal{E}\frac{\partial f_1}{\partial E}\right] = \left.\frac{\partial f_0}{\partial t}\right|_{\text{coll}}. \tag{3.113a}$$

Similarly, if we multiply eq. (3.112) by cos θd(cos θ) and integrate, we find

$$\frac{\partial f_1}{\partial t} + \upsilon\left[\frac{\partial f_0}{\partial z} - q\mathcal{E}\frac{\partial f_0}{\partial E}\right] = \left.\frac{\partial f_1}{\partial t}\right|_{\text{coll}}. \tag{3.113b}$$

Equations (3.113) are two coupled partial differential equations for f_0 and f_1, the space, energy, and time-dependent coefficients for the first two Legendre polynomials. Note that instead of having to solve a four-dimensional problem (one in position space and three in momentum space), the symmetry of the problem and the use of orthogonal polynomials has reduced the problem to two dimensions, one in position space and one in energy space. When the electric field varies in two dimensions, then we lose the symmetry about the electric field and must solve for the dependence of $f(\mathbf{p})$ on azimuthal angle, ϕ. For such cases, the Legendre polynomials are replaced by spherical harmonics; the technique is more involved mathematically but conceptually the same. Hennacy et al. [3.9] and Gnudi et al. [3.10] discuss numerical solutions by Legendre polynomial and spherical harmonic expansions.

3.6.4 Device analysis

To analyze a device, the spatial gradient terms in the BTE must be retained and boundary conditions imposed. Because the electric field may depend on the free carrier density, Poisson's equation should also be solved simultaneously with the BTE. Device analysis by solving the BTE is quite a challenge computationally because the problem is multi-dimensional in both position and momentum space. The imposition of spatial boundary conditions also requires some care. The BTE is a first order equation; one spatial boundary condition can be specified, but a one-dimensional device has two contacts. To handle the contacts, we must specify one-half of the boundary condition at each one. The problem can be shown to be mathematically well-posed when the incoming flux along the boundary is specified [3.7].

Solving the BTE for a device is computationally demanding. For bulk semiconductors, the BTE can be readily solved using the relaxation time approximation or by numerical techniques like those just described. For devices, orthogonal polynomial expansions can help make the problem manageable, but there are always concerns about how many terms of the expansion are required. More rigorous techniques such as Monte Carlo simulation, discussed in Chapter 6, or scattering matrix simulations [3.1] are available, but the computational demands increase, so these techniques are primarily used for scientific studies or as benchmarks to assess simpler approaches. For device design, engineers typically use simplified approaches to avoid solving the Boltzmann equation. One widely-used technique is the balance equation approach described in Chapter 5.

3.7 Validity of the Boltzmann transport equation

Because we so often begin at the BTE, it is important to understand its limitations. The BTE is approximate because it is a single particle description of a many particle system of carriers. Correlations between carriers are not treated. Another approximation is the semi-classical treatment of carriers as particles that obey Newton's laws. Finally, the simple treatment of scattering assumes binary collisions that occur instantly in time and are localized in space. The nature of these approximations and some of their implications are briefly addressed in this section.

The distribution function defines the most probable state of the system of carriers. Such a statistical description is appropriate when the number of carriers is large (extremely small devices might contain too few carriers to justify a statistical treatment). Since carriers interact through their electric fields, correlations between carriers exist. In principle, therefore, to determine the probability that a state at **p** is occupied, we need to know how the other states are occupied. The statistical treatment of this N-particle system is in terms of an N-particle distribution function [3.4, 3.5]. For dilute concentrations, carrier–carrier correlations are weak, and the N-particle distribution function can be contracted to the one-particle distribution function, $f(\mathbf{r}, \mathbf{p}, t)$ which satisfies the BTE [3.4, 3.5]. The influence of other carriers is, however, treated directly through the self-consistent electric field.

The BTE is also approximate because it treats carriers as classical particles that obey Newton's laws. Quantum mechanics is used only to describe the collisions. That $f(\mathbf{r}, \mathbf{p}, t)$ is a classical concept is clear because it specifies both position and momentum at the same time. According to the uncertainty principle,

$$\Delta p \Delta r \geq \hbar. \tag{3.114}$$

If we assume that the spread in carrier energy is about $k_B T$, then

$$\Delta p \simeq \sqrt{2m^* k_B T},$$

so

$$\Delta r \geq \frac{\hbar}{\sqrt{2m^* k_B T}}.$$

Since the wavelength of a thermal average carrier is $\lambda_B = h/\sqrt{2m^* k_B T}$, we conclude that

$$\Delta r \gg \lambda_B. \tag{3.115}$$

The result tells us that to treat carriers as particles, it should not be necessary to localize the carrier sharply with respect to λ_B which is typically 100–200 Å at room temperature. Equation (3.115) should be satisfied when the potential varies slowly on the scale of λ_B. As demonstrated in Chapter 1, a slow variation of potential is also a prerequisite for describing carrier dynamics by Newton's Laws. When the potential varies rapidly, a wave equation must be solved to learn how the carrier wave propagates through the device.

Another limitation of the BTE arises from the second uncertainty relation:

$$\Delta E \Delta t \geq \hbar, \tag{3.116}$$

which states that a carrier must remain in a state for a long time in order to have a well-defined energy. If we take Δt to be τ, the time between collisions, and assume that $\Delta E \simeq k_B T$, then

$$\tau \gg \frac{\hbar}{k_B T}. \tag{3.117}$$

When we multiply (3.117) by the thermal average carrier velocity, a condition on $l = \upsilon\tau$, the mean distance between collisions results. Neglecting a factor of $1/\pi$, we find

$$l \gg \lambda_B, \tag{3.118}$$

which states that the mean free path must be much longer than the mean De Broglie wavelength if the BTE is to be valid. For low-mobility semiconductors, eq. (3.118) may not be satisfied.

A condition on the maximum operating frequency for which the BTE is valid can also be obtained by multiplying eq. (3.117) by 2π and inverting to find

$$\omega \ll \frac{k_B T}{h}. \tag{3.119}$$

At room temperature, we find that ω must be less than 6×10^{12} Hz, which is well above the operating frequency of today's devices. Most conventional transistors are still described reasonably well by the BTE, but with modern epitaxial and lithographic techniques, it is possible to realize devices whose active regions are comparable in size to an electron's wavelength. The mathematical analysis of such devices is based on a wave equation rather than on the BTE.

3.8 Summary

For device analysis, the BTE is frequently assumed to be the fundamental description of carrier transport. We derived the BTE from simple arguments and obtained an equation that looks simple, but the complexity hides in the collision term. Because the BTE is so difficult to solve, we often make drastic simplifications or resort to numerical techniques using computers. We discussed one exceedingly useful simplification, the relaxation time approximation, which can be used when the scattering is elastic or isotropic. A formal transport theory based on the RTA is the subject of the following chapter. We also described some numerical techniques which can be used when the RTA can't. Another numerical technique, the Monte Carlo method, is discussed in Chapter 6. But for device analysis and simulation, the direct solution to the BTE, with or without the RTA, is often computationally prohibitive so devices are usually analyzed by solving a simplified set of equations which are derived from the BTE. These moment, or balance, equations are the subject of Chapter 5.

References and further reading

The Boltzmann transport equation is derived and applications are discussed in

3.1 Wolfe, C. M., Holonyak, N. and Stillman, G. E. *Physical Properties of Semiconductors.* Prentice-Hall, Englewood Cliffs, NJ, 1989.

3.2 Harrison, W. A. *Solid State Theory*. Dover, New York, 1980, pp. 252–7.

3.3 Conwell, E. Transport: the Boltzmann equation. *Handbook on Semiconductors,* Vol. I, Chapter 10. North Holland Publishing Company, New York, 1982.

The formal, mathematical theory of the Boltzmann equation and a derivation of the BTE from the N-particle equations of motion, eqs. (3.1) and (3.2), are discussed in

3.4 Harris, S. *An Introduction to the Theory of Boltzmann Equation.* Holt, Rinehart, and Winston, New York, 1971.

3.5 Liboff, R. *Introduction to the Theory of Kinetic Equations.* John Wiley and Sons, New York, 1969.

Rode discuses numerical solutions to the BTE in

3.6 Rode, D. L. Low-field transport in semiconductors. In *Semiconductors and Semimetals*, ed. by Willardson, R. K. and Beer, A. C., Vol. 10, Academic Press, New York, 1972.

Various techniques for solving the BTE are described in

3.7 Duderstadt, J. and Martin, W. *Transport Theory.* John Wiley and Sons, New York, 1979.

3.8 Reggiani, L. General Theory. In *Hot Electron Transport in Semiconductors*, Chapter 1. Springer-Verlag, New York, 1985.

3.9 Hennacy, K. A. and Goldsman, N. A generalized Legendre polynomial/sparse matrix approach for determining the distribution function in non-polar semiconductors. *Solid-State Electronics*, **36**, 869–877, 1993. See also: Hennacy, K. A., Wu, Y.-J., Goldsman, N. and Mayergoyz, I. D. Deterministic MOSFET simulation using a generalized spherical harmonic expansion of the Boltzmann equation. *Solid-State Electronics,* **38**, 1498–95, 1995.

3.10 Gnudi, A., Ventura, D., Baccarani, G. and Odeh, F. Two-dimensional MOSFET simulation by means of a multidimensional spherical harmonics expansion of the Boltzmann transport equation. *Solid-State Electronics,* **36**, 575, 1993.

3.11 Alam, M. A., Stettler, M. A. and Lundstrom, M. S. Formulation of the Boltzmann equation in terms of scattering matrices. *Solid-State Electronics,* **36**, 263, 1993.

Problems

3.1 Answer the following questions by evaluating the equilibrium carrier density from eq. (3.6).

(a) Verify eq. (3.10) for nondegenerate semiconductors.

(b) Derive the corresponding result without assuming nondegeneracy.
Hint: Use spherical coordinates to perform the integrations and consult page 118 of *Advanced Semiconductor Fundamentals by* R. F. Pierret (Vol. VI in the Modular Series on Solid State Devices) for a review of Fermi–Dirac integrals.

3.2 Answer the following questions about the kinetic energy density.

(a) Verify eq. (3.11) for nondegenerate semiconductors in equilibrium.

(b) Derive the result corresponding to eq. (3.11) without assuming nondegeneracy.

(c) For strongly degenerate semiconductors, $\eta_C = (E_F - E_{C0})/k_B T_L \gg 0$. For these conditions,

$$\mathcal{F}_{1/2}(\eta_C) = \frac{\eta_C^{3/2}}{\Gamma(5/2)}$$

and

$$\mathcal{F}_{3/2}(\eta_C) = \frac{\eta_C^{5/2}}{\Gamma(7/2)}.$$

Derive an expression for the average kinetic energy per carrier in a strongly degenerate electron gas. Is the result higher or lower than the corresponding result for a non-degenerate electron gas?

(d) Verify eq. (3.19) for the displaced-Maxwellian distribution.

3.3 For a three-dimensional, nondegenerate electron gas, the average kinetic energy per carrier is $3k_B T_L/2$. Work out the corresponding result for a two-dimensional, nondegenerate electron gas.

3.4 Verify eq. (3.14) assuming nondegenerate, equilibrium conditions.
Hint: Evaluate each of the nine components of the tensor separately.

3.5 Show that for a nondegenerate semiconductor in equilibrium the flux directed outward along one of the three coordinate axes is

$$\frac{J_n}{(-q)} = \frac{n}{2}\sqrt{\frac{2k_B T_L}{\pi m^*}}.$$

Alternatively, $\sqrt{2k_B T_L/\pi m^*}$ is the average velocity for a hemi-Maxwellian.

3.6 The collisions that we consider do not create or destroy carriers – they simply move carriers about in momentum space. As a result,

$$\sum_{\mathbf{p}} \left.\frac{\partial f}{\partial t}\right|_{\text{coll}} = 0. \tag{P3.1}$$

(a) For the general form of the collision operator, prove the equation stated above.
(b) Prove that the relaxation time approximation to the collision operator satisfies (P3.1).

3.7 Compare eqs. (3.29b) with (3.40b) to show that the RTA is adequate to describe the out-scattering process but not, in general, the in-scattering process.

3.8 Derive the BTE for a semiconductor with a slowly varying effective mass.
Hint: First derive an expression for $d\mathbf{p}/dt$ using arguments like those in Section 1.6.

3.9 Evaluate $\langle u_z \rangle$ from eq. (3.55), and verify that the result is eq. (3.56).

3.10 Use the principle of detailed balance in equilibrium (i.e. that $\partial f_0/\partial t = 0$) and answer the following questions.

(a) Establish the following relation between the equilibrium transition rate and its inverse:

$$\frac{S_0(\mathbf{p}', \mathbf{p})}{S_0(\mathbf{p}, \mathbf{p}')} = e^{[E(\mathbf{p}')-E(\mathbf{p})]/k_B T_L}.$$

(This relation is often true away from equilibrium because $S(\mathbf{p}, \mathbf{p}')$ is determined by the scattering potentials and is relatively insensitive to the applied fields.)

(b) Use physical arguments to explain why transitions to higher energy states are less probable than transitions to lower energy states.

3.11 Figures 3.1a and 3.1b apply to nondegenerate semiconductors. Draw corresponding figures for a degenerate semiconductor.

3.12 Assume that $f(\mathbf{p}) = f_S(\mathbf{p})$ is a Maxwellian at the lattice temperature (but don't assume that $f_S = f_0$). Show that $\partial f_S/\partial t|_{\text{coll}} = 0$.

3.13 Begin with the steady-state, spatially uniform BTE,

$$(-q)\boldsymbol{\mathcal{E}} \cdot \nabla_p f = \sum_{\mathbf{p}'} S(\mathbf{p}', \mathbf{p}) f(\mathbf{p}') - S(\mathbf{p}, \mathbf{p}') f(\mathbf{p}), \tag{P3.2}$$

and derive Rode's iterative technique for low-field transport as follows:

(a) Decompose the scattering into elastic and inelastic components, $S^{\text{el}}(\mathbf{p}, \mathbf{p}')$ and $S^{\text{in}}(\mathbf{p}, \mathbf{p}')$, and let $f = f_S + f_A$. Show that eq. (P3.2) becomes

$$f_A = \hat{I}(p) - (-q)\frac{\boldsymbol{\mathcal{E}} \cdot \nabla_p f_0}{\nu_{\text{el}} + 1/\tau_{\text{in}}}, \tag{P3.3}$$

where

$$\frac{1}{\tau_{\text{in}}} = \sum_{\mathbf{p}} S^{\text{in}}(\mathbf{p}, \mathbf{p}').$$

How are ν_{el} and $\hat{I}(p)$ defined and what are they?

(b) Let $f_A = g(p)\cos\theta$, where θ is the angle between $\mathbf{p}$ and $\boldsymbol{\mathcal{E}}$. Show that eq. (P3.3) becomes

$$g(p)\cos\theta = \frac{\hat{I}(p) - (-q)\mathcal{E}\dfrac{\partial f_0}{\partial p}\cos\theta}{\nu_{\text{el}} + 1/\tau_{\text{in}}}. \tag{P3.4}$$

(c) Solve eq. (P3.4) for $g(p)$. You'll find the theorem,

$$\sum_{\mathbf{p}'} \cos\theta' A(\cos\alpha) = \cos\theta \sum_{\mathbf{p}'} \cos\alpha A(\cos\alpha),$$

useful. In this theorem, A is an arbitrary function of $\cos\alpha$, where α is the angle between $\mathbf{p}$ and $\mathbf{p}'$. (The theorem can be proved by direct integration using a coordinate system like that in Fig. 3.6.) Show that the final result is

$$g(p) = \frac{\tilde{I}(p) - (-q)\mathcal{E}\dfrac{\partial f_0}{\partial p}}{\nu_{\text{el}} + 1/\tau_{\text{in}}}, \tag{P3.5}$$

where $\tilde{I}(p)$ is defined in eq. (3.85).

(d) Explain how eq. (P3.5) is used to iteratively solve for $g(p^2)$.

(e) Show that from the solution, $g(p)$, the mobility can be found from

$$\mu_n = -\frac{\hbar}{3nm^* E}\int k^3 g(k^2)\mathrm{d}k \tag{P3.6}$$

3.14 In metals, we often assume that $\partial f_S/\partial E \simeq \delta(E - E_F)$.

(a) Explain and justify this approximation.

(b) Derive an expression for the mobility of electrons in a metal and compare the result to eq. (3.61).

3.15 Derive eq. (3.102), the steady-state path integral solution to the BTE, by letting $t_1 \rightarrow -\infty$ in eq. (3.101).
Hint: Make a change of variables to $t'' = t - \tilde{t}$.

3.16 Evaluate the rms carrier velocity for a nondegenerate semiconductor as follows:
(a) Write an expression for the rms velocity, $\upsilon_{\text{rms}} = \sqrt{\langle \upsilon^2 \rangle}$ involving sums in momentum space.
(b) Evaluate the sum to find an expression for υ_{rms}.

3.17 Write an expression for $\langle\langle \tau_f^3 \rangle\rangle$, where $\langle\langle \tau_f \rangle\rangle = \langle E\tau_f(E)\rangle / \langle E \rangle$ with E being the kinetic energy. Assume power law scattering and express your answer in terms of Γ-functions.

4 Low-field transport

A formal theory for the flow of charge and heat will be developed in this chapter. Applications of the theory will also be explored, experimental techniques discussed, and results surveyed. Our goal is to extend Ohm's law, $\mathbf{J} = \sigma\boldsymbol{\mathcal{E}}$, to include temperature gradients and to develop an analogous expression for the heat current, $\mathbf{J}_Q$. The final result has the form

$$\mathbf{J} = L_{11}\boldsymbol{\mathcal{E}} + L_{12}\nabla T_{\mathrm{L}}, \tag{4.1a}$$

$$\mathbf{J}_{\mathrm{Q}} = L_{21}\boldsymbol{\mathcal{E}} + L_{22}\nabla T_{\mathrm{L}}. \tag{4.1b}$$

We might have expected to associate applied electric fields with electric currents and temperature gradients with heat currents, but electrons transport both charge and heat, so the two flows are coupled. Our goal is to derive eqs (4.1) from the BTE and to relate the coefficients, L_{ij}, to the semiconductor's material properties.

The theoretical treatment is based on a number of simplifying assumptions. First, conduction by electrons is assumed (but a corresponding set of equations for holes could be readily developed). We also assume that the applied fields are low, so that the currents are proportional to the driving forces (the electric field and the temperature gradient). Finally, we assume the relaxation time approximation and spherical, parabolic energy bands which makes it easy to solve the BTE.

When the band structure is complex or when the RTA doesn't apply, the BTE is difficult to solve, but the form of the resulting current equations is unchanged. We can view the *coupled current equations*, (4.1a) and (4.1b), as phenomenological relations describing low-field transport in any semiconductor, but the coefficients, L_{ij}, may have to be measured or determined from numerical solutions to the BTE. We'll illustrate how useful these phenomenological relations are by applying them to a variety of experimental situations.

4.1 Low-field solution to the BTE (B = 0)

We begin by assuming that the distribution function consists of two parts: a large component shaped like the equilibrium distribution function and a small perturbation which is obtained by solving the BTE. The first component has the form

$$f_S = \frac{1}{e^{\Theta} + 1}, \tag{4.2a}$$

where

$$\Theta = [E_{C0}(\mathbf{r}, t) + E(\mathbf{p}) - F_n(\mathbf{r}, t)]/k_B T_L(\mathbf{r}, t) \tag{4.2b}$$

and is just like f_o except that the Fermi-level is replaced by the quasi-Fermi level which may vary with position; the temperature is also allowed to vary. Equation (4.2a) cannot be the correct distribution function because it is symmetric in momentum, so the average velocity is zero, and no current flows. We assume, therefore, that

$$f = f_S + f_A, \tag{4.3}$$

where f_A is a small, anti-symmetric component that we shall find by solving the BTE.

For the steady-state BTE in the relaxation time approximation, we have

$$\upsilon \cdot \nabla_r (f_S + f_A) + \mathbf{F} \cdot \nabla_p (f_S + f_A) = -\frac{f_A}{\tau_f}. \tag{4.4}$$

To simplify eq. (4.4), we assume that $f_S \gg f_A$. We also need to assume that $|\nabla_r f_S| \gg |\nabla_r f_A|$ and that $|\nabla_p f_S| \gg |\nabla_p f_A|$ which are more difficult to justify. We can demonstrate, however, that the approach is self-consistent, which means that if we make these assumptions, the solution we find satisfies the original assumptions (see homework problem 4.3). With these assumptions, eq. (4.4) simplifies to

$$\upsilon \cdot \nabla_r f_S + \mathbf{F} \cdot \nabla_p f_S = -\frac{f_A}{\tau_f}$$

to which we apply the chain rule and write

$$\upsilon \cdot \frac{\partial f_S}{\partial \Theta} \nabla_r \Theta + \mathbf{F} \cdot \frac{\partial f_S}{\partial \Theta} \nabla_p \Theta = -\frac{f_A}{\tau_f}. \tag{4.5}$$

From eq. (4.2) we obtain

$$\nabla_r \Theta = \frac{[\nabla_r E_{C0}(\mathbf{r}) - \nabla_r F_n(\mathbf{r})]}{k_B T_L(\mathbf{r})} + [E_{C0}(\mathbf{r}) + E(\mathbf{p}) - F_n] \nabla_r \left(\frac{1}{k_B T_L(\mathbf{r})} \right) \tag{4.6a}$$

and

$$\nabla_p \Theta = \upsilon / k_B T_L. \tag{4.6b}$$

The low-field solution to the BTE is obtained by inserting eqs. (4.6) in eq. (4.5) to find

$$f_A = \frac{\tau_f}{k_B T_L} (-\partial f_S / \partial \Theta) \upsilon \cdot \mathcal{F}, \tag{4.7}$$

where

$$\mathcal{F} = -\nabla_r F_n(\mathbf{r}) + T_L [E_{C0}(\mathbf{r}) + E(\mathbf{p}) - F_n(\mathbf{r})] \nabla_r \left(\frac{1}{T_L} \right) \tag{4.8}$$

is interpreted as a *generalized force*. To derive eq. (4.7) we assumed $\mathbf{F} = -\nabla_r E_{C0}(\mathbf{r})$ which applies only when magnetic fields are absent. (Magnetic fields require a somewhat different treatment which we defer to Section (4.4.) The generalized force describes the influence of gradients in the electrostatic potential, temperature, and carrier concentration on the distribution function. Observe that eq. (4.7) has the form

$$f_A = g(p) \cos \theta, \tag{4.9}$$

where θ is the angle between the carrier velocity and the generalized force. The function, g, depends only on the magnitude of $\mathbf{p}$ because we shall permit τ_f to vary with the magnitude but not direction of $\mathbf{p}$. Equation (4.9) is the general form that we assumed in Chapter 3 in order to establish the validity of the RTA.

4.2 The coupled current equations

The electric and heat current densities are evaluated from the anti-symmetric component of the distribution function. To find the electric current density (hereafter, simply the current density) we evaluate

$$\mathbf{J} = \frac{(-q)}{\Omega}\sum_{\mathbf{p}} \upsilon f_{\mathrm{A}}(\mathbf{p}). \tag{4.10}$$

For the heat current we might use the kinetic energy current,

$$\mathbf{J}_{\mathrm{W}} = \frac{1}{\Omega}\sum_{\mathbf{p}} E(\mathbf{p})\upsilon f_{A}(\mathbf{p}) \tag{4.11}$$

because heat is associated with the kinetic energy. But part of the kinetic energy is drift energy, which is due to the average motion of carriers in the applied field. A suitable definition of the heat current should include only the random component of the kinetic energy.

A proper definition of the heat current is suggested by the thermodynamic relation between the increase in internal energy, $\mathrm{d}U$, and the increase in heat, $\mathrm{d}Q$, and particle number, $\mathrm{d}N$:

$$\mathrm{d}U = \mathrm{d}Q + F_n \mathrm{d}N, \tag{4.12}$$

where F_n the quasi-Fermi potential (or electrochemical potential), is the increase in internal energy that occurs when a small number of carriers, $\mathrm{d}N$, is added at a constant temperature. When eq. (4.12) is written as

$$\mathrm{d}Q = \mathrm{d}U - F_n \mathrm{d}N,$$

it suggests that the heat current should be defined as

$$\mathbf{J}_{\mathrm{Q}} \equiv \mathbf{J}_{\mathrm{U}} - F_n \mathbf{J}_{\mathrm{N}}, \tag{4.13}$$

where $\mathbf{J}_{\mathrm{U}}$ is the flux of internal energy (the sum of potential and kinetic energy) and $\mathbf{J}_{\mathrm{N}}$ is the particle current density (that is, the carrier flux). From eq. (4.13), we obtain the desired expression for the heat current as

$$\mathbf{J}_{\mathrm{Q}} \equiv \frac{1}{\Omega}\sum_{\mathbf{p}} [E_{\mathrm{C0}}(\mathbf{p}) + E(\mathbf{p}) - F_n]\upsilon f_{\mathrm{A}}. \tag{4.14}$$

Expressions for the electric and heat current densities result by inserting eq. (4.7) for f_{A} in eqs. (4.10) and (4.14). For the electric current, we find

$$\mathbf{J} = \frac{(-q)}{\Omega k_{\mathrm{B}} T_{\mathrm{L}}}\sum_{\mathbf{p}} \upsilon(\upsilon \cdot \mathcal{F})\tau_f(p)(-\partial f_{\mathrm{S}}/\partial\Theta), \tag{4.15}$$

and for the heat current

$$\mathbf{J}_{\mathrm{Q}} = \frac{1}{\Omega k_{\mathrm{B}} T_{\mathrm{L}}}\sum_{\mathbf{p}} \upsilon(\upsilon \cdot \mathcal{F})\tau_{\mathrm{f}}(p)(-\partial f_{\mathrm{S}}/\partial\Theta)[E_{\mathrm{C0}} + E(\mathbf{p}) - F_n]. \tag{4.16}$$

Equations (4.15) and (4.16) are written in symbolic, vector notation; it will be convenient to write these equations in *indicial notation* which displays the vector components explicitly. Written in indicial notation, the dot product of two vectors is

$$\mathbf{A} \cdot \mathbf{B} = \sum_{j=1}^{3} A_j B_j \equiv A_j B_j. \tag{4.17}$$

(Recall the summation convention introduced in Chapter 3, which states that repeated indices should be summed over each of the three coordinate axes.) In indicial notation, the ith component of the current density becomes

$$J_i = \frac{(-q)}{\Omega k_B T_L} \sum_{\mathbf{p}} \upsilon_i \upsilon_j \mathcal{F}_j \tau_f(p)(-\partial f_S/\partial \Theta). \tag{4.18}$$

From eq. (4.8), we obtain the generalized force in indicial notation as

$$\mathcal{F}_j = \partial_j(-F_n) + T_L[E_{C0}(\mathbf{r}) + E(\mathbf{p}) - F_n]\partial_j(1/T_L), \tag{4.19}$$

where

$$\partial_j(\cdot) \equiv \frac{\partial}{\partial x_j}(\cdot) \quad j = 1, 2, \text{ or } 3 \quad (\text{or } j = x, y, \text{ or } z). \tag{4.20}$$

After putting eqs. (4.18) and (4.19) together we find

$$J_i = \sigma_{ij}\partial_j(F_n/q) + B_{ij}\partial_j(1/T_L) \tag{4.21}$$

where

$$\sigma_{ij} = \frac{q^2}{\Omega k_B T_L} \sum_{\mathbf{p}} \upsilon_i \upsilon_j \tau_f(p)(-\partial f_S/\partial \Theta) \tag{4.22}$$

is the conductivity tensor and

$$B_{ij} = \frac{(-q)}{\Omega k_B T_L} \sum_{\mathbf{p}} \upsilon_i \upsilon_j \tau_f(p) T_L[E_{C0}(\mathbf{r}) + E(p) - F_n](-\partial f_S/\partial \Theta). \tag{4.23}$$

Equation (4.21) shows that the driving forces responsible for electric current are gradients in the quasi-Fermi level and inverse temperature.

The current equations can be expressed in symbolic notation if we recognize the rule for matrix multiplication,

$$[A]\mathbf{X} = \sum_{j=1}^{3} A_{ij} X_j = A_{ij} X_j, \tag{4.24}$$

and use it to express eq. (4.21) as

$$\mathbf{J} = [\sigma]\nabla_r(F_n/q) + [B]\nabla_r(1/T_L). \tag{4.25}$$

For *isotropic* materials, the tensors are diagonal,

$$[\sigma] = \sigma_0[I] \tag{4.26a}$$

or

$$\sigma_{ij} = \sigma_0 \delta_{ij}, \tag{4.26b}$$

where σ_0 is a scalar, $[I]$ the identity matrix (recall that δ_{ij}, the Kronecker delta, is defined to be zero except when $i = j$).

In a very similar manner, the heat current is evaluated by using eqs. (4.7) in (4.14); the result is

$$J_{Qi} = p_{ij}\partial_j(F_n/q) + K_{ij}\partial_j(1/T_L), \tag{4.27}$$

where

$$p_{ij} = \frac{(-q)}{\Omega k_B T_L}\sum_{\mathbf{p}} v_i v_j \tau_f(p)[E_{C0}(\mathbf{r}) + E(p) - F_n](-\partial f_S/\partial\Theta) \tag{4.28}$$

and

$$K_{ij} = \frac{1}{\Omega k_B}\sum_{\mathbf{p}} v_i v_j \tau_f(p)[E_{C0}(\mathbf{r}) + E(p) - F_n]^2(-\partial f_S/\partial\Theta). \tag{4.29}$$

The coupled current equations, (4.21) and (4.27), relate the flow of charge and heat to the driving forces – gradients in the quasi-Fermi level and in the inverse temperature. The constants of proportionality are the four tensors, σ_{ij}, B_{ij}, p_{ij}, and K_{ij}. For anisotropic materials, the current and driving forces may not be parallel, but for cubic semiconductors, the tensors are diagonal and the coupled current equations become:

$$\mathbf{J} = \sigma_0 \nabla_r(F_n/q) + B_0 \nabla_r(1/T_L) \tag{4.30a}$$

and

$$\mathbf{J}_Q = p_0 \nabla_r(F_n/q) + K_0 \nabla_r(1/T_L). \tag{4.30b}$$

For this case, the transport coefficients, σ_0, B_0, p_0, and K_0 are scalars; σ_0 and K_0 are positive, and for conduction by electrons, B_0 and p_0 are negative. Equations (4.30) are much like the expected form for the coupled current equations, (4.1), but they state that the natural driving forces are gradients in the quasi-Fermi level and inverse temperature. When the carrier concentration is uniform, $\nabla_r F_n = q\mathcal{E}$, but more generally, ∇F_n includes the effects of diffusion in a concentration gradient as well as drift in the electric field.

4.3 Transport coefficients

In this section we verify that the transport tensors are diagonal for common semiconductors and work out the sums to express the transport coefficients in terms of the semiconductor's parameters, m^* and τ_f. The four tensors that describe low-field transport,

$$\begin{Bmatrix} \sigma_{ij} \\ B_{ij} \\ p_{ij} \\ K_{ij} \end{Bmatrix} = \frac{1}{\Omega}\sum_{\mathbf{p}}\left(-\frac{f_S}{\partial\Theta}\right)\tau_f(p)\frac{\upsilon_i\upsilon_j}{k_B T_L}\begin{Bmatrix} q^2 \\ (-q)T_L[E_{C0}+E(p)-F_n] \\ (-q)[E_{C0}+E(p)-F_n] \\ T_L[E_{C0}+E(p)-F_n]^2 \end{Bmatrix} \tag{4.31}$$

are similar in form, so only the first, σ_{ij}, will be treated in detail; results for the others will simply be quoted. To keep the mathematics simple, a nondegenerate semiconductor is assumed, which means

$$f_S = e^{-\Theta} \text{ and } \frac{\partial f_S}{\partial\Theta} = -f_S. \tag{4.32}$$

We'll allow τ_f to be a function of the magnitude of $\mathbf{p}$ (that is, a function of energy) and will assume spherical, parabolic energy bands centered at $\mathbf{k} = (0, 0, 0)$. With these simplifying assumptions, eq. (4.31) for σ_{ij} becomes

$$\sigma_{ij} = \frac{q^2}{k_B T_L}\frac{1}{\Omega}\sum_{\mathbf{p}}\upsilon_i\upsilon_j\tau_f(p)f_S,$$

which can be re-written as

$$\sigma_{ij} = \frac{q^2}{m^*(k_B T_L/2)}\frac{1}{\Omega}\sum_{\mathbf{p}}\frac{m^*\upsilon_i\upsilon_j}{2}\tau_f(p)f_S. \tag{4.33}$$

The only angular dependence of the integrand arises from the $\upsilon_i\upsilon_j$ factor. By integrating over the polar angle, θ, we find

$$\int_0^{\pi}\upsilon_i\upsilon_j\sin\theta d\theta = \frac{2}{3}\upsilon^2\delta_{ij},$$

so σ_{ij} is diagonal. This result can be used to express eq. (4.33) as

$$\sigma_{ij} = \frac{q^2}{m^*(3k_B T_L/2)}\frac{1}{\Omega}\sum\left(\frac{m^*\upsilon^2}{2}\right)\tau_f(p)f_S\delta_{ij}. \tag{4.34}$$

We recognize the sum as n times $E(p)\tau_f(p)$ averaged over $f_S(p)$, and $3k_B T_L/2$ is the average kinetic energy, so eq. (4.34) can be written as

$$\sigma_{ij} = nq\frac{q}{m^*}\frac{\langle E\tau_f(E)\rangle}{\langle E\rangle}\delta_{ij} = nq\frac{q\langle\langle\tau\rangle\rangle}{m^*}\delta_{ij}. \tag{4.35}$$

The final result is what we obtained in Chapter 3, Section 4 with

$$\langle\langle\tau\rangle\rangle \equiv \frac{\langle E\tau(E)\rangle}{\langle E\rangle} = \tau_0\frac{\Gamma(s+5/2)}{\Gamma(5/2)}. \tag{4.36}$$

The quantity, $\langle\langle\tau\rangle\rangle$, is a specially defined 'average' relaxation time, and the final result on the right-hand side applies to power law scattering characterized by an exponent, s. (Power law scattering was defined in eq. (3.63), and the exponents for common scattering mechanisms were listed in Table 3.2.)

The remaining transport coefficients are evaluated in a similar manner. The result for nondegenerate semiconductors are:

$$\sigma_{ij} = nq\mu_n\delta_{ij} \tag{4.37a}$$

where

$$\mu_n = \frac{q\langle\langle\tau\rangle\rangle}{m^*} \tag{4.37b}$$

with

$$\langle\langle\tau\rangle\rangle = \tau_0\frac{\Gamma(s+5/2)}{\Gamma(5/2)}. \tag{4.37c}$$

and

$$B_{ij} = \frac{k_B}{(-q)}T_L^2[\ln(N_C/n) + (s+5/2)]\sigma_{ij} \tag{4.37d}$$

$$p_{ij} = \frac{1}{T_L}B_{ij} \tag{4.37e}$$

$$K_{ij} = \frac{k_B^2 T_L^3}{q^2}[[\ln(N_C/n) + (s+5/2)]^2 + (s+5/2)]\sigma_{ij}\quad. \tag{4.37f}$$

4.3.1 Ellipsoidal energy bands

To evaluate the transport coefficients, spherical energy bands were assumed, but for many semiconductors, the energy bands are ellipsoidal, and there are several conduction band minima. The constant energy surfaces for silicon are displayed in Fig. 4.1; they are ellipsoids of revolution described by

$$E(p) = E_{C0} + \frac{(p_x - p_{xo})^2}{2m_{xx}} + \frac{(p_y - p_{yo})^2}{2m_{yy}} + \frac{(p_z - p_{zo})^2}{2m_{zz}}, \tag{4.38}$$

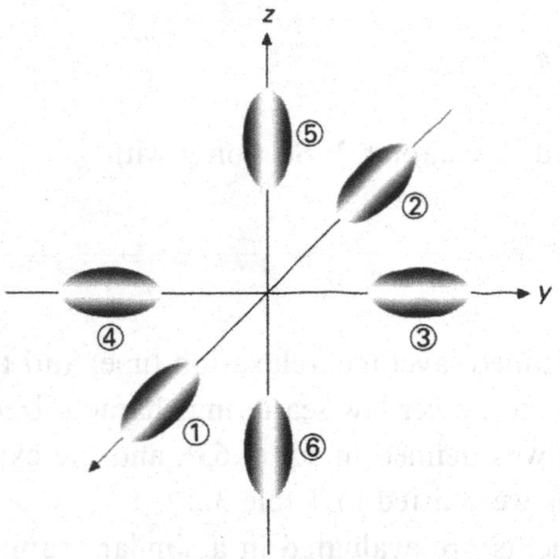

Fig. 4.1 Constant energy surfaces for the conduction band of silicon.

where $\mathbf{p}_o = (p_{xo}, p_{yo}, p_{zo})$ specifies the location of the ellipsoid's center. To find the transport coefficients for silicon, eq. (4.31) must be integrated with $E(\mathbf{p})$ given by eq. (4.38). The mathematics gets a little involved, but the final result can be had with a few simple arguments.

In and near equilibrium, carriers are evenly distributed among the six equivalent minima; to find the total conductivity, we consider the ellipsoids one at a time then add the contributions together. Consider first ellipsoid 1 whose major axis is oriented along the $\hat{x}$-axis, and assume that the generalized field is also oriented along the $\hat{x}$-axis. Carriers in this ellipsoid respond to the field with the longitudinal effective mass, m_l^*, so we expect that the contribution of this ellipsoid to the conductivity should be

$$\sigma_1 = \frac{n}{6} q(q\langle\langle\tau\rangle\rangle/m_l^*). \tag{4.39}$$

Ellipsoid 2 also responds to fields with effective mass, m_l^*, but the other four ellipsoids respond with the transverse effective mass, m_t^*. Adding the contributions from each of the six ellipsoids, we find

$$\sigma = nq(q\langle\langle\tau\rangle\rangle/m_c^*), \tag{4.40}$$

where

$$\frac{1}{m_c^*} \equiv \frac{1}{3m_l^*} + \frac{2}{3m_t^*} \tag{4.41}$$

is known as the *conductivity effective mass*. The motivation for defining this new effective mass is simply to make the result look like the expression for a simple spherical band given in eq. (4.37b). Equation (4.40) is correct for any orientation of the electric field. The conductivity of a given ellipsoid depends on the direction of the applied field, but the total conductivity is independent of direction because of the high degree of symmetry.

4.3.2 Multiple scattering mechanisms

Consider next a semiconductor in which two independent scattering mechanisms exist. From the relaxation time approximation, we write

$$\left.\frac{\partial f}{\partial t}\right|_{\text{coll}} = -\frac{f_A}{\tau_1} - \frac{f_A}{\tau_2} \equiv -\frac{f_A}{\tau_{\text{eff}}}, \tag{4.42}$$

where

$$\frac{1}{\tau_{\text{eff}}(p)} \equiv \frac{1}{\tau_1(p)} + \frac{1}{\tau_2(p)}. \tag{4.43}$$

If both τ_1 and τ_2 have a power law dependence on energy with characteristic exponents, s_1 and s_2, then

$$\tau_{\text{eff}} = \frac{\tau_{01}\tau_{02}(p^2/2m^*k_B T_L)^{s_1+s_2}}{\tau_{01}(p^2/2m^*k_B T_L)^{s_1} + \tau_{02}(p^2/2m^*k_B T_L)^{s_2}}. \tag{4.44}$$

Now if s_1 *just happens* to equal s_2, then τ_{eff} has exactly the same form as does eq. (3.63) for a single scattering mechanism. For this particularly simple case, the mathematics works out just as it did for a single scattering mechanism with the result that

$$\mu_n = \frac{q}{m^*}\left[\frac{\tau_{01}\tau_{02}}{\tau_{01} + \tau_{02}}\right]\frac{\Gamma(s+5/2)}{\Gamma(5/2)}$$

or

$$\frac{1}{\mu_n} = \frac{m^*}{q\tau_{01}}\frac{\Gamma(5/2)}{\Gamma(s+5/2)} + \frac{m^*}{q\tau_{02}}\frac{\Gamma(5/2)}{\Gamma(s+5/2)}. \tag{4.45}$$

The final result can be expressed as

$$\frac{1}{\mu} = \frac{1}{\mu_1} + \frac{1}{\mu_2}, \tag{4.46}$$

which is known as *Mathiessen's rule* and states that the mobility may be deduced from the mobility due to each mechanism acting alone. Mathiessen's rule is often used to estimate mobility when multiple scattering mechanisms are present, but it must be stressed that Mathiessen's rule applies *only when* $s_1 = s_2$. Because independent scattering mechanisms rarely have the same energy dependence, the use of Mathiessen's rule is rarely justified in practice. Nevertheless, it is commonly used in practice because it is often easy to estimate the mobility for various scattering mechanisms independently but difficult to do so when the processes occur simultaneously.

4.4 Transport in a weak magnetic field

To treat magnetic fields, the force, $(-q)\upsilon \times \mathbf{B}$, is included in the BTE, and the result is a set of four transport tensors that are functions of $\mathbf{B}$. Since the mathematics can get tedious, especially for strong magnetic fields, we shall simply outline the solution and discuss the results for weak magnetic fields. The more general problem is treated in the homework problems and in the chapter references.

We begin the analysis at eq. (4.4) but add to the BTE the force due to the magnetic field,

$$\upsilon \cdot \nabla_r f + \left[(-q)\boldsymbol{\mathcal{E}} + (-q)\upsilon \times \mathbf{B}\right] \cdot \nabla_p f = -\frac{f_A}{\tau_f}. \tag{4.47}$$

For low fields, the equation can be solved by superposition. First, we set $\mathbf{B} = \mathbf{0}$ and solve eq. (4.47). The result is just eq. (4.7) which we now denote f_A', that part of f_A due to all forces except the magnetic force. Next, the generalized force (electric field, concentration gradients and temperature gradients) is set to zero and the BTE solved for

$$f_A'' = \tau_f q(\upsilon \times \mathbf{B}) \cdot \nabla_p f. \tag{4.48}$$

The complete solution is just $f_A' + f_A''$. The temptation is to approximate $\nabla_p f$ in eq. (4.48) by $\nabla_p f_S$, but

$$\nabla_p f_S = \frac{\partial f_S}{\partial \Theta} \nabla_p \Theta = \frac{\partial f_S}{\partial \Theta} \frac{\upsilon}{k_B T}$$

and

$$(\upsilon \times \mathbf{B}) \cdot \upsilon = 0$$

so in this approximation $f_A'' = 0$. A better procedure is to approximate $\nabla_p f$ using the solution obtained by ignoring the magnetic field. That is,

$$\nabla_p f \simeq \nabla_p f_A' = \nabla_p \left[\frac{\tau_f}{k_B T_L}\left(-\frac{\partial f_S}{\partial \Theta}\right)(\upsilon \cdot \mathcal{F})\right]. \tag{4.49}$$

In eq. (4.49), the term in brackets contains several functions of energy ($\tau_f(E)$, $\partial f_S/\partial\Theta$, and $\mathcal{F}$). Consider the gradient in momentum space of a function of energy, $h(E)$,

$$\nabla_{\mathbf{p}} h(E) = \frac{\partial h}{\partial E} \nabla_{\mathbf{p}} E = \frac{\partial h}{\partial E} \upsilon,$$

so the gradient of a function of energy is proportional to the carrier velocity. If we now take the dot product with $(\upsilon \times \mathbf{B})$, we get zero because $(\upsilon \times \mathbf{B}) \cdot \upsilon = 0$. When evaluating eq. (4.49) therefore, the terms involving the gradient of a function of energy give zero when inserted into eq. (4.48). In indicial notation, eq. (4.49) is

$$\frac{\partial f'_{\mathrm{A}}}{\partial p_i} = const \times \upsilon + \frac{\tau_f}{k_{\mathrm{B}} T_{\mathrm{L}}} \left(- \frac{\partial f_{\mathrm{S}}}{\partial \Theta} \right) \frac{\partial \upsilon_j}{\partial p_i} \mathcal{F}_j. \tag{4.50a}$$

For spherical bands

$$\frac{\partial \upsilon_i}{\partial p_j} = \frac{1}{m^*} \delta_{ij}, \tag{4.50b}$$

and if eqs. (4.50a) and (4.50b) are inserted in eq. (4.48), we find

$$f''_{\mathrm{A}} = \frac{-q \tau_f^2}{m^*} \frac{1}{k_B T_L} \frac{\partial f_S}{\partial \Theta} (\upsilon \times \mathbf{B} \cdot \mathcal{F}). \tag{4.51}$$

With the vector identity $\mathbf{A} \cdot (\mathbf{B} \times \mathbf{C}) = \mathbf{B} \cdot (\mathbf{C} \times \mathbf{A})$, this equation can be rewritten as

$$f''_{\mathrm{A}} = - \frac{q \tau_f^2}{m^* k_{\mathrm{B}} T_{\mathrm{L}}} \frac{\partial f_{\mathrm{S}}}{\partial \Theta} \upsilon \cdot (\mathbf{B} \times \mathcal{F}). \tag{4.52}$$

Since the BTE has been solved, the current density can be obtained by evaluating

$$\mathbf{J} = \frac{(-q)}{\Omega} \sum_{\mathbf{p}} \upsilon (f'_{\mathrm{A}} + f''_{\mathrm{A}}). \tag{4.53}$$

The first term in eq. (4.53) was already evaluated in Section 4.3; when the generalized force consisted only of an electric field, $(\mathcal{F} = -q\boldsymbol{\mathcal{E}})$, those results were

$$\mathbf{J}' = [\sigma'] \boldsymbol{\mathcal{E}}, \tag{4.54}$$

where $[\sigma']$ is diagonal. The current density due to the additional term in eq. (4.53) arising from the **B**-field is

$$\mathbf{J}'' = \frac{1}{\Omega} \sum_{\mathbf{p}} \frac{q^3 \tau_f^2}{m^*} \frac{1}{k_{\mathrm{B}} T_{\mathrm{L}}} \left(\frac{-\partial f_{\mathrm{S}}}{\partial \Theta} \right) \upsilon [\upsilon \cdot (\mathbf{B} \times \boldsymbol{\mathcal{E}})]. \tag{4.55}$$

Before we proceed, we must explain how to write cross-products in indicial notation. To do so, we express the cross-product as

$$\mathbf{A} \times \mathbf{B} = \varepsilon_{1jk} A_j B_k \hat{x}_1 + \varepsilon_{2jk} A_j B_k \hat{x}_2 + \varepsilon_{3jk} A_j B_k \hat{x}_3$$

where ε_{ijk} is the *alternating unit tensor* defined as

$$\begin{aligned}\varepsilon_{ijk} &= +1 \text{ for } i, j, k \text{ in cyclic order } (1, 2, 3,\ \ 2, 3, 1,\ \ \text{etc}) \\ &= -1 \text{ for } i, j, k \text{ in anti-cyclic order } (3, 2, 1, 2, 1, 3 \text{ etc}) \\ &= 0 \ \ \text{otherwise}\end{aligned} \tag{4.56}$$

The ith component of the cross-product is expressed as

$$(\mathbf{A} \times \mathbf{B}) \cdot \hat{x}_i = \varepsilon_{ijk} A_j B_k. \tag{4.57}$$

Now we can write the component of **J** due to the magnetic field, eq. (4.55), in indicial notation as

$$J_i'' = \frac{1}{\Omega} \sum_{\mathbf{p}} \frac{q^3}{m^* k_B T_L} \tau_f^2 (-\partial f_S / \partial \Theta) v_i v_m \varepsilon_{mnj} B_n \mathcal{E}_j, \tag{4.58}$$

from which we can write the component of the conductivity tensor due to magnetic fields as

$$\sigma_{ij}'' = \frac{1}{\Omega} \sum_{\mathbf{p}} \frac{q^3}{m^* k_B T_L} \tau_f^2 (-\partial f_S / \partial \Theta)(v_i v_m \varepsilon_{mnj} B_n). \tag{4.59}$$

Consider first the diagonal term, σ_{11}''. The term in parentheses becomes

$$v_i v_m \varepsilon_{mnj} B_n = v_1 v_m \varepsilon_{mn1} B_n.$$

We saw in Section 4.3 that when we integrate over θ and ϕ only the $v_1 v_1$ term will be non-zero, but if $m = 1$, ε_{mn1} has a repeated index, so it is zero and $\sigma_{11}'' = 0$. Similarly, all other diagonal components of $[\sigma'']$ are zero. We conclude that the presence of a weak magnetic field does not affect the diagonal components of the conductivity tensor.

Consider next an off-diagonal term such as

$$\sigma_{12}'' = \frac{1}{\Omega} \sum_{\mathbf{p}} \frac{q^3}{m^* k_B T_L} \tau_f^2 (-\partial f_S / \partial \Theta) v_1 v_m \varepsilon_{mn2} B_n.$$

The only non-zero term permitted by the alternating unit tensor occurs for $m = 1$, $n = 3$:

$$\sigma_{12}'' = \frac{(-q) B_3}{m^*} \left\{ \frac{q^2}{k_B T_L \Omega} \sum_{\mathbf{p}} v_1 v_1 \tau_f^2 (-\partial f_S / \partial \Theta) \right\}.$$

The term in brackets is similar to the one we evaluated in Section 4.3 except for the presence of τ_f^2 instead of τ_f. The result of performing the sum is

$$\sigma_{12}'' = -\sigma_0 \mu_H B_z \tag{4.60}$$

where

$$\sigma_0 = nq\mu_n \tag{4.61}$$

and

$$\mu_{\mathrm{H}} \equiv \frac{\langle\langle\tau^2\rangle\rangle}{\langle\langle\tau\rangle\rangle^2}\mu = r_{\mathrm{H}}\mu. \tag{4.62a}$$

The *Hall factor*, r_{H}, which relates the *Hall mobility*, μ_{H}, to the *drift mobility*, μ, is given by

$$\boxed{r_{\mathrm{H}} \equiv \frac{\langle\langle\tau^2\rangle\rangle}{\langle\langle\tau\rangle\rangle^2} = \frac{\Gamma(2s+5/2)\Gamma(5/2)}{[\Gamma(s+5/2)]^2}}. \tag{4.62b}$$

The Hall factor was tabulated for common scattering mechanisms in Table 3.2.

The weak **B**-field solution to the BTE shows that the diagonal elements of the conductivity tensor are unchanged but that off-diagonal elements like eq. (4.60) are introduced. After adding the diagonal and off-diagonal elements, we find

$$J_i = \sigma_{ij}(\mathbf{B})\mathcal{E}_j, \tag{4.63}$$

where

$$\sigma_{ij}(\mathbf{B}) = \sigma_0\delta_{ij} - \sigma_0\mu_H\varepsilon_{ijk}B_k. \tag{4.64}$$

The tensor, σ_{ij}, is the *magnetoconductivity tensor*; its components are

$$[\sigma(\mathbf{B})] = \sigma_0 \begin{bmatrix} 1 & -\mu_{\mathrm{H}}B_z & +\mu_{\mathrm{H}}B_y \\ +\mu_{\mathrm{H}}B_z & 1 & -\mu_{\mathrm{H}}B_x \\ -\mu_{\mathrm{H}}B_y & +\mu_{\mathrm{H}}B_x & 1 \end{bmatrix}, \tag{4.65}$$

Alternatively, the current equation, eq. (4.63), can be expressed in symbolic notation as

$$\boxed{\mathbf{J} = \sigma_0\boldsymbol{\mathcal{E}} - \sigma_0\mu_{\mathrm{H}}\boldsymbol{\mathcal{E}} \times \mathbf{B}}, \tag{4.66}$$

Although the mathematics is more involved, the treatment of magnetic fields of arbitrary strength proceeds along similar lines. One technique is to use the distribution function we obtained for weak magnetic fields, $f_A' + f_A''$, to obtain a better approximation for $\nabla_p f$ in eq. (4.48). The result is an improved estimate, f_A'', which leads to terms proportional to B^2 in the magnetoconductivity tensor. The process can be continued iteratively to any order in B.

Strong magnetic fields are defined by the criterion,

$$\omega_c\tau_f \gg 1, \tag{4.67}$$

where

$$\omega_c = \frac{qB}{m^*} \tag{4.68}$$

is the *cyclotron frequency*, the frequency at which carriers orbit the **B**-field. The physical interpretation of eq. (4.67) is that under strong fields carriers complete several orbits before scattering, while under weak magnetic fields, they scatter many times before completing an orbit.

High magnetic fields affect both the diagonal and off-diagonal elements of the magnetoconductivity tensor. Under low magnetic fields the Hall factor approaches eq. (4.62b), and under high fields it approaches unity. For high magnetic fields applied along or perpendicular to the current, semiconductors can display *longitudinal* or *transverse magnetoresistance*. For strong magnetic fields, electron motion in the plane perpendicular to **B** is quantized into *Landau levels*, which are separated in energy by $\hbar\omega_c$. At low temperatures where $k_B T_L \ll \hbar\omega_c$, these levels have a strong influence on carrier transport [4]. Some of the high magnetic field solution techniques and transport effects are discussed in the homework problems and, more thoroughly, in the chapter references.

4.5 The phenomenological current equations

The coupled current equations,

$$J_i = \sigma_{ij}(\mathbf{B})\mathcal{E}_j + B_{ij}(\mathbf{B})\partial_j(1/T_L) \tag{4.69a}$$

and

$$J_{Qi} = p_{ij}(\mathbf{B})\mathcal{E}_j + K_{ij}(\mathbf{B})\partial_j(1/T_L) \tag{4.69b}$$

state that the electric and heat current densities are proportional to the driving forces. (The driving force, $\partial_j F_n/q$, has been replaced by the electric field because we now are going to restrict our attention to applications for which $n(\mathbf{r})$ is constant.) Although the derivation of these equations was based on a number of simplifying assumptions (e.g. the relaxation time approximation), we could have begun by *assuming* that eqs. (4.69a) and (4.69b) hold. The assumption is reasonable because fluxes should be proportional to the driving forces when the driving forces are small. From this viewpoint the equations are *phenomenological* and describe low-field transport in any semiconductor – not just those for which the simplifying assumptions hold. But the expressions we derived for the four transport tensors were based on simplifying assumptions and are not valid in general. To apply eqs. (4.69a) and (4.69b), the transport tensors must be measured or computed from a more accurate theory such as the iterative technique discussed in Chapter 3.

4.5.1 Inversion of the transport equations

The independent variables in eq. (4.69) (those under control in an experiment) are the electric field and the gradient of the inverse temperature. From an experimental point-of-view, it is considerably more convenient to force a current then measure the resulting electric field. We prefer to write transport equations as

$$\mathcal{E}_j = \rho_{jk} J_k + \alpha_{jk} \partial_k T_{\rm L} \tag{4.70a}$$

and

$$J_{{\rm Q}i} = \pi_{jk} J_k - \kappa_{jk} \partial_k T_{\rm L} \quad . \tag{4.70b}$$

That these equations find more use than their counterparts is apparent from the fact that each of the coefficients in eq. (4.70) has a name:

$\rho =$ resistivity

$\alpha =$ thermoelectric power

$\pi =$ Peltier coefficient

$\kappa =$ thermal conductivity.

To relate these parameters to those in eq. (4.69), we solve eq. (4.69a) for $\mathcal{E}_j$. The solution is easy to find when eq. (4.69a) is written in symbolic notation as

$$\mathbf{J} = [\sigma]\boldsymbol{\mathcal{E}} - \frac{[B]}{T_{\rm L}^2} \nabla_r T_L,$$

which is readily solved for

$$\boldsymbol{\mathcal{E}} = [\sigma^{-1}]\mathbf{J} + \frac{[\sigma^{-1}][B]}{T_{\rm L}^2} \nabla_r T_L. \tag{4.71}$$

Comparing eq. (4.71) with eq. (4.70a) we conclude that

$$[\rho] = [\sigma^{-1}] \tag{4.72}$$

and

$$[\alpha] = \frac{[\rho][B]}{T_{\rm L}^2}. \tag{4.73}$$

Similarly, casting eq. (4.69b) in the form of eq. (4.70b) we find that

$$[\pi] = [p][\rho] \tag{4.74}$$

and

$$[\kappa] = \frac{1}{T_{\rm L}^2}\{[K] - [p][\rho][B]\}. \tag{4.75}$$

Equations (4.72)–(4.75) express the transport parameters for the inverted equations in terms of the transport parameters derived directly from the BTE.

4.5.2 Taylor series expansions of transport tensors

Equations (4.70a) and (4.70b) are difficult to apply because it is hard to experimentally characterize the tensors for **B**-fields of arbitrary strength. We find it more convenient to work with Taylor series expansions of the tensors because the various terms in the expansions can be measured in zero, weak, or moderate strength **B**-fields. To illustrate the technique, consider eq. (4.70a) in the absence of a temperature gradient,

$$\mathcal{E}_i = \rho_{ij}(\mathbf{B})J_j, \tag{4.76}$$

and expand $\rho_{ij}(\mathbf{B})$ in a Taylor series as

$$\rho_{ij}(\mathbf{B}) = \rho_{ij}(\mathbf{0}) + \left.\frac{\partial\rho_{ij}(\mathbf{B})}{\partial B_k}\right|_{\mathbf{B}=0} B_k + \frac{1}{2}\left.\frac{\partial^2\rho_{ij}(\mathbf{B})}{\partial B_k\partial B_l}\right|_{\mathbf{B}=0} B_k B_l + \ldots \tag{4.77}$$

For notational convenience, we make several definitions:

$$\begin{aligned}\rho_{ij} &\equiv \rho_{ij}(0)\\ \rho_{ijk} &\equiv \left.\frac{\partial\rho_{ij}(\mathbf{B})}{\partial B_k}\right|_{\mathbf{B}=0}\\ \rho_{ijkl} &\equiv \frac{1}{2}\left.\frac{\partial^2\rho_{ij}(\mathbf{B})}{\partial B_k\partial B_l}\right|_{\mathbf{B}=0}\end{aligned} \tag{4.78}$$

etc.

so that eq. (4.77) can be written as

$$\rho_{ij}(\mathbf{B}) = \rho_{ij} + \rho_{ijk}B_k + \rho_{ijkl}B_kB_l + \ldots \tag{4.79}$$

When eq. (4.79) is inserted in eq. (4.76), we find that

$$\boxed{\mathcal{E}_i = \rho_{ij}J_j + \rho_{ijk}B_kJ_j + \rho_{ijkl}B_kB_lJ_j + \ldots} \tag{4.80}$$

which is the form that we shall find most useful. The first term describes electrical conductivity, the second, the Hall effect and the third, magnetoresistance. Each of the other three tensors in eq. (4.70) can be expanded in a similar manner.

4.5.3 Transport coefficients for cubic semiconductors

Before we apply the phenomenological equations to experiments, it is necessary to know something about the form of the various tensors (such as which elements are non-zero and which are equal). A very powerful technique for deducing the form of the transport tensors makes use of symmetry considerations [4.1]. When the phenomenological current equations are applied to cubic semiconductors like silicon and gallium arsenide, the tensors have a very simple form – much like those we obtained in Section 4.3. For cubic semiconductors, the conductivity tensor for $\mathbf{B} = \mathbf{0}$ is diagonal, so

$$\rho_{ij} = \rho_0 \delta_{ij} \tag{4.81}$$

where

$$\rho_0 = \frac{1}{\sigma_0}.$$

To find ρ_{ijk} we must invert the magnetoconductivity tensor, eq. (4.65), and differentiate it. The result is

$$\rho_{ijk} = \rho_0 \mu_H \varepsilon_{ijk}, \tag{4.82}$$

where ε_{ijk} is the alternating unit tensor.

To find ρ_{ijkl}, we must work with the magnetoconductivity tensor valid to second order in **B**. For cubic semiconductors, only a few of the 81 terms of this tensor are non-zero; they are:

$$\rho_{\alpha\alpha\alpha\alpha}$$

$$\rho_{\alpha\alpha\beta\beta} \tag{4.83}$$

$$\rho_{\alpha\beta\alpha\beta} = \rho_{\alpha\beta\beta\alpha}.$$

(By convention, Greek letter subscripts refer to a specific element in the tensor, and no sum over repeated indices is assumed.) The final result for cubic semiconductors is

$$\boxed{\mathcal{E}_i = \rho_0 J_i + \rho_0 \mu_H \varepsilon_{ijk} J_j B_k + \rho_{ijkl} J_j B_k B_l + \ldots} \;. \tag{4.84}$$

The remaining transport tensors are similar in form to those already quoted. For example

$$\kappa_{ij} = \kappa_0 \delta_{ij},$$

$$\kappa_{ijk} = \kappa_1 \varepsilon_{ijk},$$

and κ_{ijkl} will have non-zero components where ρ_{ijkl} does.

From the theoretical expressions for the transport tensors, eq. (4.37), and from the prescription for inverting them given in eqs. (4.72)–(4.75), we find

$$\rho_0 = 1/\sigma_0 = 1/nq\mu_n \tag{4.85a}$$

$$\alpha_0 = \frac{\rho_0 B_0}{T_L^2} = \frac{k_B}{(-q)}[\ln(N_C/n) + (s + 5/2)] \tag{4.85b}$$

$$\pi_0 = T_L \alpha_0 \tag{4.85c}$$

and

$$\kappa_0 = T_L (k_B/q)^2 (s + 5/2)\sigma_0. \tag{4.85d}$$

Although limited to nondegenerate semiconductors ($n < N_C$) with spherical and parabolic bands and to scattering mechanisms for which the RTA is valid, these expressions provide useful estimates for the transport parameters.

The thermoelectric power is negative because an n-type semiconductor was assumed; it is positive for p-type semiconductors. Since $k_B/q = 86\mu V/K$, and because $\ln(N_c/n)$ can be quite large for lightly doped semiconductors, $|\alpha_0|$ may be several millivolts per Kelvin. The close relation between the thermoelectric power and the Peltier coefficient, eq. (4.85c), applies generally and is known as the *Kelvin relation*. Finally, note that eq. (4.85d) is only the electronic contribution to the thermal conductivity; the lattice also makes a sizable contribution.

4.6 Applications of the phenomenological equations

The phenomenological equations describe currents in the presence of electric and magnetic fields and temperature gradients. Because so many effects can occur, we describe only a few simple cases to illustrate how the coupled current equations are applied to experiments. The effects are conveniently divided into three classes. *Thermoelectric* effects involve temperature gradients and electric fields, *thermomagnetic* effects involve temperature gradients and magnetic fields, and *galvanomagnetic* effects occur in the presence of electric and magnetic fields.

4.6.1 Thermoelectric effects

Thermoelectric effects relate electric fields to temperature gradients. For example, when a temperature gradient is maintained across an open-circuited semiconductor, a voltage across the two ends appears. The phenomena is know as the *Seebeck effect* and can be described by (4.70a) with $\mathbf{B} = \mathbf{0}$. If the one-dimensional semiconductor in Fig. 4.2 is open-circuited, then $J_x = 0$ and eq. (4.70a) gives

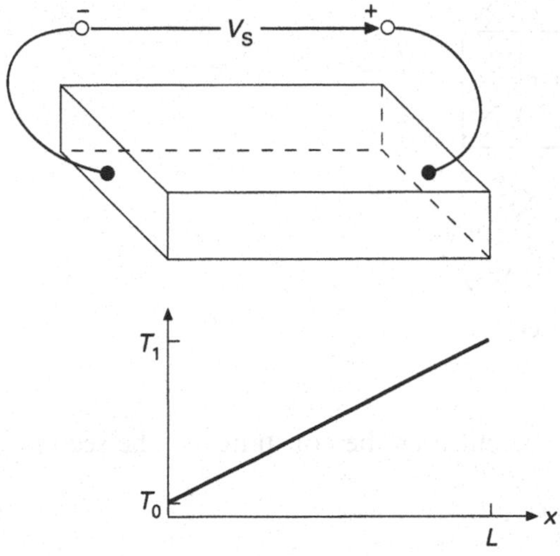

Fig. 4.2 Illustration of the Seebeck effect.

$$\mathcal{E}_x = \alpha_{xx}\frac{dT_L}{dx} = \alpha_0\frac{(T_1 - T_0)}{L},$$

where α_0 is the diagonal element of $[\alpha]$. The voltage developed across the two faces is

$$V_S = -\alpha_0 \Delta T_L. \tag{4.86}$$

The physical explanation for the Seebeck effect is that carriers diffuse down the temperature gradient; when the sample is open-circuited, they accumulate at the cold end where the resulting charge imbalance sets up an electric field that opposes the diffusion of carriers from the hot end. Since $\alpha_0 < 0$ for n-type semiconductor and $\alpha_0 > 0$ for p-type, measurement of the Seebeck voltage, typically performed in an instrument known as a *hot-point probe*, is a convenient means to determine the type of a semiconductor.

The Seebeck effect concerns electric fields produced by temperature gradients; the *Peltier effect* describes the evolution or absorption of heat caused by the flow of electric current. Consider the one-dimensional semiconductor in Fig. 4.3 which is maintained at a constant temperature T_0. According to eq. (4.70b), the heat flux is

$$J_{Qx} = \pi_{xx}J_x = \pi_0 J_x \tag{4.87}$$

where π_0 is the diagonal element of $[\pi]$. The metal contacts and the semiconductor will have different Peltier coefficients, say π_M and π_S. At contact #1 we have

$$\Delta J_{Q1} = J_{Qin} - J_{Qout} = (\pi_M - \pi_S)J_x, \tag{4.88a}$$

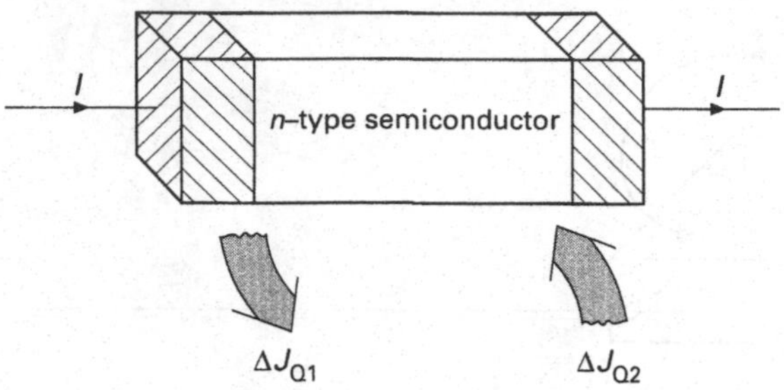

Fig. 4.3 Illustration of the Peltier effect.

since we know that the electric current must be constant. At the second contact

$$\Delta J_{\mathrm{Q2}} = J_{\mathrm{Qin}} - J_{\mathrm{Qout}} = (\pi_{\mathrm{S}} - \pi_{\mathrm{M}})J_x. \tag{4.88b}$$

If π_{S} is more negative than π_{M}, then $\Delta J_{\mathrm{Q}} = 0$ at contact #1 and $\Delta J_{\mathrm{Q}} < 0$ at contact #2. We conclude that heat must be released at contact #1 and absorbed at contact #2. The first contact will get hot, the second cold. Physically, the effect arises because electrons carry different amounts of heat in different materials, but the circuit constrains the electric current to be constant.

4.6.2 Thermomagnetic effects

Several different effects occur when a magnetic field and temperature gradient are simultaneously applied. The Seebeck effect in the presence of a $\hat{z}$-directed **B**-field can be evaluated from eq. (4.70a), which is expanded to second order in **B** as

$$\mathcal{E}_i = \rho_0 J_i + \rho_0 \mu_H \varepsilon_{ijk} J_j B_k + \rho_{ijkl} J_j B_k B_l + \alpha_0 \partial_i T_L + \alpha_1 \varepsilon_{ijk} \partial_j T_L B_k + \alpha_{ijkl} \partial_j T_L B_k B_l, \tag{4.89}$$

where we have written α_{ijk} as $\alpha_1 \varepsilon_{ijk}$. For experimental conditions which ensure that $J_i = 0$ and $\partial T_{\mathrm{L}}/\partial y = 0$, (4.89) becomes

$$\mathcal{E}_x = \alpha_0 \partial_x T_{\mathrm{L}} + \alpha_1 \varepsilon_{xxz} \partial_x T_{\mathrm{L}} B_z + \alpha_{xxzz} \partial_x T_{\mathrm{L}} B_z^2.$$

Since $\varepsilon_{xxz} = 0$,

$$\mathcal{E}_x = (\alpha_0 + \alpha_{xxzz} B_z^2) \partial T_{\mathrm{L}}/\partial x. \tag{4.90}$$

The change in thermoelectric power with magnetic field,

$$\Delta\alpha = \alpha(B) - \alpha(0) = \alpha_{xxzz} B_z^2 \tag{4.91}$$

is termed the *magneto-Seebeck effect*.

Table 4.1. *Coupled current effects in the presence of transverse magnetic field*

Effect	Experimental conditions	Result
Transverse magneto-resistance (isothermal)	$\mathbf{J} = J_x\hat{x},\ \mathbf{B} = B_z\hat{z}$ $T_L = \text{constant}$	$\mathcal{E}_x = (\rho_0 + \rho_{xxzz}B_z^2)J_x$
Transverse magneto-resistance (adiabatic)	$\mathbf{J} = J_x\hat{x},\ \mathbf{B} = B_z\hat{z}$ $J_{Qy} = 0,\ \partial T_{\mathrm{L}}/\partial x = 0$	$\mathcal{E}_x = \left(\rho_0 + \left(\rho_{xxzz} - \frac{T_L\alpha_1^2}{\kappa_0}\right)B_z^2\right)J_x$
Magneto–Seebeck (isothermal)	$\mathbf{J} = \mathbf{0},\ \mathbf{B} = B_z\hat{z}$ $\partial T_{\mathrm{L}}/\partial y = 0$	$\mathcal{E}_x = (\alpha_0 + \alpha_{xxzz}B_z^2)\frac{\partial T_{\mathrm{L}}}{\partial x}$
Magneto–Seebeck (adiabatic)	$\mathbf{J} = 0,\ \mathbf{B} = B_z\hat{z}$ $J_{Qy} = 0$	$\mathcal{E}_x = \left[\alpha_0 + \left(\alpha_{xxzz} - \frac{\alpha_1\kappa_1}{\kappa_0}\right)B_z^2\right]\frac{T_{\mathrm{L}}}{\partial x}$
Nernst (isothermal)	$\mathbf{J} = \mathbf{0},\ \mathbf{B}_z = B_z\hat{z}$ $\partial T_{\mathrm{L}}/\partial y = 0$	$\mathcal{E}_y = -\alpha_1 B_z \frac{\partial T_{\mathrm{L}}}{\partial x}$
Nernst (adiabatic)	$\mathbf{J} = 0,\ \mathbf{B}_z = B_z\hat{z}$ $J_{Qy} = 0$	$\mathcal{E}_y = -(\alpha_1 - \alpha_0\frac{\kappa_1}{\kappa_0})B_z\frac{T_L}{\partial x}$
Hall (isothermal)	$\mathbf{J} = J_x\hat{x},\ \mathbf{B} = B_{\hat{z}}$ $T_L = \text{constant}$	$\mathcal{E}_y = \rho_1 B_z J_x$
Hall (adiabatic)	$\mathbf{J} = J_x\hat{x},\ \mathbf{B} = B_z\hat{z}$ $J_{Qy} = 0$	$\mathcal{E}_y = \left[\rho_1 - \frac{T_{\mathrm{L}}\alpha_0\alpha_1}{\kappa_0}\right]B_z J_x$
Ettingshausen	$\mathbf{J} = J_x\hat{x},\ \mathbf{B} = B_z\hat{z}$ $J_{Qy} = 0\ \partial T_{\mathrm{L}}/\partial x = 0$	$\frac{\partial T_{\mathrm{L}}}{\partial y} = -\frac{\pi_1}{\kappa_0}J_x B_z$
Righi–Leduc	$\mathbf{J} = 0,\ \mathbf{B} = B_z\hat{z}$ $J_{Qy} = 0$	$\frac{\partial T_{\mathrm{L}}}{\partial y} = \frac{\kappa_1}{\kappa_0}B_z\frac{\partial T_{\mathrm{L}}}{\partial x}$

Several other thermomagnetic effects also exist; examples are the Nernst, Ettingshausen, and Righi–Leduc effects listed in Table 4.1. We examine just the first, and leave the others as exercises. For the Nernst effect, the experimental conditions dictate that no current flows. A $\hat{z}$-directed magnetic field is applied as an $\hat{x}$-directed temperature gradient. From (4.89) we find

$$\mathcal{E}_i = \alpha_0\partial_i T_{\mathrm{L}} + \alpha_1\varepsilon_{ijz}\partial_j T_{\mathrm{L}}B_z + \alpha_{ijzz}\partial_j T_{\mathrm{L}}B_z^2.$$

For $i = x$ we get the magneto-Seebeck effect, but there is also a $\hat{y}$-directed electric field given by

$$\mathcal{E}_y = -\alpha_1 B_z \partial T_L / \partial x. \tag{4.92}$$

The appearance of an electric field transverse to a temperature gradient and **B**-field is known as the *Nernst effect*; it occurs because carriers that diffuse down the temperature gradient are deflected by the magnetic field.

4.6.3 Galvanomagnetic effects

The best-known galvanomagnetic effect is the Hall effect: a magnetic field normal to the direction of current flow produces an electric field normal to both. With $\mathbf{J} = J_x \hat{x}$ and $\mathbf{B} = B_z \hat{z}$ we find that an electric field transverse to both the direction of current flow and magnetic field develops. Assuming that the sample is isothermal, eq. (4.89) gives

$$\mathcal{E}_y = -\rho_0 \mu_H J_x B_z \tag{4.93}$$

or

$$\frac{\mathcal{E}_y}{J_x B_z} \equiv R_H = -\rho_0 \mu_H. \tag{4.94}$$

The *Hall coefficient*, R_H, is easily determined from an experiment and using theoretical expressions for ρ_0 and μ_H is readily related to physical parameters as

$$R_H = \frac{r_H}{n(-q)}, \tag{4.95}$$

where the Hall factor r_H varies between 1 and 2 depending on the dominant scattering mechanism (recall eq. (4.62b) and Table 3.2). According to eq. (4.95), by measuring R_H, one can estimate the carrier density. Since the sign of R_H changes for p-type semiconductors, the sample may also be typed.

The derivation of eq. (4.94) was based on the assumption that the sample was isothermal. From an experimental perspective, it may prove easier to ensure that no heat flows across the sample. By setting $J_{Qy} = 0$ in eq. (4.70b), we find

$$0 = \pi_{yx} J_x + \pi_{yxz} J_x B_z - \kappa_{yj} \partial_j T_L - \kappa_{yjz} \partial_j T_L B_z.$$

Because $[\pi]$ and $[\kappa]$ are diagonal, and assuming the temperature can be maintained constant in the $\hat{x}$-direction, we find

$$\frac{\partial T_L}{\partial y} = \frac{\pi_{yxz}}{\kappa_0} J_x B_z = -\frac{\pi_1}{\kappa_0} J_x B_z. \tag{4.96}$$

This temperature gradient will produce an electric field by the Seebeck effect. When the field due to the thermoelectric effect is added to the Hall field, eq. (4.93), we find that

$$\mathcal{E}_y = \left[-\rho_0 \mu_{\mathrm{H}} - \frac{\alpha_0 \pi_1}{\kappa_0} \right] J_x B_z, \tag{4.97}$$

so

$$R_{\mathrm{H}}^{Q} = \left[-\rho_0 \mu_{\mathrm{H}} - \frac{\alpha_0 \pi_1}{\kappa_0} \right] \tag{4.98}$$

is the Hall coefficient under zero heat flow (adiabatic) conditions in the $\hat{y}$-direction. In an experiment, it may be difficult to know whether isothermal or adiabatic conditions apply, we don't know whether to apply eqs. (4.95) or (4.98). This uncertainty clouds most measurements.

Table 4.1 summarizes the coupled current phenomena that occur in a transverse magnetic field. The table shows that the measured result depends on whether the experiment was conducted under isothermal or adiabatic conditions. Thermomagnetic effects can also affect the accuracy of Hall effect measurements. If, for example, $\partial T_{\mathrm{L}}/\partial x \neq 0$ (perhaps due to heating caused by the Peltier effect associated with J_x) then an $\mathcal{E}_y$ component will be developed by the Nernst effect. The measured voltage is the sum of the Hall voltage and the Nernst voltage

$$V_{\mathrm{M}} = V_{\mathrm{H}} + V_{\mathrm{N}}$$

where $V_{\mathrm{H}} = -\mathcal{E}_y d$ with $\mathcal{E}_y$ from (4.93) and $V_{\mathrm{N}} = -\mathcal{E}_y d$ with $\mathcal{E}_y$ from (4.92). Notice that if both **B** and **J** are reversed, V_{H} doesn't change sign but V_{N} does (assuming that the measurement is done quickly so that $\partial T_{\mathrm{L}}/\partial x$ does not change). By averaging the two measurements, therefore, V_N can be eliminated. The Righi–Leduc and Ettingshausen effects also influence the measured voltage similarly. The influence of the Righi–Leduc effect can be eliminated by reversing **B** and **J** and averaging the measured voltages, but it is impossible to eliminate the influence of the Ettingshausen effect.

4.7 Measurement of carrier concentration and mobility

One of the most common uses of low-field transport theory is in characterizing semiconductor layers to determine the carrier concentration and mobility. Knowledge of transport theory and careful attention to detail is necessary to obtain accurate results, but the basic concepts are straightforward. For example, consider the semiconductor bar illustrated in Fig. 4.4. By forcing a current from contact A to B while applying a z-directed magnetic field, we measure the Hall voltage, V_{CE}. From eq. (4.94), we have

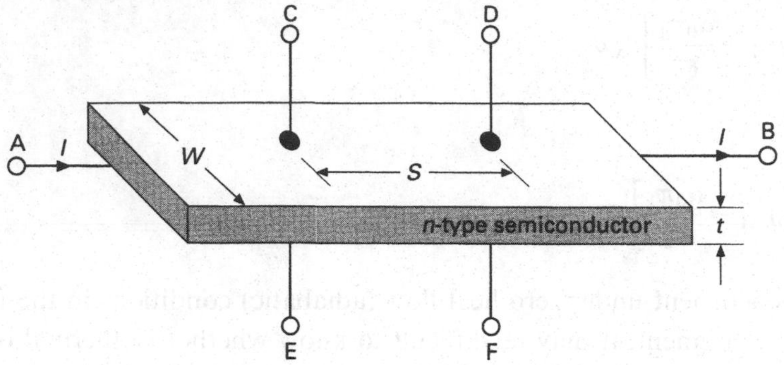

Fig. 4.4 Hall bar geometry for performing Hall effect measurements as well as resistivity measurements.

$$\rho_0 \mu_H = \frac{W}{B_z} \frac{V_{CE}}{I}, \tag{4.99a}$$

which can be solved for the carrier density to find

$$n = r_H \left[\frac{B_z I}{q W V_{CE}} \right]. \tag{4.99b}$$

This Hall effect measurement gives the carrier density, if the Hall factor is known.

To determine the carrier mobility, we need to measure the semiconductor resistivity. Using the geometry (Fig. 4.4), we force a current from contact A to B and measure the voltage from contact C to D. From this measurement, we obtain the resistivity as

$$\rho_0 = \frac{Wt}{S} \frac{V_{CD}}{I}. \tag{4.100}$$

Since the resistivity depends on both the carrier density and mobility, we can use eq. (4.99a) to solve for the mobility as

$$\mu = \frac{1}{r_H} \left[\frac{S}{t B_z} \frac{V_{CE}}{V_{CD}} \right]. \tag{4.101}$$

Equations (4.99b) and (4.101) show that to determine the drift mobility and the carrier concentration, the Hall factor must be known. If the scattering mechanisms are understood, it may be possible to estimate the Hall factor, but in most cases, the precise determination is not easy. It is common practice, therefore, to assume that the Hall factor is 1.0 when reporting results. Since the Hall factor is expected to vary from about 1.0 to 2.0, there is a corresponding uncertainty in the measured carrier density and mobility. When performing measurements, care must be taken to avoid errors from thermal gradients, misaligned Hall contacts, etc. As discussed in Section 4.6, some of these effects can be eliminated by reversing the direction of current flow and magnetic fields, remeasuring the voltage, and averaging the results. Look [4.6] discusses the uncertainties that can affect Hall effect and resistivity measurements as well as experimental techniques and data analysis.

4.7.1 Hall effect measurements in a general, 2D geometry

The semiconductor bar geometry is not always the most convenient for performing measurements. Typically, we wish to characterize transport in a thin semiconductor layer where the current flow is constrained to flow in a plane (e.g. a thin n-type layer on a p-type layer or on a semi-insulating substrate). Figure 4.5 shows a general geometry with the four contacts placed along the perimeter (the precise location of these contacts may or may not be known). Assuming that the current flow is in the x–y plane, eqs. (4.80) give

$$\mathcal{E}_x = \rho_0 J_x + (\rho_0 \mu_H B_z) J_y \tag{4.102a}$$

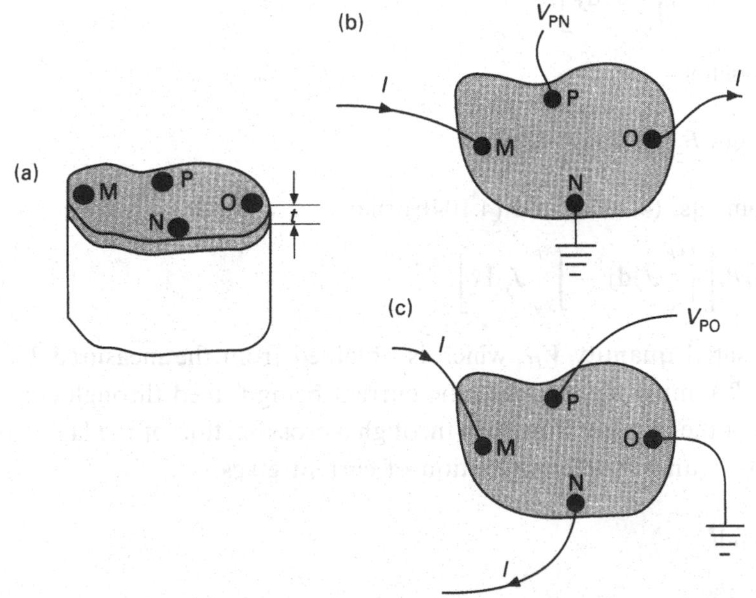

Fig. 4.5 A general two-dimensional geometry for performing Hall effect and resistivity measurements to characterize a semiconductor layer. (a) The structure consisting of a thin layer on a semi-insulating or opposite conductivity type layer. (This geometry is known as a van der Pauw pattern.) (b) A top view showing how the Hall effect is measured in the presence of a B-field directed in the z-direction. (c) A top view showing how the resistance, $R_{MN,OP}$ is measured to deduce the sheet resistance of the film. The surface of the film lies in the x–y plane with the z direction normal to the film.

and

$$\mathcal{E}_y = -(\rho_0 \mu_H B_z) J_x + \rho_0 J_y. \tag{4.102b}$$

In the presence of a z-directed B-field, the voltage measured between contacts P and N is

$$V_{PN}(B_z) = -\int_N^P \boldsymbol{\mathcal{E}} \cdot \mathrm{d}l = -\int_N^P \mathcal{E}_x \mathrm{d}x + \mathcal{E}_y \mathrm{d}y, \tag{4.103}$$

which can be written as

$$V_{PN}(B_z) = -\left[\rho_0 \int_{x_N}^{x_P} J_x \mathrm{d}x + (\rho_0 \mu_\mathrm{H} B_z) \int_{x_N}^{x_P} J_y \mathrm{d}x - (\rho_0 \mu_\mathrm{H} B_z) \int_{y_N}^{y_P} J_x \mathrm{d}y + \rho_0 \int_{y_N}^{y_P} J_y \mathrm{d}y\right]. \tag{4.104a}$$

Similarly,

$$V_{PN}(-B_z) = -\left[\rho_0 \int_{x_N}^{x_P} J_x \mathrm{d}x - (\rho_0 \mu_\mathrm{H} B_z) \int_{x_N}^{x_P} J_y \mathrm{d}x + (\rho_0 \mu_\mathrm{H} B_z) \int_{y_N}^{y_P} J_x \mathrm{d}y + \rho_0 \int_{y_N}^{y_P} J_y \mathrm{d}y\right]. \tag{4.104b}$$

Now if we define

$$V_\mathrm{H} \equiv \frac{1}{2}[V_\mathrm{PN}(+B_z) - V_\mathrm{PN}(-B_z)], \tag{4.105}$$

we find from eqs. (4.104a) and (4.104b) that

$$V_\mathrm{H} = \rho_0 \mu_\mathrm{H} B_z \left[\int_{y_N}^{y_P} J_x \mathrm{d}y - \int_{x_N}^{x_P} J_y \mathrm{d}x\right]. \tag{4.106}$$

The measured quantity V_H, which is obtained from the measured $V_\mathrm{PN}(+B_z)$ and $V_\mathrm{PN}(-B_z)$, must be related to the current being forced through contacts M and O. All of the current must pass through a cross-section of the layer connecting contacts N and P, so conservation of current gives,

$$I = t \int_\mathrm{N}^\mathrm{P} \mathbf{J} \cdot \hat{n} \mathrm{d}l, \tag{4.107}$$

where $\hat{n}$ is the normal to the path $\mathbf{d}\boldsymbol{l}$. Using

$\hat{n}\mathrm{d}l = \mathrm{d}l \times \hat{z} = \mathrm{d}y\hat{x} - \mathrm{d}x\hat{y}$

we find

$$I = t \int_\mathrm{N}^\mathrm{P} J_y \mathrm{d}x - J_x \mathrm{d}y, \tag{4.108}$$

which can be used in eq. (4.106) to obtain

$$V_{\mathrm{H}} = \left(\frac{r_{\mathrm{H}}}{qnt}\right) B_z I. \tag{4.109}$$

From eq. (4.09) we observe that the measurement determines carrier concentration per square cm, $n_{\mathrm{S}} = nt$ (within the uncertainty of the Hall factor). More generally,

$$V_{\mathrm{H}} = \left(\frac{r_{\mathrm{H}}}{qn_{\mathrm{S}}}\right) B_z I \tag{4.110}$$

where

$$n_{\mathrm{S}} = \int_0^t n(z)\mathrm{d}z. \tag{4.111}$$

We conclude that a Hall effect measurement using a general two-dimensional geometry, with four contacts placed on the boundary as illustrated in Fig. 4.5, gives the carrier concentration per cm^2 in the film. The location of the contacts does not need to be precise (but they should be small and on the boundary). There is, however, the unavoidable uncertainty in the final result associated with the Hall factor. Just as for the simple Hall bar geometry, to determine the mobility in the layer, we need to perform a resistivity measurement on the same sample.

4.7.2 Resistivity measurements in a general, 2D geometry (van der Pauw technique)

The geometry for resistivity measurements is illustrated in Fig. 4.5c, which shows that we inject a current, I, from contacts M and P and measure a voltage between contacts P and O. We define a resistance by

$$R_{\mathrm{MN,OP}} = \frac{V_{\mathrm{PO}}}{I}, \tag{4.112}$$

which describes a measurement in which the current flows from M to N and the potential drop is from P to O. Our task now is to relate this measured voltage to the resistivity of the semiconductor.

It is easier to perform the calculation in the infinite half-plane geometry shown in Fig. 4.6 rather than in the general geometry of Fig. 4.5c. After completing the calculation, we'll show that the result doesn't change when we make the geometry more general. Because we consider the semiconductor film to be a 2D sheet, the current injected at contact M spreads radially into the film with the magnitude of the radial current density being

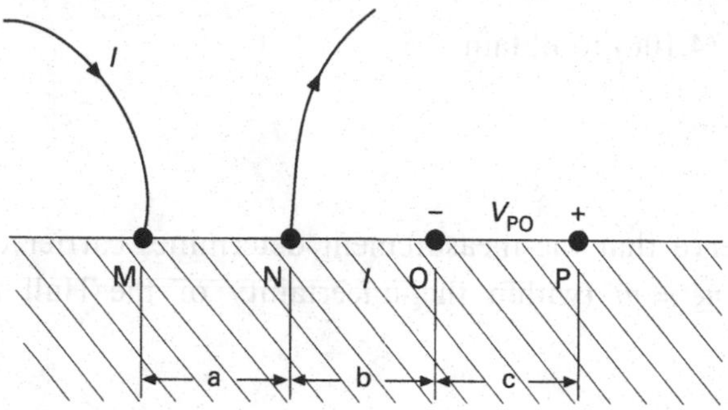

Fig. 4.6 A simplified geometry, an infinite half-plane with the four contacts at the boundary, used for establish the resistivity measurement relations for the more general geometry displayed in Fig. 4.5c.

$$J_r = \frac{I}{\pi r t}, \tag{4.113}$$

where r is the radial distance from the contact M.

If we assume that the current flow is entirely by drift in a radial electric field, then

$$\mathcal{E}_r = \frac{I\rho_0}{\pi r t}. \tag{4.114}$$

Using eq. (4.114) we can find the potential difference between any radial distance, r, and a reference location, r_0, as

$$V(r) - V(r_0) = -\frac{I\rho_0}{\pi t}\ln\left(\frac{r}{r_0}\right). \tag{4.115}$$

Now it is a simple matter to evaluate the potential at contact P,

$$V(\mathrm{P}) = -\frac{I\rho_0}{\pi t}\ln\left(\frac{a+b+c}{r_0}\right),$$

at contact O,

$$V(\mathrm{O}) = -\frac{I\rho_0}{\pi t}\ln\left(\frac{a+b}{r_0}\right),$$

and then to evaluate the potential difference as

$$V_{\mathrm{PO}} = -\frac{I\rho_0}{\pi t}\ln\left(\frac{a+b+c}{a+b}\right). \tag{4.116a}$$

So far, we have evaluated the measured potential between contacts P and O due to current flowing in at contact M. There is another contribution, with opposite sign, due to current flowing out at contact N. This contribution is

$$V'_{\mathrm{PO}} = +\frac{I\rho_0}{\pi t}\ln\left(\frac{b+c}{b}\right) \tag{4.116b}$$

The total measured potential is the difference between eqs. (4.116a) and (4.116b). From this potential difference, we evaluate the 'resistance' from eq. (4.112) as

$$R_{\mathrm{MN,OP}} = \frac{\rho_0}{\pi t}\ln\left(\frac{(a+b)(b+c)}{b(a+b+c)}\right) \tag{4.117}$$

Another 'resistance' can be measured by injecting the current into contact N and extracting it from contact O while measuring the potential between contacts M and P. Using similar arguments, we find

$$R_{\mathrm{NO,PM}} = \frac{\rho_0}{\pi t}\ln\left(\frac{(a+b)(b+c)}{ac}\right). \tag{4.118}$$

Equations (4.117) and (4.118) relate the measured 'resistances' to the resistivity of the semiconductor films in terms of the spacing of the contacts, *a*, *b*, *c*, and *d*. For some regular arrangements of the contacts, simple expressions are obtained (see homework problem 4.21). Without even knowing the location of the contacts, however, eqs. (4.117) and (4.118) give

$$\mathrm{e}^{-\frac{\pi}{R_{\mathrm{S}}}R_{\mathrm{MN,OP}}} + \mathrm{e}^{-\frac{\pi}{R_{\mathrm{S}}}R_{\mathrm{NO,PM}}} = 1, \tag{4.119}$$

where

$$R_{\mathrm{S}} = \frac{\rho_0}{t} \tag{4.120}$$

is the sheet resistance of the film. By measuring $R_{\mathrm{MN,OP}}$ and $R_{\mathrm{NO,PM}}$, eq. (4.119) can be solved either numerically or graphically to determine the sheet resistance of the film.

It may appear that the simple, infinite half-plane geometry used for these calculations has no relevance to the general geometry shown on Fig. 4.6a. Conformal mapping techniques, however, can map most geometries onto the infinite half-plane. van der Pauw [4.5] showed that if a shape meets the following conditions, it can be mapped onto an infinite half-plane:

1. $\nabla \cdot \mathbf{J} = 0$
2. $\nabla \times \mathbf{J} = 0$
3. the region is simply-connected (i.e. it contains no holes)
4. the region is homogeneous, isotropic, and of uniform thickness
5. the contacts are located on the perimeter and are point contacts.

Since these conditions are often well-approximated in practice, the van der Pauw method is widely-used in practice.

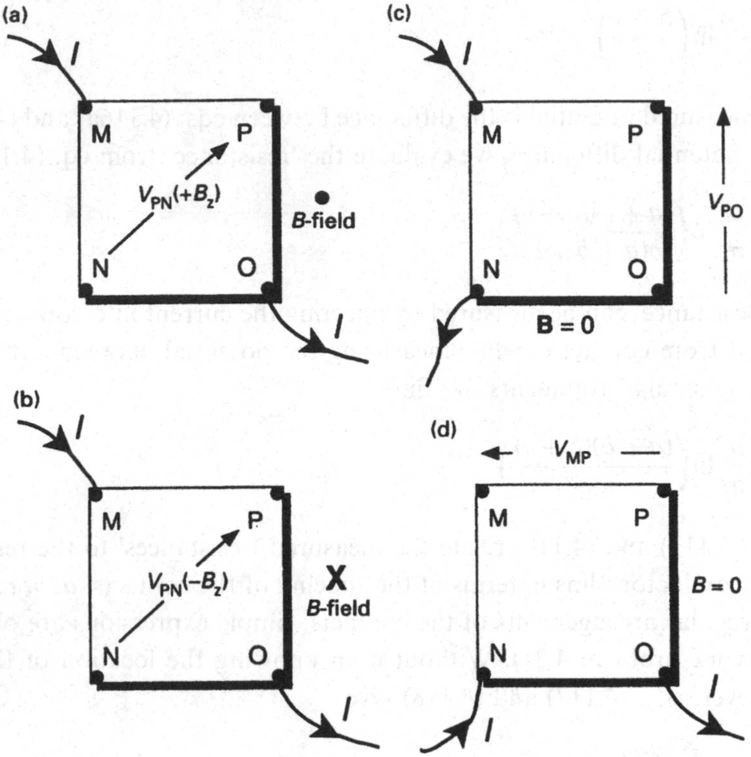

Fig. 4.7 Illustration of carrier concentration and mobility measurements using a van der Pauw pattern. (a) measurement of the Hall voltage for a $+z$-directed B-field, (b) measurement of the Hall voltage for a $-z$-directed B-field, (c) measurement of resistance, $R_{\text{MN,OP}}$, for resistivity determination, and (d) measurement of resistance, $R_{\text{NO,PM}}$.

Figure 4.7 summarizes the measurement procedure. First, a z-directed B-field is applied, a current injected into one contact and extracted from another, and a voltage measured across the other two contacts. Next, the direction of the B-field is reversed and the voltage measured. We then obtain a Hall voltage from

$$V_{\text{H}} = \frac{1}{2}[V_{\text{PN}}(+B_z) - V_{\text{PN}}(-B_z)] \tag{4.121}$$

from which we determine the carrier density per cm^2 from

$$V_{\text{H}} = \frac{r_{\text{H}}}{qn_{\text{S}}} B_z I. \tag{4.122}$$

Next, we set $B = 0$, inject a current and measure a voltage to determine $R_{\text{MN,OP}}$ and repeat the procedure to measure $R_{\text{NO,PM}}$. From these two measurements and eq. (4.119), we determine the sheet resistance of the film. Finally, from the measured Hall voltage and extracted sheet resistance, we determine the Hall mobility,

$$\mu_H = r_H \mu = \frac{V_H}{IR_S B_z}. \tag{4.123}$$

4.8 Low-field mobility of electrons in bulk Si and GaAs

The theory we've developed in this and preceding chapters generally does a good job of describing the low-field electron transport in common semiconductors. The most general theoretical approach relies on a numerical solution to the BTE and can accurately evaluate the electron mobility and other transport coefficients for a variety of semiconductors [4.5]. Hole transport is a difficult theoretical problem because of the degenerate light and heavy hole bands with their warped constant energy surfaces.

An extensive survey of the measured low-field transport properties of Si and GaAs can be found in the references [4.7–4.12]. In this section, we focus on the measured electron mobility and examine its temperature and doping dependence. Phonon scattering controls the mobility in pure semiconductors, while for doped samples, both ionized impurity and phonon scattering are important. Before we examine the measured data, we use the relaxation time approximation to establish the general features of phonon and impurity scattering.

4.8.1 Low-field mobility due to ionized impurity and phonon scattering

Because ionized impurity scattering is elastic, the relaxation time approximation is valid, and we can describe τ_f, the relaxation of $f(\mathbf{p})$ to $f_0(\mathbf{p})$ for small perturbations, by the characteristic time, τ_f. For elastic scattering, the time τ_f is precisely the momentum relaxation time. Equation (2.40) for the momentum relaxation time due to ionized impurity scattering can be expressed as

$$\tau_f(p) = \tau_m(p) = \tau_0(p)\left(\frac{E(p)}{k_B T_L}\right)^{3/2}, \tag{4.124}$$

where

$$\tau_0(p) = \frac{16\sqrt{2m^*}\pi\kappa_S^2\varepsilon_0^2}{N_1 q^4}(k_B T_L)^{3/2}\cdot\left[\ln(1+\gamma^2) - \frac{\gamma^2}{1+\gamma^2}\right]^{-1}. \tag{4.125}$$

Because the dependence of τ_0 on energy is weak, (4.124) is approximately in power law form with a characteristic exponent of $s = 3/2$. From (4.37b) and (4.37c) we find the mobility due to ionized impurity scattering as

$$\mu_{II} = \frac{q\langle\langle\tau\rangle\rangle}{m^*} = \frac{q\tau_0(\hat{E})}{m^*}\frac{\Gamma(4)}{\Gamma(5/2)}, \tag{4.126}$$

where the energy $\hat{E}$ is $(s+3/2)k_B T_L = 3k_B T_L$, which maximizes the integrand (see homework problem 4.2). The parameter, γ, which is responsible for the energy-dependence of τ_0, is evaluated at $\hat{E} = 3k_B T_L$ to obtain

$$\boxed{\gamma_{BH} = \frac{2L_D}{\hbar}\sqrt{6m^* k_B T_L}}, \tag{4.127}$$

and the mobility due to ionized impurity scattering is obtained from (4.126) and (4.127) as

$$\boxed{\mu_{BH} = \frac{128\sqrt{2\pi}\kappa_S^2\varepsilon_0^2(k_B T_L)^{3/2}}{q^3\sqrt{m^*}N_I[\ln(1+\gamma_{BH}^2) - \gamma_{BH}^2/(1+\gamma_{BH}^2)]}}. \tag{4.128}$$

As expected, eq. (4.128) shows that the mobility is approximately inversely proportional to the ionized impurity concentration. We also observe that mobility increases with temperature at $T_L^{3/2}$, which occurs because the faster moving electrons are deflected less. Observation of this temperature dependence in an experiment is often taken as the 'signature' indicating that the measured mobility is dominated by ionized impurity scattering. In the absence of screening, the Conwell–Weisskopf expression for τ_m should be used to evaluate the mobility. The result (see homework problem 4.17) should be used when the screening length is greater than half the average impurity separation.

Acoustic phonon scattering is elastic and isotropic, so the relaxation time approximation is valid and $\tau_f = \tau = \tau_m$. According to eq. (2.81) we find

$$\tau_f(p) = \tau_0\left[\frac{E(p)}{k_B T_L}\right]^{-1/2}, \tag{4.129}$$

where

$$\tau_0 = \frac{2\pi\hbar^4 c_l}{D_A^2}(2m^* k_B T_L)^{-3/2}. \tag{4.130}$$

The mobility due to ADP scattering is evaluated from

$$\mu_{AP} = \frac{q\langle\langle\tau\rangle\rangle}{m^*} = \frac{q\tau_0}{m^*}\frac{\Gamma(2)}{\Gamma(5/2)}, \tag{4.131}$$

to find

$$\boxed{\mu_{AP} = \frac{2\sqrt{2\pi} q \hbar^4 c_l}{3 D_A^2 (m^*)^{5/2} (k_B T_L)^{3/2}}} \ . \tag{4.132}$$

Equation (4.132) shows that acoustic phonon scattering by the deformation potential produces a mobility that falls with temperature as $T_L^{-3/2}$. We saw earlier that ionized impurity scattering leads to a $T_L^{+3/2}$ temperature dependence. At low temperatures, few phonons are present, so ionized impurity scattering dominates and $\mu \sim T_L^{3/2}$. At high temperatures, however, phonon scattering dominates and $\mu \sim T_L^{-3/2}$. The resulting temperature-dependent mobility, sketched in Fig. 4.8 roughly describes $\mu(T_L)$ for many semiconductors. With this background, we are ready to examine the measured low-field mobilities of silicon and gallium arsenide.

4.8.2 Low-field mobility of electrons in silicon

Measured low-field mobilities of electrons in pure (by which we mean not intentionally doped) and doped silicon are displayed in Fig. 4.9. The measured mobility varies with temperature much as expected from Fig. 4.8. For temperatures below 80 K, acoustic phonon scattering dominates, and the expected $T_L^{-3/2}$ dependence is observed. For higher temperatures, however, intervalley scattering becomes important and causes the mobility to decrease more rapidly with tem-

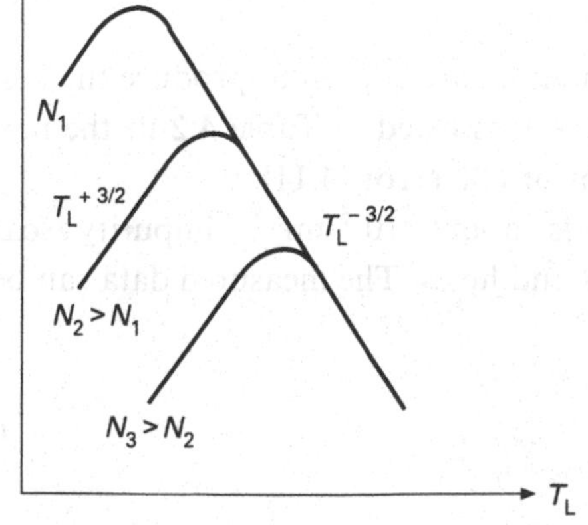

Fig. 4.8 Mobility versus lattice temperature for ionized impurity and acoustic phonon scattering.

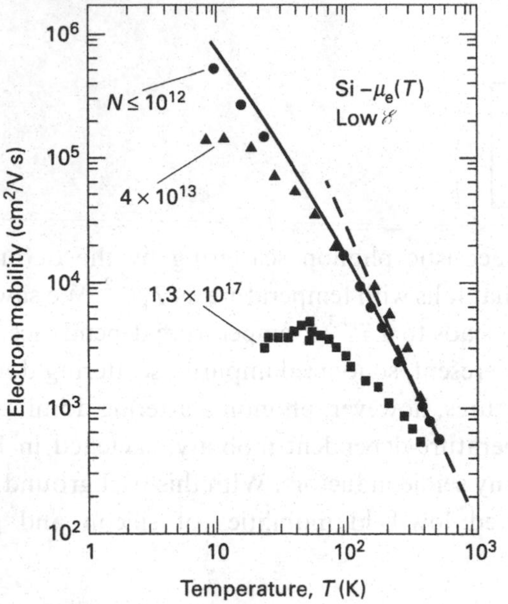

Fig. 4.9 Low-field mobility of electrons in silicon as a function of temperature. (From Jacoboni, C., Canali, C., Ottaviani, G. and Alberigi Quaranta, A. A review of some charge transport properties of silicon. *Solid-State Electronics*, **20**, 77–89, 1977. Reproduced with permission from Pergamon Press.)

perature. Near room temperature, intravalley ADP scattering and intervalley scattering are about equally frequent for electrons.

The observed temperature dependence of the low-field mobility in pure silicon can be fitted by an expression of the form [4.9]

$$\mu_0 = A/T_L^p, \tag{4.133}$$

where the constant, A, and the exponent, p, are chosen to produce the best fit to the measured results. The parameter values listed in Table 4.2 fit the measured data from 77–430 K with a maximum of 6% error [4.11].

When the doping density exceeds about $10^{16}\,\text{cm}^{-3}$, impurity scattering becomes important for both electrons and holes. The measured data can be fitted by [4.12]

$$\mu = \mu_{\text{min}} + \frac{\mu_{\text{max}} - \mu_{\text{min}}}{1 + (N/N_{\text{ref}})^\alpha}, \tag{4.134}$$

where N is the total ionized impurity concentration and the other parameters are listed in Table 4.2.

Table 4.2. *Best fitting parameters for the temperature and doping dependence of the low-field electron and hole mobilities in silicon. The parameters are those in eqs. (4.133) and (4.134) and are taken from [4.11, 4.12].*

Parameter	Electrons	Holes
A	$1.19 \times 10^6\ \mathrm{cm^2\,K^2/V\,s}$	$1.30 \times 10^8\ \mathrm{cm^2\,K^{2.2}/V\,s}$
p	2.0	2.2
μ_{max}	$1360\ \mathrm{cm^2/V\,s}$	495.0
μ_{min}	$92\ \mathrm{cm^2/V\,s}$	47.7
N_{ref}	$1.3 \times 10^{17}\ \mathrm{cm^{-3}}$	6.3×10^{16}
a	0.91	0.76

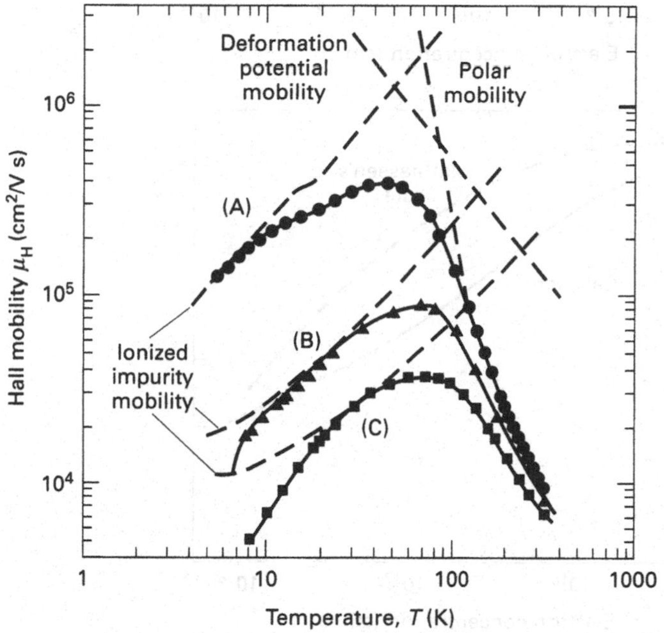

Fig. 4.10 Low-field Hall mobility of electrons in GaAs as a function of temperature. (From Stillman, G. E., Wolfe, C. M. and Dimmock, J. O. *Journal of the Physics and Chemistry of Solids*, **31**, 1199, 1970.) Sample A had an ionized impurity concentration of $7 \times 10^{13}\ \mathrm{cm^{-3}}$, for samples B and C the concentrations were higher. (Reproduced with permission from American Institute of Physics.)

4.8.3 Low-field mobility of electrons in GaAs

The low-field mobility versus temperature of electrons in GaAs is displayed in Fig. 4.10. For electrons in GaAs, the mobility is controlled by polar optical phonon scattering, but at very low temperatures, the mobility is limited by residual impurities. As the dashed lines show, the low-temperature mobility is

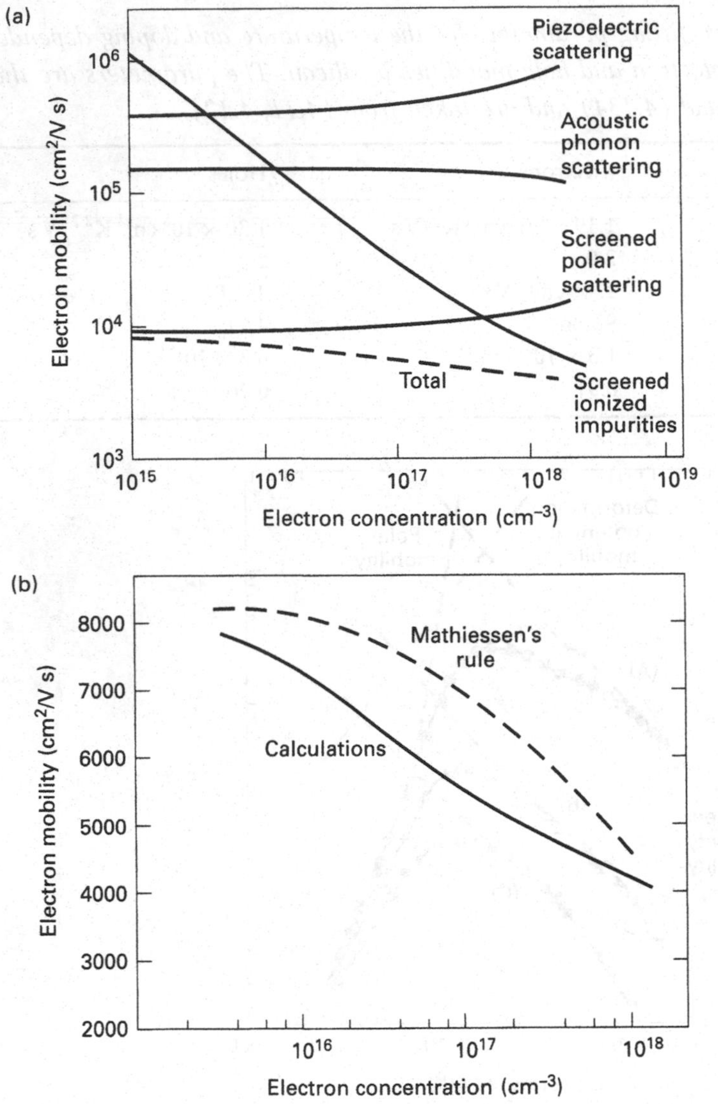

Fig. 4.11 (a) Theoretical room temperature component and total mobilities of electrons in GaAs as a function of electron concentration. (b) Comparison of the room temperature mobility of electrons in GaAs with the estimate based on Mathiessen's rule. (From Walukiewicz, W., Lagowski, L., Jastrzebski, L., Lichtensteiger, M. and Gatos, H. C. Electron mobility and free-carrier absorption in GaAs: determination of the compensation ratio, *Journal of Applied Physics,* **50**, 899–908, 1979. Reproduced with permission from American Institute of Physics.)

well-described by the Brooks–Herring theory, and the high temperature mobility can be accounted for by polar optical phonon scattering.

Figure 4.11a shows how the individual mobility components due to the various scattering processes vary with the electron concentration. For lightly doped samples, polar optical phonon scattering controls the mobility, and at heavier doping it is controlled by ionized impurity scattering. Figure 4.11b shows the total electron mobility due to all of the scattering mechanisms and compares it with the total mobility as estimated from the individual component mobilities according to Mathiessen's rule, eq. (4.46). The figure clearly shows that Mathiessen's rule must be viewed simply as a rough estimate for the mobility.

4.9 Low-field mobility of 2D carriers in Si and GaAs

In silicon and III-V field-effect transistors, the channel often consists of carriers confined in a potential well. Figure 4.12 shows a silicon metal-oxide semiconductor field-effect transistor (MOSFET) and an AlGaAs/GaAs modulation-doped field-effect transistor (MODFET). (See [8.1–8.3] for a discussion of these devices.) In the MOSFET, electrons are confined in a potential well at the oxide–silicon interface. In the MODFET, a potential well exists at the AlGaAs–GaAs heterojunction. In both cases, the depth of the potential well is controlled by the voltage on the gate. The carriers within these potential wells display quantum confinement effects and can be regarded as two-dimensional carriers. The mobility is still given by $\mu_n = q\langle\langle\tau\rangle\rangle/m^*$, but $\langle\langle\tau\rangle\rangle$ is obtained from

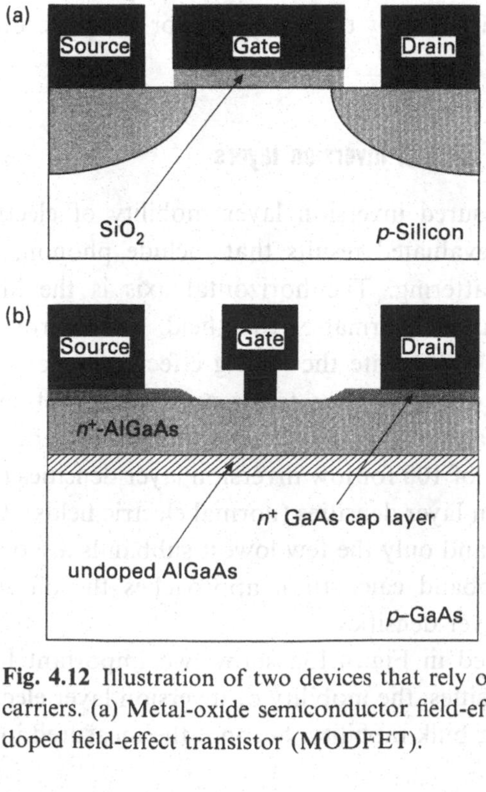

Fig. 4.12 Illustration of two devices that rely on the transport of confined, two-dimensional carriers. (a) Metal-oxide semiconductor field-effect transistor (MOSFET). (b) Modulation-doped field-effect transistor (MODFET).

the scattering rate of 2D carriers as discussed in Chapter 2, Section 2.11. Because of the different elastic properties of silicon and silicon dioxide, the phonon dispersion curve may be modified, and stresses at the oxide–silicon interface may change the phonon deformation potentials. The first effect is thought to be small, but the second leads to a lower mobility for inversion layer electrons. Because of the similar material properties of GaAs and AlGaAs, these effects are small in AlGaAs–GaAs heterostructures.

In addition to changes in the scattering rate due to the two-dimensional nature of the confined carriers, additional scattering mechanisms also occur for transport along an interface. Carriers may scatter from charges in the oxide or in the AlGaAs as well as from charges in the depletion layers of the silicon and GaAs bulk regions. The oxide charge may be a contaminant, such as sodium, or a fixed charge at the oxide–silicon interface. For the MODFET, the charges are likely to be dopants in the AlGaAs. An undoped spacer layer of AlGaAs is frequently inserted at the AlGaAs–GaAs interface in order to spatially separate the dopants from the carriers and thereby reduce the charged impurity scattering rate. Carriers may also scatter from roughness at the oxide–silicon or AlGaAs–GaAs interface. For Si, oxidation tends to produce a rough interface, so this scattering component is substantial, but AlGaAs–GaAs epitaxy produces very smooth interfaces so that surface roughness scattering is not as important. Finally, because of the high carrier densities in the inversion layer, it is important to include the suppression of scattering by the $(1-f)$ factor which accounts for final state filling [recall eq. (2.3a)].

4.9.1 Low-field transport of electrons in SiO_2–Si inversion layers

Figure 4.13a compare the measured inversion layer mobility of electrons in silicon with some numerically evaluated results that include phonon, ionized impurity, and oxide charge scattering. The horizontal axis is the inversion layer density, which increases as the normal electric field, which provides the confining potential, increases. To illustrate the strong effects of inter-subband scattering, two calculations are displayed; one treats only the lowest subband, and the second includes all subbands up to $10k_BT$ above the lowest (the number of subbands can be on the order of 100 for low inversion layer densities (normal electric fields). For high inversion layer densities (normal electric fields), the subband energy splittings increase, and only the few lowest subbands are occupied. This explains why the one subband calculation approaches the all subband calculation for high inversion layer densities.

The measured results displayed in Fig. 4.13a show two important features. First, at low inversion layer densities, the mobility of inversion layer electrons is about 65% of the corresponding bulk mobility. Second, the measured inversion

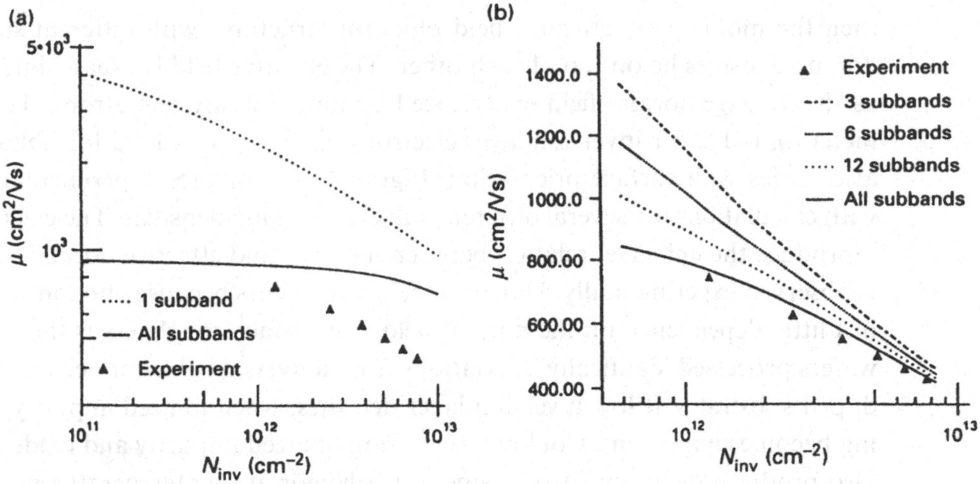

Fig. 4.13 The calculated inversion layer mobility versus inversion layer density for electrons in silicon. The lines are calculated results and the points are the measured mobility for lightly doped silicon. (a) The calculated phonon-limited mobility. (b) The calculated mobility including both phonon and surface roughness scattering. (From Jungemann et al. [4.13].) (Reproduced with permission of Pergamon Press.)

layer mobility shows a strong decrease at high inversion layer densities. Recall that surface roughness scattering was not included in these calculations, so the lower mobility cannot be due to that. The all-subband calculation matches the measured results at low inversion layer densities if the acoustic deformation potential is increased from its value in the bulk. The increase is thought to be due to interfacial stress at the oxide–silicon interface and is, therefore, a parameter that varies as the oxidation procedures vary. The calculations do not, however, show the strong decrease in mobility at high inversion layer densities that the measured results display.

Figure 4.13b shows that when surface roughness scattering is included in the calculation, the observed decrease in mobility at high inversion layer densities can be reproduced. As discussed in Chapter 2, Section 2.12, surface roughness scattering depends on the magnitude and spatial correlation length of the roughness and is, therefore, dependent on the oxidation conditions used to grow the SiO_2.

In spite of the fact that the inversion layer mobility is sensitive to processing specifics (processing conditions may affect surface roughness and stress at the oxide–silicon interface), a type of universal behavior has been observed. When the measured mobility is plotted as a function of the effective normal electric field,

$$\mathcal{E}_{\text{eff}} = \left(\frac{q}{\kappa_S \varepsilon_0}\right)[N_{\text{depl}} + \eta N_{\text{inv}}], \tag{4.135}$$

then the mobility vs. effective field plots for structures with different substrate doping densities lie on top of each other. The effective field is usually interpreted as the average normal field experienced by inversion layer electrons. The parameter, η, is 1/2 for inversion layer electrons on (100) Si and 1/3 for holes, but it also varies with surface orientation. Figure 4.14 compares experimental results with calculations for several different substrate doping densities. The calculations reproduce the universal relation between mobility and effective normal field that is observed experimentally. Universal behavior occurs because phonon scattering has little dependence on the normal field and surface roughness is the same for wafers processed identically. Deviations from universal behavior occur in heavily doped substrates at low inversion layer densities, when ionized impurity scattering becomes important. Coulomb scattering (ionized impurity and oxide charge) also produces deviations from universal behavior at low temperatures, and low normal fields.

Finally, we examine the temperature dependence of the inversion layer mobility in Fig. 4.15. The mobility increases as the temperature is lowered because of the decreasing importance of phonon scattering. The low-temperature mobility also displays a steeper decrease with normal electric field in comparison to the room temperature mobility. This occurs because at room temperature, phonon scattering, which has a weak dependence on normal field, is substantial, while at low temperatures, surface roughness scattering, which has a strong dependence on normal field, dominates. At very low temperatures and low normal field, ionized impurity scattering causes the mobility to drop.

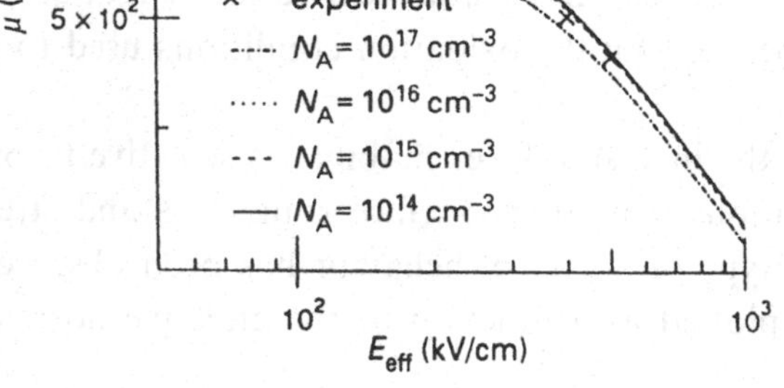

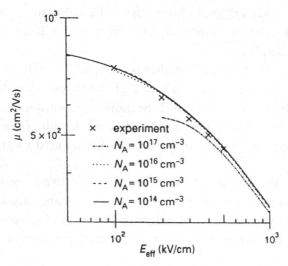

Fig. 4.14 The calculated inversion layer mobility versus normal electric field for four different substrate dopings. Experimental results are also displayed. (From Jungemann et al. [4.13].) (Reproduced with permission of Pergamon Press.)

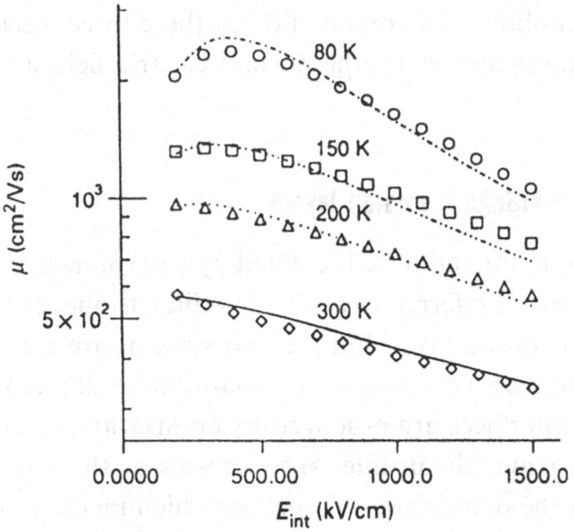

Fig. 4.15 The calculated inversion layer mobility versus normal electric field for four different temperatures. Experimental results are also displayed. (From Jungemann et al. [4.13].) (Reproduced with permission of Pergamon Press.)

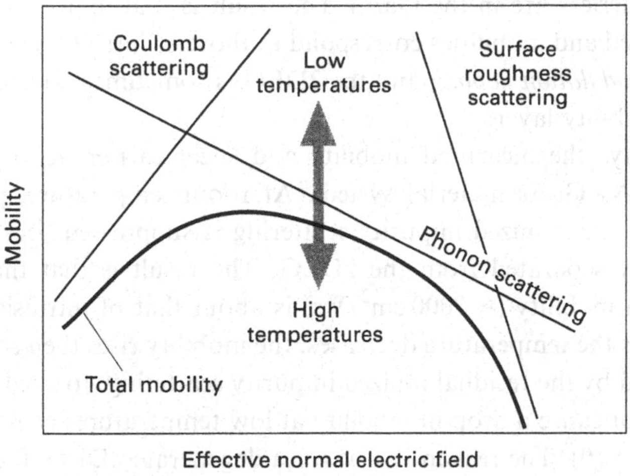

Fig. 4.16 Illustration of the normal field (or inversion layer density) dependence of the inversion layer mobility. (From Takagi et al. [4.14].) (© 1994 IEEE)

Figure 4.16 summarizes the conceptual picture of inversion layer transport. Coulomb scattering (ionized impurities in the depletion region and charges in the oxide) dominates at low normal fields (where the inversion layer densities are too small to screen the charges) and at low temperatures (where phonon scattering is reduced). Surface roughness scattering has a strong dependence on the normal field, because the strong fields confine carriers close to the interface where the scattering occurs. Phonon scattering has the strongest temperature dependence.

The overall inversion layer mobility is a combination of these three mechanisms, and their relative contributions are set by the normal electric field and by the temperature.

4.9.2 Transport of electrons in GaAs-AlGaAs inversion layers

Carriers can also be confined in potential wells created by heterojunctions. In III-V systems, the carriers are usually referred to as the two-dimensional electron gas (2DEG) rather than as an inversion layer, but the two systems are conceptually very similar. There are, however, very important quantitative differences. For III-V material systems, the interfaces are produced by epitaxy and can be much smoother than that of the oxide–silicon interface. Stresses at the interface are also smaller, so increases in the deformation potential, which increases the scattering rate, are minimal. The result is that the mobility of the 2DEG can be much greater than the corresponding bulk mobility, rather than much less as in the SiO_2/Si material system. In a bulk semiconductor dopants are necessary to produce carriers, but in an *n*-AlGaAs/*i*-GaAs system, the dopants are in the AlGaAs while the carriers are in the GaAs. The result is that ionized impurity scattering is minimized and mobilities correspond to those of intrinsic GaAs. The effect is known as *modulation doping*, and the 2DEG is sometimes referred to as the high electron mobility layer.

Figure 4.17 displays the measured mobility and sheet carrier density for a 2DEG in the AlGaAs–GaAs material system. At room temperature, phonon scattering dominates, but ionized impurity scattering is suppressed because the dopants are spatially separated from the 2DEG. The result is that the room temperature electron mobility ($\approx 7000\,\mathrm{cm^2/V\,s}$) is about that of intrinsic GaAs (recall Fig. 4.11a). As the temperature decreases, the mobility rises then saturates at a value determined by the residual ionized impurity scattering. Ionized impurity scattering does not cause a drop in mobility at low temperatures as it does in the bulk (recall Fig. 4.10). The reason is that for a degenerate 2DEG, the impurities are strongly screened and the average thermal velocity in the degenerate limit varies little with temperature. The result is that ionized impurity scattering in the degenerate limit shows little variation with temperature. Note that very high mobilities can be achieved at very low temperatures. Mobilities well above $10^6\,\mathrm{cm^2/V\,s}$ can be achieved.

Figure 4.18 shows how the undoped AlGaAs spacer layer thickness affects the mobility and sheet carrier density. Thin spacer layers reduce the separation between the 2DEG and the ionized impurities and, therefore, lower the mobility. With the proper spacer layer thickness, this sample displays mobility of over $2 \times 10^6\,\mathrm{cm^2/V\,s}$. It may come as a surprise that the mobility falls when the spacer layer is very large. This occurs because the inversion layer density is reduced,

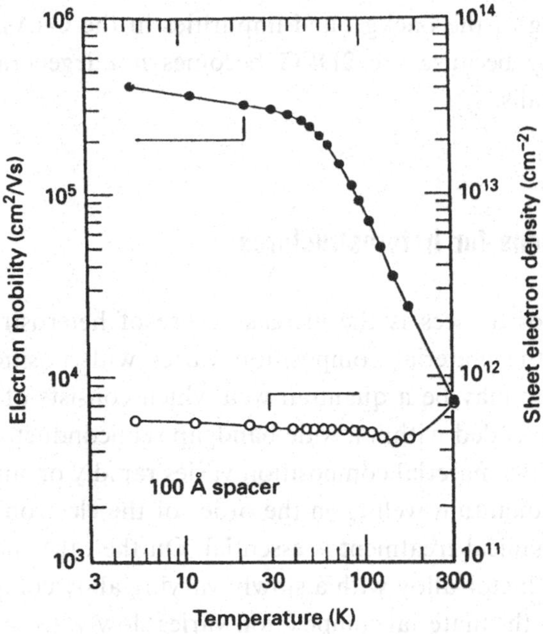

Fig. 4.17 The measured temperature dependence of the 2DEG mobility and density in an *n*-AlGaAs–GaAs heterostructure. (From Mori, Y., Nakamura, F. and Wantabe, N. *Journal of Applied Physics*, **60**, 334–337, 1986.) (Reproduced with permission of American Institute of Physics.)

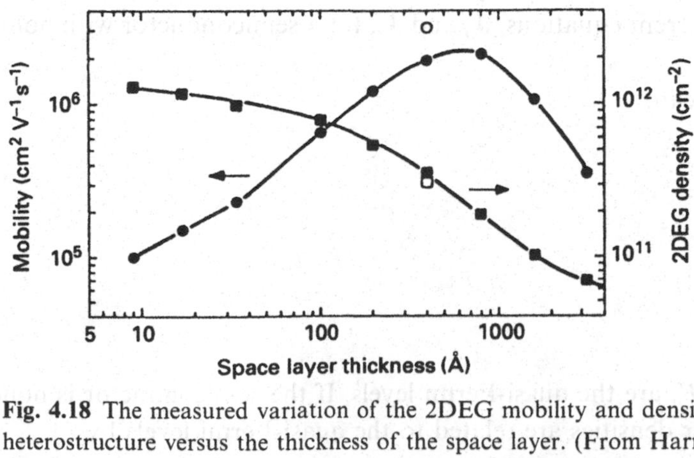

Fig. 4.18 The measured variation of the 2DEG mobility and density in an *n*-AlGaAs/GaAs heterostructure versus the thickness of the space layer. (From Harris, Foxon, Barnham, Lacklison, Hewett, and White. *Journal of Applied Physics*, **61**, 1219–1221, 1987.) (Reproduced with permission of American Institute of Physics.)

which reduces the screening of the background impurities in the GaAs. It also lowers the thermal velocity because the 2DEG becomes nondegenerate. The result is that the mobility falls.

4.10 Low-field transport equations for heterostructures

One trend in semiconductor devices is the increasing use of heterostructures, semiconductors in which the material composition varies with position. For example, the heterostructure may be a quantum well which consists of a small bandgap semiconductor imbedded within a wide bandgap semiconductor (recall Section 1.3.2). In this case, the material composition varies rapidly on an atomic scale and the width of the quantum well is on the order of the electron's wavelength, so a quantum mechanical treatment is essential. On the other hand, the material may be a semiconductor alloy with a slowly varying alloy composition (recall Section 1.3.3). When the material composition varies slowly, we should be able to describe transport semiclassically. That is our objective in this section, to derive semiclassical transport equations for semiconductor heterostructures when the fields and quasi-fields are low. When the fields are high, the transport equations can be derived from moments of the BTE, as discussed in Chapter 5, Section 5.8. The reader may wish to review heterostructure fundamentals in Chapter 1, Section 1.3.3.

4.10.1 Phenomenological derivation of the heterostructure drift-diffusion equation

If we restrict ourselves to near-equilibrium conditions, it is easy to derive the hole and electron current equations, $\mathbf{J}_p$ and $\mathbf{J}_n$, for a semiconductor with nonuniform composition.

To begin, recall:

$$J_{pz} = p\mu_p \frac{\mathrm{d}F_p}{\mathrm{d}z} \tag{4.136a}$$

and

$$J_{nz} = n\mu_n \frac{dF_n}{dz}, \tag{4.136b}$$

where F_p and F_n are the quasi-Fermi levels. If the semiconductor is nondegenerate, the carrier densities are related to the quasi-Fermi levels by

$$p = N_v(z)\mathrm{e}^{(E_{v0} - F_p)/kT} \tag{4.137a}$$

and

$$n = N_c(z)e^{(F_n - E_{C0})/kT}. \tag{4.137b}$$

From eqs. (4.137a) and (4.137b), we obtain

$$F_p = E_{V0} - kT\log\left(\frac{p}{N_v}\right) \tag{4.138a}$$

and

$$F_n = E_{C0} + kT\log\left(\frac{n}{N_c}\right). \tag{4.138b}$$

Consequently

$$\frac{dF_p}{dz} = \frac{dE_{V0}}{dz} - kT\left[\frac{1}{p}\frac{dp}{dz} - \frac{1}{N_c}\frac{dN_v}{dz}\right] \tag{4.139a}$$

and

$$\frac{dF_n}{dz} = \frac{dE_{C0}}{dz} + kT\left[\frac{1}{n}\frac{dn}{dz} - \frac{1}{N_c}\frac{dN_c}{dz}\right]. \tag{4.139b}$$

Using the above in (4.136a) and (4.136b) we get

$$J_{pz} = p\mu_p\left[\frac{dE_{V0}}{dz} + \frac{kT}{N_v}\frac{dN_v}{dz}\right] - kT\mu_p\frac{dp}{dz} \tag{4.140a}$$

$$J_{nz} = n\mu_n\left[\frac{dE_{C0}}{dz} - \frac{kT}{N_c}\frac{dN_c}{dz}\right] + kT\mu_n\frac{dn}{dz}. \tag{4.140b}$$

These expressions look like drift–diffusion equations – especially if we use the Einstein relation

$$\frac{D_p}{\mu_p} = \frac{D_n}{\mu_n} = \frac{kT}{q}. \tag{4.141}$$

To complete the derivation, we must express dE_{C0}/dz and dE_{V0}/dz in terms of material parameters and the electrostatic potential, $V(z)$. From our band model for a nonuniform semiconductor, eqs. (1.46a) and (1.46b), and eqs. (4.140a) and (4.140b), we find

$$J_{pz} = -pq\mu_p\left[\frac{d}{dz}\left\{V(z) + \frac{\chi(z)}{q} + \frac{E_G(z)}{q}\right\} - \frac{kT}{q}\frac{1}{N_v}\frac{dN_v}{dz}\right] - qD_p\frac{dp}{dz} \tag{4.142a}$$

$$J_{nz} = -nq\mu_n\left[\frac{d}{dz}\left\{V(z) + \frac{\chi(z)}{q}\right\} + \frac{kT}{q}\frac{1}{N_c}\frac{dN_c}{dz}\right] + qD_n\frac{dn}{dz}. \tag{4.142b}$$

In examining these results, we see that compositional variations introduce additional terms into the drift-diffusion equations. The drift current (the part in { } brackets) includes both electric and quasi-electric fields. We also see a conventional diffusion current, but there is another term involving the gradient of the effective densities of states. It is not clear whether to write this *density of states effect* as a drift current (because it is proportional to the carrier concentration) or as a diffusion current (because it involves a gradient). Physically, this term arises because carriers tend to diffuse in the direction of increasing density of states, because there are more states available for the random walk. In practice, however, the biggest effect is drift in the quasi-electric fields, and the density of states effect is rather small.

4.10.2 Derivation of the heterostructure drift-diffusion equation from the BTE

A transport equation for heterostructure can also be derived by solving the Boltzmann equation in the relaxation time approximation. We begin with

$$\upsilon_z \frac{\partial f}{\partial z} + \left\{ -q\mathcal{E}_z + \frac{d\chi}{dz} \right\} \frac{\partial f}{\partial p_z} - \frac{p^2}{2} \frac{\partial}{\partial z} \left(\frac{1}{m^*} \right) \frac{\partial f}{\partial p_z} = -\frac{f_A}{\tau_f}, \tag{4.143}$$

which is the steady-state Boltzmann equation, eq. (3.22), but with the equation of motion, dp_z/dt as given by eq. (1.95b) for a semiconductor with a slowly varying effective mass. We then solve for f_A by approximating f on the left-hand side by f_S to find,

$$f_A = q\tau_f[\mathcal{E}_z - d(\chi/q)/dz] \frac{\partial f_S}{\partial p_z} - \tau_f \upsilon_z \frac{\partial f_S}{\partial z} + \frac{p^2}{2} \frac{d(1/m^*)}{dz} \frac{\partial f_S}{\partial p_z}. \tag{4.144}$$

To evaluate eq. (4.144), assume near-equilibrium conditions,

$$f_S = e^{-(E_{C0} - p^2/2m^* - F_n)/k_B T_L}. \tag{4.145}$$

When eq. (4.145) is inserted in eq. (4.144), we find

$$f_A = \frac{\tau_f}{k_B T_L} \upsilon_z f_S \frac{dF_n}{dz}, \tag{4.146}$$

which, after inserting in eq. (4.10) gives

$$J_{nz} = n\mu_n \frac{dF_n}{dz}. \tag{4.147}$$

Equation (4.146) was where we started when deriving the near-equilibrium transport equation for heterostructures, eq. (4.136b).

4.11 Summary

The low-field transport theory developed in this chapter was based on the relaxation time approximation. When acoustic phonon or ionized impurity scattering dominates, the RTA is valid and the theory applies, but for semiconductors such as GaAs, polar optical phonon scattering often dominates, so the RTA cannot be used. For such cases, numerical solutions to the BTE can be used (see the article by Rode [4.7] for a discussion of one such technique). Nevertheless, we can view the final result of our simple theory, the coupled current equations, as phenomenological descriptions of low-field current flow in any semiconductor. The specific expressions we developed for the various transport tensors, however, are not valid generally; they must be measured or computed from a more accurate theory. The phenomenological couple current equations are readily applied to experimental situations They describe the thermoelectric, thermomagnetic, and galvanomagnetic effects that occur in semiconductors. Our discussion of the temperature-dependence of the carrier mobility in silicon and gallium arsenide showed that the temperature dependence of the bulk mobility could be understood in terms of the increasing strength of phonon scattering and the decreasing influence of ionized impurity scattering as the temperature increases. For the inversion layer mobility, surface roughness scattering also has to be considered as well as the effects of carrier degeneracy and stresses at the interface. Finally, we showed how low-field transport equations can be generalized to treat transport in heterostructures, when the material composition varies slowly.

References and further reading

Thorough treatments of low-field transport theory can be found in the texts listed below. The treatment presented in this chapter closely follows that of Smith, Janek, and Adler.

4.1 Smith, A. C., Janek, J. and Adler, R. *Electronic Conduction In Solids*. McGraw-Hill, New York, 1965.

4.2 Wolfe, C. M., Holonyak, N. and Stillman, G. E. *Physical Properties of Semiconductors*. Prentice-Hall, Englewood Cliffs, NJ, 1989.

4.3 Conwell, E. Transport: the Boltzmann equation. In *Handbook on Semiconductors*, Vol. I. North Holland Publishing Co., 1982, p. 513.

The influence of quantum effects under very strong magnetic fields on transport is discussed by Seeger. He also discusses the influence of the valence band structure on the transport of holes.

4.4 Seeger, K., *Semiconductor Physics*, 3rd edn. Springer-Verlag, New York, 1985.

Experimental techniques for measuring carrier densities and mobilities are discussed in

4.5 van der Pauw, L. J. A method of measuring specific resistivity and Hall effect of discs of arbitrary shape. *Phillips Research Reports* **13**, 1–9, 1958.

4.6 Look, D. C. *Electrical Characterization of GaAs Materials and Devices.* John Wiley and Sons, New York, 1989.

Low-field transport characteristics of silicon and GaAs are discussed in

4.7 Rode, D. L. Low-field electron transport. In *Semiconductors and Semimetals.* ed. by Willardson, R. K. and Beer, A. C. Vol. 10. Academic Press, New York, 1975, pp. 1–89.

4.8 Walukiewicz, W., Lagowski, L., Jastrzebski, L., Lichtensteiger, M. and Gatos, H. C. Electron mobility and free-carrier absorption in GaAs: determination of the compensation ratio. *Journal of Applied Physics,* **50**, 899–908, 1979.

4.9 Jacoboni, C., Canali, C., Ottaviani, G. and Alberigi Quaranta, A. A review of some charge transport properties of silicon. *Solid-State Electronics,* **20**, 77–89, 1977.

4.10 Wiley, J. D. Mobility of holes in III-V compounds. In *Semiconductors and Semimetals,* ed. by Williardson, R. K. and Beer, A. C. Academic Press, New York, 1975.

4.11 Ali Omar, M. and Reggiani, L. Drift and diffusion of charge carriers in silicon and their empirical relation to the electric field. *Solid-State Electronics,* **30**, 693–7, 1987.

4.12 Baccarani, G. and Ostoja, P. Electron mobility empirically related to the phosphorus concentration in silicon. *Solid-State Electronics,* **18**, 579–80, 1975.

The mobility of two-dimensional electrons in the Si/SiO_2 system is discussed in

4.13 Jungemann, Chr., Emunds, A. and Engl, W. L. Simulation of linear and nonlinear electron transport in homogeneous silicon inversion layers. *Solid-State Electronics,* **36**, 1529–40, 1993.

4.14 Takagi, S., Toriumi, A., Iwase, M. and Tango, H. On the universality of inversion layer mobility in Si MOSFET's. *IEEE Transactions on Electron Devices,* **41**, 2357–68, 1994.

Problems

4.1 Verify the results eqs. (4.37d), (4.37e), and (4.37f) for the tensors, B_{ij}, p_{ij}, and K_{ij}.

4.2 The integrand in the numerator of the expression for $\langle\langle\tau\rangle\rangle$, eq. (4.36), is of the form

$$\int_0^\infty e^{-p^2/2m^* k_B T_L} \tau_0 \left(\frac{p^2}{2m^* k_B T_L}\right)^s p^4 dp.$$

Assume that τ_0 is a constant, change variables to integrate over energy, and show that the integrand peaks at

$$\hat{E} = (s + 3/2) k_B T_L.$$

For ionized impurity scattering, τ_0 varies with energy, but the variation is slow so it can be moved outside the integral and evaluated at $E = \hat{E}$ when computing $\langle\langle\tau\rangle\rangle$. This technique is used to evaluate the Brooks–Herring mobility, eq. (4.128).

4.3 For a thermal average electron, show that $|\nabla_r f_S| \gg |\nabla_r f_A|$. Assume n-type silicon with $\mathcal{E} = 100$ V/cm.

4.4 Show that eq. (4.30a) reduces to the conventional drift–diffusion equation under isothermal conditions.

4.5 Obtain an expression relating the carrier density in silicon to the quasi-Fermi level. Approach the problem as follows:

(a) Show that eq. (4.38) can be written as

$$E(p') = E_{C0} + \frac{p'^2}{2m_0},$$

where

$$p'^2 = p_x'^2 + p_y'^2 + p_z'^2,$$

and

$$p_x' = \sqrt{\frac{m_0}{m_{xx}}}(p_x - p_{xo}), \text{ etc.}$$

This change of variables effectively stretches the coordinate axes, so that $E(p)$ appears to be spherical.

(b) Since the coordinate axes have been stretched, the number of states in a volume, $dp_x'dp_y'dp_z'$ differs from the number in a volume, $dp_x dp_y dp_z$. Show that these volumes are related by

$$dp_x'dp_y'dp_z' = \sqrt{\frac{m_0^3}{m_{xx}m_{yy}m_{zz}}}dp_x dp_y dp_z.$$

(c) Show that the carrier density due to a single ellipsoid,

$$n_1 = \frac{1}{\Omega}\sum_{\mathbf{p}} f_S,$$

(where the sum is over the states near the center of the ellipsoid) evaluates to

$$n_1 = \frac{(2\pi k_B T_L)^{3/2}}{4\pi^3\hbar^3}\sqrt{m_l^* m_t^{*2}}e^{(F_n - E_{C0})/k_B T_L}.$$

(d) Add the contributions for the six ellipsoids, obtain an expression for n, and show that the result is

$$n = N_C e^{(F_n - E_{C0})/k_B T_L},$$

where

$$N_C = \frac{(2\pi k_B T_L m_d^*)^{3/2}}{4\pi^3\hbar^3}$$

and

$$m_d^* \equiv 6^{2/3}(m_l^* m_t^{*2})^{1/3}.$$

Like m_c^*, the *density of states effective mass*, m_d^*, is defined so that the result has the same form as the result for spherical bands.

4.6 Compute σ_{ij} for silicon and verify the results obtained in Section 4.3. Approach the problem as follows:

(a) Use the change of variables described in problem 4.5 and show that when the sum in eq. (4.31) is converted to an integral over the primed coordinate system, the conductivity for a single ellipsoid is

$$\sigma_{\alpha\alpha} = \frac{(2\pi k_B T_L)^{3/2}}{4\pi^3\hbar^3}\sqrt{m_{xx}^* m_{yy}^* m_{zz}^*}e^{(F_n - E_{C0})/k_B T_L}\frac{q^2\langle\langle\tau\rangle\rangle}{m_{\alpha\alpha}}.$$

(b) Show that when the contributions for all six equivalent minima are added, the result is eq. (4.40).

4.7 Use arguments similar to those in Section 4.3 to obtain an expression for the conductivity effective mass in germanium.

4.8 Show that $\langle\langle\tau^2\rangle\rangle = \tau_o^2\Gamma(2s+5/2)/\Gamma(5/2)$ and verify eq. (4.62b) for the Hall factor, r_H.

4.9 Obtain an expression for σ_{ij} without assuming that the semiconductor is non-degenerate. Assume spherical, parabolic energy bands and power law scattering.

4.10 Verify the result stated in Table 4.1 for the Ettingshausen effect, and provide a simple, physical explanation for the effect.

4.11 Evaluate the isothermal longitudinal magneto-resistance for a cubic crystal. Let **B** and **J** be $\hat{x}$-directed and find the apparent resistivity, $\mathcal{E}_x/J_x = \rho_A$. Find an expression for $\Delta\rho = \rho_A(B_x) - \rho_A(0)$.

4.12 For a cubic crystal aligned along the coordinate axes, evaluate the isothermal planar Hall effect. Let **J** be $\hat{x}$-directed and find $\mathcal{E}_y$ when **B** lies in the x–y plane oriented at 45° with respect to the x-axis. Show that $R_p \equiv \mathcal{E}_y/J_xB^2 = \rho_{\alpha\beta\alpha\beta}$.

4.13 Assume that the electric current is zero and that $\partial T_L/\partial y = 0$. Show that the isothermal (in the $\hat{y}$-direction) thermal conductivity κ_T is magnetic-field dependent. Assume a cubic crystal aligned along the coordinate axes and define κ_T by

$$\kappa_T \equiv -\frac{J_{Qx}}{\partial T_L/\partial x}.$$

(Assume that **B** is z-directed.)

4.14 Evaluate the adiabatic (in the $\hat{y}$-direction) electrical resistivity for cubic crystals. Assume $J_y = \partial T_L/\partial x = 0$ and $\mathbf{B} = B_z\hat{z}$ as before, but now assume $J_{Qy} = 0$ instead of $\partial T_L/\partial y = 0$. Evaluate $\rho_Q \equiv \mathcal{E}_x/J_x$. Show that $\rho_Q = \rho_0$ for $\mathbf{B} = \mathbf{0}$ but that the transverse magnetic-field dependence differs from the isothermal result.

4.15 Estimate ρ_0, π_0, and κ_0 for n-type silicon doped with 10^{14} carriers/cm^3. Assume that acoustic phonon scattering dominates and that $T_L = 300\,\text{K}$. (You may use measured data for the electron mobility.)

4.16 Assume that the transport tensors are scalars as in eq. (4.30), and obtain expressions for the corresponding parameters in the inverted equations. Show that the results are analogous to eqs. (4.72)–(4.75).

4.17 Use the Conwell–Weisskopf treatment of ionized impurity scattering and show that the mobility is

$$\boxed{\mu_{CW} = \frac{128\sqrt{2\pi}\kappa_S^2\varepsilon_0^2(k_BT_L)^{3/2}}{q^3\sqrt{m^*}N_I[\ln(1+\gamma_{CW}^2)]}}\,.$$

Explain how to determine when to use the Brooks–Herring approach and when to use the Conwell–Weisskopf approach.

4.18 Compute the electric current in the presence of a magnetic field using a simple particle approach based on the equation of motion for an average particle:

$$\frac{d\mathbf{p}}{dt} = -(q)\boldsymbol{\mathcal{E}} + (-q)v \times \mathbf{B} - \frac{\mathbf{p}}{\tau},$$

Approach the problem as follows:

(a) Write the current as

$$\mathbf{J} = (-q)nv = (-q)n\frac{\mathbf{p}}{m^*},$$

assume steady-state conditions, and show from the equation of motion that the steady-state current is

$$\mathbf{J} = \frac{nq^2\tau}{m^*}\boldsymbol{\mathcal{E}} - \frac{(-q)\tau}{m^*}\mathbf{B} \times \mathbf{J}.$$

(b) To find the current, we must solve a vector equation of the form

$$\mathbf{c} = \mathbf{a} + \mathbf{b} \times \mathbf{c},$$

which has a solution,

$$\mathbf{c} = \frac{\mathbf{a} + \mathbf{b} \times \mathbf{a} + (\mathbf{a} \cdot \mathbf{b})\mathbf{b}}{1 + b^2}.$$

Use this result to show that

$$\mathbf{J} = \frac{nq^2}{m^*}\left\langle\left\langle\frac{\tau}{1+\omega_c^2\tau^2}\right\rangle\right\rangle\boldsymbol{\mathcal{E}} - \frac{n(-q)^3}{(m^*)^2}\left\langle\left\langle\frac{\tau^2}{1+\omega_c^2\tau^2}\right\rangle\right\rangle\mathbf{B} \times \boldsymbol{\mathcal{E}}$$
$$+ \frac{nq^4}{(m^*)^3}\left\langle\left\langle\frac{\tau^3}{1+\omega_c^2\tau^2}\right\rangle\right\rangle(\boldsymbol{\mathcal{E}} \cdot \mathbf{B})\mathbf{B},$$

where

$$\omega_c = \frac{qB}{m^*}$$

is the cyclotron frequency.

(c) Demonstrate that when a magnetic field is applied parallel to the electric field, $\mathbf{J} = \sigma_0\boldsymbol{\mathcal{E}}$, which means that there is no *longitudinal magnetoresistance*. This result applies to semiconductors with a simple, spherical energy band centered at $\mathbf{k} = (0, 0, 0)$. It also applies to ellipsoidal bands if the magnetic field is directed along the major or minor axis of all of the ellipsoids.

(d) Consider the effect of a magnetic field applied transverse to the current flow by letting $\mathbf{J}$ lie in the $\hat{x}$–$\hat{y}$ plane,

$$J_x = \sigma_{xx}\mathcal{E}_x + \sigma_{xy}\mathcal{E}_y$$
$$J_y = \sigma_{xy}\mathcal{E}_x + \sigma_{yy}\mathcal{E}_y.$$

Assume that $\mathbf{B} = B\hat{z}$ and that $J_y = 0$, and show that an electric field transverse to the direction of current flow develops. Show that

$$\mathcal{E}_y = \frac{\sigma_{xy}}{\sigma_{xx}^2 + \sigma_{xy}^2}J_x \equiv \frac{r_H}{(-q)n}B_zJ_x.$$

Show also that the Hall factor for arbitrary strength magnetic fields is

$$r_{\mathrm{H}} = \frac{\left\langle\left\langle \dfrac{\tau^2}{(1+\omega_c^2\tau^2)} \right\rangle\right\rangle}{\left\langle\left\langle \dfrac{\tau}{1+\omega_c^2\tau^2} \right\rangle\right\rangle^2 + \omega_c^2 \left\langle\left\langle \dfrac{\tau^2}{1+\omega_c^2\tau^2} \right\rangle\right\rangle^2}$$

and that r_{H} approaches eq. (4.62b) for weak magnetic fields and unity for strong ones.

4.19 It is also possible to solve the BTE directly for arbitrary strength magnetic fields. The solution begins by assuming a solution of the form

$$f_{\mathrm{A}} = \frac{q\tau_{\mathrm{f}}}{k_{\mathrm{B}} T_{\mathrm{L}}} \frac{\partial f_{\mathrm{s}}}{\partial \Theta} \upsilon \cdot \mathbf{G}$$

which is similar to eq. (4.7) except that **G** is an unknown vector which must be determined.

(a) Solve the BTE and show that **G** is given by the vector equation,

$$\mathbf{G} = \boldsymbol{\mathcal{E}} + \frac{q\tau_{\mathrm{f}}}{m^*}(\mathbf{B} \times \mathbf{G}).$$

(You may find the vector identity, $\mathbf{G} \cdot (\upsilon \times \mathbf{B}) = \upsilon \cdot (\mathbf{B} \times \mathbf{G})$, useful.)

(b) Solve the vector equation to find

$$\mathbf{G} = \frac{\boldsymbol{\mathcal{E}} + \omega_c \tau_{\mathrm{f}} \left(\dfrac{\mathbf{B}}{B} \times \boldsymbol{\mathcal{E}}\right) + (\omega_c \tau_{\mathrm{f}})^2 \left(\dfrac{\mathbf{B}}{B} \cdot \boldsymbol{\mathcal{E}}\right) \dfrac{\mathbf{B}}{B}}{1 + (\omega_c \tau_{\mathrm{f}})^2}.$$

Hint: See problem 4.18b.

(c) Explain how one uses this solution to find $\sigma_{ij}(\mathbf{B})$, and show that when **B** is aligned along $\hat{z}$,

$$\sigma_{xx} = \frac{nq^2}{m^*} \left\langle\left\langle \frac{\tau_{\mathrm{f}}}{1+\omega_c^2 \tau_{\mathrm{f}}^2} \right\rangle\right\rangle.$$

4.20 Repeat the derivation of the Hall effect in a general geometry, but use eq. (4.80) and work to second order in the magnetic field, instead of first order as done in eqs. (4.102a) and (4.102b). Show that magnetoresistance has no effect on the measurement.

4.21 Consider a van der Pauw pattern with the contacts at the four corners of a square, as in Fig. 4.7. Show that the sheet resistance of the film is

$$R_{\mathrm{S}} = \left(\frac{\pi}{\ln 2}\right)\frac{V}{I},$$

where I is the current forced through two of the contacts and V is the voltage measured across the other two.

4.22 The sheet resistance of a semiconductor film is often measured by a 'four-point probe'. As shown in Fig. P4.1, it consists of four probes in a line placed in the center of a semi-infinite layer. A current, I, is forced through the outer two probes and a voltage, V is measured across the inner two. Using methods similar to those in Section 4.7.2, develop an expression which relates the measured voltage to the semiconductor resistivity.

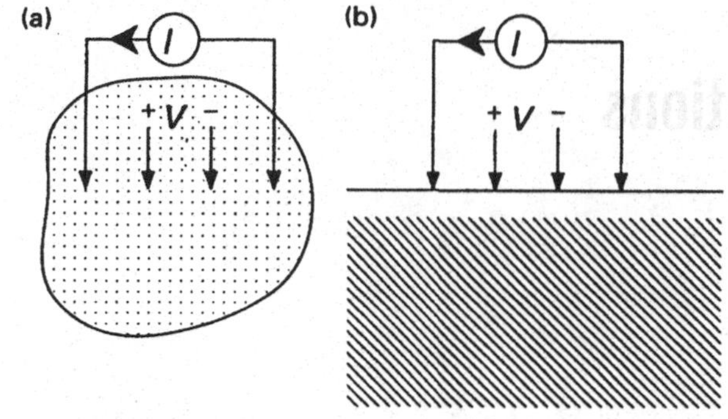

Fig. P4.1 Geometry for a four-point probe measurement of semiconductor resistivity. (a) Top view of the probes on a semi-infinite plane. (b) Cross-section showing the layer being measured and substrate beneath it.

5 Balance equations

Because it is so difficult to solve the Boltzmann transport equation (BTE) directly, simpler approaches are often adopted when analyzing, designing, and optimizing devices. The use of balance, or conservation, equations which are derived from the BTE is a common approach. Balance equations have a very clear physical interpretation. For example, the electron continuity equation,

$$\frac{\partial n}{\partial t} = -\nabla \cdot \mathbf{F}_n + G_n - R_n, \tag{5.1}$$

states that the net rate of increase of average carrier density at a specified location and time, $n(\mathbf{r}, t)$, is given by the rate per unit volume at which carriers are flowing in (the negative divergence of the electron flux, $\mathbf{F}_n$) plus the rate per unit volume of electron creation, G_n (due to optical or avalanche generation, for example) minus the rate per unit volume at which electrons disappear (by recombining with a hole or defect). Figure 5.1 illustrates this conservation law schematically. Balance equations for the average carrier momentum and energy density can also be formulated and expressed as continuity equations in the form of eq. (5.1). Such equations find wide application in device analysis. The familiar drift–diffusion equation, for example, is a simplified form of the momentum balance equation.

This chapter begins by introducing a mathematical prescription for generating balance equations directly from the BTE. Balance equations for the average

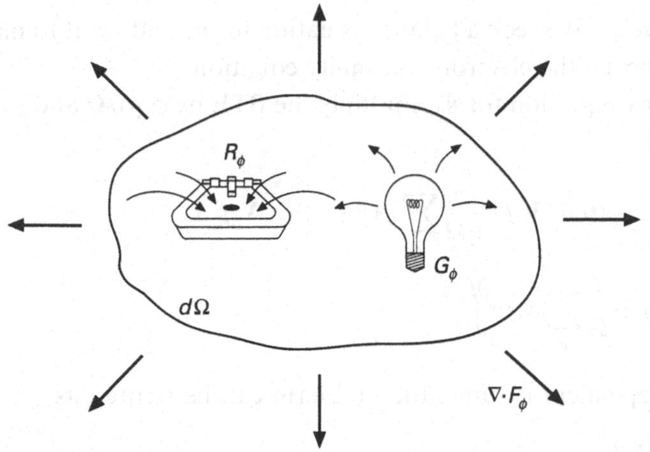

Fig. 5.1 Illustration of a balance equation.

carrier density, momentum density, and energy density are then formulated. An infinite number of such equations can be generated; they are useful when only a few are required to describe a device accurately. During the course of this discussion, we'll derive the drift–diffusion equation which continues to serve as the basis for most device analysis. An important objective is to clearly establish the limitations of the drift–diffusion equation because it is losing its validity as device dimensions shrink. In the drift–diffusion approach, the unknowns are the average carrier density and velocity. When the average carrier energy is also needed, another balance equation can be solved. Although the balance equation approach is conceptually straightforward, many different sets of balance equations can be formulated, depending on the specific approach and the simplifying assumptions. Because such equations are so widely used, it is important to clearly understand their relation to the BTE.

5.1 Derivation from the Boltzmann equation

To evaluate the total carrier momentum, we should weight the momentum of a state by the probability that the momentum state is occupied then sum over all states. Similarly, the total value of a quantity associated with the function of momentum, $\phi(\mathbf{p})$, is the weighted sum

$$n_\phi(\mathbf{r}, t) \equiv \frac{1}{\Omega}\sum_{\mathbf{p}} \phi(\mathbf{p}) f(\mathbf{r}, \mathbf{p}, t). \tag{5.2}$$

By appropriately selecting $\phi(\mathbf{p})$, n_ϕ may represent, for example, the carrier density, momentum, or the energy density. (For these examples, $\phi(\mathbf{p}) = 1$,

$\mathbf{p}$ and $E(\mathbf{p})$ respectively.) We seek a balance equation for n_ϕ and want to make the result look like eq. (5.1), the electron continuity equation.

To find the balance equation for n_ϕ, multiply the BTE by $\phi(\mathbf{p})/\Omega$ and sum over momentum, to find

$$\frac{1}{\Omega}\sum_{\mathbf{p}}\phi(\mathbf{p})\frac{\partial f}{\partial t}+\frac{1}{\Omega}\sum_{\mathbf{p}}\phi(\mathbf{p})\upsilon\cdot\nabla_r f+\frac{1}{\Omega}\sum_{\mathbf{p}}\phi(\mathbf{p})(-q)\boldsymbol{\mathcal{E}}\cdot\nabla_p f$$
$$=\frac{1}{\Omega}\sum_{\mathbf{p}}\phi(\mathbf{p})s(\mathbf{r},\mathbf{p},t)+\frac{1}{\Omega}\sum_{\mathbf{p}}\phi(\mathbf{p})\left.\frac{\partial f}{\partial t}\right|_{\text{coll}}. \tag{5.3}$$

Because $\phi(\mathbf{p})$ is independent of time, the first term can be written as

$$\frac{1}{\Omega}\sum_{\mathbf{p}}\phi(\mathbf{p})\frac{\partial f}{\partial t}=\frac{\partial n_\phi(\mathbf{r},t)}{\partial t}. \tag{5.4}$$

Similarly, for the second term, the spatial gradient can be moved outside the sum to write

$$\frac{1}{\Omega}\sum_{\mathbf{p}}\phi(\mathbf{p})\upsilon\cdot\nabla_r f=\nabla\cdot\mathbf{F}_\phi, \tag{5.5}$$

where

$$\mathbf{F}_\phi(\mathbf{r},t)\equiv\frac{1}{\Omega}\sum_{\mathbf{p}}\phi(\mathbf{p})\upsilon f \tag{5.6}$$

is the flux associated with the quantity n_ϕ. (For example, if n_ϕ is the carrier density, then F_ϕ is the carrier flux. If n_ϕ is the carrier energy, then F_ϕ is the energy flux.)

The third term in eq. (5.3) can be re-expressed as

$$(-q)\boldsymbol{\mathcal{E}}\cdot\sum_{\mathbf{p}}\phi(\mathbf{p})\nabla_p f=(-q)\boldsymbol{\mathcal{E}}\cdot\sum_{\mathbf{p}}\nabla_p(\phi f)+q\boldsymbol{\mathcal{E}}\cdot\sum_{\mathbf{p}}f\nabla_p\phi,$$

which may be simplified by noting that the first term on the right-hand side sums to zero because $f(\mathbf{r},\mathbf{p},t)$ approaches zero rapidly for large p. (Although this assumption is standard, it may have to be questioned for extremely high electric fields.) By defining a 'generation rate', G_ϕ, we can write

$$(-q)\boldsymbol{\mathcal{E}}\cdot\frac{1}{\Omega}\sum_{\mathbf{p}}\phi(\mathbf{p})\nabla_p f=-G_\phi, \tag{5.7}$$

where

$$G_\phi(\mathbf{r},t)\equiv(-q)\boldsymbol{\mathcal{E}}\cdot\frac{1}{\Omega}\sum_{\mathbf{p}}\nabla_p\phi(\mathbf{p}). \tag{5.8}$$

Because the electric field increases carrier momentum, it also increases n_ϕ, and for this reason, the term involving the electric field has been labeled as a generation rate.

In addition to the increase of n_ϕ caused by the electric field, there is also an increase (or decrease) of n_ϕ caused by the creation (or recombination) of carriers by the source (or sink) terms, $s(\mathbf{r}, \mathbf{p}, t)$. This contribution to the balance equation is

$$S_\phi(\mathbf{r}, t) \equiv \frac{1}{\Omega} \sum_{\mathbf{p}} \phi(\mathbf{p}) s(\mathbf{r}, \mathbf{p}, t). \tag{5.9}$$

Collisions destroy momentum and should, therefore, produce a 'recombination' term in the balance equation. Because collisions oppose deviations from equilibrium, we define the rate of loss of n_ϕ to be proportional to the deviation from equilibrium, $n_\phi = n_\phi^o$, or

$$R_\phi = \frac{-1}{\Omega} \sum_{\mathbf{p}} \phi(\mathbf{p}) \left.\frac{\partial f}{\partial t}\right|_{\text{coll}} \equiv \left\langle\left\langle \frac{1}{\tau_\phi} \right\rangle\right\rangle [n_\phi(r, t) - n_\phi^0(\mathbf{r}, t)], \tag{5.10}$$

where $\langle\langle 1/\tau_\phi \rangle\rangle$ is the *ensemble relaxation rate*. Equation (5.10) should not be confused with the relaxation time approximation; it simply *defines* the quantity $\langle\langle 1/\tau_\phi \rangle\rangle$; no approximation is involved.

To find an explicit expression for $\langle\langle 1/\tau_\phi \rangle\rangle$, we expand the collision term (assuming nondegenerate conditions) to find

$$\sum_{p} \phi(\mathbf{p}) \left.\frac{\partial f}{\partial t}\right|_{\text{coll}} = \sum_{\mathbf{p}} \sum_{\mathbf{p}'} \phi(\mathbf{p}) S(\mathbf{p}', \mathbf{p}) f(\mathbf{p}') - \phi(\mathbf{p}) S(\mathbf{p}, \mathbf{p}') f(\mathbf{p}),$$

which can be simplified by interchanging the dummy indices $\mathbf{p}$ and $\mathbf{p}'$ in the first term, then reversing the order of summation. The result is

$$\sum_{\mathbf{p}} \phi(\mathbf{p}) \left.\frac{\partial f}{\partial t}\right|_{\text{coll}} = \sum_{\mathbf{p}} f(\mathbf{p}) \phi(\mathbf{p}) \sum_{\mathbf{p}'} \left(\frac{\phi(\mathbf{p}')}{\phi(\mathbf{p})} - 1 \right) S(\mathbf{p}, \mathbf{p}') \equiv - \sum_{\mathbf{p}} \frac{f(\mathbf{p}) \phi(\mathbf{p})}{\tau_\phi(\mathbf{p})}, \tag{5.11}$$

where

$$\frac{1}{\tau_\phi(\mathbf{p})} \equiv \sum_{\mathbf{p}'} \left(1 - \frac{\phi(\mathbf{p}')}{\phi(\mathbf{p})} \right) S(\mathbf{p}, \mathbf{p}') \tag{5.12}$$

is the out-scattering rate associated with ϕ. This characteristic rate is just the transition rate from $\mathbf{p}$ to $\mathbf{p}'$ weighted by the fractional change in ϕ summed over all states, $\mathbf{p}'$, to which the carrier can scatter.

The characteristic rate, $1/\tau_\phi(\mathbf{p})$, describes only the out-scattering process, but changes in n_ϕ are the result of both in-scattering and out-scattering. The effect of scattering on n_ϕ is described by the ensemble relaxation rate, $\langle\langle 1/\tau_\phi \rangle\rangle$, which is found by equating eqs. (5.10) and (5.11) to obtain

$$\left\langle\left\langle \frac{1}{\tau_\phi} \right\rangle\right\rangle \equiv \frac{\frac{1}{\Omega}\sum_{\mathbf{p}} f(\mathbf{p})\phi(\mathbf{p})/\tau_\phi(\mathbf{p})}{[n_\phi(\mathbf{r},t) - n_\phi^0(\mathbf{r},t)]}. \tag{5.13}$$

Note that $1/\tau_\phi(\mathbf{p})$ depends only on the scattering physics as described by $S(\mathbf{p}, \mathbf{p}')$, but $\langle\langle 1/\tau_\phi \rangle\rangle$ depends both on the scattering physics and on how the carriers are distributed in momentum. (Under degenerate conditions, however, both the in- and out-scattering rates depend on the distribution function.) Finally, it is important to note that according to eq. (5.13), the ensemble relaxation time, $1/\langle\langle 1/\tau_\phi \rangle\rangle$ is not $\langle\langle \tau_\phi \rangle\rangle$.

Finally, after collecting the results, eqs. (5.4), (5.5), (5.7), (5.9), and (5.10), and inserting them in eq. (5.3), we obtain the desired balance equation for the quantity n_ϕ:

$$\boxed{\frac{\partial n_\phi(\mathbf{r},t)}{\partial t} = -\nabla \cdot \mathbf{F}_\phi + G_\phi - R_\phi + S_\phi}\ . \tag{5.14a}$$

Although this section has been somewhat mathematical, the final result, eq. (5.14a), has the clear physical interpretation illustrated in Fig. 5.1. Any balance equation we seek can now be obtained with the appropriate choice of $\phi(\mathbf{p})$. The mathematical prescription for generating the various terms in the balance equation is summarized below.

Density: $$n_\phi(\mathbf{r},t) \equiv \frac{1}{\Omega}\sum_{\mathbf{p}} f\phi(\mathbf{p}) \tag{5.14b}$$

Associated flux: $$\mathbf{F}_\phi(\mathbf{r},t) \equiv \frac{1}{\Omega}\sum_{\mathbf{p}} \mathbf{v} f\phi(\mathbf{p}) \tag{5.14c}$$

Field generation rate: $$G_\phi(\mathbf{r},t) \equiv (-q)\boldsymbol{\mathcal{E}} \cdot \frac{1}{\Omega}\sum_{\mathbf{p}} f\nabla_p\phi(\mathbf{p}) \tag{5.14d}$$

Scattering recombination rate: $$R_\phi(\mathbf{r},t) \equiv \left\langle\left\langle \frac{1}{\tau_\phi} \right\rangle\right\rangle [n_\phi(\mathbf{r},t) - n_\phi^0(\mathbf{r},t)] \tag{5.14e}$$

Associated out-scattering rate: $$\frac{1}{\tau_\phi(\mathbf{p})} \equiv \sum_{\mathbf{p}'}\left(1 - \frac{\phi(\mathbf{p}')}{\phi(\mathbf{p})}\right) S(\mathbf{p},\mathbf{p}') \tag{5.14f}$$

Particle gen-rec rate: $$S_\phi(\mathbf{r},t) \equiv \frac{1}{\Omega}\sum_{\mathbf{p}} \phi(\mathbf{p}) s(\mathbf{r},\mathbf{p},t) \tag{5.14g}$$

Ensemble relaxation rate: $$\left\langle\left\langle \frac{1}{\tau_\phi} \right\rangle\right\rangle \equiv \frac{\frac{1}{\Omega}\sum_{\mathbf{p}} f(\mathbf{r},\mathbf{p},t)\phi(\mathbf{p})/\tau_\phi(\mathbf{p})}{[n_\phi(\mathbf{r},t) - n_\phi^0(\mathbf{r},t)]}. \tag{5.14h}$$

5.2 Characteristic times

The ensemble relaxation rates describe the change of $n_\phi(\mathbf{r}, t)$ due to collisions. From eqs. (5.11) and (5.13) we find

$$\left.\frac{\partial n_\phi}{dt}\right|_{\text{coll}} = \frac{1}{\Omega}\sum_{\mathbf{p}} \phi(\mathbf{p})\left.\frac{\partial f}{\partial t}\right|_{\text{coll}} \equiv -\left\langle\left\langle\frac{1}{\tau_\phi}\right\rangle\right\rangle[n_\phi - n_\phi^0]. \tag{5.15}$$

The characteristic rate, $\langle\langle 1/\tau_\phi\rangle\rangle$, describes how collisions tend to drive the collective variable, n_ϕ, to its equilibrium value, n_ϕ^0. By selecting $\phi(\mathbf{p}) = p_i$, we find $n_\phi = P_i$ and

$$\left.\frac{dP_i}{dt}\right|_{\text{coll}} = -\left\langle\left\langle\frac{1}{\tau_m}\right\rangle\right\rangle P_i, \tag{5.16}$$

where $P_i = nm^* \upsilon_{di}$ is the average momentum density of the ensemble (υ_d is the average velocity of the ensemble). The quantity, $\langle\langle 1/\tau_m\rangle\rangle$, is an 'average' momentum relaxation rate of the ensemble and includes both in-scattering and out-scattering contributions. In Section 5.5, we will discuss how $\langle\langle 1/\tau_m\rangle\rangle$ relates to $\langle\langle \tau_f\rangle\rangle$, the 'average' time that appeared in the relaxation time approximation solution to the BTE.

By selecting $\phi(\mathbf{p}) = E(\mathbf{p})$, we find $n_\phi = W$, the kinetic energy density, and

$$\left.\frac{dW}{dt}\right|_{\text{coll}} = \left\langle\left\langle\frac{1}{\tau_{\text{E}}}\right\rangle\right\rangle(W - W^0), \tag{5.17}$$

where $W = nu$ (with u being the average kinetic energy per carrier) is the average kinetic energy density of the ensemble, and $\langle\langle 1/\tau_{\text{E}}\rangle\rangle$ is the ensemble energy relaxation rate. Both $\langle\langle 1/\tau_m\rangle\rangle$ and $\langle\langle \tau_{\text{E}}\rangle\rangle$ are evaluated from the prescription, eq. (5.14h).

It is important to reiterate that eqs. (5.16) and (5.17) are exact – no relaxation time approximation is assumed. But to evaluate the ensemble relaxation rates from the prescription, the distribution function must be known. To avoid solving the BTE, the ensemble relaxation rates are often assumed to be constant, or to depend only upon the average carrier energy. When this approach is taken, eqs. (5.16) and (5.17) comprise a type of relaxation time approximation.

5.2.1 The out-scattering rates

According to the prescription, eq. (5.14h), $\langle\langle 1/\tau_\phi\rangle\rangle$ appears to be an average of $1/\tau_\phi(\mathbf{p})$ over $\phi(\mathbf{p})f(\mathbf{p})$. For $\phi(\mathbf{p}) = p_z$, $1/\tau_\phi = 1/\tau_m$, where

$$\frac{1}{\tau_m(\mathbf{p})} = \sum_{\mathbf{p}'} S(\mathbf{p}, \mathbf{p}')[1 - p_z'/p_z]. \tag{5.18}$$

When $\phi(\mathbf{p}) = E(\mathbf{p})$, $1/\tau_\phi = 1/\tau_{\mathrm{E}}$ where

$$\frac{1}{\tau_{\mathrm{E}}(\mathbf{p})} \equiv \sum_{\mathbf{p}'} S(\mathbf{p}, \mathbf{p}')\left[1 - \frac{E(\mathbf{p}')}{E(\mathbf{p})}\right]. \tag{5.19}$$

Equations (5.18) and (5.19) should be recognized as the momentum and energy relaxation rates introduced in Chapter 2. As was illustrated in Fig. 2.1, these characteristic times describe the rate of loss of momentum or energy of a beam of electrons injected into a semiconductor; the probability of in-scattering is small.

Figure 5.2 illustrates the difference between the out-scattering and ensemble momentum relaxation times. As shown in Fig. 5.2a, the momentum relaxation rate due to out-scattering is a known function of momentum and is determined solely by the scattering processes. (We are assuming a non-degenerate semiconductor. Recall also that we computed $1/\tau_m(\mathbf{p})$ for several scattering processes in Chapter 2.) To determine how the total momentum of the ensemble relaxes, however, we need to know both the distribution function and the out-scattering

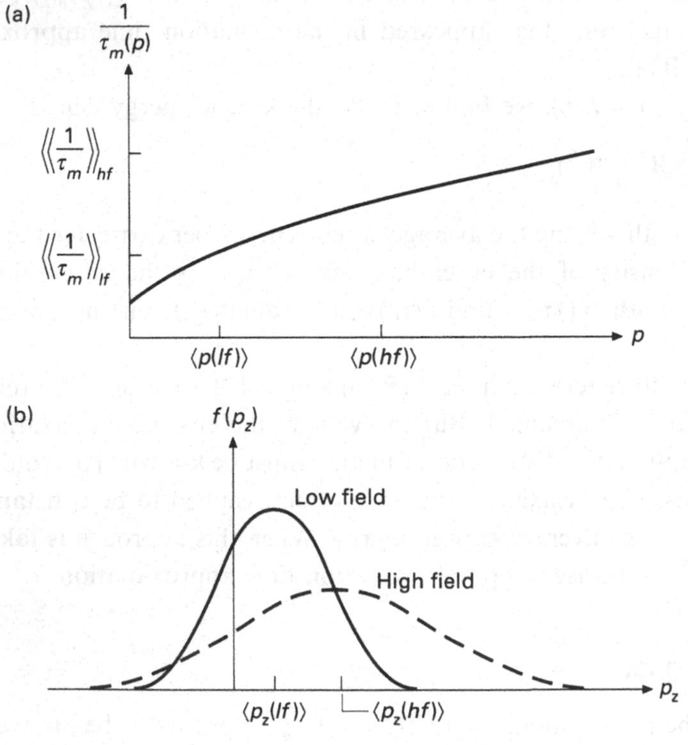

Fig. 5.2 (a) The momentum relaxation rate versus p. The ensemble relaxation rates for the two distribution functions displayed in Fig. 5.2b are also indicated. (b) The distribution functions under low and high applied fields versus the $\hat{z}$ component of **p**. The average momenta, $\langle p_z(lf)\rangle$ and $\langle p_z(hf)\rangle$ are indicated.

rate. Representative distribution functions under low and high applied fields are sketched in Fig. 5.2b. If the average momenta under low and high fields are p_{lf} and p_{hf} respectively, then we find that the ensemble relaxation rates are somewhat greater than $1/\tau_m(p_{lf})$ and $1/\tau_m(p_{hf})$ respectively. The reason is that $1\tau_m(p)$ generally increases with energy and the prescription, eq. (5.14h) weights the higher momentum states more.

5.3 The balance equations

Only a few of the balance equations have a simple physical interpretation. Fortunately, these few are sufficient for analyzing many devices. In this section, we derive the balance equations that are commonly needed for analyzing devices.

5.3.1 The carrier density balance equation

To derive the balance equation for carrier density, n, we let $\phi(\mathbf{p}) = 1$, so that $n_\phi = n$ and find from eq. (5.14c),

$$\mathbf{F}_\phi = \frac{1}{\Omega}\sum_{\mathbf{p}} v f = n v_d = \frac{\mathbf{J}_n}{(-q)}, \tag{5.20}$$

where v_d is the average electron velocity, and $\mathbf{J}_n/(-q)$ is the electron flux. The general relations, eq. (5.14), also show that both G_ϕ and R_ϕ are zero (because $\nabla_p \phi$ and $1/\tau_\phi$ are both zero). Physically, this occurs because the electric field and scattering merely rearrange carriers in momentum space; the scattering processes we consider don't generate carriers. After assembling these results and inserting them in the general balance equation, (5.14a), we find

$$\boxed{\frac{\partial n}{\partial t} = \frac{1}{q}\nabla \cdot \mathbf{J}_n + \mathcal{S}_n}, \tag{5.21}$$

the familiar electron continuity equation. The term, $\mathcal{S}_n$ is a particle generation–recombination rate and plays the role of $(G_n - R_n)$ in eq. (5.1).

5.3.2 The momentum balance equation

A balance equation for the $\hat{z}$-directed component of the momentum density results by setting $\phi(\mathbf{p}) = p_z$. From eq. (5.14b) we find

$$n_\phi \equiv \frac{1}{\Omega}\sum_{\mathbf{p}} p_z f = P_z = n m^* v_{dz}, \tag{5.22}$$

where P_z is the $\hat{z}$-component of $\mathbf{P}$, the total momentum density, and υ_{dz} is the $\hat{z}$-component of the average carrier velocity. The flux associated with n_ϕ must be a 'flux of momentum' and is obtained from eq. (5.14c) as

$$\mathbf{F}_\phi = \frac{1}{\Omega}\sum_{\mathbf{p}} \boldsymbol{\upsilon} p_z f, \tag{5.23a}$$

which is

$$F_{\phi i} = \frac{1}{\Omega}\sum_{\mathbf{p}} \upsilon_i p_z f \equiv 2W_{iz} \tag{5.23b}$$

in indicial notation. Note that the product of velocity and momentum is related to the kinetic energy.

The electric field 'generates' momentum because it accelerates carriers in the direction of the field. From eq. (5.14d) we find

$$G_\phi = (-q)n\mathcal{E}_z. \tag{5.24}$$

On the other hand, collisions with the lattice randomize momentum and lead to the loss term, R_ϕ. From eq. (5.14e) we find that momentum is lost by collisions at the rate

$$R_\phi = \left\langle\left\langle \frac{1}{\tau_m} \right\rangle\right\rangle P_z. \tag{5.25}$$

After collecting these results and inserting them in eq. (5.14a), we find the momentum balance equation as

$$\frac{\partial P_z}{\partial t} = -\frac{\partial}{\partial x_i}(2W_{iz}) + n(-q)\mathcal{E}_z - \left\langle\left\langle \frac{1}{\tau_m} \right\rangle\right\rangle P_z, \tag{5.26}$$

where we have written out the divergence of the momentum flux in indicial notation. Similar equations can be written for the other two components of $\mathbf{P}$, so in general

$$\frac{\partial P_j}{\partial t} = -\frac{\partial}{\partial x_i}(2W_{ij}) + n(-q)\mathcal{E}_j - \left\langle\left\langle \frac{1}{\tau_m} \right\rangle\right\rangle P_j. \tag{5.27a}$$

According to eq. (5.27a), the rate of increase of momentum is the rate at which momentum flows into the volume plus the rate at which momentum is generated by the field minus the rate at which collisions with the lattice destroy momentum. An explicit source term is absent in our momentum balance equation because we've assumed that $s(\mathbf{r}, \mathbf{p}, t)$ creates carriers with randomly directed momenta. In symbolic notation, the momentum balance equation becomes

$$\boxed{\frac{\partial \mathbf{P}}{\partial t} = -2\nabla \cdot \overleftrightarrow{W} + n(-q)\boldsymbol{\mathcal{E}} - \left\langle\left\langle \frac{1}{\tau_m} \right\rangle\right\rangle \mathbf{P}}\,, \tag{5.27b}$$

where

$$W_{ij} \equiv \frac{1}{2\Omega} \sum_{\mathbf{p}} \upsilon_i p_j f \tag{5.28}$$

defines the (ij)th component of the tensor $\overleftrightarrow{W}$ and

$$\nabla \cdot \overleftrightarrow{W} \cdot \hat{x}_j \equiv \frac{\partial}{\partial x_i} W_{ij} \tag{5.29}$$

defines the dot product of a tensor (note that the dot product of a tensor is a vector). The trace of this tensor has a clear physical meaning,

$$W_{ii} = \sum_i \frac{1}{\Omega} \sum_{\mathbf{p}} \frac{1}{2} m^* \upsilon_i^2 f(\mathbf{p}) = W = nu, \tag{5.30}$$

where W is the average kinetic energy density and u is the average kinetic energy per carrier.

For simple spherical, parabolic energy bands, the current density can be obtained directly from the momentum density as

$$\mathbf{J}_n = (-q)n\upsilon_d = (-q)\frac{\mathbf{P}}{m^*}. \tag{5.31}$$

It is a simple matter, therefore, to obtain an equation for the electric current density from the momentum balance equation,

$$\boxed{\frac{\partial \mathbf{J}_n}{\partial t} = \frac{-2(-q)\nabla \cdot \overleftrightarrow{W}}{m^*} + \frac{q^2 n \boldsymbol{\mathcal{E}}}{m^*} - \left\langle\left\langle \frac{1}{\tau_m} \right\rangle\right\rangle \mathbf{J}_n} \tag{5.32}$$

We'll show in Section 5.7 that with appropriate simplifications, eq. (5.32) reduces to a drift–diffusion equation.

5.3.3 The energy balance equation

When $\phi(\mathbf{p}) = E(\mathbf{p})$ is selected, a balance equation for

$$n_\phi = \frac{1}{\Omega} \sum_{\mathbf{p}} E(\mathbf{p}) f = W, \tag{5.33}$$

which is the kinetic energy density, results. For this case, the associated flux is

$$\mathbf{F}_\phi = \frac{1}{\Omega}\sum_{\mathbf{p}} \upsilon E(\mathbf{p}) f = \mathbf{F}_{\mathrm{W}}, \tag{5.34}$$

an energy flux. Energy is supplied to the carriers by the electric field. From eq. (5.14d) we find the generation rate as

$$G_\phi = (-q)\boldsymbol{\mathcal{E}} \cdot \frac{1}{\Omega}\sum_{\mathbf{p}} [\nabla_p E(\mathbf{p})] f = (-q)\boldsymbol{\mathcal{E}} \cdot \upsilon_d = \mathbf{J}_n \cdot \boldsymbol{\mathcal{E}} \tag{5.35}$$

(recall that $\nabla_p E(\mathbf{p})$ is the carrier velocity). As expected, the input power density, $\mathbf{J}_n \cdot \boldsymbol{\mathcal{E}}$, is responsible for increasing the energy density of the carriers.

Applied fields increase carrier energy, but energy is lost by collisions with the lattice. The rate of energy loss is obtained from eq. (5.14e) as

$$R_\phi = \left\langle\left\langle \frac{1}{\tau_{\mathrm{E}}} \right\rangle\right\rangle (W - W^0), \tag{5.36}$$

where W^0 is the energy density in thermal equilibrium.

Putting these results together, we obtain the energy balance equation as

$$\boxed{\frac{\partial W}{\partial t} = -\nabla \cdot \mathbf{F}_{\mathrm{W}} + \mathbf{J}_n \cdot \boldsymbol{\mathcal{E}} - \left\langle\left\langle \frac{1}{\tau_E} \right\rangle\right\rangle (W - W^0) + \mathcal{S}_{\mathrm{E}}} \; . \tag{5.37}$$

According to eq. (5.37), the rate of increase of carrier energy, $\partial W/\partial t$, is due to energy flowing in $(-\nabla \cdot \mathbf{F}_{\mathrm{W}})$, to the field accelerating carriers $(\mathbf{J}_n \cdot \boldsymbol{\mathcal{E}})$, or to the explicit generation of carriers as described by $\mathcal{S}_{\mathrm{E}}$. Energy is lost by collisions with the lattice at the rate, $\langle\langle 1/\tau_{\mathrm{E}} \rangle\rangle (W - W^0)$.

5.3.4 Discussion

Using the prescription developed in Section 5.1, we've derived three balance equations. We began by letting $\phi(\mathbf{p}) = 1$ and derived a balance equation for carrier density, which contained two unknowns, carrier density and current density. We then sought a balance equation for current density (or momentum), but the resulting balance equation contained the tensor $\overleftrightarrow{W}$ whose trace we interpreted as the kinetic energy density. The balance equation for carrier energy introduced the energy flux. As discussed in homework problem 5.6, a balance equation for the energy flux can be written as

$$\boxed{\frac{\partial \mathbf{F}_{\mathrm{W}}}{\partial t} = -\nabla \cdot \overleftrightarrow{X} + \frac{(-q)W}{m^*}\boldsymbol{\mathcal{E}} + 2(-q)\frac{\boldsymbol{\mathcal{E}} \cdot \overleftrightarrow{W}}{m^*} - \left\langle\left\langle \frac{1}{\tau_{FW}} \right\rangle\right\rangle \mathbf{F}_{\mathrm{W}}} \; , \tag{5.38}$$

where $\overleftrightarrow{X}$ is a tensor defined as

$$X_{ij} \equiv \frac{1}{\Omega}\sum_{\mathbf{p}} \upsilon_i \upsilon_j E(\mathbf{p}) f, \tag{5.39}$$

but now we have another unknown tensor to deal with. No matter how many balance equations we write, they always contain one more unknown than the number of equations. The solution to this infinite set of balance (or moment) equations is the solution to the BTE itself.

To summarize, the first four balance equations derived from the BTE are

(i) *carrier density:*

$$\frac{\partial n}{\partial t} = \frac{1}{q}\nabla \cdot \mathbf{J}_n + S_n \tag{5.21}$$

(ii) *momentum density:*

$$\frac{\partial \mathbf{P}}{\partial t} = -2\nabla \cdot \overleftrightarrow{W} + n(-q)\boldsymbol{\mathcal{E}} - \left\langle\left\langle \frac{1}{\tau_m} \right\rangle\right\rangle \mathbf{P} \tag{5.27b}$$

(iii) *energy density:*

$$\frac{\partial W}{\partial t} = -\nabla \cdot \mathbf{F}_\mathrm{W} + \mathbf{J}_n \cdot \boldsymbol{\mathcal{E}} - \left\langle\left\langle \frac{1}{\tau_\mathrm{E}} \right\rangle\right\rangle (W - W^0) + S_\mathrm{E} \tag{5.37}$$

(iv) *energy flux:*

$$\frac{\partial \mathbf{F}_W}{\partial t} = -2\nabla \cdot \overleftrightarrow{X} + \frac{(-q)W}{m^*}\boldsymbol{\mathcal{E}} - \left\langle\left\langle \frac{1}{\tau_\mathrm{FW}} \right\rangle\right\rangle \mathbf{F}_\mathrm{W}. \tag{5.38}$$

Notice that the first and third balance equations describe the conservation of a quantity, carrier density or kinetic energy, and that the second and fourth describe the flow of a quantity, mass or kinetic energy. We reiterate that these equations were derived from the BTE without simplifying approximations. To make use of them, however, the hierarchy of balance equations has to be truncated and simplifying assumptions are necessary in order to express the relaxation times in terms of macroscopic quantities such as the average kinetic energy rather than in terms of the unknown distribution function. Before discussing the application of these equations, however, we reformulate the balance equations in a more convenient form.

5.4 Carrier temperature and heat flux

The carriers in a semiconductor comprise a gas with a temperature, T_C, that may differ from the lattice temperature, T_L. Electric fields can increase the carrier energy so $T_\mathrm{C} > T_\mathrm{L}$, but carriers exchange energy with the lattice which tends to equalize the two temperatures. The flow of kinetic energy consists, in part, of a flow of heat. For many applications, we find it convenient to work with balance equations formulated in terms of the carrier temperature and heat flow, which are the subjects of this section.

5.4.1 Carrier temperature

To begin, examine the kinetic energy density:

$$W_{ii} = \frac{1}{2\Omega}\sum_{\mathbf{p}} p_i \upsilon_i f = \frac{m^*}{2\Omega}\sum_{\mathbf{p}} \upsilon_i^2 f = \frac{nm^*}{2}\langle \upsilon^2 \rangle. \tag{5.40}$$

Remember the implied sum over repeated indices, and recall that the brackets, $\langle\cdot\rangle$, denote an average over the distribution function and that the double brackets, $\langle\langle\cdot\rangle\rangle$, are reserved for the specially defined, ensemble averages. The carrier velocity,

$$\upsilon = \upsilon_d + \mathbf{c}, \tag{5.41}$$

has an average component, υ_d, due to the applied field, and a random component, $\mathbf{c}$, due to collisions. Using eq. (5.41) in eq. (5.40) we find

$$W_{ii} = \sum_i nm^* \upsilon_{di}^2 + \frac{1}{2}nm^*\langle c_i^2 \rangle = \frac{1}{2}nm^*\upsilon_d^2 + \frac{1}{2}nm^*\langle c^2 \rangle \tag{5.42}$$

(because $\langle c_i \rangle = 0$ by definition). The first term represents the *drift energy*,

$$W_{\text{drift}} = \frac{1}{2}nm^*\upsilon_d^2, \tag{5.43}$$

and the second term is due to random, thermal motion; its magnitude is

$$W_{\text{thermal}} = \frac{1}{2}m^* n\langle c^2 \rangle, \tag{5.44}$$

which, for an ideal gas is $3/2nk_BT$ (one third of which is associated with each degree of freedom). Equation (5.44) suggests that we define the carrier temperature by

$$\boxed{\frac{3}{2}nk_{\text{B}}T_{\text{C}} \equiv \frac{1}{2}m^* n\langle c^2 \rangle}\ , \tag{5.45}$$

so that the kinetic energy of the ensemble consists of a drift component, associated with the average motion of the ensemble, and a thermal component associated with the random component of the velocity.

Figure 5.3 illustrates the difference between drift and thermal kinetic energy. The low-field distribution function displayed in Fig. 5.3a is characterized by a temperature equal to the lattice temperature, T_{L}, because the car-

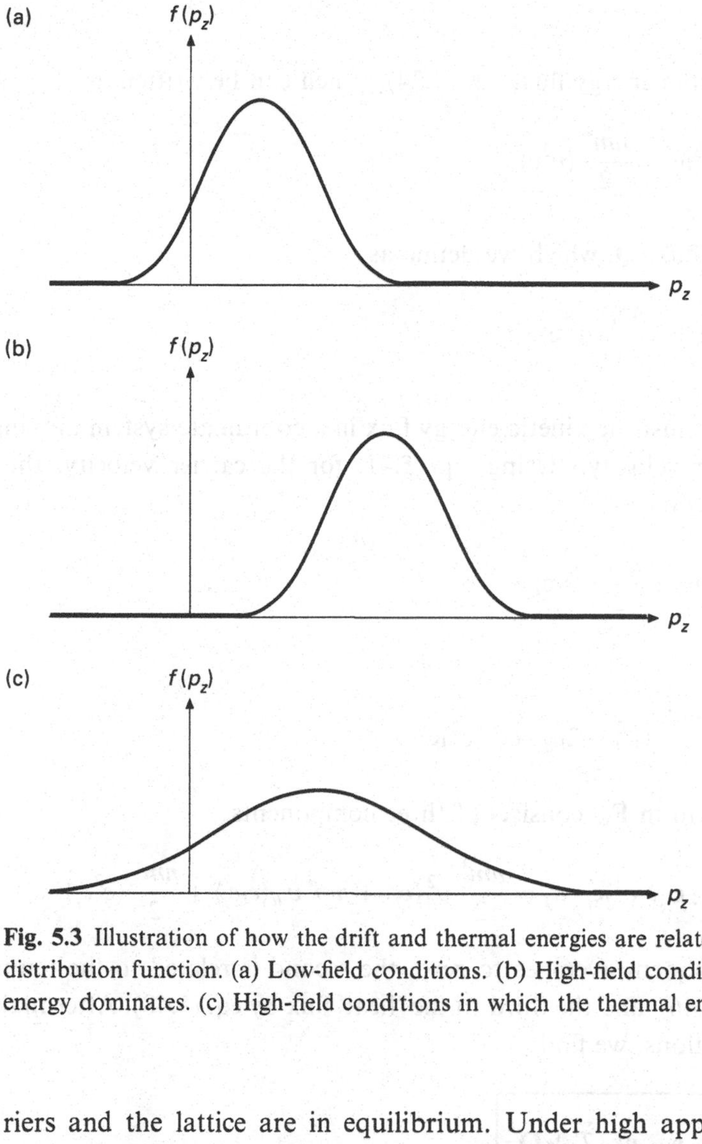

Fig. 5.3 Illustration of how the drift and thermal energies are related to the shape of the distribution function. (a) Low-field conditions. (b) High-field conditions in which the drift energy dominates. (c) High-field conditions in which the thermal energy dominates.

riers and the lattice are in equilibrium. Under high applied fields the kinetic energy of the distribution increases; Figure 5.3b shows a high-field distribution for which the additional kinetic energy is largely drift energy. This situation can occur in devices shortly after the application of a high field, but for uniform or slowly varying fields, scattering produces a large random component to the velocity, and the kinetic energy is mostly thermal as illustrated in Fig. 5.3c.

Guided by eq. (5.45), a temperature tensor can now be defined as

$$\boxed{\frac{1}{2}nk_B T_{ij} \equiv \frac{1}{2}m^* n\langle c_i c_j \rangle} \,. \tag{5.46}$$

5.4.2 Heat flux

Consider next the energy flux, eq. (5.34), which can be written as

$$\mathbf{F}_{\mathrm{W}} = \frac{m^*}{2\Omega}\sum_{\mathbf{p}} \upsilon^2 \boldsymbol{\upsilon} f = \frac{nm^*}{2}\langle \upsilon^2 \boldsymbol{\upsilon}\rangle, \tag{5.47}$$

and the 'heat' flux, $\mathbf{Q}$, which we define as

$$\mathbf{Q} \equiv \frac{m^*}{2\Omega}\sum_{\mathbf{p}} c^2 \mathbf{c} f = \frac{nm^*}{2}\langle c^2 \mathbf{c}\rangle. \tag{5.48}$$

(The flux, $\mathbf{Q}$, is just the kinetic energy flux in a coordinate system moving at the average carrier velocity.) Using eq. (5.41) for the carrier velocity, the kinetic energy flux becomes

$$\mathbf{F}_{\mathrm{W}} = \frac{nm^*}{2}\langle \upsilon^2\rangle \boldsymbol{\upsilon}_d + \frac{nm^*}{2}\langle \upsilon^2 \mathbf{c}\rangle \tag{5.49}$$

or

$$\mathbf{F}_{\mathrm{W}} = W\boldsymbol{\upsilon}_d + \frac{nm^*}{2}\langle (\upsilon_d^2 + 2\boldsymbol{\upsilon}_d \cdot \mathbf{c} + c^2)\mathbf{c}\rangle. \tag{5.50}$$

The second term in $\mathbf{F}_{\mathrm{W}}$ consists of three components,

$$\frac{nm^*}{2}\langle (\upsilon_d^2 + 2\boldsymbol{\upsilon}_d \cdot \mathbf{c} + c^2)\mathbf{c}\rangle \cdot \hat{x}_j = \frac{nm^*}{2}\upsilon_d^2\langle c_j\rangle + nm^*\upsilon_{di}\langle c_i c_j\rangle + \frac{nm^*}{2}\langle c^2 c_j\rangle. \tag{5.51}$$

The first component averages to zero, the second is related to the temperature through eq. (5.46), and the third to the 'heat' flux by eq. (5.48). After making use of these definitions, we find

$$\boxed{\mathbf{F}_{\mathrm{W}} = W\boldsymbol{\upsilon}_d + \boldsymbol{\upsilon}_d \cdot nk_{\mathrm{B}}\overleftrightarrow{T} + \mathbf{Q}}\,. \tag{5.52}$$

Equation (5.52) can be understood physically by viewing the electrons as an ideal gas. Recalling that $\mathcal{P}\Omega = Nk_{\mathrm{B}}T$ for an ideal gas tells us that $\mathcal{P} = nk_{\mathrm{B}}T$ is the pressure of the electron gas. If we consider moving a small volume element through the gas, the first term in eq. (5.52) is simply the energy density within the volume element times the velocity at which it moves. The second term is the velocity of the volume element times the pressure of the electron gas which represents the work to push the volume element through the gas. The third term describes the loss of energy from this volume element by the flow of heat out.

5.4.3 The meaning of Q in the hydrodynamic flow equations

The balance equations we have derived are known as the *hydrodynamic equations* because of their similarity to the equations commonly used to analyze fluid flow. The flux, **Q**, arises when we express the energy flux in terms of carrier temperature rather than average kinetic energy directly. Equation (5.48) shows that **Q** is a flux in a coordinate system moving at the average carrier velocity. Since it is associated with the random component of the carrier velocity, it seems reasonable to associate **Q** with the flow of thermal energy, or heat. Indeed, the quantity, **Q**, is commonly termed the heat flux, but it is reasonable to ask how it relates to the heat flux we defined in eq. (4.13).

In Chapter 4 we defined a heat flux under near-equilibrium conditions using thermodynamic arguments. The result, eq. (4.30b), had a Peltier component which accounted for the heat carried by the particles as they flow, and a diffusive component associated with heat flowing down a temperature gradient. To derive a thermodynamically sound balance equation for the heat flux, we should begin by recalling that the increase in heat is $T_C \mathrm{d}S$, where T_C is the carrier temperature, and $\mathrm{d}S$ is the increase in their entropy. The heat current, therefore, should be related to flow of entropy by

$$\mathbf{J}_Q = T_C \mathbf{J}_S \tag{5.53}$$

where the entropy flux is given by

$$\mathbf{J}_S = \frac{1}{\Omega}\sum_{\mathbf{p}} \upsilon\{f(\mathbf{p})\ln f(\mathbf{p}) - [1 - f(\mathbf{p})]\ln[1 - f(\mathbf{p})]\}. \tag{5.54}$$

Entropy is a measure of disorder. Notice that if all states are either empty ($f(\mathbf{p}) = 0$) or full ($f(\mathbf{p}) = 1$), then, as expected, the entropy is zero. For a derivation of eq. (5.54), consult Chapter 7 of Smith, Janek and Adler [4.1].

Under near-equilibrium conditions, the general prescription for the heat flux, eq. (5.53), reduces to eq. (4.13). Equation (5.48) for the vector **Q**, however, does not. In a bulk semiconductor, the vector **Q** actually points in a direction opposite to the direction of carrier flow! (If it were the heat flux, heat would have to be carried with the particles.) Although it may be difficult to associate the vector, **Q**, with a physical quantity, that does not prevent us from solving the balance equations for physically meaningful quantities such as the average carrier density, velocity, and energy.

5.4.4 The momentum and energy balance equations

Having introduced the carrier temperature, T_C, and a related vector, **Q**, we can reformulate the balance equations in terms of them. Guided by eq. (5.42), we write the tensor, W_{ij}, as

$$W_{ij} = \frac{1}{2} nm^* v_{di} v_{dj} + \frac{1}{2} nk_B T_{ij} \tag{5.55}$$

or

$$W_{ij} = K_{ij} + \frac{nk_B T_{ij}}{2}, \tag{5.56}$$

where

$$K_{ij} \equiv \frac{nm^*}{2} v_{di} v_{dj}. \tag{5.57}$$

Equation (5.56) decomposes $\overleftrightarrow{W}$ into a component, $\overleftrightarrow{K}$, associated with the average motion of carriers, and other, $\overleftrightarrow{T}$, which is determined by the random component of the carriers' velocity. After inserting eqs. (5.56) and (5.27b), we obtain the momentum balance equation as

$$\boxed{\frac{\partial \mathbf{P}}{\partial t} = -\nabla \cdot (2\overleftrightarrow{K} + nk_B \overleftrightarrow{T}) + (-q)n\boldsymbol{\mathcal{E}} - \left\langle\left\langle \frac{1}{\tau_m} \right\rangle\right\rangle \mathbf{P}}\,. \tag{5.58}$$

With the aid of eq. (5.52), we can express the energy balance equation, eq. (5.37), as

$$\boxed{\frac{\partial W}{\partial t} = -\nabla \cdot (W\boldsymbol{v}_d + \mathbf{Q} + nk_B \boldsymbol{v}_d \cdot \overleftrightarrow{T}) + \boldsymbol{\mathcal{E}} \cdot \mathbf{J}_n - \left\langle\left\langle \frac{1}{\tau_E} \right\rangle\right\rangle (W - W^0) + \mathcal{S}_E}\,. \tag{5.59}$$

Equations (5.58) and (5.59) are the momentum and energy balance equations written in terms of the carrier temperature. They are mathematically equivalent to eqs. (5.27b) and (5.37), which are the corresponding equations expressed in terms of the carrier energy.

5.5 The displaced Maxwellian approximation

Before the balance equations can be solved, several simplifying assumptions are necessary. The hierarchy of equations must be truncated, some of the terms simplified, and expressions for the ensemble relaxation times developed. One approach is to make some simplifying assumptions and to extract parameters from measured results and from Monte Carlo simulations (Monte Carlo simulation is the subject of Chapter 6, and measured results are discussed in Chapter 7). A second approach is to guess the form of the distribution function and then use the balance equations to solve for the parameters in this functional form. The most commonly used guess is the *displaced* (or *drifted*) *Maxwellian*,

$$f(\mathbf{p}) \sim \exp\left[-\left|\mathbf{p} - m^* \upsilon_d\right|^2 / 2m^* k_B T_C\right] \tag{5.60}$$

which contains two parameters, the drift velocity, υ_d, and the carrier temperature, T_C. We might have expected the distribution function to be strongly distorted in the direction of the applied field which is accelerating the carriers. To achieve the symmetrical distribution function described by eq. (5.60), frequent collisions to randomize the momentum gained in the direction of the field must occur. As discussed in Chapter 3, Section 3.3.1, if the carrier density is high enough, collisions between carriers will randomize momentum and energy and lead to a Maxwellian distribution. A carrier density in excess of $10^{16}/\text{cm}^3$ is typically required [5.4]. Consequently we may expect the distribution function to be Maxwellian in the channel of a MOSFET where the carrier density is very high but not in the space-charge region of bipolar transistors where the carrier density is low.

5.5.1 Balance equations in the displaced Maxwellian approximation

Let's consider how the second form of the balance equations, those involving carrier temperature and heat flux, are simplified when a displaced Maxwellian distribution function is assumed. First, the temperature tensor for a displaced Maxwellian can be shown to be diagonal (see homework problem 5.4),

$$T_{ij} = T_C \delta_{ij}. \tag{5.61}$$

More complex situations can occur, however. For example, different temperatures might be associated with motion parallel and perpendicular to the applied field. For complex band structures and arbitrarily oriented fields, the tensor may even contain off-diagonal elements. For diagonal tensors, however, the divergence of the tensor becomes the gradient of a scalar,

$$\nabla \cdot \overleftrightarrow{T} = \nabla T_C, \tag{5.62}$$

so the momentum balance equation, eq. (5.58), can be simplified to

$$\frac{\partial P_j}{\partial t} = -\frac{\partial (n m^* \upsilon_{di} \upsilon_{dj})}{\partial x_i} + \frac{\partial (n k_B T_C)}{\partial x_j} + n(-q)\mathcal{E}_j - \left\langle\left\langle \frac{1}{\tau_m} \right\rangle\right\rangle P_j. \tag{5.63}$$

To solve the momentum balance equation, the carrier temperature must be known. Under low applied fields, we can assume that the carrier temperature is the temperature of the lattice. Under high applied fields, however, the carrier temperature must be found by solving the energy balance equation. But the energy balance equation introduces the 'heat' flux, **Q**. If the **Q** is evaluated from its definition, eq. (5.48), using the displaced Maxwellian distribution function, eq. (5.60), we find that $\mathbf{Q} = \mathbf{0}$ (see homework problem 5.10). If the distribution function is truly a displaced Maxwellian, then $\mathbf{Q} = \mathbf{0}$, but the heat flux is often important in devices [5.3], so it is better to abandon the displaced Maxwellian when evaluating $\mathbf{Q}$ and to describe it phenomenologically. Since we know that heat flows down a temperature gradient, it seems reasonable to approximate $\mathbf{Q}$ by

$$\mathbf{Q} = -\kappa \nabla T_C, \tag{5.64}$$

where κ is the thermal conductivity. This equation is approximate; it ignores the Peltier effect discussed in Chapter 4, and the linear form of the equation applies only when the temperature gradient is gentle. More importantly, it is based on assigning a physical interpretation to $\mathbf{Q}$ that we know is incorrect. Nevertheless, eq. (5.64) is a plausible way to terminate the hierarchy of equations and is preferable to simply ignoring heat conduction. With eq. (5.64) and the use of a diagonal temperature tensor, the energy balance equation, (5.59), becomes

$$\frac{\partial W}{\partial t} = -\nabla \cdot (W\upsilon_d - \kappa \nabla T_C + nk_B T_C \upsilon_d) + \mathbf{J}_n \cdot \boldsymbol{\mathcal{E}} - \left\langle\left\langle \frac{1}{\tau_E} \right\rangle\right\rangle (W - W_0) + \mathcal{S}_E. \tag{5.65}$$

We now have a complete set of simplified balance equations, eqs. (5.21), (5.63), and (5.65) in a form suitable for solving for the unknowns, $n(\mathbf{r})$, $\upsilon(\mathbf{r})$, and $T_C(\mathbf{r})$. When spatial variations are confined to the $\hat{z}$-direction and no sources or sinks for carriers are present, the simplified balance equations for electron transport reduce to:

$$\frac{\partial n}{\partial t} = \frac{1}{q}\frac{\partial J_n}{\partial z}, \tag{5.66a}$$

$$\frac{\partial P_z}{\partial t} = \frac{\partial (nm^* \upsilon_{dz}^2 + nk_B T_C)}{\partial z} + n(-q)\mathcal{E}_z - \left\langle\left\langle \frac{1}{\tau_m} \right\rangle\right\rangle P_z, \tag{5.66b}$$

$$\frac{\partial W}{\partial t} = -\frac{\partial}{\partial z}\left[(W + nk_B T_C)\upsilon_{dz} - \kappa \frac{\partial T_C}{\partial z}\right] + J_{nz}\mathcal{E}_z - \left\langle\left\langle \frac{1}{\tau_E} \right\rangle\right\rangle (W - W^0), \tag{5.66c}$$

along with the constitutive relations,

$$P_z = nm^* \upsilon_{dz} \tag{5.66d}$$

and

$$\frac{W}{n} = \frac{1}{2} m^* \upsilon_{dz}^2 + \frac{3}{2} k_B T_C. \tag{5.66e}$$

These equations comprise a set of three equation in the three unknowns, $n(z, t)$, $\upsilon_{dz}(z, t)$, and $T_C(z, t)$ – providing that we can approximate the ensemble relaxation times in terms of these unknowns.

5.5.2 Relaxation times in the displaced Maxwellian approximation

To evaluate the ensemble momentum relaxation rate,

$$\left\langle\left\langle\frac{1}{\tau_m}\right\rangle\right\rangle = \frac{\sum_{\mathbf{p}} p_z f/\tau_m(\mathbf{p})}{\sum_{\mathbf{p}} p_z f}, \tag{5.67}$$

it is convenient to expand the displaced Maxwellian as

$$f(\mathbf{p}) = \mathrm{e}^{-(p^2 - 2m^*\mathbf{p}\cdot\upsilon_d + m^{*2}\upsilon_d^2)/2m^*k_B T_C} \tag{5.68}$$

because the last term, which represents the drift energy, can be ignored for low fields and for uniform high fields, and the second term can be assumed small and expanded to obtain

$$f(\mathbf{p}) = e^{-p^2/2m^*k_B T_C}(1 + \mathbf{p}\cdot\upsilon_d/k_B T_C) = f_S + f_A. \tag{5.69}$$

(The validity of these assumptions will be established in Chapter 7, Section 2.) The displaced Maxwellian has been factored into two components, a symmetric component, f_S, which is even in momentum and an anti-symmetric component, f_A, which is odd. If $\tau_m(\mathbf{p})$ is an even function of momentum, only the odd component of f contributes to eq. (5.67), and we find

$$\left\langle\left\langle\frac{1}{\tau_m}\right\rangle\right\rangle = \frac{\sum_{\mathbf{p}} p_z(\mathbf{p}\cdot\upsilon_d)f_S(p)/\tau_m(p)}{\sum_{\mathbf{p}} p_z(\mathbf{p}\cdot\upsilon_d)f_S(p)}. \tag{5.70}$$

After dividing numerator and denominator by $2m^*$ and assuming a one-dimensional problem ($\upsilon_d = \upsilon_{dz}\hat{z}$), eq. (5.70) becomes

$$\left\langle\left\langle\frac{1}{\tau_m}\right\rangle\right\rangle = \frac{\sum_{\mathbf{p}} \frac{p_z^2}{2m^*} f_S(p)/\tau_m(p)}{\sum_{\mathbf{p}} \frac{p_z^2}{2m^*} f_S(p)}. \tag{5.71}$$

Due to the assumed spherical symmetry of the problem, the same result is obtained for any choice of direction, so using $E(p) = p^2/2m^*$, we can write

$$\left\langle\left\langle\frac{1}{\tau_m}\right\rangle\right\rangle = \frac{\sum_{\mathbf{p}} E(p)f_S(p)/\tau_m(p)}{\sum_{\mathbf{p}} E(p)f_S(p)} = \frac{\langle E(p)/\tau_m(p)\rangle}{\langle E(p)\rangle}, \tag{5.72}$$

where the averages are over f_S.

For power-law scattering,

$$\tau_m(E) = \tau_0(E/k_B T_C)^s, \tag{5.73}$$

and eq. (5.72) evaluates to

$$\left\langle\left\langle \frac{1}{\tau_m} \right\rangle\right\rangle = \frac{1}{\tau_0} \frac{\Gamma(5/2 - s)}{\Gamma(5/2)}. \tag{5.74}$$

To underscore the fact that the ensemble relaxation times depend on the assumed distribution function, note that if we use, instead, $f(\mathbf{p})$ as given by the relaxation time approximation solution to the BTE, we find

$$\left\langle\left\langle \frac{1}{\tau_m} \right\rangle\right\rangle = \frac{1}{\tau_0} \frac{\Gamma(5/2)}{\Gamma(5/2 + s)} \tag{5.75}$$

(see homework problem, 5.13). For the same power law scattering characteristic, we find different ensemble relaxation times for the two different assumed distribution functions. Obviously, the best answer is the one for which the assumed distribution function corresponds most closely to the actual distribution function under the conditions of interest.

5.5.3 Discussion

Equations like (5.66) are frequently the starting point for device analysis. Reference [5.3] describes the application of these equations to silicon devices. We formulated eqs. (5.66) by simplifying the general equations assuming a displaced Maxwellian distribution function. In practice, the approach is not so clean. The temperature tensor is assumed to be diagonal (which can be justified for a displaced Maxwellian distribution function) and the relaxation times are extracted from Monte Carlo simulations for which the carrier distributions are highly non-Maxwellian. There are many issues to address when applying balance equations to devices, and other approaches to consider as well. Because such equations are so widely used, a clear understanding of these issues is essential, so we return to the subject again in Chapter 8.

5.6 Stratton's approach

The set of simplified balance equations as presented in eqs. (5.66a)–(5.66e) represents one formulation of the macroscopic transport equations. Many variations on this basic theme exist, but one popular approach is somewhat different [5.5, 5.6]. Recall the first and third balance equations expressed continuity of carrier density and energy and that the second and fourth were flow equations. For the flow equations, Stratton worked directly with the BTE using a relaxation time approximation [5.5]. The key difference, then, is that the so-called hydrodynamic transport equations use a macroscopic relaxation time which describes the ensemble of carriers while Stratton uses a microscopic relaxation time that describes how individual carriers scatter.

Stratton's approach essentially solves the BTE in the relaxation time approximation as we did in Chapters 3 and 4, but it removes the restriction to low electric fields and near-equilibrium conditions. We begin by decomposing the carrier distribution into symmetric and anti-symmetric components in momentum,

$$f(\mathbf{p}) = f_S(p) + f_A(\mathbf{p}), \tag{5.76}$$

which can always be done. The next step is to write the collision integral as

$$\left.\frac{\partial f}{\partial t}\right|_{\text{coll}} = \left.\frac{\partial f_S}{\partial t}\right|_{\text{coll}} + \left.\frac{\partial f_A}{\partial t}\right|_{\text{coll}} \approx \left.\frac{\partial f_A}{\partial t}\right|_{\text{coll}} = -\frac{f_A}{\tau_f}. \tag{5.77}$$

Equation (5.77) is the same assumption we used to solve the BTE under near-equilibrium conditions. The difference is that we are now considering more general conditions where the carriers may be near or very far from equilibrium. The symmetric component of the carrier distribution will not be an equilibrium Maxwellian. If it is a heated Maxwellian, then the assumption that $\partial f_S/\partial t|_{\text{coll}} = 0$ can be justified. As demonstrated in Chapter 3, we also require that the dominant scattering mechanisms be either elastic or isotropic.

Stratton's approach begins with simplifying assumptions about the scattering integral. The heated Maxwellian assumption requires that the carrier density be high, and the assumption of isotropic or elastic scattering applies to nonpolar semiconductors like silicon. In the balance equation approach, we formulated the equations without making simplifying assumptions, but to solve the equations, they must be simplified. As a result, we shall find that the hydrodynamic and Stratton equations, as they are actually solved, look very similar.

Using the microscopic relaxation time approximation as given in eq. (5.77), we write the steady-state BTE as

$$\upsilon_z \frac{\partial f}{\partial z} - q\mathcal{E}_z \frac{\partial f}{\partial p_z} = -\frac{f_A}{\tau_f}. \tag{5.78}$$

We can approximate f on the LHS of eq. (5.78) by f_S and solve for f_A to obtain

$$f_A = q\tau_f \mathcal{E}_z \frac{\partial f_S}{\partial p_z} - \tau_f \upsilon_z \frac{\partial f_S}{\partial z}. \tag{5.79}$$

Now that we have solved the BTE for $f(\mathbf{p})$, we can evaluate the fluxes directly. For the carrier flux (or, rather the current density), we have

$$J_{nz} = \frac{(-q)}{\Omega} \sum_{\mathbf{p}} \upsilon_z f_A(\mathbf{p}), \tag{5.80}$$

which, using eq. (5.79) becomes

$$J_{nz} = \frac{q}{\Omega}\sum_{\mathbf{p}}\{q\tau_f(-\partial f_S/\partial p_z)\upsilon_z\mathcal{E}_z + \tau_f\upsilon_z^2(\partial f_S/\partial z)\}. \tag{5.81}$$

For the assumed Maxwellian distribution,

$$-\frac{\partial f_S}{\partial p_z} = \left(\frac{\upsilon_z}{k_B T_C}\right) f_S, \tag{5.82}$$

which can be inserted in eq. (5.81) to find

$$J_{nz} = \frac{q}{\Omega}\sum_{\mathbf{p}}\left\{\frac{q\tau_f\upsilon_z^2 f_S}{k_B T_C}\mathcal{E}_z + \tau_f\frac{\partial(\upsilon_z^2 f_S)}{\partial z}\right\}. \tag{5.83}$$

To evaluate eq. (5.83), note that because of the symmetry of the assumed Maxwellian, $\langle\upsilon_x^2\rangle = \langle\upsilon_y^2\rangle = \langle\upsilon_z^2\rangle = \langle\upsilon^2\rangle/3$. Note also that $\langle E\rangle = 3k_B T_C/2$. Using these expressions, eq. (5.83) becomes

$$J_{nz} = \frac{q}{\Omega}\sum_{\mathbf{p}}\left\{\frac{q\tau_f\upsilon^2 f_S}{2\langle E\rangle}\mathcal{E}_z + \frac{\tau_f}{3}\frac{\partial(\upsilon^2 f_S)}{\partial z}\right\}. \tag{5.84}$$

Stratton assumed that τ_f was position-independent. This is true for phonon scattering, but if the doping density varies with position, then the ionized impurity relaxation time will vary with position. Nevertheless, if we follow Stratton and assume that τ_f is position-independent, then eq. (5.84) becomes

$$J_{nz} = \frac{q}{\Omega}\sum_{\mathbf{p}}\frac{q}{m^*\langle E\rangle}\left\{\left(\frac{m^*\upsilon^2}{2}\right)(\tau_f f_S)\mathcal{E}_z + \frac{k_B T_C}{q}\frac{\partial[(m^*\upsilon^2/2)\tau_f f_S]}{\partial z}\right\}, \tag{5.85}$$

which, finally, we can write as

$$J_{nz} = nq\mu_n\mathcal{E}_z + \frac{\partial(k_B T_C\mu_n n)}{\partial z}, \tag{5.86}$$

where

$$\mu_n \equiv \frac{q}{m^*}\frac{\langle E\tau_f\rangle}{\langle E\rangle}. \tag{5.87}$$

If we make another definition,

$$D_n = \frac{k_B T_C}{q}\mu_n, \tag{5.88}$$

then we can write Stratton's current equation as

$$J_{nz} = nq\mu_n\mathcal{E}_z + q\frac{\partial(D_n n)}{\partial z}, \tag{5.89}$$

(but recall that we have assumed that τ_f is spatially independent).

The energy flux equation is derived by a similar procedure. Beginning with

$$F_{Wz} = \frac{1}{\Omega}\sum_{\mathbf{p}} E(\mathbf{p})\upsilon_z f_A \tag{5.90}$$

and using eq. (5.79), we obtain

$$F_{\mathrm{Wz}} = -C_e\left\{n\mu_n k_{\mathrm{B}} T_{\mathrm{C}} \mathcal{E}_z + \frac{\partial(n\mu_n k_{\mathrm{B}}^2 T_{\mathrm{C}}^2/q)}{\partial z}\right\}, \tag{5.91}$$

where

$$C_e = (5/2 + s) \tag{5.92}$$

for the power law scattering. Alternatively, Stratton's energy flux can be expressed as

$$F_{\mathrm{Wz}} = k_{\mathrm{B}} T_{\mathrm{C}} C_e\left(\frac{J_{nz}}{(-q)}\right) - \kappa_n \frac{\partial T_{\mathrm{C}}}{\partial z} \tag{5.93}$$

where

$$\kappa_n = nq\mu_n\left(\frac{k_B}{q}\right)^2 C_e T_{\mathrm{C}}. \tag{5.94}$$

(Homework problems 5.5 and 5.20 ask you to fill in the steps of the derivation and to relate Stratton's equation to the hydrodynamic equations)

Stratton's approach provides an alternative set of flow equations for describing transport. In practice, both forms are used. Stratton's equations will be compared to the hydrodynamic flow equations in Section 5.7.

5.7 Drift-diffusion equations

The simplified hydrodynamic equations, or Stratton's equations form a closed set of equations useful for analyzing transport in bulk semiconductors or in devices. Electrical engineers, however, are used to treating carrier transport with the even simpler, drift–diffusion equation,

$$\mathbf{J}_n = nq\mu_n \boldsymbol{\mathcal{E}} + qD_n \nabla n. \tag{5.95}$$

Our objectives in this section are to understand the origin of the drift–diffusion equation and the assumptions that limit its validity.

5.7.1 Derivation from the momentum balance equation

Starting at the momentum balance equation as stated in eq. (5.32),

$$\mathbf{J}_n + \frac{1}{\langle\langle 1/\tau_m\rangle\rangle}\frac{\partial \mathbf{J}_n}{\partial t} = \frac{q^2 n}{m^*\langle\langle 1/\tau_m\rangle\rangle}\boldsymbol{\mathcal{E}} + \frac{2q}{m^*\langle\langle 1/\tau_m\rangle\rangle}\nabla\cdot\overleftrightarrow{W},$$

we define a carrier mobility as

$$\mu_n \equiv \frac{q}{m^* \langle\langle 1/\tau_m \rangle\rangle}. \tag{5.96}$$

We can normally assume that the current does not vary appreciably over a momentum relaxation time, $1/\langle\langle 1/\tau_m \rangle\rangle$, to obtain

$$\mathbf{J}_n = nq\mu_n \boldsymbol{\mathcal{E}} + 2\mu_n \nabla \cdot \overleftrightarrow{W}. \tag{5.97}$$

According to eq. (5.55), the tensor $\overleftrightarrow{W}$ contains components due to the average (drift) velocity and due to the random, thermal energy. To simplify eq. (5.97), ignore the drift velocity components of $\overleftrightarrow{W}$ and assume that the temperature tensor is diagonal, so

$$W_{ij} = \frac{nk_{\mathrm{B}}T_{\mathrm{C}}}{2}\delta_{ij} = \frac{W}{3}\delta_{ij}, \tag{5.98}$$

where

$$W = \frac{3}{2}nk_{\mathrm{B}}T_{\mathrm{C}} \tag{5.99}$$

is the kinetic energy density (assuming again that the drift energy is small). When eq. (5.99) is inserted in eq. (5.97), we find

$$\boxed{\mathbf{J}_n = nq\mu_n \boldsymbol{\mathcal{E}} + \frac{2}{3}\mu_n \nabla W}\ . \tag{5.100}$$

Equation (5.100) looks like a drift–diffusion equation, but it demonstrates that diffusion is associated with gradients in the kinetic energy density which can result from variations in either the carrier density or average kinetic energy per carrier. After expanding the gradient and writing the result in terms of the carrier temperature, the current equation becomes

$$\boxed{\mathbf{J}_n = q\mu_n n\boldsymbol{\mathcal{E}} + qD_n \nabla n + qS_n \nabla T_{\mathrm{C}}}\ , \tag{5.101}$$

and we have our drift–diffusion equation. In eq. (5.101), we have defined two new parameters:

$$D_n \equiv \frac{k_{\mathrm{B}}T_{\mathrm{C}}}{q}\mu_n \tag{5.102}$$

is the diffusion coefficient, and

$$S_n = n\mu_n\left(\frac{k_{\mathrm{B}}}{q}\right) \tag{5.103}$$

is the Soret coefficient.

Equation (5.101) shows that the drift–diffusion equation is valid when gradients in carrier temperature are gentle. There are some additional assumptions, frequencies much less than the momentum relaxation rate, a diagonal temperature tensor, neglect of the drift energy, but these are generally well satisfied, especially when ∇T_C is small. Note, in particular, that there is no restriction to low-field, near-equilibrium conditions; drift–diffusion equations under high-field conditions when gradients in carrier temperature are not large.

5.7.2 Comparison to Stratton's equation

Both Stratton's approach and the balance equations yield similar, drift–diffusion-like current equations, but the differences are interesting. Stratton's approach gives

$$\mathbf{J}_n = nq\mu_n\boldsymbol{\mathcal{E}} + q\nabla(D_n n), \tag{5.104}$$

which should be compared with eq. (5.101), the hydrodynamic result. [Recall than when deriving Stratton's equation, we assumed that the microscopic relaxation time did not vary with position. Equation (5.104) has to be modified when τ_f is position-dependent, but the result still does not agree with eq. (5.01)]. The essential difference between the two approaches is that the mobility (or diffusion coefficient) appears inside the gradient in Stratton's approach and outside it in the hydrodynamic approach.

It should not surprise us that the two approaches give different results; they are based on different assumptions. But even when the underlying physical assumptions are the same (i.e. isotropic or elastic scattering so that a relaxation time can be assumed and sufficient carrier-carrier scattering so that the distribution is Maxwellian) the two approaches still give different forms of the resulting equations. This occurs because Stratton's approach uses a microscopic relaxation time while the hydrodynamic approach is formulated in terms of a macroscopic, ensemble relaxation time. When the underlying physical problem justifies the microscopic relaxation time approximation and a Maxwellian shape for f_S, then the two approaches give the same final result. In practice, comparison of the two approaches is clouded by the fact that both methods use experimental data and Monte Carlo simulations to extract parameters. We return to this issue in Chapter 8 where the use of macroscopic transport equations for device analysis is discussed.

5.7.3 The carrier mobility

To make use of the current equation, the mobility, diffusion coefficient, and carrier temperature must be known. Under low fields, the carrier temperature is that of the lattice, but for high fields, we need to solve the energy balance equation to find T_C. To evaluate μ_n, the distribution function must be known or approximated. For low fields, however, the perturbation in $f(\mathbf{p})$ is proportional to the electric field, so the mobility becomes a material-dependent parameter (see homework problem 5.11). For high applied fields in bulk semiconductors, there is a one-to-one correspondence between the electric field and the distribution function, so we can view μ_n as a field-dependent, material parameter. For small devices the applied fields are also high, but the distribution function depends on the electric field throughout the device and must be found by solving the BTE. The drift–diffusion equation loses its simplicity when applied to a small device because μ_n and D_n depend on the device structure and applied bias as well as the material properties of the semiconductor.

To illustrate the dependence of the mobility on the distribution function, consider a heavily doped semiconductor for which ionized impurity scattering dominates and the RTA applies. Evaluating $\langle\langle 1/\tau_m \rangle\rangle$ from eq. (5.75), we find

$$\mu_n(RTA) = \frac{q\tau_0}{m^*}\frac{\Gamma(5/2+s)}{\Gamma(5/2)}. \tag{5.105}$$

If, however, there is also a high density of electrons, then electron–electron scattering may dominate. Scattering of electrons by electrons may seem unimportant because the total momentum of the electron ensemble is unaffected, but it does influence the mobility indirectly by its effect on the distribution function. For high electron–electron scattering rates, a displaced Maxwellian results, and $\langle\langle 1/\tau_m \rangle\rangle$ must be evaluated from eq. (5.74). The corresponding mobility is

$$\mu_n(DM) = \frac{q\tau_0}{m^*}\frac{\Gamma(5/2)}{\Gamma(5/2-s)}. \tag{5.106}$$

For ionized impurity scattering, $s = 3/2$, and we find

$$\mu_n(RTA) \simeq 3.4\mu_n(DM), \tag{5.107}$$

so the assumed distribution function can have a large effect on the mobility.

5.8 Balance equations for heterostructures

One trend in semiconductor devices is the increasing use of heterostructures, semiconductors in which the material composition varies with position. For example, the heterostructure may be a quantum well which consists of a small bandgap semiconductor embedded within a wide bandgap semiconductor. In this case, the material composition varies rapidly on an atomic scale and the width of the quantum well is on the order of the electron's wavelength, so a quantum mechanical treatment is essential. On the other hand, the material may be a semiconductor alloy with a slowly varying alloy composition. When the material composition varies slowly, we should be able to describe transport semiclassically. That is our objective in this section, to derive semiclassical transport equations for semiconductor heterostructures. The reader may wish to review heterostructure fundamentals in Chapter 1, Section 1.3.3 and the derivation of the drift–diffusion equation for heterostructures in Chapter 4, Section 4.10.

The equation of motion for an electron in momentum space,

$$\frac{d\mathbf{p}}{dt} = \mathbf{F}_e = (-q)\boldsymbol{\mathcal{E}}(\mathbf{r}), \tag{5.108}$$

has to be modified in a heterostructure as given by eq. (1.95b),

$$\frac{d\mathbf{p}}{dt} = -\nabla E_{C0} - \nabla E(\mathbf{p}). \tag{5.109}$$

In this equation, the first term represents the actual force on the electron, $\mathbf{F}_e = -\nabla E_{C0}$, where the E_{C0} is the electron's potential energy (the bottom of the conduction band). In the second term in eq. (5.109), $E(\mathbf{p})$ represents the electron's kinetic energy. Using eq. (1.46a) for the bottom of the conduction band and assuming simple energy bands, we obtain

$$\frac{d\mathbf{p}}{dt} = (-q)\boldsymbol{\mathcal{E}} + \nabla_r \chi - \frac{p^2}{2}\nabla_r\left(\frac{1}{m^*}\right), \tag{5.110}$$

which leads to the BTE for heterostructures,

$$\frac{\partial f}{\partial t} + \upsilon \cdot \nabla_r f + \left\{(-q)\boldsymbol{\mathcal{E}} + \nabla_r \chi\right\} \cdot \nabla_p f + \left\{-\frac{p^2}{2}\nabla_r\left(\frac{1}{m^*}\right)\right\} \cdot \nabla_p f = \left.\frac{\partial f}{\partial t}\right|_{\text{coll}}. \tag{5.111}$$

The derivation of balance equations starting from eq. (5.111) now proceeds with the procedure of Section 5.1. The appearance of the quasi-electric field causes no problems, we simply add the quasi-electric field to the actual field in the final result. The balance equation will have an additional term associated with the position-dependent effective mass to the balance equation, and the term involving the gradient of the associated flux will change.

Beginning with the second term in the BTE, we multiply by $\phi(\mathbf{p})$ and sum over momentum to find

$$\sum_{\mathbf{p}} \phi(p)\upsilon \cdot \nabla_r f = \nabla_r \cdot \sum_{\mathbf{p}} \phi(p)\frac{\mathbf{p}}{m^*} f - \nabla_r\left(\frac{1}{m^*}\right)\sum_{\mathbf{p}} \phi(p)\mathbf{p} f, \tag{5.112}$$

which can be re-expressed as

$$\sum_{\mathbf{p}} \phi(p) v \cdot \nabla_r f = \nabla_r \cdot F_\phi + \left(\frac{1}{m^*}\nabla_r m^*\right) \cdot F_\phi. \tag{5.113}$$

Multiplying the fourth term in the BTE by $\phi(\mathbf{p})$ and summing over the Brillouin zone, we find

$$M = -\sum_{\mathbf{p}} \phi(\mathbf{p}) \frac{p^2}{2} \frac{\partial(1/m^*)}{\partial z} \frac{\partial f}{\partial p_z} = +\left(\frac{1}{m^{*2}}\right) \frac{\partial m^*}{\partial z} \sum_{\mathbf{p}} \frac{\phi(\mathbf{p}) p^2}{2} \frac{\partial f}{\partial p_z}, \tag{5.114}$$

which can be written as

$$M = \left(\frac{1}{m^{*2}}\right) \frac{\partial m^*}{\partial z} \sum_{\mathbf{p}} \frac{\partial}{\partial p_z} \left(\frac{\phi(\mathbf{p}) p^2 f}{2}\right) - \left(\frac{1}{m^{*2}}\right) \frac{\partial m^*}{\partial z} \sum_{\mathbf{p}} \frac{\partial}{\partial p_z} \left(\frac{\phi(\mathbf{p}) p^2}{2}\right) f. \tag{5.115}$$

The first term integrates to zero if f goes to zero at the boundaries of the Brillouin zone, so the additional term in the balance equation is

$$M = -\left(\frac{1}{m^{*2}}\right) \frac{\partial m^*}{\partial z} \sum_{\mathbf{p}} \frac{\partial}{\partial p_z} \left(\frac{\phi(\mathbf{p}) p^2}{2}\right) f. \tag{5.116}$$

The general prescription for a balance equation for heterostructures is:

$$\frac{\partial n_\phi(\mathbf{r}, t)}{\partial t} = -\nabla \cdot \mathbf{F}_\phi + \left(\frac{1}{m^*}\nabla_r m^*\right) \mathbf{F}_\phi + G_\phi + M - R_\phi + \mathcal{S}_\phi \tag{5.117a}$$

Density: $$n_\phi(\mathbf{r}, t) \equiv \frac{1}{\Omega} \sum_{\mathbf{p}} f \phi(\mathbf{p}) \tag{5.117b}$$

Associated flux: $$\mathbf{F}_\phi(\mathbf{r}, t) \equiv \frac{1}{\Omega} \sum_{\mathbf{p}} v f \phi(\mathbf{p}) \tag{5.117c}$$

Field generation rate: $$G_\phi(\mathbf{r}, t) \equiv ((-q)\mathcal{E} + \nabla_r \chi) \cdot \frac{1}{\Omega} \sum_{\mathbf{p}} f \nabla_p \phi(\mathbf{p}) \tag{5.117d}$$

Effective mass term: $$M \equiv -\left(\frac{1}{m^{*2}}\right) \nabla_r m^* \cdot \sum_{\mathbf{p}} \nabla_p \left(\frac{\phi(\mathbf{p}) p^2}{2}\right) f \tag{5.117e}$$

Scattering recombination rate: $$R_\phi(\mathbf{r}, t) \equiv \left\langle\left\langle \frac{1}{\tau_\phi} \right\rangle\right\rangle [n_\phi(\mathbf{r}, t) - n_\phi^0(\mathbf{r}, t)] \tag{5.117f}$$

Associated out-scattering rate: $$\frac{1}{\tau_\phi(\mathbf{p})} \equiv \sum_{\mathbf{p}'} \left(1 - \frac{\phi(\mathbf{p}')}{\phi(\mathbf{p})}\right) S(\mathbf{p}, \mathbf{p}') \tag{5.117g}$$

Particle gen-rec rate:	$S_\phi(\mathbf{r}, t) \equiv \frac{1}{\Omega} \sum_{\mathbf{p}} \phi(\mathbf{p}) s(\mathbf{r}, \mathbf{p}, t)$	(5.117h)
Ensemble relaxation rate:	$\left\langle\left\langle \frac{1}{\tau_\phi} \right\rangle\right\rangle \equiv \frac{\frac{1}{\Omega}\sum_{\mathbf{p}} f(\mathbf{r}, \mathbf{p}, t)\phi(\mathbf{p})/\tau_\phi(\mathbf{p})}{[n_\phi(\mathbf{r}, t) - n_\phi^0(\mathbf{r}, t)]}.$	(5.117h)

It's now a straightforward matter to derive a momentum balance equation for heterostructures. Proceeding as in Section 5.3.2, we find

$$\begin{aligned}\frac{\partial P_z}{\partial t} = &-\frac{\partial}{\partial x_i}(2W_{iz}) + \left(\frac{1}{m^*}\frac{\partial m^*}{\partial x_i}\right) W_{iz} + n\left\{(-q)\mathcal{E}_z + \frac{\partial \chi}{\partial z}\right\} \\ &- \left(\frac{1}{m^*}\frac{\partial m^*}{\partial z}\right) W - \left(\frac{2}{m^*}\frac{\partial m^*}{\partial z}\right) W_{zz} - \left\langle\left\langle \frac{1}{\tau_m} \right\rangle\right\rangle P_z,\end{aligned} \tag{5.118}$$

which can be simplified by assuming a Maxwellian distribution,

$$W = 3W_{zz} = \frac{3}{2} n k_B T_C. \tag{5.119}$$

By also assuming time variations slow on the scale of the momentum relaxation time, we arrive at a current equation of the form

$$\mathbf{J}_n = nq\mu_n\left[\boldsymbol{\mathcal{E}} - \nabla(\chi/q)\right] + \mu_n \nabla(nk_B T_C) - \mu_n(3nk_B T_C/2)\frac{\nabla m^*}{m^*}. \tag{5.120}$$

The first two terms in this current equation represent drift in the electric and quasi-electric fields, the second diffusion in concentration and temperature gradients, and the third term arises from the density of states effect. When the temperature is uniform, eq. (5.120) reduces to the expected result, eq. (4.141b)

5.9 Summary

By solving the BTE, we learn how electrons are distributed in momentum space as a function of location within the device. By solving the balance equations, we learn considerably less: the average number of electrons, their average momentum, and their average energy. This information, however, often suffices for analyzing the performance of a device, and it is far simpler to solve the balance equations than it is to solve the BTE directly. Much of present-day device analysis and simulation consists of solving balance equations. The accuracy of the solution depends on how the infinite series of balance equations is truncated, how the resulting equations are simplified, and how the relaxation times are defined. We return to these issues in Chapter 8 where the application of macroscopic transport equations to device analysis is discussed.

The drift–diffusion equation serves as the cornerstone of semiconductor device analysis, so our discussion of how it is derived was especially important. To put the momentum balance equation in drift–diffusion form, three assumptions were necessary. First, temporal variations were assumed to occur in a time much longer than the momentum relaxation time. Second, the drift component of the kinetic energy density was assumed to be negligible. And finally, a diagonal temperature tensor was assumed. Conventional drift–diffusion equations are based on even more assumptions. First, the thermoelectric effect is assumed to be small, which means that the field is either low or if high, that it is uniform. Second, the transport parameters, μ_n and D_n are assumed to be material-dependent but device-independent. This also necessitates low, or uniform high applied fields so that the transport parameters are either independent of field or else depend only on the local electric field. Modern devices contain both low and high-field regions, and the spatial variations are rapid. Many of the interesting phenomena that occur under such conditions will be described in Chapter 8.

References and further reading

Liboff discusses the derivation of balance, or hydrodynamic, equations for the BTE in

5.1 Liboff, R. L., *Kinetic Equations.* Gordon and Breach, New York, 1971.

Various forms of the balance equations for multi-valley semiconductors are described in

5.2 Blotekjaer, K. Transport equations for electrons in two-valley semiconductors. *IEEE Transactions on Electronic Devices,* **ED-17**, 38–47, 1970.

In the following paper, Baccarani and Wordeman simplify the balance equations and solve them for silicon devices.

5.3 Baccarani, G. and Wordeman, M. R. An investigation of steady-state velocity overshoot in silicon. *Solid-State Electronics,* **28**, 407–16, 1985.

The validity of the displaced Maxwellian approximation to the high-field distribution function is discussed in

5.4 Conwell, E. High field transport in semiconductors. In *Solid State Physics, Suppl.* 9. Ed. by Seitz, F., Turnbull, D. and Ehrenreich, H. Academic Press, New York, 1967.

Stratton's approach for deriving the current and energy flux equations is discussed in

5.5 Stratton, R. Diffusion of hot and cold electrons in semiconductor barriers. *Physical Review,* **126**(6), 2002–13, 1962.

5.6 Chen, D., Kan, E. C., Ravaioli, U., Shu, C.-W. and Dutton, R. W. An improved energy transport model including nonparabolicity and non-Maxwellian distribution effects. *IEEE Electronic Device Letters,* **13**, 26–8, 1992.

For a treatment of electron transport in compositionally nonuniform semiconductors, consult

5.7 Marshak, A. H. and van Vliet, K. M. Electrical currents in solids with position-dependent band structure. *Solid-State Electronics,* **21**, 417–28, 1978.

5.8 Marshak, A. H. and van Vliet, K. M. Carrier densities and emitter efficiency in degenerate materials with position-dependent band structure. *Solid-State Electronics,* **21**, 429–34, 1978.

5.9 Van Vliet, C. M., Nainaparampil, J. J. and Marshak, A. H. Ehrenfest derivation of the mean forces acting in materials with non-uniform band structure: a canonical approach. *Solid-State Electronics,* **38**, 217–23, 1995.

Problems

5.1 Show that the sum, $\sum_{\mathbf{p}} \nabla_p(\phi f)$, is indeed zero as argued in Section 5.1. Be sure to state the conditions on $\phi(\mathbf{p})$ and $f(\mathbf{p}, \mathbf{r}, t)$ that are required. **Hint**: Convert the sum to an integral in rectangular coordinates.

5.2 Derive expressions for $1/\tau_\phi(\mathbf{p})$ and for $\langle\langle 1/\tau_\phi\rangle\rangle$ analogous to eq. (5.14f) but valid for degenerate semiconductors.

5.3 Evaluate the thermal energy from the definition, eq. (5.44), assuming a displaced Maxwellian distribution function, and show that the result is eq. (5.45).

5.4 (a) Evaluate the tensor, W_{ij}, for a displaced Maxwellian distribution function and show that the result is

$$W_{ij} = \frac{nm^*}{2}\upsilon_{di}\upsilon_{dj} + \left[\frac{nk_{\mathrm{B}}T_{\mathrm{C}}}{2}\right]\delta_{ij}.$$

(b) When drift energy is negligibly small show that $W_{ij} = (W/3)\delta_{ij}$ where $W = 3nk_{\mathrm{B}}T_{\mathrm{C}}/2$ is the kinetic energy density.

These results demonstrate that a displaced Maxwellian with negligible drift energy satisfies the assumptions made in Section 5.5 to derive a drift–diffusion equation.

5.5 Stratton writes the steady-state current and energy flux equations for electrons as

$$J_{nz} = nq\mu_n\mathcal{E}_z + q\mathrm{d}(D_n n)/\mathrm{d}z \tag{P5.1}$$

and

$$S(T_{\mathrm{e}}) = -\kappa(T_{\mathrm{e}})\mathrm{d}T_e/\mathrm{d}z - (J_{nz}/q)\delta(T_{\mathrm{e}})k_{\mathrm{B}}T_e. \tag{P5.2}$$

$$J_{nz}\cdot\mathcal{E}_z = nB(T_{\mathrm{e}}) + \mathrm{d}S(T_{\mathrm{e}})/\mathrm{d}z, \tag{P5.3}$$

(see [5.5]). In these equations, $S(T_{\mathrm{e}})$ is the energy flux, $\delta(T_{\mathrm{e}})$ is described as 'the average kinetic energy transported per electron arising from the current flow', and $B(T_{\mathrm{e}})$ is the average rate of energy loss per carrier to the lattice.

Beginning with the hydrodynamic transport equations, eqs. (5.32), (5.37), and (5.52), recast them in the form of eqs. (P5.1), (P5.2), and (P5.3). You should assume that $Q_z = -\kappa_n(T_{\mathrm{e}})\mathrm{d}T_{\mathrm{e}}/\mathrm{d}z$ and should carefully list all assumptions that you make. Obtain expressions for $B(T_{\mathrm{e}})$ and $\delta(T_{\mathrm{e}})$. Under what conditions does $\delta(T_{\mathrm{e}}) = 5/2$? This exercise shows that the hydrodynamic flow equations and Stratton's equations, though derived from very different approaches, can be cast in very similar forms.

5.6 To derive the balance equation for the $\hat{z}$ component of the energy flux, eq. (5.38), we set $\phi(\mathbf{p}) = \upsilon_z E(\mathbf{p})$. Answer the following questions about the energy flux balance equation.

(a) Assume spherical and parabolic energy bands to show that the energy flux balance equation is

$$\frac{\partial F_{Wz}}{dt} = -\frac{\partial X_{iz}}{\partial x_i} + \frac{(-q)\mathcal{E}_z W}{m^*} + \frac{2(-q)\mathcal{E}_i W_{iz}}{m^*} - \left\langle\left\langle \frac{1}{\tau_{FW}} \right\rangle\right\rangle F_{Wz}, \tag{P5.4}$$

which, in symbolic notation is

$$\frac{\partial \mathbf{F}_W}{dt} = -\nabla \cdot \overleftrightarrow{X} + \frac{(-q)\boldsymbol{\mathcal{E}}W}{m^*} + 2(-q)\frac{\boldsymbol{\mathcal{E}} \cdot \overleftrightarrow{W}}{m^*} - \left\langle\left\langle \frac{1}{\tau_{FW}} \right\rangle\right\rangle \mathbf{F}_W. \tag{P5.5}$$

The tensor, $\overleftrightarrow{X}$, is defined as

$$X_{ij} \equiv \frac{1}{\Omega} \sum_{\mathbf{p}} v_i v_j E(\mathbf{p}) f. \tag{P5.6}$$

(b) Assuming that $f(\mathbf{p})$ is a Maxwellian, evaluate the elements of X_{ij} to show that

$$X_{ij} = \frac{5}{3} \frac{k_B T_C}{m^*} W \delta_{ij}.$$

5.7 According to Widiger et al. (*IEEE Transactions on Electronic Devices*, **ED-32**, 1092–1102, 1985) the energy flux balance equation can be written as

$$\frac{1}{\langle\langle 1/\tau_{FW} \rangle\rangle} \frac{\partial \mathbf{F}_W}{\partial t} + \mathbf{F}_W = -\mu_E W \boldsymbol{\mathcal{E}} - \nabla(D_E W).$$

(a) Begin with eq. (5.38) and derive this expression. List all assumptions that are required.
(b) Show that $\mu_E = \alpha\mu$ and obtain an expression for α. Simplify your expression for power law scattering (i.e., $\tau = \tau_0(E(p)/k_B T)^s$).
(c) Obtain an 'Einstein relation' for D_E and μ_E.

5.8 Derive the energy balance equation for a semiconductor with a position-dependent effective mass.

5.9 Begin at the definition of the ensemble momentum relaxation rate for a Maxwellian, eq. (5.72), and verify the result for power law scattering, eq. (5.74).

5.10 Evaluate the heat flux from (5.48) using a displaced Maxwellian distribution function. **Hint**: sum over the peculiar momentum, $\mathbf{p}_c = m^*\mathbf{c}$, rather than over $\mathbf{p}$, and exploit the symmetry of the integrand. Show that the heat flux for a displaced Maxwellian is identically zero.

5.11 Assume that the low-field distribution function has the form

$$f(\mathbf{p}) = f_S(p) + f_A(\mathbf{p})$$

where f_S is symmetric in $\mathbf{p}$ and $f_A = g(\mathbf{p}) \cdot \mathcal{E}_z$ is antisymmetric. The function, $g(\mathbf{p})$, is material-dependent. Show that the mobility is a material-dependent number that is independent of the electric field.

5.12 Compare $\tau_m(p)$ as given by eq. (5.18) with $\tau_f(p)$ as given by eqs. (3.69) and (3.75).
(a) Show that $\tau_f(p) = \tau_m(p)$ for elastic scattering.
(b) Show that $\tau_f(p) = \tau_m(p)$ for isotropic scattering.

5.13 Assume that the distribution function is given by eq. (3.51) which was derived from the relaxation time approximation. Assuming power law scattering, answer the following.
(a) Derive an expression for $\langle\langle 1/\tau_m \rangle\rangle$ and show that the result is eq. (5.75).

(b) Assume isotropic scattering, and derive an expression for $\langle\langle 1/\tau_E \rangle\rangle$. Explain the significance of the result.

5.14 The drift–diffusion equation, eq. (5.101), is sometimes written as

$$\mathbf{J}_n = q\mu_n n\boldsymbol{\mathcal{E}} + q\nabla(D_n n).$$

What assumption is required in order to write eq. (5.101) in this form?

5.15 Begin with the general prescription, eq. (5.14), and derive a balance equation for $nk_B T_C$ where T_C is the carrier temperature. Approach the problem as follows:
(a) Determine the appropriate $\phi(\mathbf{p})$.
(b) Evaluate $\mathbf{F}_\phi$, and express the result in terms of T_C, υ_d, and the heat flux, $\mathbf{Q}$.
(c) Evaluate G_ϕ.
(d) Write out the complete balance equation.

5.16 Derive a balance equation for the tensor, W_{ij}. You can check your result by comparing it with Eq. (6) in Bringer, A and Schon, G. *Journal of Applied Physics*, **64**, 2447–2455, 1988.

5.17 Begin with the general prescription, eq. (5.14), and derive a balance equation for the carrier flux. Use this result to derive a drift–diffusion equation without assuming spherical, parabolic energy bands.

5.18 Let $\phi = \delta(\mathbf{p} - \mathbf{p}_o)$ and derive the corresponding balance equation. Explain the significance of n_ϕ and of the balance equation.

5.19 Assume a hypothetical, two-valley semiconductor with the valleys labeled 1 and 2. (These may represent the lower, Γ-valley and the equivalent, L-valleys in GaAs, for example.) Also assume isotropic intervalley scattering and that the average intervalley scattering rate from valley 1 to 2 is $\langle 1/\tau_{12} \rangle$ and from valley 2 to 1, $\langle 1/\tau_{21} \rangle$. Derive balance equations for this multi-valley semiconductor, and answer the following.
(a) Show that the carrier balance equation for valley 1 is

$$\frac{\partial n_1}{\partial t} = \frac{1}{q}\nabla \cdot \mathbf{J}_{n1} + \left\langle \frac{1}{\tau_{21}} \right\rangle n_2 - \left\langle \frac{1}{\tau_{12}} \right\rangle n_1.$$

(b) Show that the momentum balance equation for valley 1 is

$$\frac{\partial \mathbf{J}_{n1}}{\partial t} = \frac{-2(-q)\nabla \cdot \overleftrightarrow{W}_1}{m^*} + q^2 n_1 \frac{\boldsymbol{\mathcal{E}}}{m^*} - \left\langle \frac{1}{\tau_{m1}} \right\rangle \mathbf{J}_{n1}.$$

(c) Finally, show that the energy balance equation is

$$\frac{\partial W_1}{\partial t} = -\nabla \cdot \mathbf{F}_{W_1} + \mathbf{J}_{n1} \cdot \boldsymbol{\mathcal{E}} - \left\langle \frac{1}{\tau_{E1}} \right\rangle (W_1 - W_1^0) + \left\langle \frac{1}{\tau_{21}} \right\rangle W_2 - \left\langle \frac{1}{\tau_{12}} \right\rangle W_1.$$

5.20 Using Stratton's approach, derive an expression for the energy flux, eq. (5.91). Compare your answer to the result obtained from the balance equation approach.

5.21 In this chapter we have generally assumed simple, parabolic energy bands. Derive the current equation for a semiconductor with an energy-dependent effective mass. Proceed as follows.
(a) Beginning with the steady-state Boltzmann equation,

$$\upsilon_z \frac{\partial f}{\partial z} - q\mathcal{E}_z \frac{\partial f}{\partial p_z} = \left.\frac{\partial f}{\partial t}\right|_{\text{coll}}$$

multiply by p_z/Ω and sum over momentum to show:

$$\frac{1}{\Omega}\sum_{\mathbf{p}} p_z \frac{\partial f}{\partial t}\bigg|_{\text{coll}} = nq\mathcal{E}_z + 2\frac{\partial(nu_{zz})}{\partial z}. \tag{P5.7}$$

How is u_{zz} defined?

(b) Show that the collision integral can be written as

$$\frac{1}{\Omega}\sum_{\mathbf{p}} p_z \frac{\partial f}{\partial t}\bigg|_{\text{coll}} = -n\langle m^*(E)\upsilon_z/\tau_m\rangle. \tag{P5.8}$$

(c) Using the results from parts (a) and (b), show that the current equation is

$$J_{nz} = nq\mu_n\mathcal{E}_z + 2\mu_n\frac{\partial(nu_{zz})}{\partial z}$$

where

$$\mu_n \equiv \frac{q\langle\upsilon_z\rangle}{\langle m^*\upsilon_z/\tau_m\rangle}. \tag{P5.9}$$

It is also possible to remove the assumption of spherical energy bands. See Bandyopadhyay, S., et al., *IEEE Transactions on Electron Devices,* **34**, pp. 392–399, 1987.

6 Monte Carlo simulation

In Chapter 3 we introduced the Boltzmann Transport Equation (BTE) as an alternative to calculating the position and momentum versus time for each carrier within a device. The BTE is usually very difficult to solve; it is much easier to simulate the trajectories of individual carriers as they move through a device under the influence of electric fields and random scattering forces. Since each path is determined by choosing random numbers (properly distributed to reflect the probabilities of the various scattering events) the technique is a game of chance which has become known as Monte Carlo simulation. If the number of simulated trajectories is large enough, the average results are a good approximation to the average behavior of the carriers within a real device. In many cases, Monte Carlo simulation is the most accurate technique available for simulating transport in devices; it is frequently the standard against which the validity of simpler approaches is gauged.

Much of our understanding of high-field transport in bulk semiconductors and in devices has been obtained through Monte Carlo simulation, so it is important to understand the basics of the method. Because it directly mimics the physics, an understanding of the technique is also useful for the insight it affords. This chapter's emphasis is on the underlying principles of the Monte Carlo technique and on how the results of a Monte Carlo simulation are interpreted. (For the important details involved in actually implementing a Monte Carlo simulation program, consult the chapter references.) The chapter begins by discussing how to simulate carrier trajectories in semiconductors. Approaches for applying the method to bulk transport and to transport in devices are then described.

To make our discussion concrete, we define a 'model semiconductor' chosen to roughly approximate the behavior of silicon at room temperature. (In Section 6.5 we discuss how to generalize the approach to use a numerical table of the full band structure.) The equivalent conduction band minima are assumed to be spherical and parabolic with an effective mass ratio of unity. Four types of electron scattering processes will be considered; the first is ADP scattering in the form of eq. (2.84), in the elastic limit, with a numerical value of

$$\frac{1}{\tau_1} = (2 \times 10^{13})\sqrt{E(p)/q}. \tag{6.1}$$

The second is equivalent intervalley scattering by phonon absorption and is described by

$$\frac{1}{\tau_2} = (1.5 \times 10^{13)}\sqrt{[E(p)/q] + 0.050}, \tag{6.2}$$

and the third is equivalent intervalley scattering by phonon emission,

$$\frac{1}{\tau_3} = (1 \times 10^{14})\sqrt{[E(p)/q] - 0.050}, \tag{6.3}$$

which applies when $E(p) > 0.050\,\text{eV}$. Notice that both eqs. (6.2) and (6.3) are in the form of eq. (2.86). Finally, we include ionized impurity scattering in the Conwell–Weisskopf formulation as stated in homework problem 2.2,

$$\frac{1}{\tau_4} = (1 \times 10^{13})\left(\frac{N_{\mathrm{I}}\,(\mathrm{cm}^{-3})}{10^{16}}\right)^{1/3} \sqrt{E(p)/q}. \tag{6.4}$$

For this model semiconductor, the total scattering rate versus energy, $\Gamma(p)$, is the sum of the rates due to these four independent processes,

$$\Gamma(p) = \sum_{i=1}^{4} \frac{1}{\tau_i(p)}, \tag{6.5}$$

and is plotted versus energy in Fig. 6.1a. In Fig. 6.1b the contributions to Γ from each of the four individual processes are displayed. We shall refer frequently to these figures as the simulation algorithm is described.

6.1 Particle simulation

Figure 6.2 shows the trajectory of a representative electron moving under the influence of an electric field. (A bulk semiconductor is assumed for now; the

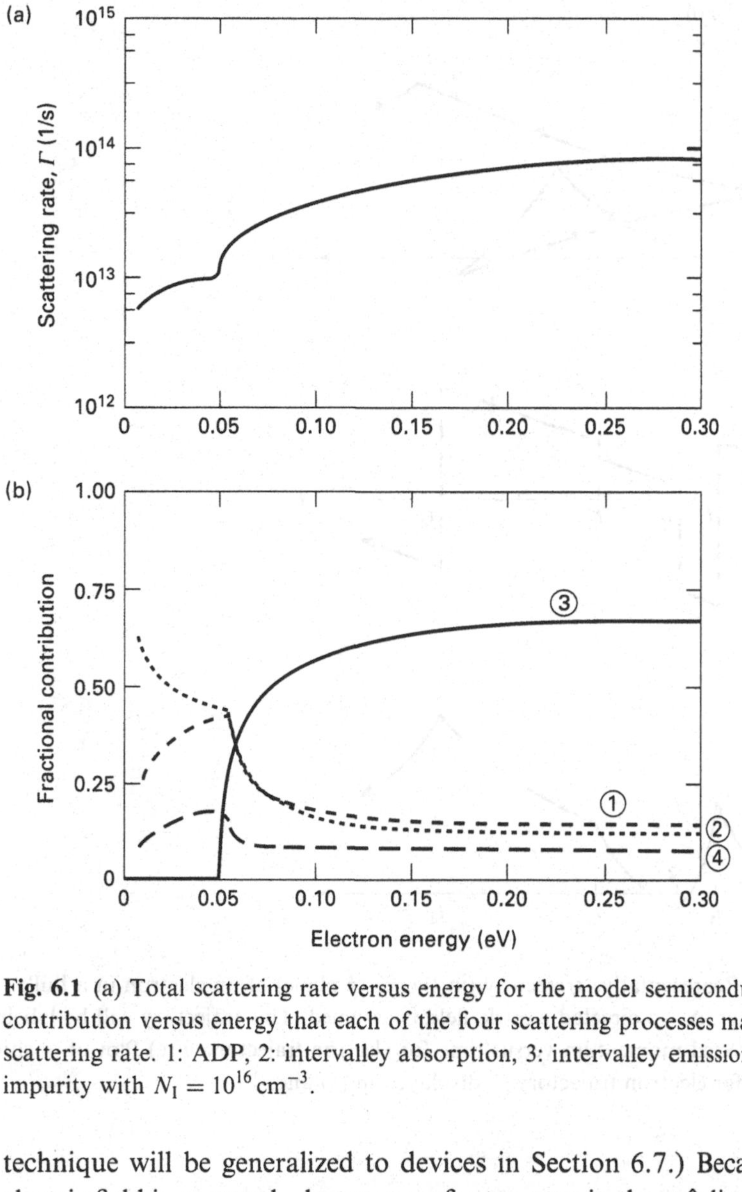

Fig. 6.1 (a) Total scattering rate versus energy for the model semiconductor. (b) Fractional contribution versus energy that each of the four scattering processes makes to the total scattering rate. 1: ADP, 2: intervalley absorption, 3: intervalley emission, and 4: ionized impurity with $N_I = 10^{16}\,\text{cm}^{-3}$.

technique will be generalized to devices in Section 6.7.) Because a $-\hat{z}$-directed electric field is assumed, electrons prefer to move in the $+\hat{z}$ direction according to

$$\mathbf{F}_e = \frac{d\mathbf{p}}{dt} = (-q)\boldsymbol{\mathcal{E}} \tag{6.6}$$

and frequently scatter off phonons and impurities. Because the duration of a collision is typically much shorter than the duration of the *free-flight* between collisions, collisions are treated as instantaneous events. Figure 6.2b shows a representative electron trajectory in momentum space. Between collisions, the electron's momentum increases with the time according to eq. (6.6), but scattering produces an instant change in the electron's momentum.

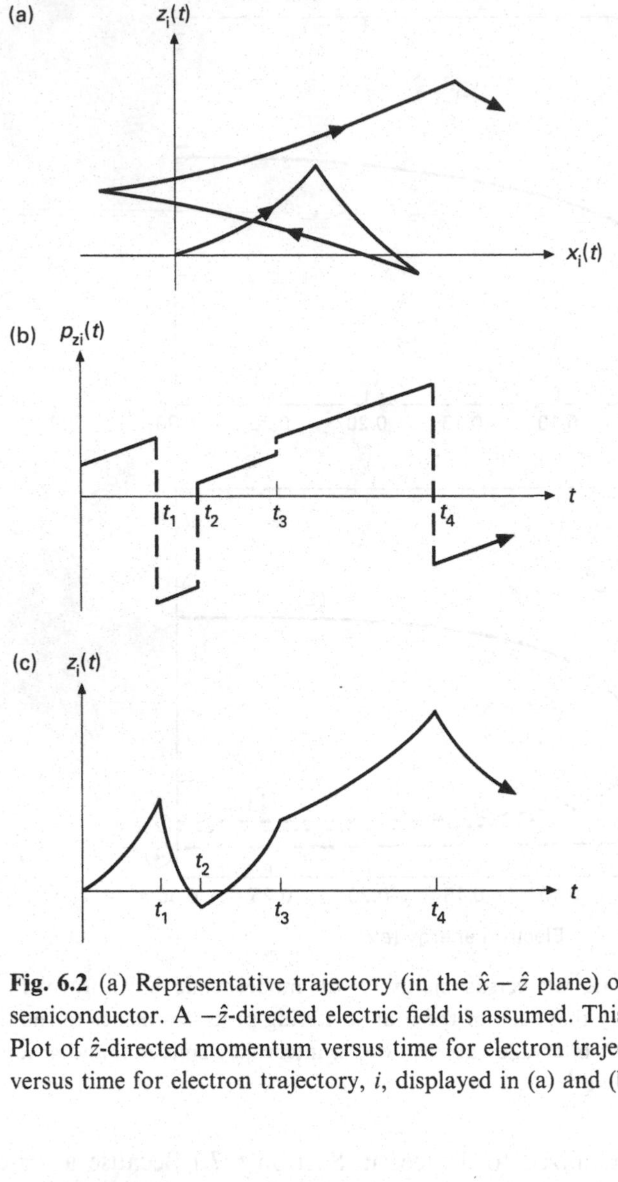

Fig. 6.2 (a) Representative trajectory (in the $\hat{x} - \hat{z}$ plane) of an electron in a bulk semiconductor. A $-\hat{z}$-directed electric field is assumed. This trajectory will be labeled, i. (b) Plot of $\hat{z}$-directed momentum versus time for electron trajectory, i (c) Plot of position, z, versus time for electron trajectory, i, displayed in (a) and (b).

The carrier's position varies with time according to

$$\mathbf{r}(t) = \mathbf{r}(0) + \int_0^t \upsilon(t')\mathrm{d}t'. \tag{6.7}$$

Figure 6.2c shows $z(t)$ for the trajectory displayed in Fig. 6.2b. Notice that collisions produce instantaneous changes in momentum but not in position. The reason is that the change in position is $\Delta r = \upsilon \Delta t$, where υ is the average velocity during the collision, and Δt is the duration of the collision. Because we assume that Δt is zero, no change in position can occur.

To simulate a single free flight and scattering event, a sequence of four random numbers is generated. The first specifies the free-flight duration during which the carrier moves in accordance with Newton's laws. At the end of the free flight, the carrier's position and momentum are updated according to eqs. (6.6) and (6.7). Next, a random number is generated to identify the scattering event that terminated the free flight. The last step is to determine the final state after scattering. The energy may increase or decrease due to phonon absorption or emission, and the direction of β will change. Two more random numbers are then generated to specify the polar and azimuthal angles after scattering. (As discussed in Section 6.5, final state selection is more involved when full, numerical energy bands are used.) A flow chart for this basic algorithm is shown in Fig. 6.3.

In the next few sections, we discuss each of the components of a Monte Carlo simulation and describe how the random numbers are chosen to mimic the physics. In Section 6.7 we'll explain how the technique is applied to devices.

6.2 Free flight

After moving for a time, t, under the influence of a $\hat{z}$-directed electric field, the electron's momentum and position are obtained from eqs. (6.6) and (6.7) as

$$\begin{aligned} p_x(t) &= p_x(0) \\ p_y(t) &= p_y(0) \\ p_z(t) &= p_z(0) + (-q)\mathcal{E}_z t \end{aligned} \tag{6.8}$$

and

$$\begin{aligned} x(t) &= x(0) + \frac{p_x(0)}{m^*}t \\ y(t) &= y(0) + \frac{p_y(0)}{m^*}t \\ z(t) &= z(0) + \frac{E(t) - E(0)}{(-q)\mathcal{E}_z}, \end{aligned} \tag{6.9}$$

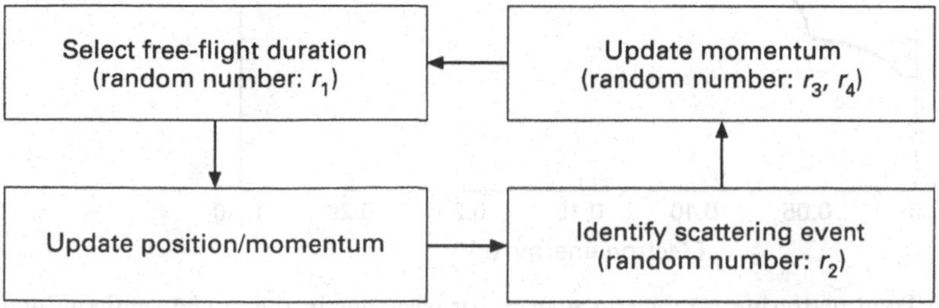

Fig. 6.3 Algorithm for Monte Carlo simulation of carrier trajectories.

where, because parabolic energy bands are assumed,

$$E(t) = \frac{p^2(t)}{2m^*}. \tag{6.10}$$

We assume that the field, $\mathcal{E}_z$, is nearly constant for the duration of the free flight. The first question to consider is: how long should the free flight continue – or what is the time of the next collision? The duration of the free flight is directly related to the scattering rate – the higher the scattering rate the shorter the average free flight.

The rate at which carriers scatter is

$$\Gamma(p) = \sum_{i=1}^{k} \frac{1}{\tau_i(p)}, \tag{6.11}$$

where the sum includes all the k scattering mechanisms ($k = 4$ for our model semiconductor). Because Γ varies with energy and the carrier's energy changes with time as it moves in the field, Γ is a function of time. For now, we simplify the problem by approximating $\Gamma(t)$ by a constant, Γ_0 (see Fig. 6.4). This approximation greatly simplifies the mathematics, and a clever trick will enable us to apply the result to real semiconductors.

Consider an ensemble consisting of carriers that have not undergone a collision since $t = 0$. Since each carrier has a scattering rate of Γ_0, the time rate of change of the collision-free carrier density, n_{CF}, is

$$\frac{dn_{CF}}{dt} = -\Gamma_0 n_{CF} \tag{6.12a}$$

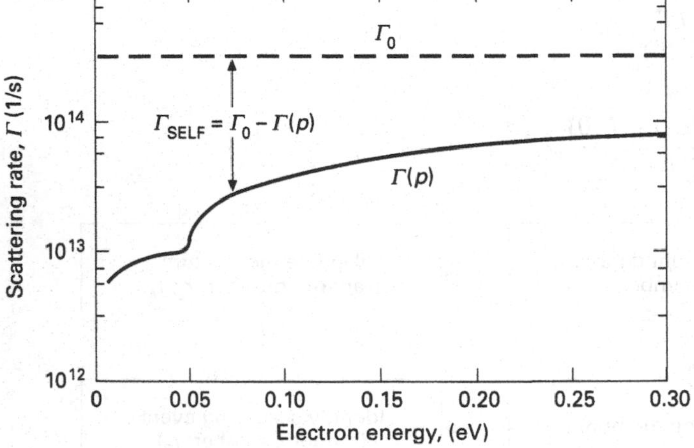

Fig. 6.4 Total scattering rate versus energy for electrons in the model semiconductor. An approximation that assumes $\Gamma(p) = \Gamma_0$ is also displayed.

which can be solved to find

$$n_{CF}(t) = n_{CF}(0)e^{-\Gamma_0 t}. \tag{6.12b}$$

The probability that an electron survives until time t, without scattering is, therefore,

$$\frac{n_{CF}(t)}{n_{CF}(0)} = e^{-\Gamma_0 t}. \tag{6.13}$$

The probability that a carrier undergoes its *first* collision between t and $t + dt$ is the scattering rate times the probability that it survives until time t, or

$$\mathcal{P}(t)dt = \Gamma_0 e^{-\Gamma_0 t}dt. \tag{6.14}$$

The free-flight durations must be selected at random in accordance with the probability distribution function specified by eq. (6.14). We need to ensure that the probability of selecting a random number between r and $r + dr$ is equal to the probability of selecting a collision time between t and $t + dt$; that is

$$\mathcal{P}(r)dr = \mathcal{P}(t)dt. \tag{6.15}$$

For a random number generator that produces random numbers uniformly distributed between 0 and 1, $\mathcal{P}(r) = 1$ and

$$dr = \Gamma_0 e^{-\Gamma_0 t}dt$$

which can be integrated to find

$$\int_0^{r_c} dr = \Gamma_0 \int_0^{t_c} e^{-\Gamma_0 t}dt,$$

or

$$r_c = 1 - e^{-\Gamma_0 t_c}. \tag{6.16}$$

The randomly distributed collision times are, therefore, related to these computer-generated random numbers by

$$\boxed{t_c = -\frac{1}{\Gamma_0}\ln(r_1)}\,, \tag{6.17}$$

where $r_1 = 1 - r_c$. (Note that r_1 is also a random number uniformly distributed between 0 and 1.) Figure 6.5a shows a histogram of 10 000 free-flight durations computed from eq. (6.17) with the aid of a pseudo-random number generator. A value of $\Gamma_0 = 3 \times 10^{14}\ s^{-1}$ was assumed. As expected, the histogram approximates the probability distribution, eq. (6.14), and the average duration of a free flight is close to $1/\Gamma_0$.

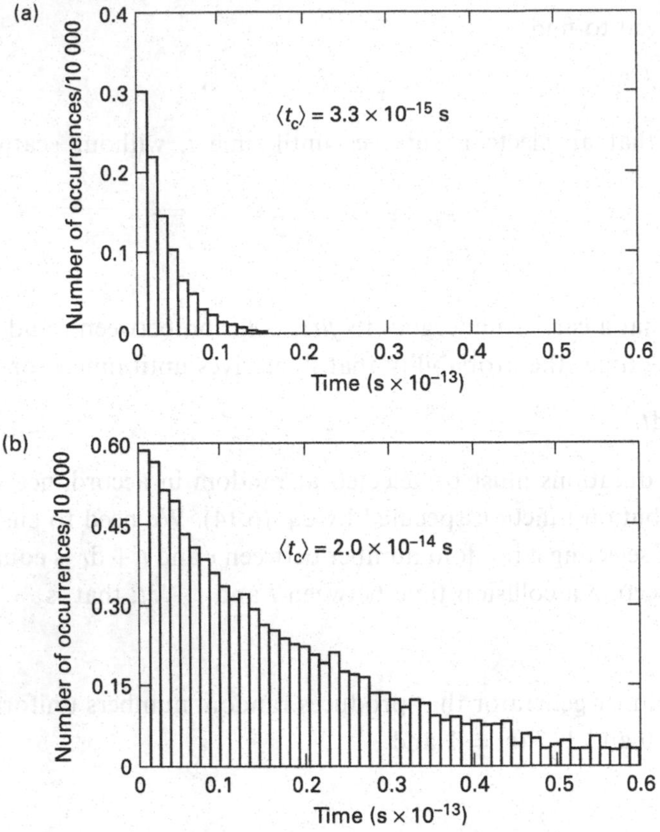

Fig. 6.5 Histogram of (a) 10 000 free flight durations for a semiconductor with $\Gamma(p) = \Gamma_0$ and (b) 10 000 real free flight durations (those terminated by a real scattering event) for the model semiconductor. The times were computed from eq. (6.17) assuming $\Gamma_0 = 3 \times 10^{14}\,\text{s}^{-1}$ by summing the times for successive random numbers until a real scattering event was detected. The electron's kinetic energy was assumed to be 0.15 eV. The times were computed from eq. (6.17).

Although eq. (6.17) is a simple prescription for generating free flight durations, it is not valid for real semiconductors in which $\Gamma(p)$ varies with energy. Fortunately there is a simple expedient, we just add to the k real scattering mechanisms under consideration a fictitious one called *self-scattering*. As Fig. 6.4 shows,

$$\frac{1}{\tau_{k+1}} = \Gamma_{\text{self}}(p) = \Gamma_0 - \Gamma(p), \tag{6.18}$$

where $\Gamma(p)$ includes only the real scattering mechanisms. Note that $\Gamma_{\text{self}}(p)$ is energy dependent as displayed in Fig. 6.4 and that Γ_0 (which is usually specified by the simulation program's user) *must* be greater than $\Gamma(p)$ over the energies occurring during simulation to ensure that Γ_{self} is always positive.

With the addition of self-scattering, the total scattering rate is constant, so eq. (6.17) now applies, but we must be certain that the fictitious scattering mechanism introduced does not alter the problem. Real scattering events alter the carrier's momentum, but when a self-scattering event occurs we do not change the carrier's momentum. Self-scattering does not affect the carrier's trajectory – it simply makes the scattering rate constant so that eq. (6.17) applies. The histogram of free flight durations in Fig. 6.5a includes both the real and the fictitious scattering events. When a free flight is terminated by a fictitious scattering event, a new random number is generated and the free flight continues. Figure 6.5b shows a histogram of 10 000 real free flights for carriers with $E(p) = 0.5\,\text{eV}$. Note that the average duration of a real free flight is $1/\Gamma(p)$ – not $1/\Gamma_0$.

Figure 6.6 shows the momentum–space and position–space trajectories for an electron with the real and fictitious scattering events indicated. Several self-scattering events typically precede each real event. The price for the simplicity of using eq. (6.17) to generate free flight durations is an added computational burden due to the increased number of scattering events, most of which do not affect the carrier's trajectory at all. (Techniques to minimize the number of self-scattering events and thereby maximize computational efficiency are

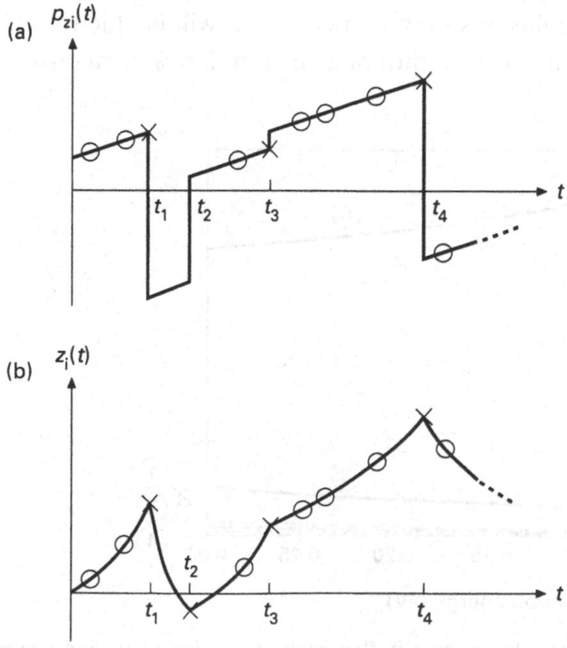

Fig. 6.6 (a) Momentum versus time for an electron trajectory. (b) Position versus time for the electron trajectory. Fictitious scattering events are identified by an 'O', real events by an '×'

described in Jacoboni and Reggiani [6.1].) Our next task is to learn to separate the real and fictitious scattering events.

6.3 Identification of the scattering event

After selecting the duration of the free flight using the prescription, eq. (6.17), the carrier's momentum, position, and energy are updated at time t_c^- according to eqs. (6.8), (6.9), and (6.10). Collisions alter the carrier's momentum, but each mechanisms does so differently. To update the momentum at t_c^+, we must first identify the scattering event that terminated the free flight and determine whether it was real or fictitious.

As Fig. 6.1b illustrates, the contribution that each individual scattering mechanism makes to the total scattering rate varies considerably with energy. Since we have now added a $(k+1)$st scattering mechanism, the contribution of self-scattering must also be included. By using eqs. (6.1)–(6.4) for the real processes and eq. (6.18) for the fictitious process, we construct the new version of Fig. 6.1b displayed in Fig. 6.7. Because the carrier's energy at the end of the free flight is known, the probabilities of the various events can be read directly from the figure. As a specific example, let's assume that $E(t_c^-) = 0.15\,\text{eV}$, then Fig. 6.7 shows that 83.4% of the collisions such carriers suffer will be due to self-scattering, 10.5% will occur by intervalley phonon emission, 2.6% by acoustic phonon

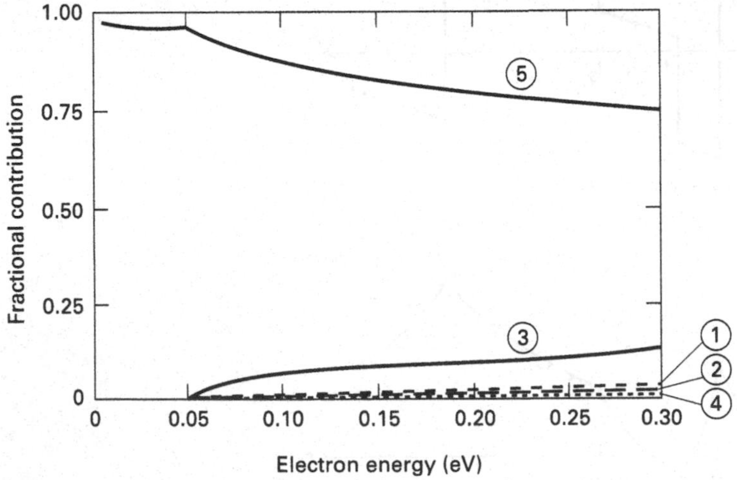

Fig. 6.7 Fractional contribution versus energy that each of the five scattering processes makes to the total scattering rate. The first four processes are those defined for the model semiconductor in eqs. (6.1)–(6.4), and the fifth is self-scattering as specified by eq. (6.18). A value of $\Gamma_0 = 3.0 \times 10^{14}\ \text{s}^{-1}$ is assumed.

scattering, 2.2% by intervalley phonon absorption, and 1.3% by ionized impurity scattering. By adding up the various contributions in this order, we obtain the graph shown in Fig. 6.8a. Selection of a random number, r_2, uniformly distributed from zero to one locates a region in the graph and identifies the scattering event.

The mathematical description of the identification procedure is to select mechanism l, if

$$\frac{\sum_{i=1}^{\ell-1} \frac{1}{\tau_i(p)}}{\Gamma_0} \leq r_2 < \frac{\sum_{i=1}^{\ell} \frac{1}{\tau_i(p)}}{\Gamma_0} \qquad l = 1, 2, 3, \ldots k+1. \tag{6.19}$$

The procedure consists of determining the carrier's energy just before the collision, constructing a bar graph like that in Fig. 6.8a, choosing a random number, r_2, and locating it within the bar graph to identify the scattering event. (This procedure was used to separate the real from the fictitious scattering events in Fig. 6.5a in order to produce Fig. 6.5b.) When prescription (6.19) is applied to 10 000 scattering events in the model semiconductor, the distribution of scattering events displayed in Fig. 6.8b results. As expected, the distribution of scattering events identified by random number r_2, is very near the physical distribution summarized in Fig. 6.8a.

6.4 Selecting a final state after scattering

Before beginning a new free flight, the carrier's state (its location in momentum space) is updated to reflect the effect of scattering. In general, both the magnitude and orientation of the carrier's momentum are altered by scattering. For spherical, parabolic energy bands, the magnitude of the carrier's momentum just after scattering is

$$p(t_c^+) = p' = \sqrt{2m^*[E(t_c^-) + \Delta E]}, \tag{6.20}$$

where ΔE is the change in energy associated with the particular scattering event selected by random number r_2. For elastic scattering, $\Delta E = 0$, and for inelastic scattering it is typically a phonon energy. Because there is a unique ΔE associated with each scattering event, random number r_2 also determines the magnitude of the carrier's momentum after scattering, but to update the orientation of $\mathbf{p}$, two more random numbers must be selected.

When updating the orientation of $\mathbf{p}$, it is convenient to work in a coordinate system in which the $\hat{z}$ axis is directed along the initial momentum $\mathbf{p}$. The new coordinate system $(\hat{x}_r, \hat{y}_r, \hat{z}_r)$ is obtained by rotating the $(\hat{x}, \hat{y}, \hat{z})$ system by an

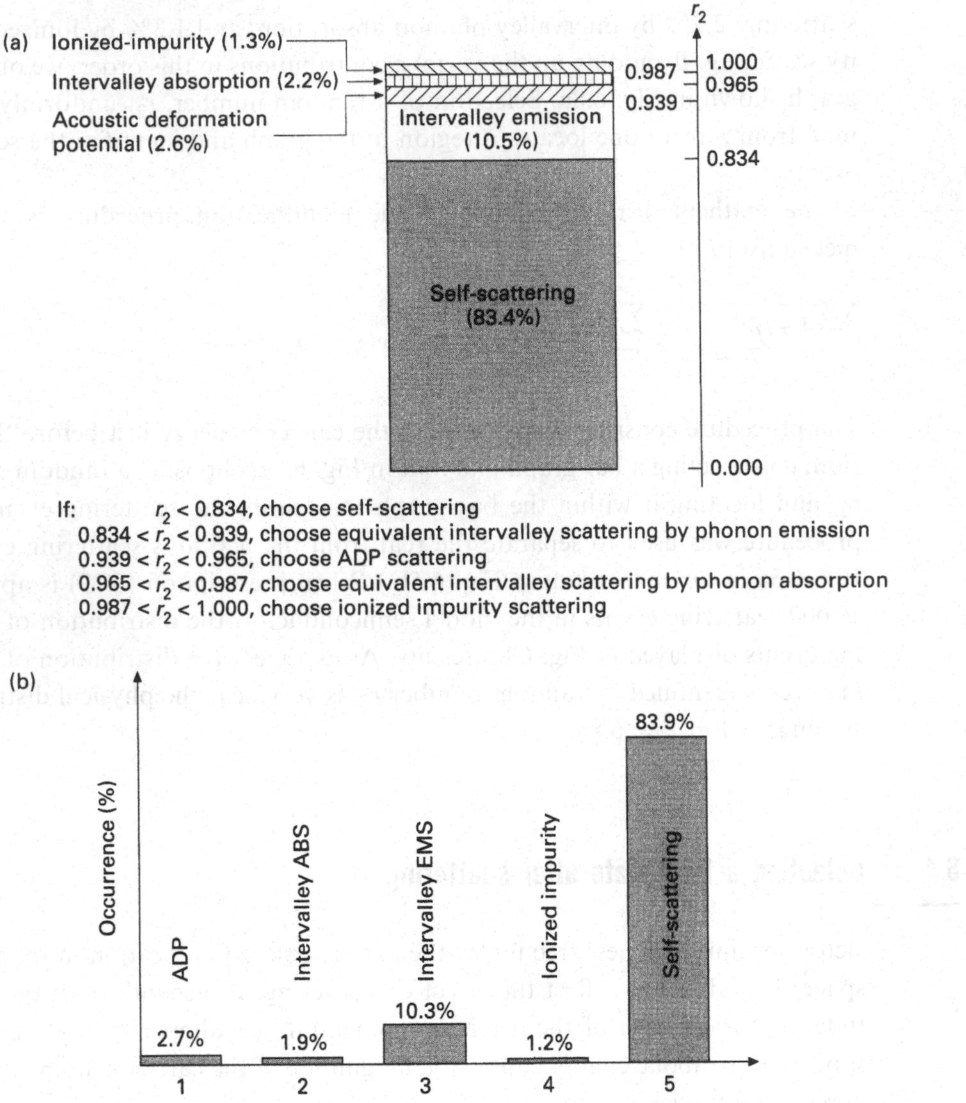

Fig. 6.8 (a) Illustration of the procedure for identifying a scattering event. An electron with kinetic energy of 0.15 eV in the model semiconductor is assumed. The various contributions were obtained from Fig. 6.7. (b) Result of identifying 10 000 scattering events for electrons with 0.15 eV kinetic energy in the model semiconductor. Note that the results are distributed very nearly as expected from Fig. 6.7.

angle ϕ about the $\hat{z}$ axis, then θ about $\hat{y}$ as illustrated in Fig. 6.9. The probability that $\mathbf{p}'$ lies between azimuthal angle β and $\beta + \mathrm{d}\beta$ is found by evaluating

$$\mathcal{P}(\beta)\mathrm{d}\beta = \frac{\mathrm{d}\beta \int_0^\infty \int_0^\pi S(\mathbf{p}, \mathbf{p}') \sin\alpha \mathrm{d}\alpha p'^2 \mathrm{d}p'}{\int_0^{2\pi} \mathrm{d}\beta \int_0^\infty \int_0^\pi S(\mathbf{p}, \mathbf{p}') \sin\alpha \mathrm{d}\alpha p'^2 \mathrm{d}p'}. \tag{6.21a}$$

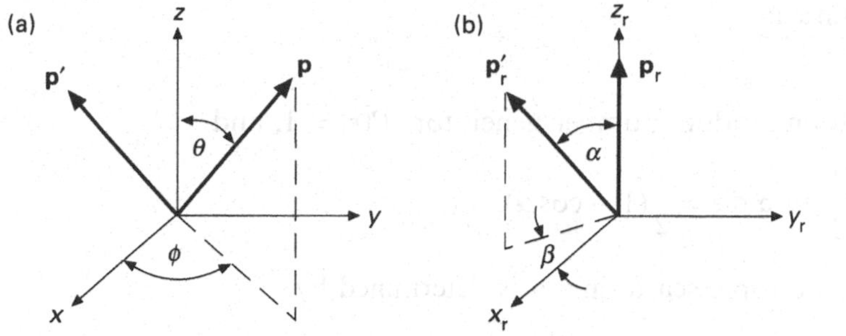

Fig. 6.9 (a) Scattering event in the $(\hat{x}, \hat{y}, \hat{z})$ coordinate system. The incident momentum is $\mathbf{p}$ and the scattered momentum, $\mathbf{p}'$. (b) The same scattering event in the rotated coordinate system, $(\hat{x}_r, \hat{y}_r, \hat{z}_r)$, which is obtained by rotating $(\hat{x}, \hat{y}, \hat{z})$ by an angle of ϕ about the $\hat{z}$-axis, then θ about the $\hat{y}$-axis. In the rotated system, the incident momentum is $\mathbf{p}_r$ and the scattered momentum $\mathbf{p}_r'$.

(Do not confuse the azimuthal angle, β, with the phonon wavevector, β.) Because our simple treatment of scattering makes the transition rate independent of β, the integration over β in the denominator can be performed directly, and we find

$$\mathcal{P}(\beta)\mathrm{d}\beta = \mathrm{d}\beta/2\pi, \tag{6.21b}$$

which states that the azimuthal angle is uniformly distributed between 0 and 2π. The azimuthal angle after scattering is specified by a third random number, r_3, according to

$$\boxed{\beta = 2\pi r_3}\,. \tag{6.22}$$

If r_3 is uniformly distributed from zero to one, then β will be uniformly distributed from 0 to 2π.

The prescription for selecting the polar angle α is slightly more involved because $S(\mathbf{p}, \mathbf{p}')$ may depend on α (recall that ionized impurity and polar optical phonon scattering favor small angle scattering). By analogy with eq. (6.21a), we find

$$\mathcal{P}(\alpha)\mathrm{d}\alpha = \frac{\sin\alpha\mathrm{d}\alpha \int_0^\infty \int_0^{2\pi} S(\mathbf{p}, \mathbf{p}')\mathrm{d}\beta p'^2 \mathrm{d}p'}{\int_0^{2\pi} \int_0^\infty \int_0^\pi [S(\mathbf{p}, \mathbf{p}')\sin\alpha\,\mathrm{d}\alpha]\mathrm{d}\beta p'^2 \mathrm{d}p'}. \tag{6.23}$$

Consider an isotropic scattering mechanism like acoustic phonon scattering for which $S(\mathbf{p}, \mathbf{p}') = C_{\mathrm{AP}}\delta(E' - E)/\Omega$, as in eq. (2.6). Because C_{AP} is independent of α, eq. (6.23) gives

$$\mathcal{P}(\alpha)\mathrm{d}\alpha = \frac{\sin\alpha\,\mathrm{d}\alpha}{2}.$$

The angle, α, is specified by a fourth random number according to

$$\mathcal{P}(r)dr = \frac{\sin\alpha\, d\alpha}{2}.$$

For a uniform random number generator, $\mathcal{P}(r) = 1$, and

$$\int_0^{r_4} dr = \frac{1}{2}\int_0^{\alpha} \sin\alpha\, d\alpha = \frac{1}{2}(1 - \cos\alpha),$$

so that for isotropic scattering, α is determined by

$$\boxed{\cos\alpha = 1 - 2r_4}\,. \tag{6.24}$$

Figures 6.10a and 6.10b show histograms of 10 000 azimuthal and polar scattering angles selected according to prescriptions, eqs. (6.22) and (6.24). Note that β is uniformly distributed between 0 and 2π, but α is not uniformly distributed between 0 and π (rather, it is $\cos\alpha$ that is uniformly distributed between -1 and $+1$).

For anisotropic scattering, small angle deflections are most probable. The procedure for selecting the polar angle begins with eq. (6.23), but the appropriate $S(\mathbf{p}, \mathbf{p}')$ must be used. An example calculation for ionized impurity scattering follows.

Example: polar angle selection for ionized impurity scattering

Ionized impurity scattering is described in the Conwell–Weisskopf approach as

$$S(\mathbf{p}, \mathbf{p}') = 0 \qquad (\text{for } \alpha \le \alpha_{\min})$$

and

$$S(\mathbf{p}', \mathbf{p}) = \frac{C_{CW}\delta(E' - E)}{\sin^4\left(\frac{\alpha}{2}\right)} \qquad (\text{for } \alpha > \alpha_{\min})$$

where C_{CW} was given in eq. (2.36b). In this case, the probability of scattering between α and $\alpha + d\alpha$, eq. (6.23), becomes

$$\mathcal{P}(\alpha)d\alpha = \frac{\left(\sin\alpha/\sin^4\frac{\alpha}{2}\right)d\alpha}{\int_{\alpha_{\min}}^{\pi}\left(\sin\alpha/\sin^4\frac{\alpha}{2}\right)d\alpha}.$$

The procedure leading to eq. (6.24) can now be followed to obtain

$$\cos\alpha = 1 - \frac{1 - \cos\alpha_{\min}}{1 - r_4\left[1 - \frac{1}{2}(1 - \cos\alpha_{\min})\right]}, \tag{6.25}$$

which relates the fourth random number to the polar angle for ionized impurity scattering. Figure 6.10c shows a histogram of 10 000 randomly selected polar angles for ionized impurity scattering. The contrast with Fig. 6.10b is sharp – ionized impurity scattering strongly favors small angle deflections.

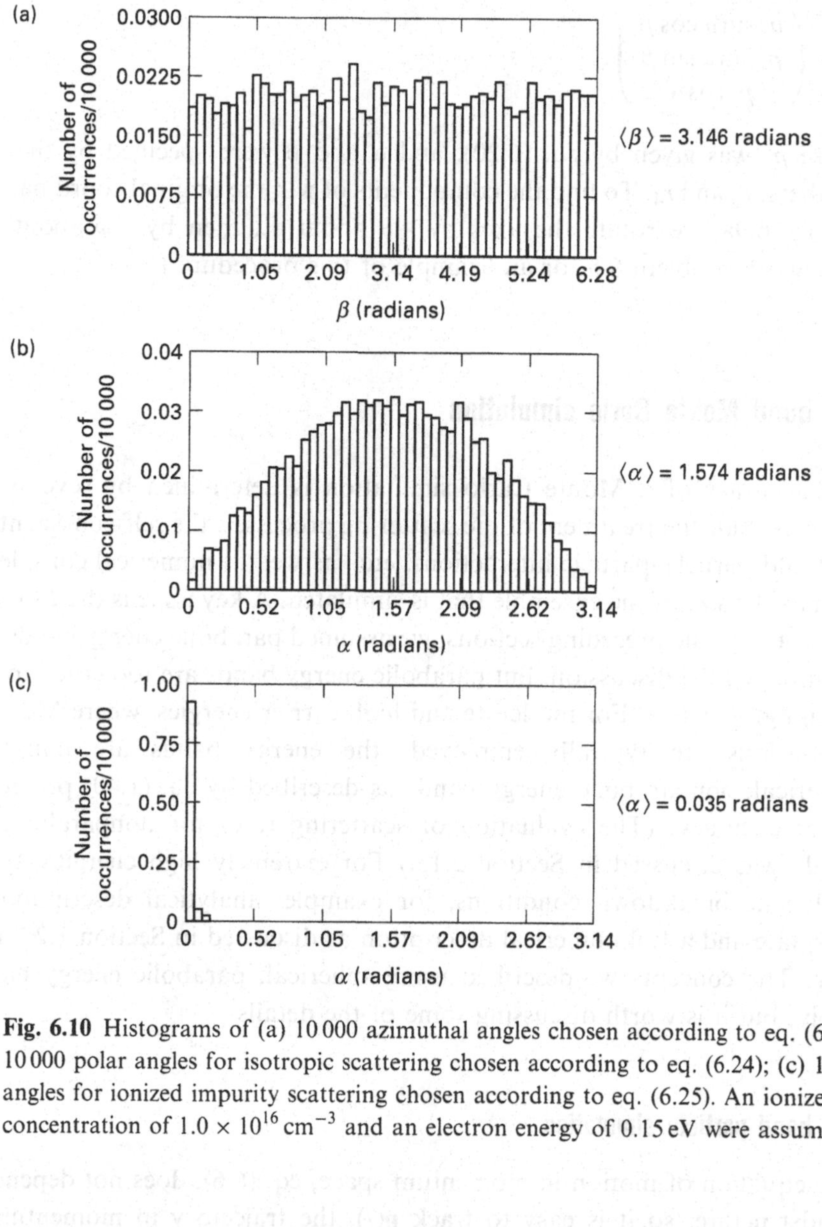

Fig. 6.10 Histograms of (a) 10 000 azimuthal angles chosen according to eq. (6.22); (b) 10 000 polar angles for isotropic scattering chosen according to eq. (6.24); (c) 10 000 polar angles for ionized impurity scattering chosen according to eq. (6.25). An ionized impurity concentration of 1.0×10^{16} cm^{-3} and an electron energy of 0.15 eV were assumed.

Having determined the orientation of $\mathbf{p}'$ in the rotated system, we now need to determine it in the original coordinate system. (Note that $\mathbf{p}$ and $\mathbf{p}'$ do not change when the coordinate system is rotated, but their components as measured along the coordinate axes do. When the initial and scattered momenta are labeled, $\mathbf{p}_r$ and $\mathbf{p}_r'$, their components are measured in the rotated system.) In the rotated coordinate system, the initial momentum, $\boldsymbol{p}_r$, was directed along the $\hat{z}_r$ axis as illustrated in Fig. 6.9b. After scattering, the components of $\mathbf{p}_r'$ in the rotated coordinate system are

$$\mathbf{p}_r' = \begin{pmatrix} p' \sin\alpha \cos\beta \\ p' \sin\alpha \sin\beta \\ p' \cos\alpha \end{pmatrix}, \tag{6.26}$$

where p' was given by eq. (6.20), and α and β were specified by the random numbers, r_3 and r_4. To find the components of $\mathbf{p}$ in the original coordinate system of Fig. 6.9a, we rotate the axes by $-\theta$ about $\hat{y}_r$, then by $-\phi$ about $\hat{z}_r$. (See homework problem 6.5 for an example of this procedure.)

6.5 Full band Monte Carlo simulation

The accuracy of a Monte Carlo simulation is determined by several factors. These include the treatment of the scattering processes, the self-consistent electric field and particle–particle interactions, etc., as well as numerical considerations, such as the size of the ensemble that is simulated. A key issue is the $E(\mathbf{p})$ description used. In the preceding sections, we assumed parabolic energy bands because it simplified the discussion, but parabolic energy bands are too crude to provide meaningful results. For moderate and high carrier energies, where Monte Carlo simulations are typically employed, the energy bands are nonparabolic. Spherical, nonparabolic energy bands as described by eq. (1.40) provide much better accuracy. (The evaluation of scattering rates for nonparabolic energy bands was discussed in Section 2.13.) For extremely high energies, typical of avalanche breakdown conditions, for example, analytical descriptions aren't adequate and a full numerical description as discussed in Section 1.2.2 is necessary. The concepts we described using spherical, parabolic energy bands still apply, but it is worth discussing some of the details.

6.5.1 Full band particle simulation

The equation of motion in momentum space, eq. (6.6), does not depend on the bandstructure, so it is easy to track $\mathbf{p}(t)$, the trajectory in momentum space. Equation (6.8) can still be used to update $\mathbf{p}$ after a short timestep. The energy can also be easily evaluated from the numerical table describing $E(\mathbf{p})$. Tracking particles in real space, however, is more involved. The real space trajectory is still described by eq. (6.7), but the carrier velocity is numerically evaluated from

$$v[\mathbf{p}(t)] = \nabla_p E[\mathbf{p}(t)]. \tag{6.27}$$

Equation (6.9) can be still used to update the carrier's position after a short timestep if $\mathbf{p}/m^*$ is replaced by the group velocity.

6.5.2 Full band scattering events

In a full band simulation, the electron–phonon scattering rate for a carrier at state **p** is evaluated from eq. (2.61) using eq. (2.71c) for A_β to find

$$\frac{1}{\tau(\mathbf{p}_i)} = \frac{1}{\Omega}\sum_{\beta}\left(\frac{\pi}{\rho\omega_\beta}\right)\left|I(\mathbf{p}',\mathbf{p})\right|^2\left|K_\beta\right|^2(N_\beta + 1/2 \mp 1/2)\delta(E' - E \mp \hbar\omega_\beta)\delta_{\mathbf{p}',\mathbf{p}\pm\hbar\boldsymbol{\beta}}, \tag{6.28a}$$

where $|I(\mathbf{p}',\mathbf{p})|$ is the overlap integral involving the Bloch functions [recall eq. (1.116)], and $|K_\beta|^2$ is the Fourier component of the perturbing potential which connects the initial and final states [recall eq. (2.58)]. The perturbing potential is anisotropic for electrostatic scattering processes, which favor small angle events. The phenomenological deformation potentials are usually taken to be constants, but rigorous calculations show that they can depend on the initial and final states [6.9]. The full dispersion curve which describes ω_β is usually used rather than assuming that $\omega = \omega_0$ for optical phonons and $\omega = \beta v_S$ for acoustic phonons. Finally, note that the sum is over the phonon wavevector, $\boldsymbol{\beta}$, which by momentum conservation is equivalent to summing over the final electron states, $\mathbf{p}'$.

Electrons in semiconductors may populate several different energy bands (recall Fig. 1.9), and electron–phonon scattering may occur from several different phonons (longitudinal, transverse, acoustic, or optical as shown in Fig. 1.27). The electron–phonon scattering rate should, therefore, be written as $1/\tau_{\eta,\nu}(\mathbf{p})$, where η refers to the type of phonon and v to the band the electron resides in before scattering. Equation (6.28a) becomes

$$\frac{1}{\tau_{\eta,\nu}(\mathbf{p}_i)} = \frac{1}{\Omega}\sum_{\nu',\boldsymbol{\beta}}\left(\frac{\pi}{\rho\omega_{\eta,\beta}}\right)\left|I(\nu',\mathbf{p}';\nu,\mathbf{p})\right|^2\left|K_{\eta,\nu'}(\boldsymbol{\beta})\right|^2$$
$$\times\,(N_{\eta,\beta} + 1/2 \mp 1/2)\delta(E'_\nu - E_\nu \mp \hbar\omega_{\eta,\beta})\delta_{\mathbf{p}',\mathbf{p}\pm\hbar\boldsymbol{\beta}}, \tag{6.28b}$$

which includes a sum over all of the final energy bands that the electron may occupy.

Because the bandstructure is tabulated throughout the Brillouin zone (in a set of cubes in **k**-space, for example), the sum in eq. (6.28) can be carried out directly to evaluate the scattering rate. By adding the contributions of all important mechanisms, a table of scattering rate versus energy like that displayed in Fig. 2.26 is constructed. Self-scattering, the selection of free-flight times, and identification of the free-flight terminating scattering events all proceed just as they do for simple energy bands.

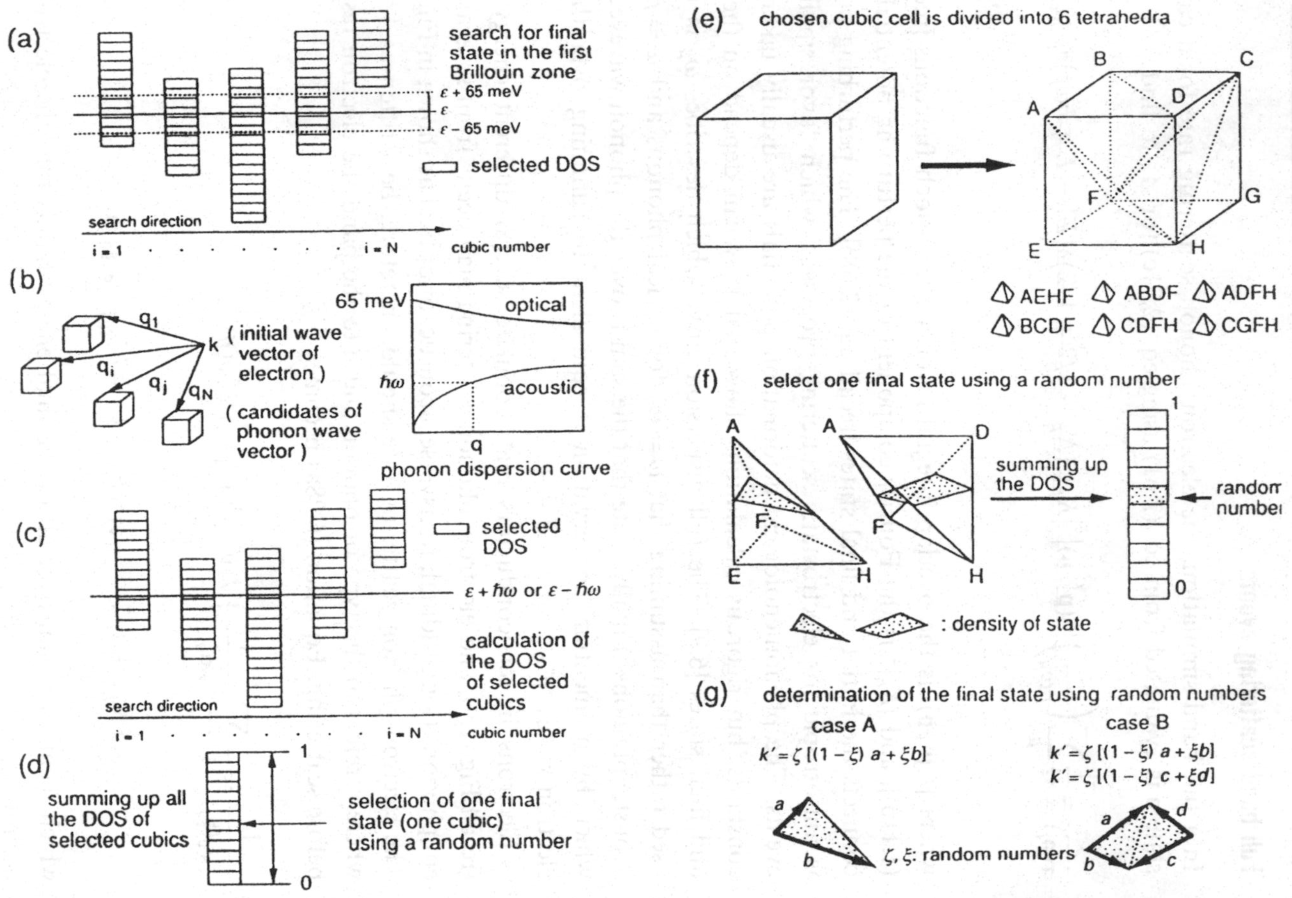

Fig. 6.11 Illustration of how the final state is selected in a full-band Monte Carlo simulation. (From [6.9].) (Reproduced with permission of American Institute of Physics.)

6.5.3 Selection of the final state

The last step in the procedure is to select a final state, and this is the most involved step in a full band Monte Carlo simulation. A final state that conserves energy and momentum must be selected. For simple bands, this was done analytically, but for full bands it involves a search throughout the Brillouin zone. Figure 6.11 summarizes the final state selection procedure.

The first step is a 'coarse search' throughout the discretized Brillouin zone to identify all cubes that may contain states that conserve energy and momentum. If, for example, the electron energy is E, and the maximum phonon energy for any transition is 65 meV, then all cubes that contain energies in the range $E \pm 65\,\text{meV}$ are located (Fig. 6.11a). The next step is a 'fine search'; each cube is subdivided into fine cubes, and each fine cube is examined. The difference between the momentum of each possible final state and the initial state momentum gives $\hbar\beta$, and the phonon dispersion curve gives the corresponding phonon energy (Fig. 6.11b). Only cubes containing fine cubes that satisfy energy and momentum conservation are retained (Fig. 6.11c).

After identifying all potential final state cubes, the density of states for each cube can be computed and a total sum computed. If the scattering process depends on β, each term in the sum is weighted appropriately. By selecting a random number between zero and one and comparing it with the normalized, running sum density of states, one state can be selected (Fig. 6.11d). Finally, since each cube spans a significant volume of momentum space, it is necessary to identify the constant energy plane within the cube and to choose two random numbers to select a specific momentum within the cube. For simple energy bands, this process of selecting a final state that conserves energy and momentum was done by analytically integrating the δ-functions expressing energy and momentum conservation as discussed in Section 2.6.

6.6 Monte Carlo simulation for bulk semiconductors

Having described how Monte Carlo techniques are used to simulate trajectories of individual electrons, we are ready to describe how transport is simulated. For a bulk semiconductor, with a uniform electric field, the technique is simple, as illustrated in Fig. 6.12. We begin with an electron whose momentum is specified (the actual value does not matter; because the simulation will run to $t \to \infty$, the initial condition is not important). A 'clock' with time intervals of ΔT, which is generally small compared to the average time between collisions, is then started, and the electron trajectory is tracked by Monte Carlo methods. Every ΔT sec, the state of the electron is examined and its velocity and energy are recorded.

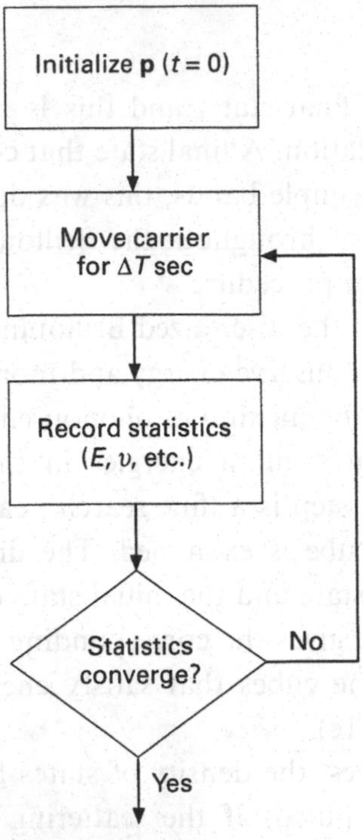

Fig. 6.12 The basic algorithm used for simulating electron transport in a bulk semiconductor. Note that scattering events may occur during the time, ΔT, for which carriers are moved.

(Alternatively, we could use the collisions as the clock and record the electron's velocity and energy just before each collision.) As the process continues, a statistical estimate for the average electron velocity and energy is generated. The process terminates when the statistical error in the estimates is acceptable. To generate an energy distribution, we define a set of bins with finite width in energy. At each sampling time, we note the energy bin the electron resides in, then add to a counter for that bin. The final population of the bins gives an estimate for the energy distribution function. Note that the magnitude of the energy distribution at a particular energy is proportional to the time an electron spends in that energy bin. We'll find this concept useful when we talk about devices too.

As we will discuss in Chapter 7, Monte Carlo simulations of electron transport in bulk semiconductors provide a good description of high-field transport in bulk semiconductors. The spatially varying electric field and the boundary conditions make the simulation of devices somewhat more involved. Nevertheless, the central element, the tracking of carrier trajectories, is done by the same methods.

6.7 Monte Carlo simulation for devices

Having discussed how to generate carrier trajectories by the Monte Carlo method, we are now ready to apply the technique to devices. Two different approaches are in use; in the first, which we'll term 'ensemble Monte Carlo', the trajectories of an ensemble of particles within a device are followed in 'parallel' in time. In the second, or 'incident flux', approach, carriers are injected from a contact and followed one at a time as they traverse the device. This technique works best for steady-state simulations. We begin with the ensemble Monte Carlo technique, the most widely-used method for simulating devices.

6.7.1 Ensemble Monte Carlo

To discuss the ensemble Monte Carlo method, consider the two-dimensional MOSFET sketched in Fig. 6.13, which is divided into cells by a numerical grid. Each of the cells is first populated with carriers, then the position and momentum of each carrier is tracked as it moves from cell to cell under the influence of the electric field and scattering potentials. It is convenient to begin the simulation in thermodynamic equilibrium where the solution is known. Each cell is then populated with carriers whose momentum is randomly selected from a Maxwellian (or Fermi–Dirac) distribution. Under charge-neutral conditions, the total number of free carriers within the device must equal the total number of ionized donors within the device, but to keep the computational burden manage-

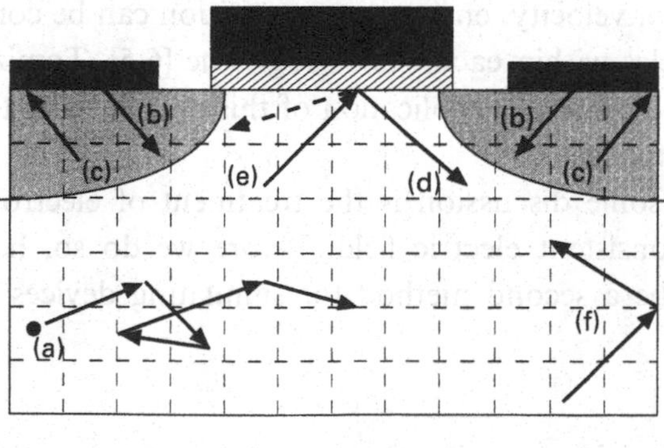

Fig. 6.13 (a) Sketch of a two-dimensional silicon–metal–oxide semiconductor field-effect transistor (MOSFET). A numerical grid which divides the device into cells is also shown. Monte Carlo techniques track the carriers as they move within the device (a). Carriers can enter (b) and exit (c) the device from ohmic contacts, and they can reflect specularly (d) or diffusively (e) from the Si/SiO_2 interface. Carriers that leave through a line of symmetry (f) are reflected.

able, the number of simulated electrons must be limited – 10 000 or so is typical. Since there are far more dopants within a typical device, each electron is treated as a *superelectron* for charge-assignment purposes and represents the charge of many electrons. The charge on each superelectron is simply

$$Q = -q\frac{N}{N_{\text{sim}}}, \tag{6.29}$$

where N is the number of charges in the device, and N_{sim} is the number of superelectrons used in the simulation.

To simulate devices, the boundaries must be treated properly. Ohmic contacts are often assumed to be perfect absorbers, so carriers that reach them simply exit the device. To maintain space–charge neutrality at the contact, carriers are injected as needed. For two-dimensional devices, the noncontacted free surfaces are commonly treated as reflecting boundaries for carriers. For field-effect transistors, roughness at the surface of the channel can cause scattering. A simple approach is to treat some fraction of the encounters with the surface as specular scattering events and the remainder as diffuse scattering events. The specific fraction is selected to match transport measurements.

After the device has been populated with electrons, the bias is applied, and the simulation begins. A short timestep, ΔT is defined, and the momentum and position of each particle are tracked by the techniques discussed earlier. At the end of the timestep (typically tens of femtoseconds), thermal electrons are injected from the contacts to maintain space-charge neutrality there. Poisson's equation is then solved to update the electric field. The whole process can be repeated as illustrated in Fig. 6.14 to simulate the transient response of the device; eventually the steady-state is achieved. At any time during the simulation, the average carrier density, velocity, energy versus position can be computed by averaging over the particles within each slab. Moglestue [6.5], Tomizawa [6.7], and Kunikiyo et al. [6.9] discuss the application of this approach to the simulation of two-dimensional transistors.

One issue that merits some discussion is the treatment of electron–electron scattering and the self-consistent electric field. Before we do so, however (in Section 6.7.3), we describe a second method for simulating devices by Monte Carlo techniques.

6.7.2 Incident flux approach

An alternative approach consists of viewing the contacts as sources which continually inject fluxes of carriers into the device. The basic idea, as illustrated in Fig. 6.15, begins with a guess for the self-consisted conduction band profile, $E_C(z)$. The approach consists of selecting a carrier at random from the injected

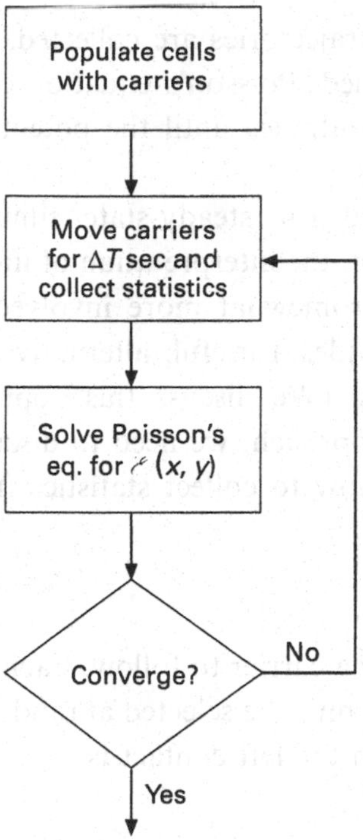

Fig. 6.14 Algorithm for time-dependent, ensemble Monte Carlo simulation of devices. Carriers are moved for ΔT seconds by Monte Carlo techniques.

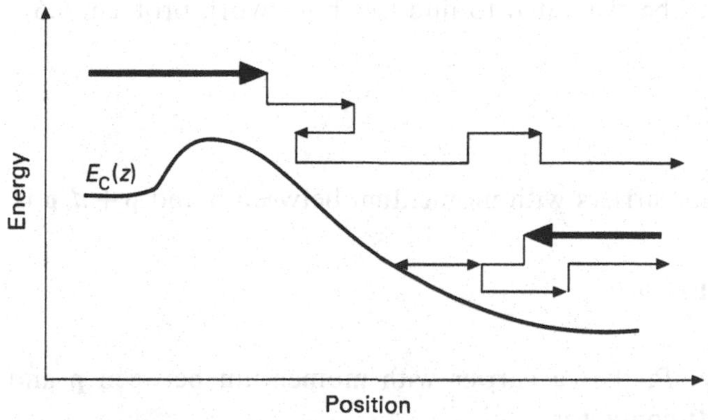

Fig. 6.15 Basic concepts of incident flux Monte Carlo simulation. The heavy lines indicate the carriers injected from the contacts.

flux and following it until it exits through the same, or another contact. Another carrier is then selected and the process repeated to simulate an ensemble of trajectories. Carriers are followed one at a time, in 'series' rather than in 'parallel' as in the many-particle approach. While the carrier is in the device, statistics are collected in order to extract from the ensemble of trajectories the collective variables of interest. When enough trajectories are collected, an estimate of the carrier profile, $n(z)$, can be obtained. Poisson's equation can then be solved to update $E_C(z)$, and the process continues until the potential converges.

The incident flux approach is best suited for steady-state simulations. Compared to ensemble Monte Carlo simulation, the interpretation of individual trajectories in term of collective variables is somewhat more involved in the incident flux approach, but the approach provides a useful, alternative way to think about steady-state transport in devices. (We discuss this approach in Chapter 9.) To understand the incident flux approach, we need to discuss how to select carriers from the incident flux and how to collect statistics from the carrier trajectories.

Injecting electrons from the contacts

The incident flux simulation begins by selecting a carrier to follow. Each contact injects a flux of carriers into the device; carriers must be selected at random from this incoming flux. The total flux incident from the left contact is

$$\frac{J_{\mathrm{L}}}{(-q)} = \frac{1}{\Omega}\sum_{\mathbf{p},p_z>0}\frac{J_L(\mathbf{p})}{(-q)} = \frac{1}{\Omega}\sum_{\mathbf{p},p_z>0}\frac{p_z}{m^*}f_{\mathrm{L}}(\mathbf{p}), \tag{6.30}$$

where $f_{\mathrm{L}}(\mathbf{p})$ is the Fermi function in the left contact. For a nondegenerate contact region, the sum can be evaluated to find (see homework problem 6.6)

$$\frac{J_{\mathrm{L}}}{(-q)} = N_{\mathrm{D}}\sqrt{\frac{k_{\mathrm{B}}T_{\mathrm{L}}}{2\pi m^*}}. \tag{6.31}$$

The flux of injected carriers with momentum between $\mathbf{p}$ and $\mathbf{p}+d^3\mathbf{p}$ is

$$\frac{J_L(\mathbf{p})}{(-q)}\mathrm{d}^3\mathbf{p} = \frac{p_z}{m^*}f_L(\mathbf{p})\mathrm{d}^3\mathbf{p}, \tag{6.32}$$

so the probability, $\mathcal{P}$, that a carrier with momentum between $\mathbf{p}$ and $\mathbf{p}+d^3\mathbf{p}$ enters from the left contact is

$$\mathcal{P}(\mathbf{p})d^3\mathbf{p} = \frac{J_L(\mathbf{p})d^3\mathbf{p}}{J_L} \sim \frac{p_z}{m^*}f_L(\mathbf{p}). \tag{6.33}$$

If the contact region is nondegenerate, then

$$f_L(p) = \exp\left[-(E_{\mathrm{CL}} + p^2/2m^* - F_{n\mathrm{L}})/k_{\mathrm{B}}T_{\mathrm{L}}\right] \tag{6.34}$$

(where $F_{n\mathrm{L}}$ is the Fermi level in the left contact), and eq. (6.33) can be written as

$$\mathcal{P}(\mathbf{p})\mathrm{d}^3\mathbf{p} = \mathcal{P}(p_x)\mathrm{d}p_x\mathcal{P}(p_y)\mathrm{d}p_y\mathcal{P}(p_z)\mathrm{d}p_z \sim \left[\exp(-p_x^2/2m^*k_BT_L)\mathrm{d}p_x\right]$$
$$\times \left[\exp(-p_y^2/2m^*k_BT_L)\mathrm{d}p_y\right]\left[\frac{p_z}{m^*}\exp(-p_z^2/2m^*k_BT_L)\,\mathrm{d}p_z\right], \tag{6.35}$$

which states that p_x and p_y are normally distributed, but the distribution for p_z is *velocity weighted* towards larger values of p_z because such carriers enter from the contact more often.

The entering electrons must be selected at random in accordance with the probability distributions specified by eq. (6.35). The momentum of the entering carrier is specified by three random numbers, r_x, r_y, and r_z which determine p_x, p_y, and p_z. Since random number generators with normal distributions are widely available, p_x and p_y are easily selected. To select p_z, we must ensure that

$$\mathcal{P}(r)\mathrm{d}r = \mathcal{P}(p_z)\mathrm{d}p_z, \tag{6.36}$$

where $\mathcal{P}(p_z)$ is the probability distribution function for p_z as given in eq. (6.35) and $\mathcal{P}(r)$ is the probability distribution function of the random number generator. For a uniform random number generator, $\mathcal{P}(r) = 1$, and

$$\mathrm{d}r = \frac{\frac{p_z}{m^*}\mathrm{e}^{-p_z^2/2m^*k_\mathrm{B}T}\mathrm{d}p_z}{\int_0^\infty \frac{p_z}{m^*}\mathrm{e}^{-p_z^2/2m^*k_\mathrm{B}T}\mathrm{d}p_z}, \tag{6.37}$$

where the denominator is a normalizing factor to ensure that the probability of finding a carrier with p_z somewhere between zero and infinity is unity. Both sides of eq. (6.37) can be integrated to find

$$\int_0^{r_z}\mathrm{d}r = \frac{\int_0^{p_z} \frac{p_z}{m^*}\mathrm{e}^{-p_z^2/2m^*k_\mathrm{B}T}\mathrm{d}p_z}{\int_0^\infty \frac{p_z}{m^*}\mathrm{e}^{-p_z^2/2m^*k_\mathrm{B}T}\mathrm{d}p_z}$$

or

$$p_z = \sqrt{-2m^*k_\mathrm{B}T\ln r_z}, \tag{6.38}$$

which specifies the z-component of the carrier's momentum from the random number, r_z. Figure 6.16 shows histograms of p_x, p_y, and p_z as determined by random numbers, r_x ,r_y, and r_z for 10 000 carriers. Along $\hat{x}$ and $\hat{y}$ the velocities are distributed almost normally with an average velocity 1000 times smaller than in the $\hat{z}$ direction. The finite values for $\langle v_x \rangle$ and $\langle v_y \rangle$ are a consequence of the statistical nature of the Monte Carlo method. (Statistical noise poses problems for Monte Carlo simulation under moderate or low fields where the average velocities are comparable to or less than the noise.)

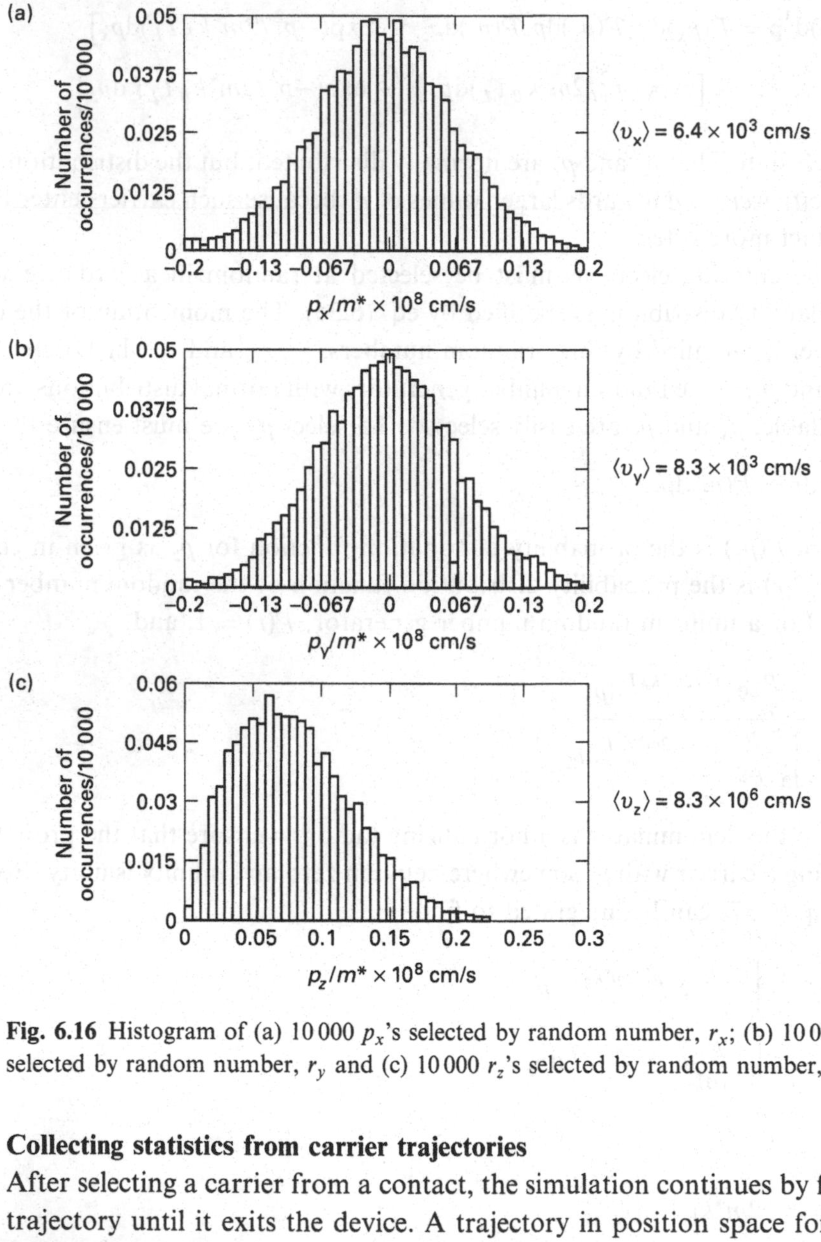

Fig. 6.16 Histogram of (a) 10 000 p_x's selected by random number, r_x; (b) 10 000 p_y's selected by random number, r_y and (c) 10 000 r_z's selected by random number, r_z.

Collecting statistics from carrier trajectories

After selecting a carrier from a contact, the simulation continues by following its trajectory until it exits the device. A trajectory in position space for one representative carrier is displayed in Fig. 6.17a. The result of the simulation process is a whole collection of such trajectories, $z_i(t)$ and $\mathbf{p}_i(t)$ where i runs from one to N_{sim} for the N_{sim} carriers selected from the contacts. By collecting statistics from these trajectories, the values of collective variables such as the average carrier density can be deduced as described next.

We begin by asking for the average carrier density within each slab within the device. If we consider only one trajectory i, then either the carrier is in the slab 'j' or it is not, so the carrier density near $z = z_j$ *due to this single carrier* is

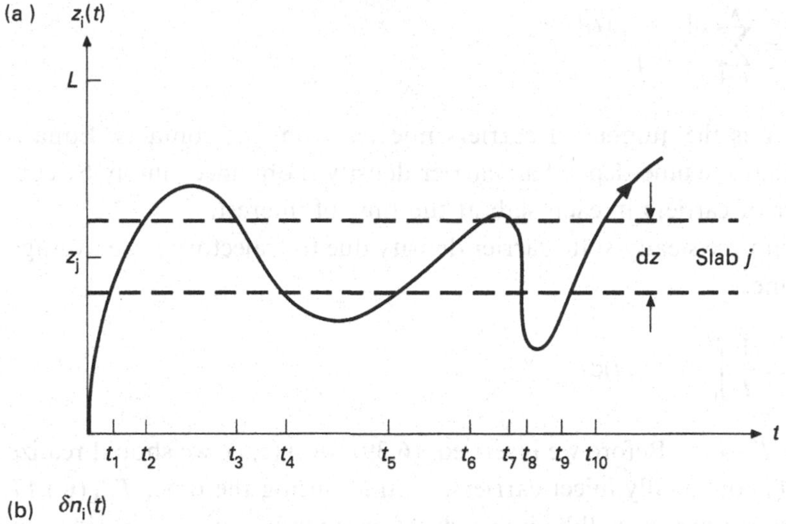

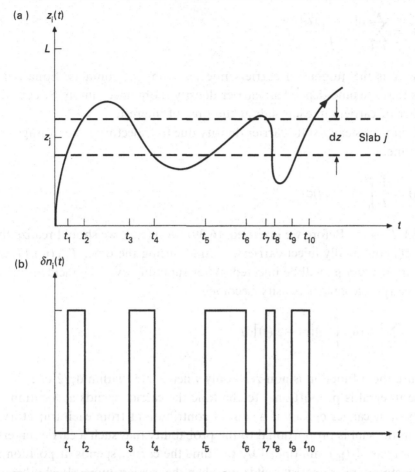

Fig. 6.17 (a) Representative trajectory of a single carrier through a device. (b) Carrier density versus time at slab 'j' due to a single carrier trajectory.

$$\delta n_i(z_j, t) = \frac{1}{A\,\mathrm{d}z}$$

when the carrier is in the jth slab; it is zero otherwise. (Here A is the cross-sectional area of the device and $\mathrm{d}z$ the width of the slabs.) The carrier density in slab 'j' versus time due to this single trajectory is shown in Fig. 6.17b and is mathematically described by

$$\delta n_i(z_j, t) = \frac{\Delta[z_j - z_i(t)]}{A}, \tag{6.39}$$

where $\Delta(z) = 1/\mathrm{d}z$ when $-\mathrm{d}z/2 < z < \mathrm{d}z/2$ and zero otherwise. Equation (6.39) gives the contribution of trajectory 'i' to the carrier density at $z = z_j$. To find the total, time-dependent carrier density, we add the contributions from each injected carrier to find

$$n(z_j, t) = \sum_{i=1}^{N} \frac{\Delta[z_j - z_i(t)]}{A}, \tag{6.40}$$

where N is the number of carriers injected from the contacts. Equation (6.40) states that the time-dependent carrier density is obtained simply by counting the number of carriers in each slab at the time of interest.

To find the steady-state carrier density due to trajectory i, we average $\delta n_i(z_j, t)$ over time,

$$\delta n_i(z_j) = \frac{1}{T}\int_0^T \delta n_i(z_j, t)\mathrm{d}t, \tag{6.41}$$

and let $T \to \infty$. Before we insert eq. (6.39) for $n_i(z, t)$, we should realize that the contacts continually inject carriers, so that during the time, $TJ_{\mathrm{L}}(\mathbf{p}_i)AT$ carriers with momentum $\mathbf{p}_i$, will be injected. After summing over the incident momenta, the steady-state carrier density becomes

$$n(z_j) = \sum_{\mathbf{p}_i} J_{\mathrm{L}}(\mathbf{p}_i)\int_0^T \Delta\left[z_j - z_i(t)\right]\mathrm{d}t. \tag{6.42}$$

Because the Δ-function is non-zero only when $z_i(t)$ is within $\mathrm{d}z/2$ of z_j, the value of the integral is proportional to the time the carrier spends at location z_j. The steady-state carrier density is a sum of contributions from each trajectory. Each term in the sum is proportional to the probability that such a carrier enters from the contacts, $J_L(\mathbf{p}_i)$, multiplied by the time the carrier spends at position z. The steady-state carrier density builds up when the carriers move slowly through the device because the contacts are constantly injecting carriers.

To evaluate the steady-state carrier density, a carrier is first selected at random from one of the contacts as described above, then its trajectory is followed until it leaves the device. While the carrier is within the device, its location is sampled every ΔT seconds, and the slab it is in is recorded. The number of times a carrier appears in a slab is proportional to the time spent in that slab, as illustrated in Fig. 6.18. The process is repeated for a second carrier, and the results are added to those for the first carrier. The simulation continues until it is determined that the shape of the $n(z)$ profile has converged to its steady-state value. To obtain a profile of the average velocity or kinetic energy versus position, we simply multiply the time spent in the slab by the velocity or energy of the carrier. Similarly, the distribution function, $f(p_k, z_j)$ is proportional to the time a carrier spends in the momentum bin at p_k while it is simultaneously in the position bin at z_j. By these means, the value of any collective variable of interest can be obtained from the collection of carrier trajectories computed by incident flux Monte Carlo simulation.

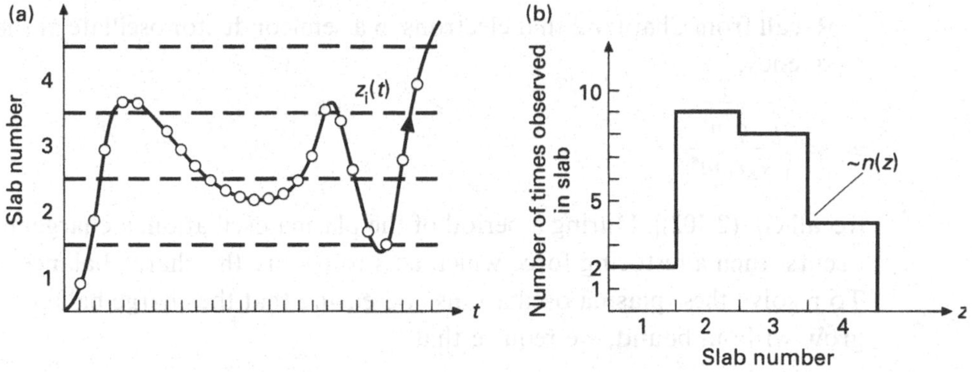

Fig. 6.18 (a) Carrier trajectory showing spatial bins and sampling times. (b) Histogram showing time spent in each bin.

6.7.3 Treatment of Coulomb effects in Monte Carlo simulation

The treatment of electron–electron scattering and space-charge effects (recall Chapter 2, Section 10) are key issues in Monte Carlo simulation. For ultra-small devices, the number of carriers and dopants is small, and the Coulomb forces between each discrete charge and all others can be evaluated directly. These *molecular dynamic simulations* treat Coulomb effects exactly, within a classical framework, but the computational burden is large when the number is charges is large. For large devices, the mobile and fixed charges can be treated as a continuous charge density with Poisson's equation giving the self-consistent electrostatic field. In general, however, we need to treat the short-range effects of discrete charges (binary electron–electron scattering) as well as the long-range space charge effects described by Poisson's equation.

Typical practice in Monte Carlo simulations treats binary, electron–electron scattering of an electron with its neighbors within a radius, R, that is roughly a Debye length. (At longer distances, screening smooths the charge distribution so that discrete charges are not sensed.) The effect of these more distant electrons is described indirectly through the self-consistent electrostatic field obtain from Poisson's equation. Fischetti and Laux [6.8] describe one implementation in a Monte Carlo simulation program.

Coulomb effects also influence the timestep, ΔT, between Poisson solutions in a Monte Carlo simulation. Of course, electrons cannot be allowed to move too far, or the electric field will change rapidly during the timestep. The criterion is

$$\Delta T < \frac{h}{\upsilon_{\max}}, \tag{6.43}$$

where h is the size of a spatial cell, and $\upsilon_{\max}$ is the maximum carrier velocity (typically $\approx 10^8$ cm/s). But a much more stringent limitation exists.

Recall from Chapter 2 that electrons in a semiconductor oscillate at the plasma frequency,

$$\omega_p = \sqrt{\frac{q^2 n}{\kappa_S \varepsilon_0 m^*}} \tag{6.44}$$

[recall eq. (2.102)]. During a period of the plasma oscillation, a charge imbalance occurs, then a restoring force which acts to restore the charge balance is set up. To resolve these plasma oscillations and ensure that the charge balance does not grow without bound, we require that

$$\Delta T < \frac{0.5}{\omega_p}. \tag{6.45}$$

In the heavily doped contact regions of a device, $n \approx 10^{20}\ \text{cm}^{-3}$, $\omega_p^{-1} \approx 1\text{fs}$, which means that extremely short timesteps are necessary.

6.8 Monte Carlo simulation and the BTE

In the previous section we presented a prescription for finding the distribution function by Monte Carlo simulation, but we also know that $f(\mathbf{r}, \mathbf{p}, \mathbf{t})$ can be found by solving the BTE directly. In this section we demonstrate that Monte Carlo simulation is a numerical technique for solving the BTE. (Actually, if Coulomb effects are carefully treated, then Monte Carlo simulation can also treat carrier–carrier correlations and, therefore, provides more detail than the Boltzmann equation which ignores such correlations.)

To begin, we need a definition for $f(z, p, t)$. We begin by letting the width of each slab in a one-dimensional device approach zero, so that the carrier density due to the single trajectory, i, becomes

$$\delta n_i(\mathbf{r}, t) = \delta[\mathbf{r} - \mathbf{r}_i(t)], \tag{6.46}$$

which is the mathematical definition of carrier density due to a single trajectory. Similarly, we find

$$\delta f_i(\mathbf{r}, \mathbf{p}, t) = \delta[\mathbf{r} - \mathbf{r}_i(t)]\delta[\mathbf{p} - \mathbf{p}_i(t)], \tag{6.47}$$

which is the contribution to the distribution function due to the single trajectory. To find how $f(\mathbf{r}, \mathbf{p}, \mathbf{t})$ evolves with time, we apply the chain rule to eq. (6.45) to obtain

$$\frac{\partial \delta f_i}{\partial t} = \nabla_{r_i}\delta f_i \cdot \frac{d\mathbf{r}_i}{dt} + \nabla_{p_i}\delta f_i \cdot \frac{d\mathbf{p}_i}{dt}, \tag{6.48}$$

where

$$\frac{d\mathbf{r}_i}{dt} = v_i \tag{6.49}$$

and

$$\frac{d\mathbf{p}_i}{dt} = (-q)\mathcal{E}(\mathbf{r}_i) + \left.\frac{\partial \mathbf{p}_i}{\partial t}\right|_{\text{coll}}, \tag{6.50}$$

because momentum can change for two reasons – fields may accelerate carriers or collisions may occur.

At this point we should stress that **r**, **p**, and t are the *independent variables* which we would like to appear in the final equation but $\mathbf{r}_i(t)$ and $\mathbf{p}_i(t)$, the *dependent variables*, refer to the time-dependent position and momentum of a single trajectory. From eq. (6.47) we observe that $\nabla_{r_i} f = -\nabla_r f$ and $\nabla_{p_i} f = -\nabla_p f$ so eq. (6.48) becomes

$$\frac{\partial \delta f_i}{\partial t} = -\nabla_r \delta f_i \cdot v_i + q\mathcal{E}(\mathbf{r}_i) \cdot \nabla_p \delta f_i + \left.\frac{\partial \mathbf{p}_i}{\partial t}\right|_{\text{coll}} \cdot \nabla_{p_i} \delta f_i.$$

The final term is the rate of change of δf_i with momentum times the rate at which momentum changes with time *due to collisions* and is therefore $\partial \delta f_i / \partial t|_{\text{coll}}$. We also note that $\delta(\mathbf{r} - \mathbf{r}_i)$ is non-zero only when $\mathbf{r} = \mathbf{r}_i$, so when summed over all trajectories,

$$\frac{\partial \delta f_i}{\partial t} + \nabla_r \delta f_i \cdot v - q\mathcal{E}(\mathbf{r}) \cdot \nabla_p \delta f_i = \left.\frac{\partial \delta f_i}{\partial t}\right|_{\text{coll}}$$

gives

$$\boxed{\frac{\partial f}{\partial t} + \nabla_r f \cdot v - q\mathcal{E}(\mathbf{r}) \cdot \nabla_p f = \left.\frac{\partial f}{\partial t}\right|_{\text{coll}}}, \tag{6.51}$$

and we have succeeded in showing that the distribution function obtained by Monte Carlo simulation satisfies the BTE.

6.9 Summary

In this chapter we showed how to simulate the motion of carriers through a device as they respond to the applied and built-in potentials and to the random scattering potentials. The average effects of the various scattering processes are simulated by properly selecting random numbers. The technique is appealing because it directly mimics the physical processes that occur during transport and because very accurate solutions to the BTE can be obtained by relatively simple means. This chapter focused on the basic concepts central to Monte Carlo simulation – how a particle trajectory is simulated and how the technique is applied to devices. To apply Monte Carlo simulation to real semiconductors and obtain accurate results, several details must be carefully considered. For example, energy bands are rarely spherical and parabolic, and carriers may scatter to other valleys. (For such cases, we must keep track of the valley in which the carrier resides in addition to tracking its position and momentum.) The chapter references describe how the Monte Carlo technique is applied to silicon and gallium arsenide.

Although the Monte Carlo method is often the most accurate method available for simulating devices, it does suffer from some limitations. First, the statistical uncertainty in the results should be obvious from the histograms presented in this chapter. For a typical semiconductor under low applied fields, the expected drift velocity can be smaller than the statistical noise. The statistical error shrinks as the number of simulated carriers, N_{sim}, increases, but it only drops as $1/\sqrt{N_{sim}}$, so large numbers of carriers have to be simulated. Monte Carlo simulation is not well-suited for analyzing low-field transport, and most devices do contain some low-field regions. Other difficulties include carrier injection over a large energy barrier and electron-hold recombination. Because these processes are statistically unlikely, the simulation times can be long.

Because of such limitations, devices are frequently analyzed by approximate techniques based on the collective variable view point rather than by direct Monte Carlo simulation. The most commonly used approach is to solve the balance equations introduced in Chapter 5. Quite often, however, Monte Carlo simulation is used to evaluate the ensemble relaxation times needed for the balance equations and to assess the validity of the simplifying approximations that are necessary when using balance equations.

References and further reading

Application of Monte Carlo simulation to group IV and III-V semiconductors is discussed in the following reviews.

6.1 Jacoboni, C. and Reggiani, L. The Monte Carlo method for the solution of charge transport in semiconductors with applications to covalent materials. *Reviews of Modern Physics*, **55**, 645–705, 1983.

6.2 Reggiani, L. *Hot-Electron Transport in Semiconductors*, In *Topics in Applied Physics*, Vol. 58, Chapter 2. Springer-Verlag, New York, 1985.

6.3 Fawcett, W., Boardman, A. D. and Swain, S. Monte Carlo determination of electron transport in gallium arsenide. *Journal of the Physics and Chemistry of Solids*, **31**, 1963–90, 1970.

6.4 Price, P. J. Monte Carlo calculation of electron transport in solids. *Semiconductors and Semimetals*, **14**, 249–334, 1979.

6.5 Moglestue, C. A self-consistent Monte-Carlo particle model to analyze semiconductor microcomponents of any geometry. *IEEE Transactions on Computer-Aided Design*, **CAD-5**, 326–45, 1986.

The classic reference for simulating the flow of particles is

6.6 Hockney, R. W. and Eastwood, J. W. *Computer Simulation Using Particles*. McGraw-Hill, New York, 1981.

This book describes how a Monte Carlo semiconductor device simulation is implemented in a computer program.

6.7 Tomizawa, K. *Numerical Simulation of Submicron Semiconductor Devices*. Artech House, Boston, 1993.

'Full band' Monte Carlo simulation is described in

6.8 Fischetti, M. V. and Laux, S. E. Monte Carlo analysis of electron transport in small semiconductor devices. *Physics Review B*, **38**, 9721–45, 1988.

6.9 Kunikiyo, T., Takenaka, M., Kamakura, Y., Yamaji, M., Mizuno, H., Morifuji, M. et al. A Monte Carlo simulation of anisotropic electron transport in silicon using full band structure and anisotropic impact–ionization model. *Journal of Applied Physics*, **75**, 297–312, 1994.

Problems

6.1 If free flights are distributed according to eq. (6.14), show that the average duration of a free flight is $\langle t_c \rangle = 1/\Gamma_0$. **Hint:**

$$\langle t_c \rangle = \int_0^\infty t\mathcal{P}(t)\mathrm{d}t \Big/ \int_0^\infty \mathcal{P}(t)\mathrm{d}t.$$

6.2 Examine the computer-generated free flight times as follows.

(a) Write a computer program to select scattering times from eq. (6.17) for $\Gamma_0 = 3 \times 10^{14}$/s.

(b) How many selections are required to make that average time between collisions correct to within 10%?

6.3 Derive an expression analogous to eq. (6.14) for the probability that a carrier's first collision occurs between t and $t + \mathrm{d}t$, but *do not assume that the scattering rate, Γ, is constant.*

6.4 Derive an expression for the polar angle, α, after scattering that is valid for ionized impurity scattering in the Brooks–Herring formulation.

6.5 As discussed in Section 6.4, updating the carrier momentum after scattering is most easily accomplished in a rotated coordinate system. The following questions concern the operations required to rotate, update, then rotate back.

(a) The rotated $\hat{x}$ axis $\hat{x}_r$ is related to the original $\hat{x}$ axis by

$$\hat{x}_r = Y_\theta Z_\phi \hat{x}$$

where Y_θ describes a rotation of θ about the $\hat{y}$-axis, and Z_ϕ describes a rotation of ϕ about the $\hat{z}$-axis. Show that

$$Y_\theta = \begin{pmatrix} \cos\theta & 0 & \sin\theta \\ 0 & 1 & 0 \\ -\sin\theta & 0 & \cos\theta \end{pmatrix}$$

and that

$$Z_\phi = \begin{pmatrix} \cos\phi & -\sin\phi & 0 \\ \sin\phi & \cos\phi & 0 \\ 0 & 0 & 1 \end{pmatrix}.$$

(b) Assume elastic scattering with an incident momentum, $\mathbf{p} = (p/\sqrt{2}, p/\sqrt{2}, 0)$. If $\alpha = \pi$ and $\beta = \pi/2$, find $\mathbf{p}'$ after scattering.

6.6 Answer the following questions about the flux injected from a contact.

(a) Evaluate the sum in eq. (6.30) and show that the result for a nondegenerate semiconductor is eq. (6.31).

(b) Find the corresponding result for a degenerate contact.

6.7 Compare the accuracy of the average velocity of the computer-selected injected flux by answering the following:

(a) Derive an expression for $\langle v_z \rangle$ for carriers injected in the $\hat{z}$-direction from a contact. Assume that $m^* = m_0$.

(b) Write a computer program and use it to select 1000 values of p_z according to eq. (6.38).

(c) How close is the average, computer-generated $\langle v_z \rangle$ to the exact value? How many more selections are required to reduce this error by a factor of two?

6.8 Assume that the p_x momentum component of an injected flux is distributed according to $P(p_x)dp_x = \exp[-p_x^2/2m^* k_B T_L]dp_x$ and that a random number generator which generates normally distributed with a standard deviation of 1.0 is available. How would you use this random number generator to select p_x?

6.9 Consider electrons flowing through the device structure sketched below. The electrons are each injected with $p_z = p_0$; the injected current is J_0. Assume that the structure is short, so that scattering does not occur. The electron's potential energy is $U(z) = -q\mathcal{E}z$.

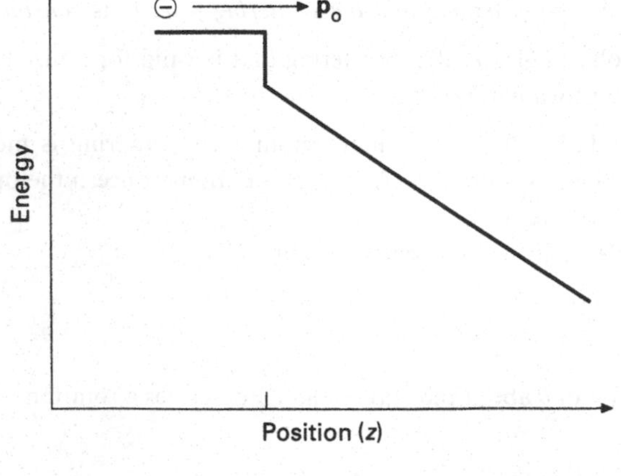

(a) Find an expression for the carrier density at $z = 0$.

(b) Assume that the current is constant, $J(z) = J_0$, and show that the steady-state carrier density is

$$n(z) = \frac{n(0)}{\sqrt{1 + 2m^* U(z)/p_0^2}}.$$

(c) Use an expression analogous to eq. (6.40),

$$n(z) = \sum_{p_i} J_L(\mathbf{p}_i) \int_0^T \delta[z - z_i(t)]\mathrm{d}t$$

to derive the result of part (b).

6.10 Write a computer program to compute a table of the self-scattering rate versus energy for the model semiconductor. Let the maximum energy be 1.0 eV and assume that $\Gamma_0 = 3 \times 10^{14}\,\mathrm{s}^{-1}$.

6.11 Write a computer program to compute the relative percentages of the various scattering events for the model semiconductor. (Assume that $\Gamma_0 = 3 \times 10^{14}\,\mathrm{s}^{-1}$.)

(a) Assume $E = 0.15\,\mathrm{eV}$ and thereby verify the results in Fig. 6.8.

(b) Repeat for $E = 0.30\,\mathrm{eV}$.

7 High-field transport in bulk semiconductors

Steady-state, low-field transport in bulk semiconductors is described by the drift–diffusion equation,

$$J_{nz} = nq\mu_n\mathcal{E}_z + qD_n\frac{\mathrm{d}n}{\mathrm{d}z},$$

where μ_n and D_n are *material-dependent* transport parameters related by the Einstein relation. The parameters are independent of the field because the low-field perturbation in the distribution function is directly proportional to the electric field (recall homework problem 5.11). Under high fields, however, the distribution function becomes a nonlinear function of the field, so μ_n and D_n become *field-dependent* parameters. In devices, the situation is even more complex because the distribution function has a nonlocal dependence on the electric field. In this chapter, our focus is on steady-state, high-field transport in homogenous semiconductors. In the following chapter, we discuss the interesting and important effects that can occur in devices when the applied fields vary rapidly in time and space.

The chapter begins in Section 7.1 with a qualitative description of the general features of high-field transport in bulk semiconductors. The electron temperature approach for solving the momentum and energy balance equations to obtain the field-dependent mobility and diffusion coefficient is introduced in Section 7.2. Although limited in accuracy, this approach is useful for the insight it provides. The application of Monte Carlo simulation to accurately evaluate the field-dependent transport coefficients is then described. Some experimental results for the high-field transport of electrons in bulk Si and GaAs are examined in Section 7.4, and high-field transport of confined carriers is examined in Section 7.5.

7.1 General features of high-field transport

By assuming steady-state conditions and that the energy-related tensor, $\overleftrightarrow{W}$, is diagonal, the electron current is obtained from eq. (5.97) as

$$J_{nz} = nq\mu_n\mathcal{E}_z + 2\mu_n\frac{q\,dW_{zz}}{dz}, \tag{7.1}$$

where

$$\mu_n = \frac{q}{m^*\langle\langle 1/\tau_m\rangle\rangle} \tag{7.2}$$

is the electron mobility and

$$W_{zz} \equiv \tfrac{1}{2}n\langle p_z v_z\rangle = nu_{zz} \tag{7.3}$$

is (for simple bands) the average kinetic energy density associated with the degree of freedom along the $\hat{z}$-axis (u_{zz} is the average kinetic energy component per electron). The second term in eq. (7.1) can be expanded into a diffusion term, which involves the concentration gradient, and a thermoelectric term, which involves the kinetic energy gradient. A bulk semiconductor in which the average kinetic energy per carrier is uniform (although it may be higher than in equilibrium) is assumed, so we find

$$J_{nz} = nq\mu_n(\mathcal{E})\mathcal{E}_z + qD_n(\mathcal{E})\frac{dn}{dz}, \tag{7.4}$$

where

$$\frac{D_n(\mathcal{E})}{\mu_n(\mathcal{E})} = \frac{2u_{zz}(\mathcal{E})}{q}. \tag{7.5}$$

The mobility and diffusion coefficient are determined by the distribution function, but in bulk semiconductors, there is a one-to-one correspondence between the applied field and the distribution function, so we can speak of field-dependent mobilities and diffusion coefficients. Equation (7.4) describes how a pulse of electrons moves through a bulk semiconductor under the influence of a high applied field. The average velocity of the pulse is determined by the mobility, and the spatial spread of the pulse increases with time as $\sqrt{D_n t}$. The *longitudinal diffusion coefficient* defined by eq. (7.5) describes the spread of the pulse along the direction of the field. In the transverse plane, the spread is controlled by the *transverse diffusion coefficient* which may be different because u_{zz} may not equal

u_{xx} or u_{yy}. Our objective is to understand, both qualitatively and quantitatively, how μ_n and D_n vary with the electric field in common semiconductors.

The free carriers and the semiconductor lattice are two separate systems which exchange energy via phonon scattering. Under low fields, the carriers and lattice are in equilibrium, so $u_{zz} = k_B T_L/2$. Under high fields, however, the carriers gain energy from the field, so their kinetic energy (or, loosely speaking, temperature) can be higher than that of the lattice. High-field transport is commonly termed *hot-carrier* transport because the temperature of the carriers exceeds the temperature of the lattice. The difference in energy of the two systems is determined by the rate at which they exchange energy through phonon scattering.

These ideas are illustrated by the bucket analogy shown in Fig. 7.1. The liquid inside the bucket represents the free carriers, while that outside represents the

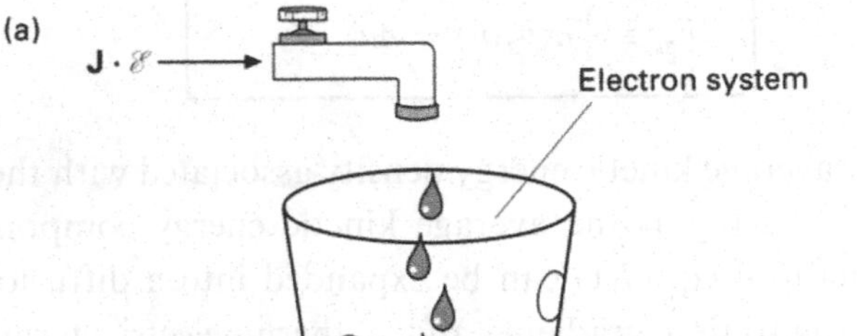

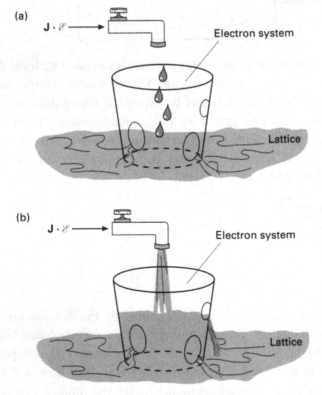

Fig. 7.1 Bucket analogy to illustrate the balance of energy between the free carriers and lattice. The faucet represents the energy input from the field, the holes in the bucket represent the phonon scattering processes, and the height of the liquid is a measure of the kinetic energy density. (a) The applied field is low, so the carriers and lattice are in equilibrium. (b) The applied field is high, so the kinetic energy of the carriers exceeds that of the lattice.

lattice. The height of the liquid corresponds to the kinetic energy (or temperature), and the holes in the bucket represent the phonon scattering processes through which the two systems interact. Under low fields, the energy input from the field (which is represented by the faucet) is slow, so scattering (i.e. holes in the bucket) ensures that the energies of the carriers and lattice are nearly equal. Under high fields, however, the energy input is rapid, so the fluid inside the bucket rises until a new balance is achieved. As Fig. 7.1b illustrates, the high-field carrier energy exceeds that of the lattice.

With this background, the general features of high-field, electron transport in silicon and gallium arsenide, as summarized in Fig. 7.2, are readily explained. In covalent semiconductors such as silicon, the conduction band consists of a set of

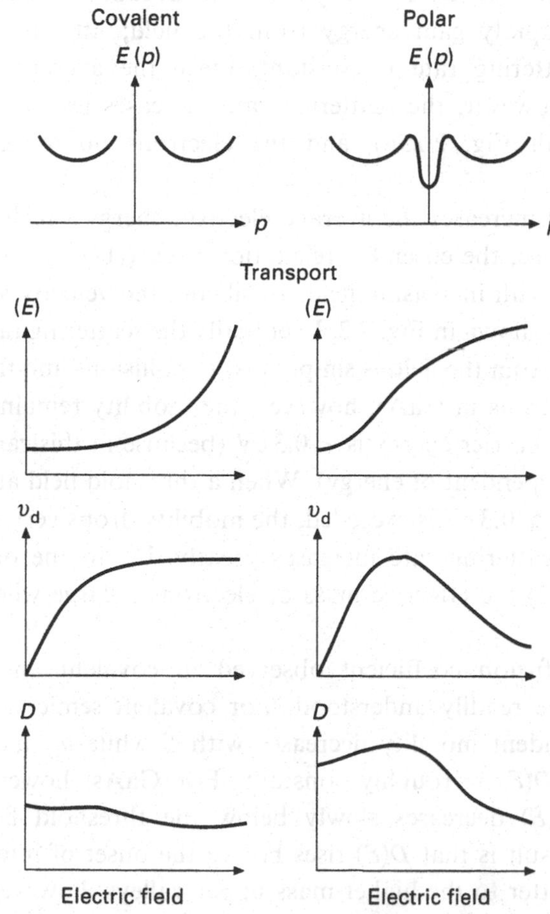

Fig. 7.2 General features of high-field electron transport in covalent and polar semiconductors. (From Jacobini, C. and Reggiani, L. *Advances in Physics*, **28**, 493–553, 1979. Reproduced with permission from Taylor and Frances.)

equivalent minima. (Higher states do exist and can be populated as discussed in Section 7.3.3.) Gallium arsenide and similar polar semiconductors have a low mass valley centered at $\mathbf{p} = 0$ and a set of higher mass, Si-like, equivalent upper valleys which lie within a few tenths of an electron volt of the lowest valley. These upper valleys become populated at modest fields and profoundly influence carrier transport.

As we illustrated in Fig. 2.25a, the dominant energy-relaxation mechanism for silicon is equivalent intervalley scattering whose rate increases with carrier energy. The average electron energy slowly increases with field as shown in Fig. 7.2 because much of the energy gained from the field is lost to the lattice in collisions. For GaAs, however, the situation is much different. When the carrier energy is greater than the optical phonon energy ($\sim 0.032\,\text{eV}$) and less than the Γ to L-valley separation ($\simeq 0.3\,\text{eV}$), scattering is predominantly by polar optical phonon emission, and the rate is nearly constant. Because their effective mass is small, electrons rapidly gain energy from the field; little is lost by collisions because the scattering rate is constant. Once the average carrier energy exceeds $\simeq 0.3\,\text{eV}$, however, the scattering rate increases greatly due to intervalley scattering (recall Fig. 2.25b), and the electrons no longer gain energy as rapidly.

Because the applied field increases the average electron energy which causes the scattering rate to increase, the ensemble relaxation rate, $\langle\langle 1/\tau_m \rangle\rangle$, increases and the mobility decreases with increasing field. In silicon, the velocity vs. field characteristic is sublinear as shown in Fig. 7.2. Eventually the scattering rate is so high that any further input from the field is simply lost to collisions and the drift velocity saturates. For electrons in GaAs, however, the mobility remains quite high as long as the average carrier energy is $< 0.3\,\text{eV}$ (because in this range the scattering rate is nearly independent of energy). When a threshold field at which the average electron energy $\simeq 0.3\,\text{eV}$ is exceeded, the mobility drops very rapidly for two reasons: (1) the scattering rate increases greatly due to the onset of intervalley scattering and (2) the effective mass of electrons is large when they reside in the upper valleys.

The field-dependent diffusion coefficient observed in covalent and polar semiconductors can also be readily understood. For covalent semiconductors like silicon, the field-dependent mobility decreases with $\mathcal{E}$ while u_{zz} increases, so, as eq. (7.5) predicts, $D(\mathcal{E})$ is roughly constant. For GaAs, however, u_{zz} increases with $\mathcal{E}$ while $\mu(\mathcal{E})$ decreases slowly below the threshold field for intervalley transfer. The result is that $D(\mathcal{E})$ rises before the onset of intervalley transfer. When carriers scatter to the higher mass upper valleys, however, their energy increases less rapidly with field while, at the same time, their mobility plummets, so above the threshold field, the diffusion coefficient begins to decrease with field.

7.2 Velocity saturation

The tendency of the average carrier velocity to saturate in a bulk semiconductor under high electric fields is one of the most important high-field transport effects. In the next section, we'll use an electron temperature approach to compute the velocity versus electric field characteristic for electrons in Si and show that the drift velocity does, indeed, approach a limiting value at high electric fields. In this section, we present a simple derivation of the saturated velocity to provide some physical insight into why the velocity saturates.

To begin, recall that scattering rates are generally proportional to the density of final states. For scattering by optical phonon emission, eq. (2.86) gives

$$\frac{1}{\tau} \propto g(E - \hbar\omega_o). \tag{7.6}$$

Optical phonon scattering is isotropic, so

$$\frac{1}{\tau_m} = \frac{1}{\tau}, \tag{7.7}$$

and the energy relaxation time due to optical phonon emission is

$$\tau_E = \left(\frac{E}{\hbar\omega_o}\right)\tau_m. \tag{7.8}$$

According to eq. (7.1), the drift velocity is

$$\upsilon_d = \mu\mathcal{E} = \frac{q\langle\tau_m\rangle}{m^*}\mathcal{E}. \tag{7.9}$$

Note that to keep the notation simple, we write the average momentum relaxation time as $\langle\tau_m\rangle$ rather than as $1/\langle\langle 1/\tau_m\rangle\rangle$. The final result will be uncertain by a statistical factor of the order of unity. (The assumption of a constant effective mass is also unrealistic, but sufficient to illustrate the underlying physics.)

For a uniform semiconductor, the energy balance equation, eq. (5.66c), becomes

$$J_{nz}\mathcal{E}_z = n\frac{(u - u_o)}{\langle\tau_E\rangle}, \tag{7.10}$$

which can be solved for

$$u = u_o + \frac{\langle\tau_E\rangle\langle\tau_m\rangle}{m^*}(q\mathcal{E}_z)^2 \tag{7.11}$$

Under high field conditions, $u \gg u_o$, so we can use eq. (7.8), in eq. (7.11) to find

$$\langle \tau_m \rangle = \sqrt{\frac{\hbar\omega_o m^*}{q\mathcal{E}_z}}. \tag{7.12}$$

Here we are assuming that eq. (7.8), which applied to the energy-dependent, microscopic relaxation times, also applies to the average relaxation times.

Finally, if we insert eq. (7.12) into eq. (7.9), we find

$$\boxed{\upsilon_{dz} = \upsilon_{sat} = \sqrt{\frac{\hbar\omega_o}{m^*}}}\ , \tag{7.13}$$

which is a constant, independent of the electric field.

According to this simple derivation, the electron velocity saturates because the momentum relaxation time varies as $1/\mathcal{E}_z$ under high-field conditions. The $1/\mathcal{E}_z$ dependence arises from energy balance considerations and the close connection between momentum and energy relaxation for optical phonon emission. The effective mass should be interpreted as an average effective mass for the hot carriers, not that at the energy band minima. Its value is likely to be similar for energetic electrons in various semiconductors. Optical phonon energies do vary, however, and eq. (7.13) indicates that its value should affect the saturated velocity. High optical phonon energies lead to efficient energy relaxation, which keeps the electron ensemble near to equilibrium where the velocity versus field characteristic is nearly linear.

Si, Ge and SiC are covalent semiconductors with different optical phonon energies. For Si, $\hbar\omega_o = 0.063\,\text{eV}$ and $\upsilon_{sat} \approx 1 \times 10^7$ cm/s. For Ge, $\hbar\omega_o = 0.037$, and as expected from eq. (7.12), the saturation velocity is lower ($\upsilon_{sat} \approx 0.6 \times 10^7$ cm/s). Silicon carbide has a much higher optical phonon energy ($\hbar\omega_o = 0.12\,\text{eV}$), and the corresponding saturation velocity is also much higher than Si ($\upsilon_{sat} \approx 1.5 \times 10^7$ cm/s).

This simple derivation gives some indication as to why the carrier velocity saturates at high electric fields and indicates the trends for different semiconductors. There is much that it misses, for example, it does not explain why the saturated velocity increases as the temperature decreases. To address such equations, and to compute the complete velocity versus electric field characteristic, we need a more sophisticated theory, such as Monte Carlo simulation or the electron temperature approach discussed in the next section.

7.3 The electron temperature approach

The correct way to evaluate the field-dependent transport coefficients is to solve for the distribution function then find the mobility and diffusion coefficient from eqs. (7.2) and (7.5). Numerical treatments such as the Monte Carlo or iterative techniques are generally required, but simpler, analytical approaches are often useful for the qualitative insight they provide. In this section, we use the simplified balance equations presented in Section 5.5, along with the assumption that the distribution function has a displaced Maxwellian form,

$$f \simeq e^{-|\mathbf{p}-m^*\upsilon_d|^2/2m^*k_BT_e}, \tag{7.14}$$

to compute the field-dependent mobility and diffusion coefficient. As we discussed in Section 5.5, electron–electron scattering tends to produce a Maxwellian distribution function, so eq. (7.14) can be justified when the electron density is sufficient to ensure that electron–electron scattering dominates. A density of at least 10^{17} cm^{-3} is typically required.

In Section 5.5.2 we showed that the argument of the displaced Maxwellian could be expanded into three terms, the second term assumed small, and the third term neglected to obtain

$$f = e^{-(p^2-2m^*\mathbf{p}\cdot\upsilon_d+m^{*2}\upsilon_d^2)/2m^*k_BT_e} \simeq e^{-p^2/2m^*k_BT_e}\left(1+\frac{\mathbf{p}\cdot\upsilon_d}{k_BT_e}\right) = f_S + f_A. \tag{7.15}$$

One can show (see homework problem 7.2) that the simplification described by eq. (7.15) is valid when the drift energy is small compared to the thermal energy. To examine the validity of this assumption, consider electrons in silicon with an applied field of 50 kV/cm. Under such conditions, Monte Carlo simulations show that $u \simeq 0.20$ eV and that $\upsilon_d \simeq 10^7$ cm/s (refer ahead to Fig. 7.4a). The drift energy component,

$$u_{\text{drift}} = \tfrac{1}{2}m^*\upsilon_d^2 \simeq 0.03\,\text{eV},$$

is only a small fraction of the total kinetic energy, so eq. (7.15) should be a good approximation to eq. (7.14). For uniform fields, whether they be high or low, we usually find that eq. (7.15) is valid, but in devices, where spatial variations are strong, the drift energy can be significant.

Under the uniform high fields we are considering, the carriers' kinetic energy is mostly thermal so

$$u_{zz} \simeq \frac{k_BT_e}{2}, \tag{7.16}$$

and the Einstein relation, eq. (7.5), assumes the particularly simple form,

$$\boxed{D_n \simeq \frac{k_BT_e}{q}\mu_n} \tag{7.17}$$

The drifted Maxwellian approach is also known as the electron temperature model because the kinetic energy is measured by the electron temperature.

We should stress that the drifted Maxwellian is simply a reasonable guess for the form of the distribution function – one that considerably simplifies the mathematics of the theory. But distribution functions that are far from Maxwellian occur frequently in practice. For example, polar optical phonon scattering (often the dominant mechanism in GaAs) favors small angle scattering, so the momentum gained from the field is not effectively randomized but rather focused in the direction of the applied field. The use of a drifted Maxwellian for transport in GaAs can produce sizable errors.

7.3.1 Solution by balance equations

Because our assumed form for the distribution function contains parameters related to momentum and energy, we use the momentum and energy balance equations to solve for them. From eq. (5.66b) under steady-state, spatially uniform conditions, we find

$$\upsilon_{dz} = \frac{(-q)}{m^* \langle\langle 1/\tau_m \rangle\rangle} \mathcal{E}_z, \tag{7.18}$$

and from the energy balance equation, eq. (5.66c), we find

$$\frac{T_e}{T_L} = 1 + \frac{2q^2}{3k_B T_L m^*} \frac{1}{\langle\langle 1/\tau_m \rangle\rangle \langle\langle 1/\tau_E \rangle\rangle} \mathcal{E}_z^2. \tag{7.19}$$

To actually solve eqs. (7.18) and (7.19), the ensemble relaxation times must be specified. According to the prescription, eq. (5.14h), these times depend on the distribution function, so they are functions of υ_{dz} and T_e. The neglect of the drift energy, however, leads to a further simplification; we showed in Section 5.5.2 that when the drifted Maxwellian is approximated by eq. (7.15) the ensemble relaxation times have the especially simple form

$$\langle\langle \frac{1}{\tau_m} \rangle\rangle = \frac{\langle E(p)/\tau_m(p) \rangle}{\langle E(p) \rangle}, \tag{7.20}$$

where $<\cdot>$ denotes an average over $f_S(p)$, the symmetric component of the distribution function. For power law scattering with a characteristic exponent, s, eq. (7.20) simplifies to

$$\langle\langle \frac{1}{\tau_m} \rangle\rangle = \frac{1}{\tau_0} \frac{\Gamma(5/2 - s)}{\Gamma(5/2)}. \tag{7.21}$$

A similar result can be obtained for $\langle\langle 1/\tau_E \rangle\rangle$.

To summarize, the electron temperature approach to high-field transport theory consists of first identifying the scattering mechanism that controls momentum relaxation, then evaluating $\langle\langle 1/\tau_m \rangle\rangle$. The mechanism responsible for energy relaxation is then identified (it may be different from the one that controls momentum relaxation), and $\langle\langle 1/\tau_E \rangle\rangle$ is computed. Finally, the momentum and energy balance equations, eqs. (7.18) and (7.19), are solved for υ_{dz} and T_e.

7.3.2 The hot carrier mobility

An expression for $\langle\langle 1/\tau_m \rangle\rangle$ is required, but we can't simply replace T_L in the expressions obtained in Chapter 2 by T_e because part of the temperature dependence is due to phonons whose number is determined by the lattice temperature. The technique will be illustrated by considering two examples; for the first, momentum relaxation is controlled by acoustic deformation potential (ADP) scattering and for the second by ionized impurity (II) scattering.

Example: Electron temperature dependent mobility for ADP scattering.

As shown in Chapter 2 [recall eq. (2.84)],

$$\tau_m(p) = \frac{\pi\hbar^4 c_\ell}{\sqrt{2}(m^*)^{3/2} D_A^2} \frac{1}{k_B T_L} E^{-1/2} \tag{7.22}$$

when ADP scattering dominates. The term, $1/k_B T_L$, arose because the number of phonons was determined according to equipartition. Equation (7.22) can be rewritten as

$$\tau_m(p) = \left[\frac{\pi\hbar^4 c_\ell}{\sqrt{2}(m^*)^{3/2} D_A^2} \frac{1}{(k_B T_L)^{3/2}}\right] \times \sqrt{\frac{T_L}{T_e}} \left(\frac{E}{k_B T_e}\right)^{-1/2} = \tau_0 \times \left(\frac{E}{k_B T_e}\right)^{-1/2}, \tag{7.23}$$

which is in power law form with $s = -1/2$. The constant, τ_0, differs from the low-field result by the factor $\sqrt{T_L/T_e}$, so

$$\langle\langle 1/\tau_m \rangle\rangle = \langle\langle 1/\tau_m \rangle\rangle^o \sqrt{\frac{T_e}{T_L}}, \tag{7.24}$$

where $\langle\langle 1/\tau_m \rangle\rangle^o$ is the ensemble, low-field momentum relaxation rate due to ADP scattering. Alternatively, in terms of mobility we have

$$\boxed{\mu_n(T_e) = \mu_n^o \sqrt{\frac{T_L}{T_e}} (ADP)}\,, \tag{7.25}$$

where $\mu_n^o = q/m^* \langle\langle 1/\tau_m \rangle\rangle^o$ is the low-field mobility due to ADP scattering. Equation (7.25) shows that when ADP scattering dominates, the mobility decreases as the electron temperature increases.

When other types of scattering dominate, the functional dependence of $\mu_n(T_e)$ will differ from eq. (7.25). Only one more example, $\mu_n(T_e)$ when ionized impurity scattering dominates, will be considered.

Example: Electron temperature dependent mobility for II scattering.

Recall from Chapter 2 that for ionized impurity scattering

$$\tau_m(p) = CE^{3/2}, \tag{7.26}$$

where the constant C is independent of temperature and nearly independent of energy. After rewriting eq. (7.26) as

$$\tau_m(p) = \left[C(k_B T_L)^{3/2}\right] \times \left(\frac{T_e}{T_L}\right)^{3/2} \times \left(\frac{E}{k_B T_e}\right)^{3/2}, \tag{7.27}$$

we can readily express the ensemble relaxation time as

$$\langle\langle 1/\tau_m \rangle\rangle = \langle\langle 1/\tau_m \rangle\rangle^o \left(\frac{T_L}{T_e}\right)^{3/2} \tag{7.28}$$

or $$\mu_n = \mu_n^o \left(\frac{T_e}{T_L}\right)^{3/2} \quad \text{(II)} . \tag{7.29}$$

Equation (7.29) should be contrasted with eq. (7.25). When ionized impurity scattering dominates, the mobility increases as the carriers become hot; the velocity versus field characteristic can be super-linear.

Figure 7.3 summarizes the expected mobility versus carrier temperature characteristic for acoustic deformation potential and ionized impurity scattering. For GaAs, the mobility is usually determined by polar optical phonon (POP) scattering. While the calculation of $\langle\langle 1/\tau_m \rangle\rangle$ for this mechanism is tedious [7.9], recall that the POP scattering rate decreases with increasing energy. As a consequence, $\mu(T_e)$ should increase with T_e when POP scattering dominants.

7.3.3 The energy relaxation time

To convert the mobility versus electron temperature characteristic to a mobility versus electric field characteristic, we need to relate the electron temperature to the applied field. To do so, we solve the energy balance equation, but the energy relaxation time must first be expressed in terms of the electron temperature. The first step is to identify the scattering mechanism that controls energy relaxation, which may be different from the one that controls momentum relaxation.

For electrons in silicon, energy relaxation is typically a consequence of equivalent intervalley scattering (acoustic deformation potential scattering is approxi-

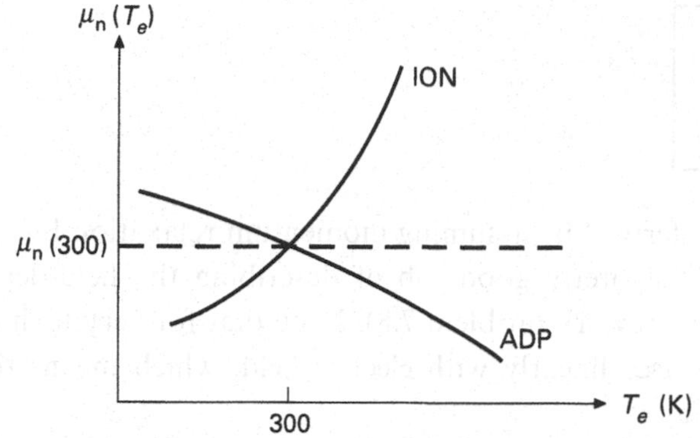

Fig. 7.3 Expected mobility versus carrier temperature characteristic for acoustic deformation potential (ADP) and ionized impurity (ION) scattering.

mately elastic, except at low temperatures). The energy relaxation rate works out to be [7.11]

$$\langle\langle \frac{1}{\tau_E} \rangle\rangle \simeq \frac{2}{3} \frac{C}{k_B T_L} \sqrt{\frac{T_L}{T_e}}, \tag{7.30}$$

where $C \simeq 10^{-8}$ watts for silicon. Using the energy balance equation, (7.19), and assuming momentum relaxation by acoustic phonons, we find

$$\frac{T_e}{T_L} = 1 + \frac{q\mu_n^o}{C} \mathcal{E}_z^2, \tag{7.31}$$

where μ_n^o is the low-field mobility due to ADP scattering. After defining a critical field by

$$\mathcal{E}_{CR} \equiv \sqrt{\frac{C}{q\mu_n^o}}, \tag{7.32}$$

which has a numerical value of $\simeq 7 \times 10^3$ V/cm for silicon, eq. (7.31) becomes

$$\frac{T_e}{T_L} = 1 + (\mathcal{E}_z/\mathcal{E}_{cr})^2, \tag{7.33}$$

which can be used to compute the field-dependent electron temperature.

To determine the field dependence of the mobility, we insert eq. (7.33) in eq. (7.25) to find

$$\boxed{\mu_n(\mathcal{E}) = \frac{\mu_n^o}{\sqrt{1+(\mathcal{E}/\mathcal{E}_{cr})^2}}} . \tag{7.34}$$

Although eq. (7.34) was derived by assuming momentum relaxation by acoustic phonons, it usually does a pretty good job of describing the field-dependent mobility of silicon (see homework problem 7.8). Note that for very high electric fields, the mobility decreases linearly with electric field, which means that the drift velocity saturates at

$$\upsilon_{\text{sat}} = \mu_n^o \mathcal{E}_{cr}. \tag{7.35}$$

For lightly doped silicon, eq. (7.35) gives $\upsilon_{sat} \simeq 10^7$ cm/s – in close agreement with the measured value.

We should also be able to find the high-field diffusion coefficient by multiplying the field dependent mobility by $k_B T_e/q$. According to eqs. (7.33) and (7.34), the electron temperature approach predicts that $D(E)$ should increase with field. In this respect, the electron temperature model *does not* agree with experiments. The measured diffusion coefficient for electrons in silicon decreases slowly with field as was indicated in Fig. 7.2.

The electron temperature model is much more difficult to apply to multi-valley semiconductors such as GaAs because balance equations for each valley are required and because carrier transfer between valleys must be accounted for. In addition, POP scattering tends to dominate in GaAs and focuses the distribution function along the applied field, so that $f(\mathbf{p})$ is not well-approximated by a drifted Maxwellian. Nevertheless, the general features of high-field transport in GaAs can be described with an electron temperature model (refer to the text by Nag [7.8]).

7.4 The Monte Carlo approach

By using Monte Carlo simulation, the simplifying approximations inherent to the electron temperature model can be removed. The shape of the distribution function can be computed, so there is no need to assume a simple form such as the displaced Maxwellian. Details of the energy band structure and scattering mechanisms can also be treated. Monte Carlo simulations are generally in excellent agreement with the experimental results to be reviewed in the following sections. Jacoboni and Reggiani [7.5] and Fawcett et al. [7.6] describe the application of Monte Carlo techniques to electron transport in bulk silicon and GaAs.

7.4.1 Monte Carlo simulation of high-field electron transport in intrinsic Si

Figures 7.4–7.6 summarize Monte Carlo simulations of high-field electron transport in intrinsic silicon. (These simulations used a simple, nonparabolic, ellipsoidal description of the energy bands and the scattering mechanisms described in reference [7.5].) The drift velocity and average kinetic energy per carrier are plotted versus electric field in Fig. 7.4a. Note that the velocity versus field characteristic becomes sub-linear above about 10 kV/cm, which is also the field at which the kinetic energy increase becomes significant. This field, above which hot carrier effects are important, is roughly the critical field defined in eq. (7.32). Note also that the drift velocity saturates at about 10^7 cm/s.

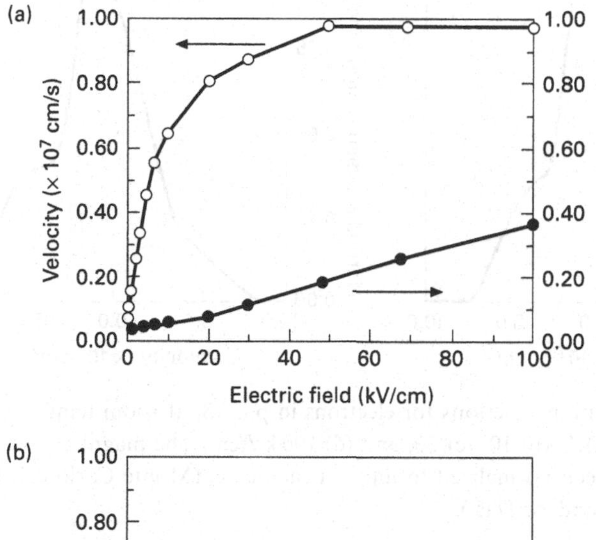

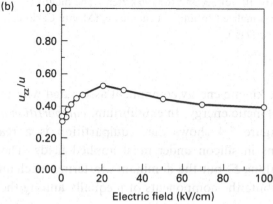

Fig. 7.4 (a) Average velocity and kinetic energy of electrons in pure silicon at 300 K versus electric field. The field is oriented along a ⟨100⟩ direction. (b) Ratio of $\hat{z}$-directed kinetic energy component to the total kinetic energy ($u_{zz} = u$) versus electric field for electrons in pure silicon. The field is oriented along a ⟨100⟩ direction. (Monte Carlo calculations courtesy of M. A. Stettler and A. Das.)

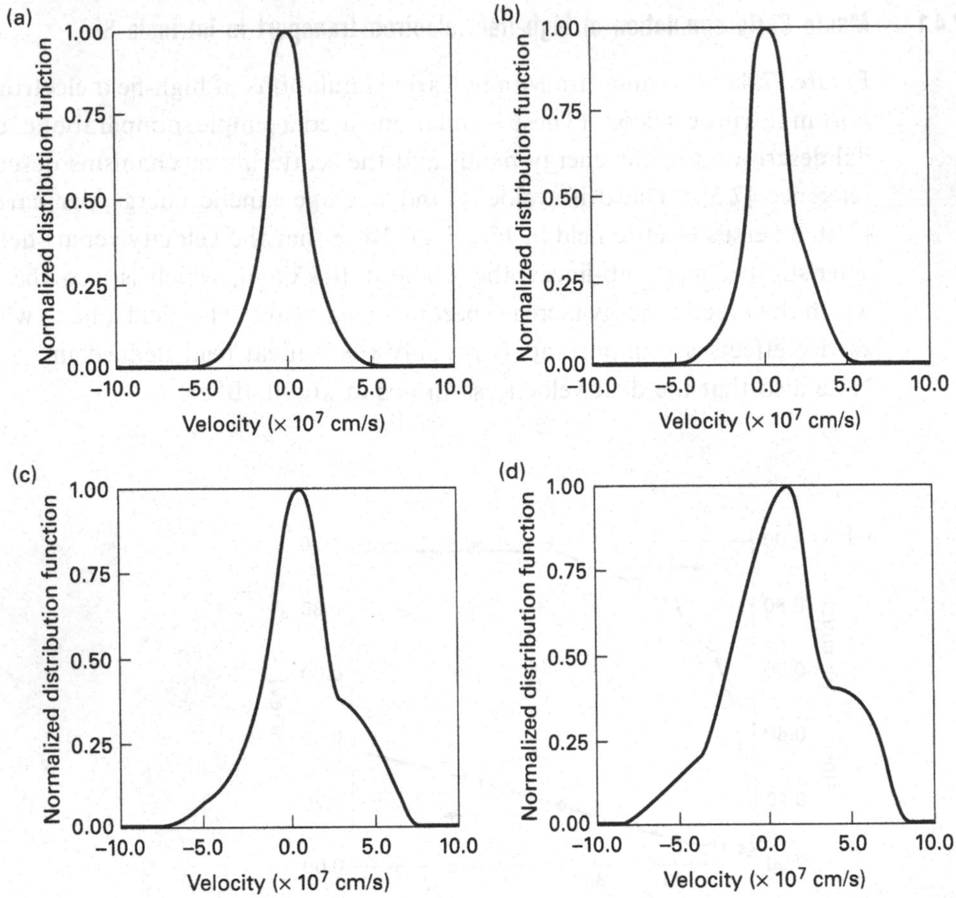

Fig. 7.5 Computed distribution functions for electrons in pure Si at room temperature for applied electric fields of: (a) 1, (b) 10, (c) 50, and (d) 100 kV/cm. The magnitude of the distribution function has been normalized to unity in each case. (Monte Carlo calculations courtesy of M. A. Stettler and A. Das.)

Figure 7.4b compares the kinetic energy component associated with motion in the $\hat{z}$-direction to the total kinetic energy. In equilibrium, *equipartition of energy* applies, and $u_{zz} = u/3$. Figure 7.4 shows that equipartition is a reasonable approximation for electrons in silicon under most applied fields. The electric field does tend to align **p** along $\boldsymbol{\mathcal{E}}$, but the dominant scattering mechanisms are isotropic and tend to distribute the components of **p** equally among the coordinate axes.

Figure 7.5, which shows the computed distribution functions for four different applied fields, illustrates the detail afforded by Monte Carlo simulation. For modest fields, the computed distribution function is approximately Maxwellian, but as the field increases, the peak of the distribution shifts to

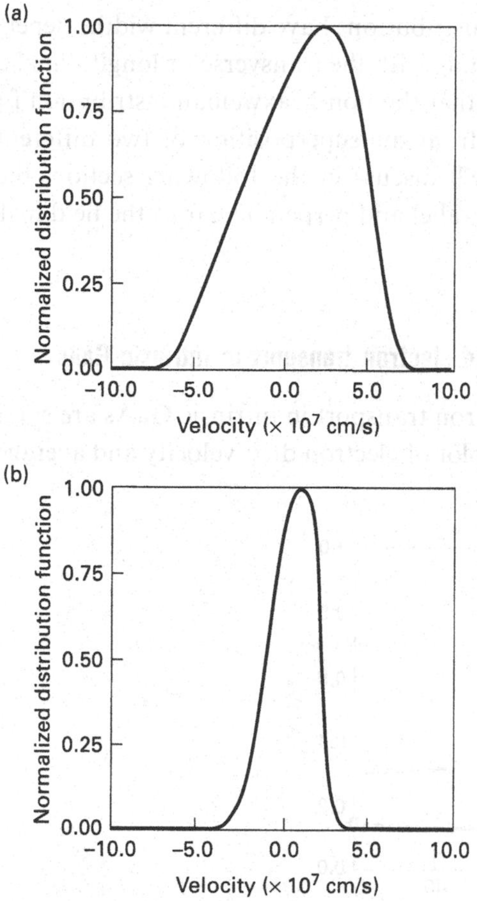

Fig. 7.6 Normalized distribution function for (a) electrons residing in the four valleys for which the major axes of the constant energy ellipsoids are oriented perpendicular to the applied field; (b) electrons residing in the two valleys for which the major axes of the constant energy ellipsoids are oriented parallel to the applied field. Pure silicon and an applied field of 50 kV/cm are assumed. The field is oriented along a ⟨100⟩ direction. (Monte Carlo calculations courtesy of M. A. Stettler and A. Das.)

positive velocities, and it broadens, which indicates that the carrier temperature is increasing. It is also apparent that the distribution function is only roughly Maxwellian. The reason is that carriers residing in the four ellipsoids with major axis perpendicular to the field respond with the transverse effective mass while those in the other two ellipsoids respond with the heavier, longitudinal effective mass.

Figure 7.6 shows separate distribution functions for carriers in the four valleys for which the major axes of the constant energy ellipsoids are perpendicular to the applied field and for the two whose ellipsoids are oriented along the electric field. The results show a roughly Maxwellian distribution of electrons for each of the two types of valleys, but the distributions have different widths depending on whether the electrons are responding with the transverse or longitudinal effective mass. These results demonstrate that the non-Maxwellian distribution functions plotted in Fig. 7.5 largely arise from the superposition of two different nearly Maxwellian distributions. As we'll discuss in the following section, the differences between valleys oriented parallel and perpendicular to the field make high-field transport anisotropic in Si.

7.4.2 Monte Carlo simulation of high-field electron transport in intrinsic GaAs

Monte Carlo simulations of electron transport in intrinsic GaAs are summarized in Figs. 7.7–7.9. Figure 7.7a is a plot of electron drift velocity and average kinetic

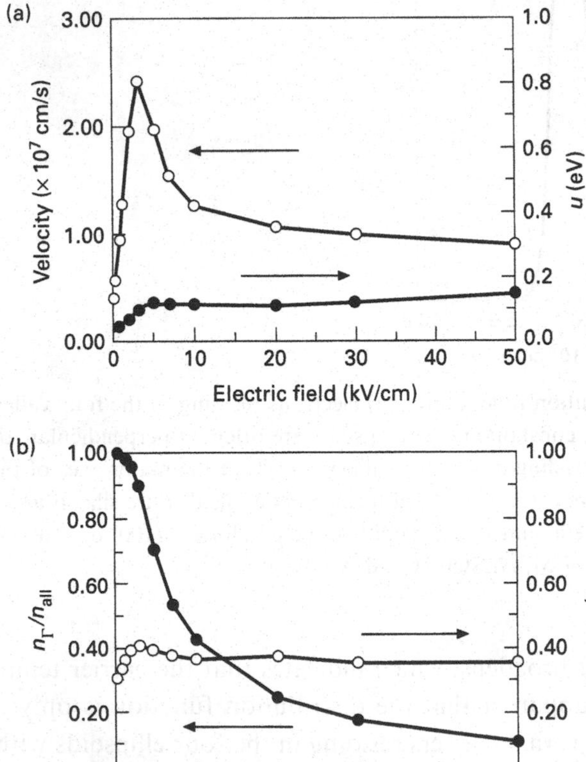

Fig. 7.7 (a) Average velocity and kinetic energy of electrons in pure gallium arsenide at 300 K versus electric field. (b) The steady-state, fractional population of the Γ-valley versus electric field. Also plotted is the ratio of $\hat{z}$-directed kinetic energy component to the total kinetic energy, (u_{zz}/u), versus electric field. (Monte Carlo calculations courtesy of M. A. Stettler and A. Das.)

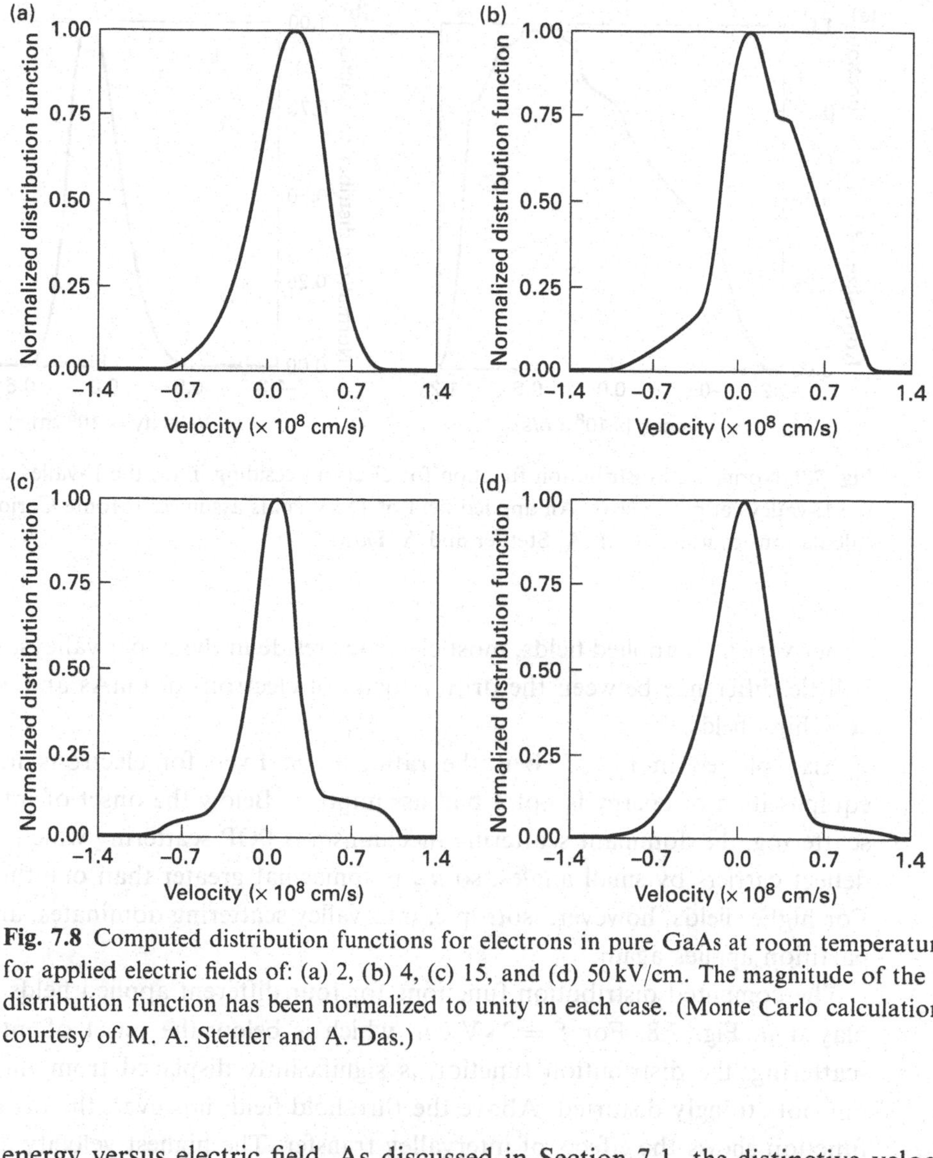

Fig. 7.8 Computed distribution functions for electrons in pure GaAs at room temperature for applied electric fields of: (a) 2, (b) 4, (c) 15, and (d) 50 kV/cm. The magnitude of the distribution function has been normalized to unity in each case. (Monte Carlo calculations courtesy of M. A. Stettler and A. Das.)

energy versus electric field. As discussed in Section 7.1, the distinctive velocity versus field characteristic for electrons in GaAs is a result of intervalley transfer. Under low applied fields, electrons reside in the small effective mass, Γ valley where their mobility is high. When their kinetic energy exceeds $\simeq 0.3$ eV, the intervalley separation energy, they begin to transfer to the high-mass, L valleys. Figure 7.7a shows that the peak velocity occurs when the average kinetic energy of the carriers is still well below that required for intervalley transfer.

The steady-state, Γ-valley occupation factor is plotted versus electric field in Fig. 7.7.b. This figure shows again that the distinctive velocity-field characteristic of GaAs is determined by the intervalley transfer of electrons. For low fields, the mobility of electrons in GaAs is much higher than that for electrons in Si, but

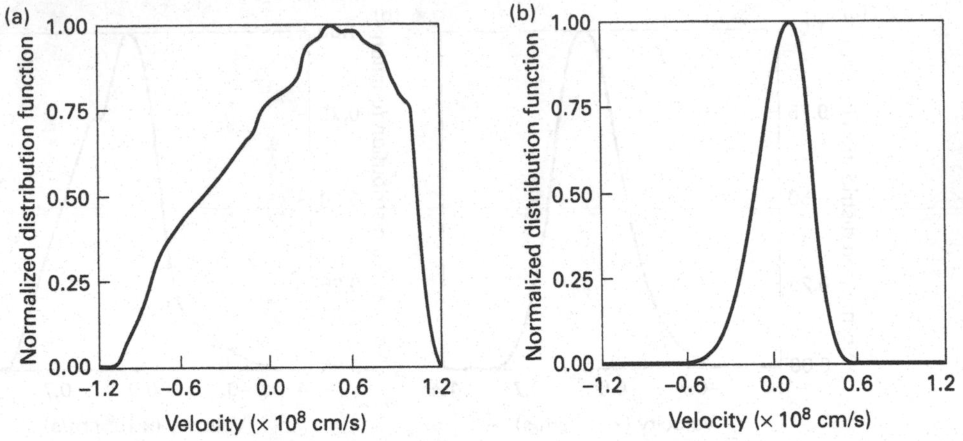

Fig. 7.9 Normalized distribution function for electrons residing in (a) the Γ-valley and (b) the L-valleys of pure GaAs. An applied field of 15 kV/cm is assumed. (Monte Carlo calculations courtesy of M. A. Stettler and A. Das.)

under very high applied fields, most electrons reside in the upper valleys, so there is little difference between the drift velocity of electrons in GaAs and Si under very high fields.

Also plotted in Fig. 7.7b is the ratio, u_{zz}/u. Even for electrons in GaAs, equipartition of energy is not a bad assumption. Below the onset of intervalley scattering, the dominant scattering mechanism is POP scattering which tends to deflect carriers by small angles, so u_{zz} is somewhat greater than one-third of u. For higher fields, however, isotropic, intervalley scattering dominates, and equipartition applies again.

The computed distribution functions for four different applied fields are displayed in Fig. 7.8. For $\mathcal{E} = 2\,\text{kV/cm}$, which is below the onset of intervalley scattering, the distribution function is significantly displaced from the origin, but not strongly distorted. Above the threshold field, however, the distribution function shows the effects of intervalley transfer. The highest velocity electrons tend to be located in the Γ-valley and the slower ones in the L-valleys.

Because the interpretation of Fig. 7.8 is clouded by the fact that electrons from two different valleys are superimposed, we plot in Fig. 7.9 the distribution functions for the two valleys separately assuming an electric field of 15 kV/cm. Since randomizing intervalley scattering dominates, the upper valley electrons display a Maxwellian distribution, but the Γ-valley electrons have a distinctly non-Maxwellian distribution. The focusing of the distribution along the applied field which results from the anisotropic POP scattering is clear.

Figures 7.8 and 7.9 clearly show why the electron temperature model must be used with caution when applied to GaAs. The basic approximation, the use of a

Maxwellian distribution function, is simply not adequate in many cases. Monte Carlo simulation is usually the method of choice when the highest accuracy is demanded. It also imposes a computational burden, however, so the simpler approaches based on the use of balance equations continue to be widely used for analyzing and designing devices.

7.4.3 Full band Monte Carlo simulation of high-field electron transport

For very high electric fields, a simple description of the energy bands no longer suffices, and a full numerical description of $E(\mathbf{p})$ becomes necessary (recall Sections 1.2.2 and 2.14.3). So-called full band Monte Carlo simulations have been developed and applied to transport in a variety of semiconductors (see [6.8], [6.9], and [7.16]). Full band simulations are essential for treating impact ionization and other hot-carrier effects involving energetic carriers.

Figure 7.10 shows the average energy versus electric field for electrons in Si and GaAs as computed by full band Monte Carlo simulation. These results should be compared with Figs. 7.4 and 7.7 which used a simple, nonparabolic description of the energy bands. The full band results extend to electric fields an order of magnitude higher, and they show that the average carrier energy can approach 1 eV.

Figure 7.11 shows the computed electron distribution functions versus energy for electrons in silicon. Under very high electric fields, the distributions are highly non-Maxwellian, and energy states as high as 4–5 eV above the bottom of the conduction band are occupied. Figure 7.12 shows how the electrons are distributed in momentum space. At 10 kV/cm, electrons are still localized to the vicinity of the six equivalent X-valley minima. By 100 kV/cm, they have noticeably spread out and at 1 MV/cm, they occupy the entire Brillouin zone. The results make it clear that a simple treatment of the energy bands fails at very high electric fields.

7.5 Electron transport in bulk Si and GaAs

The measured characteristics of high-field transport in Si and GaAs display the general features characteristic of electron transport in covalent and polar semiconductors as described in Section 7.1. Theoretical treatments based on the electron temperature approach generally reproduce the qualitative features as measured, and Monte Carlo simulations are usually in good quantitative agreement. As for low fields, high-field hole transport is a more difficult theoretical problem because of the complex valence band structure. Measured high-field

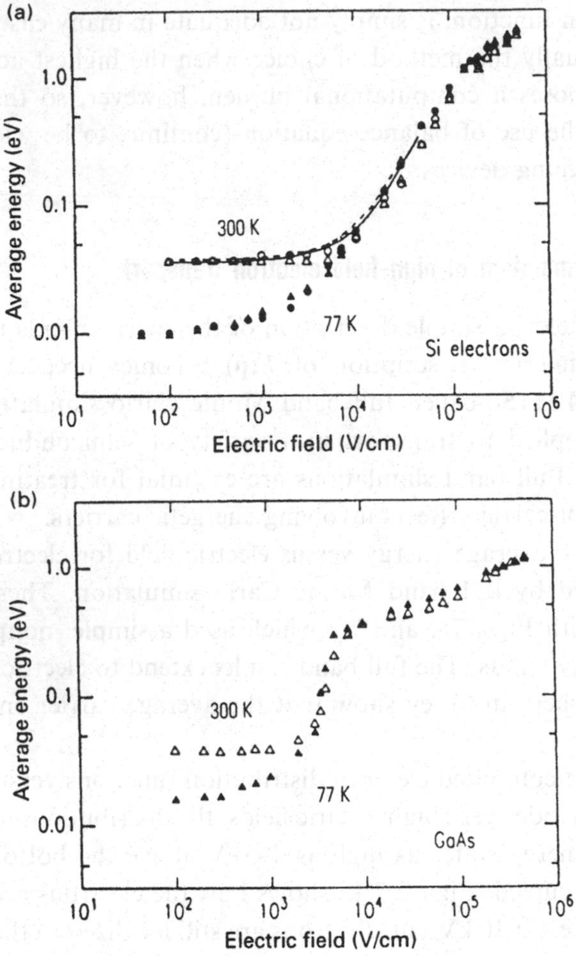

Fig. 7.10 Average electron kinetic energy versus electric field as evaluated by full band Monte Carlo simulation for (a) silicon and (b) GaAs. (From [7.16].)

characteristics are reviewed in several comprehensive surveys [7.1–7.3]. In this section, we examine a few of the key results.

7.5.1 High-field electron transport in silicon

Figure 7.13a shows the measured velocity versus field characteristics for electrons in Si at 300, 77, and 80 K. In contrast to low-field transport, high-field transport is seen to be anisotropic. The measured field-dependent mobility can be fit by [7.2]

$$\mu_n(\mathcal{E}) = \frac{\mu_n^o}{[1 + (\mathcal{E}/\mathcal{E}_{cr})^{\beta}]^{1/\beta}}, \tag{7.36}$$

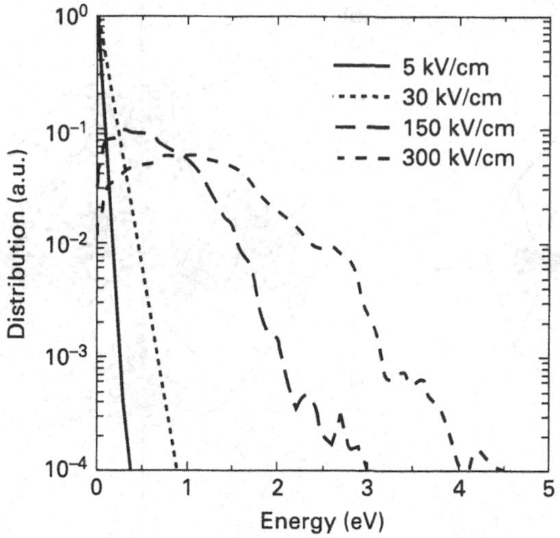

Fig. 7.11 Electron energy distribution in bulk silicon at several different electric fields. (From Kunikiyo, T. et al., *Journal of Applied Physics*, **75**, 297–312, 1994.) (Reproduced with permission of American Institute of Physics.)

which is motivated by the theoretical result, eq. (7.36), but the parameters β and $\mathcal{E}_{cr}$ are viewed as fitting parameters. For a $\langle 111 \rangle$-oriented field in high-purity Si [7.2],

$$\beta = 2.57 \times 10^{-2} \times T_{\mathrm{L}}^{1.55} \tag{7.37a}$$

and

$$\mathcal{E}_{cr} = 1.01 \times T_{\mathrm{L}}^{1.55}. \tag{7.37b}$$

As Fig. 7.13b shows, the saturated drift velocity increases as the temperature decreases. This result is in agreement with theoretical expectations such as eq. (7.35). (Because of the increased saturation velocity at low temperatures, device speed might be improved with cooling.) The observed temperature-dependent saturation velocity can be fit by [7.2]

$$\upsilon_{\mathrm{sat}} = \frac{2.4 \times 10^7}{1 + 0.8 e^{T_{\mathrm{L}}/600}} \ \mathrm{cm/s}. \tag{7.38}$$

Finally, we note that the measured high-field, longitudinal diffusion coefficient, D_n, behaves as sketched in Fig. 7.2. Except for the saturation that occurs at very high fields, the measured $D_n(\mathcal{E})$ can be fit by [7.3]

$$D_n = \frac{D_n^o}{[1 + (\mathcal{E}/\mathcal{E}_{cr})^{\beta}]^{(\beta-1)}}. \tag{7.39}$$

The anisotropic conduction displayed in Fig. 7.13a is a consequence of the ellipsoidal, multi-valley band structure of silicon. For the two field orientations

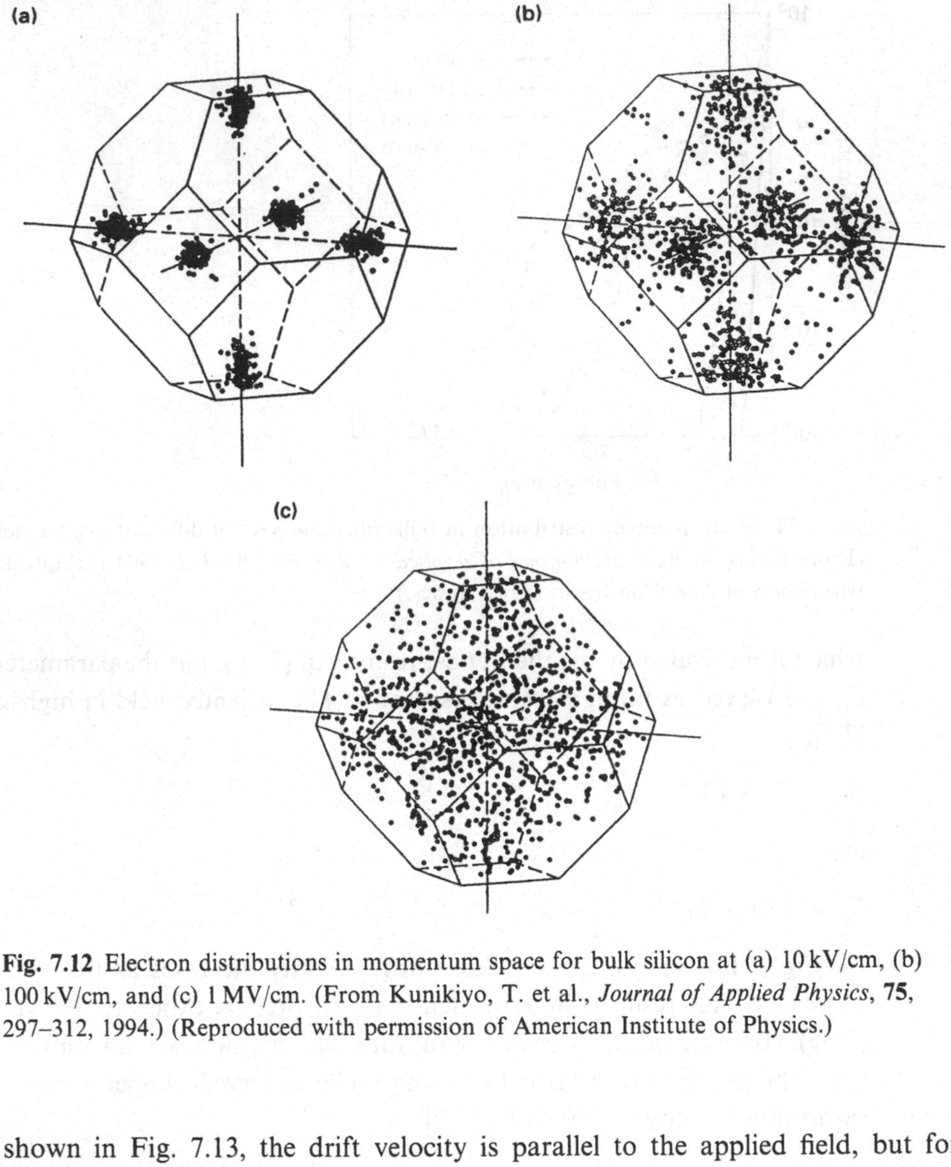

Fig. 7.12 Electron distributions in momentum space for bulk silicon at (a) 10 kV/cm, (b) 100 kV/cm, and (c) 1 MV/cm. (From Kunikiyo, T. et al., *Journal of Applied Physics*, **75**, 297–312, 1994.) (Reproduced with permission of American Institute of Physics.)

shown in Fig. 7.13, the drift velocity is parallel to the applied field, but for arbitrary orientations, υ_d and $\mathcal{E}$ may not even be parallel. The anisotropy of υ_d can be understood with reference to Fig. 7.14. Within a single ellipsoid, transport is anisotropic because carriers respond to applied fields along the major axis with a heavier effective mass (m_ℓ^*) than when the field is along a minor axis (m_t^*). The measured current, which is the sum of the contributions from each ellipsoid, is isotropic at low fields because the ellipsoids are equally populated, so these effects average out. Consider, however, a high-field oriented along the $\hat{y}$-axis (a $\langle 010 \rangle$ direction). Electrons in the four ellipsoids in the $\hat{x}$–$\hat{z}$ plane respond to the field with a small effective mass, m_t^*. Carriers in these ellipsoids are rapidly heated and the intervalley scattering rate increases. Electrons in the other two

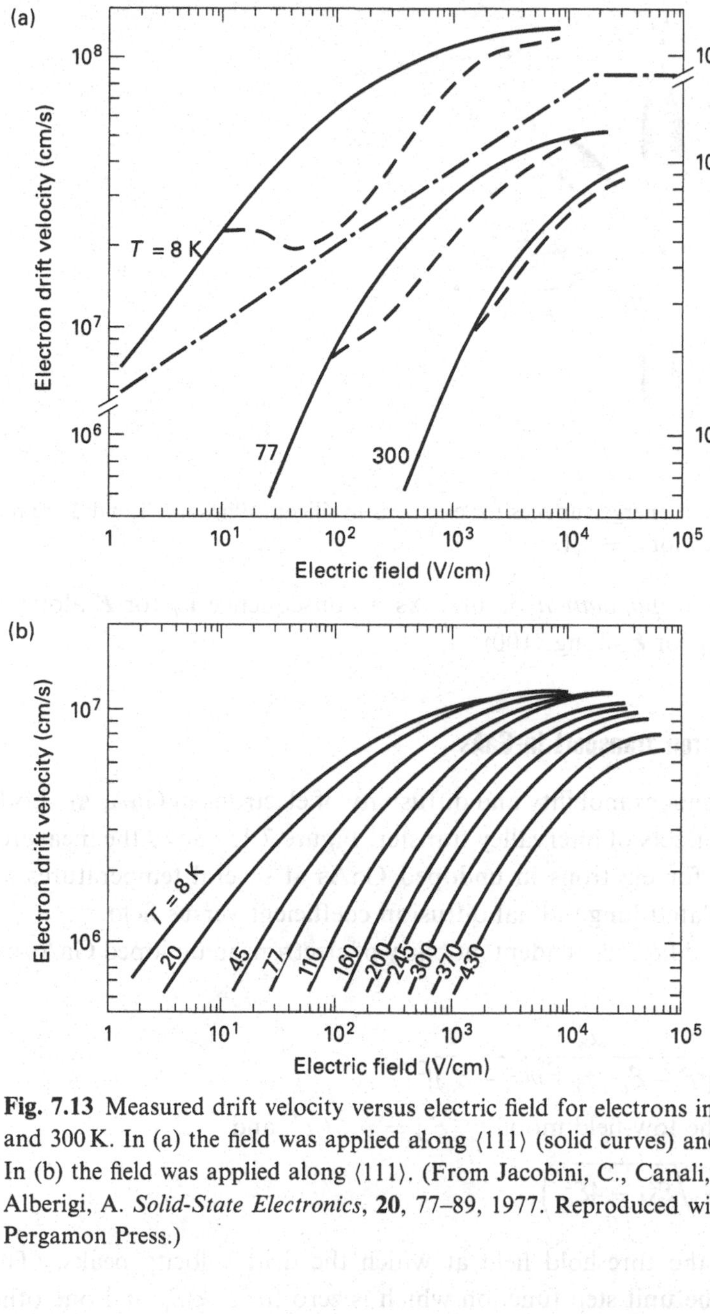

Fig. 7.13 Measured drift velocity versus electric field for electrons in silicon at $T_L = 8, \sim 77$ and 300 K. In (a) the field was applied along $\langle 111 \rangle$ (solid curves) and $\langle 100 \rangle$ (dashed curves). In (b) the field was applied along $\langle 111 \rangle$. (From Jacobini, C., Canali, C., Ottaviani, G., and Alberigi, A. *Solid-State Electronics*, **20**, 77–89, 1977. Reproduced with permission from Pergamon Press.)

ellipsoids respond with m_ℓ^*, a heavier effective mass, and, since they are not heated as much, they don't scatter out as rapidly. The result is that the two ellipsoids whose major axes are oriented along the field are more heavily populated at high fields. Since these carriers respond with the heaviest effective mass, the mobility is lower than expected. When the field is oriented along a $\langle 111 \rangle$ direction, each ellipsoid is oriented the same with respect to the electric field and

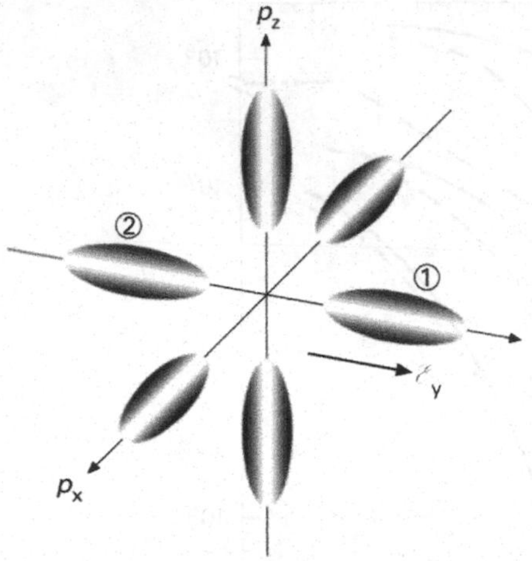

Fig. 7.14 Constant energy surfaces for electrons in silicon. Ellipsoids 1 and 2 are most heavily populated for $\mathcal{E} = \mathcal{E}_y \hat{y}$.

no such *valley repopulation* occurs. As a consequence v_d for $\boldsymbol{\mathcal{E}}$ along $\langle 111 \rangle$ is higher than v_d for $\boldsymbol{\mathcal{E}}$ along $\langle 100 \rangle$.

7.5.2 High-field electron transport in GaAs

The field-dependent mobility and diffusivity of electrons in GaAs are both dominated by the effects of intervalley transfer. Figure 7.15 shows the measured $v_d(\mathcal{E})$ characteristic for electrons in undoped GaAs at several temperatures and Fig. 7.16 the simulated longitudinal diffusion coefficient versus field.

The observed field-dependent mobility of electrons in undoped GaAs can be fit by [7.4]

$$\mu_n(\mathcal{E}) = \frac{\mu_n^o}{[1 + [(\mathcal{E} - \mathcal{E}_o)/\mathcal{E}_{cr}]^2 u(\mathcal{E} - \mathcal{E}_o)]^2}, \tag{7.40}$$

where μ_n^o is the low-field mobility, $\mathcal{E}_{cr} = v_{sat}/\mu_n^o$, and

$$\mathcal{E}_o = \frac{1}{2}\left(\mathcal{E}_M + \sqrt{\mathcal{E}_M^2 - 4\mathcal{E}_{cr}^2}\right), \tag{7.41}$$

where $\mathcal{E}_M$ is the threshold field at which the drift velocity peaks. (The term, $u(\mathcal{E} - \mathcal{E}_o)$ is the unit step function which is zero for $\mathcal{E} < \mathcal{E}_o$ and one otherwise.) The best fit to the measured velocity versus field characteristic is obtained by setting $\mathcal{E}_M = 3.2\,\text{kV/cm}$ and $v_{sat} = 8 \times 10^6$ cm/s. The temperature-dependence of v_{sat} displayed in Fig. 7.15 can be summarized by [7.4]

$$v_{sat} = 1.2 \times 10^7 - (1.5 \times 10^4)T_L\,\text{cm/sec}. \tag{7.42}$$

As for electrons in Si, the saturation velocity increases upon cooling.

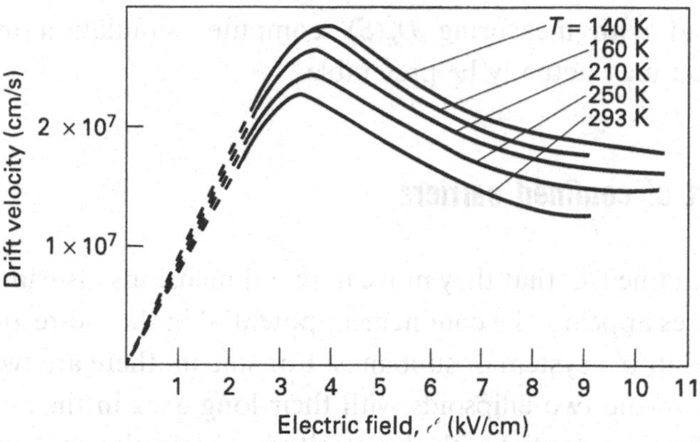

Fig. 7.15 Measured drift velocity versus electric field for electrons in undoped GaAs at several temperatures. (From Ruch, J. G. and Kino, G. S. *Physics Review*, **174**, 921–931, 1968. Reproduced with permission from American Institute of Physics.)

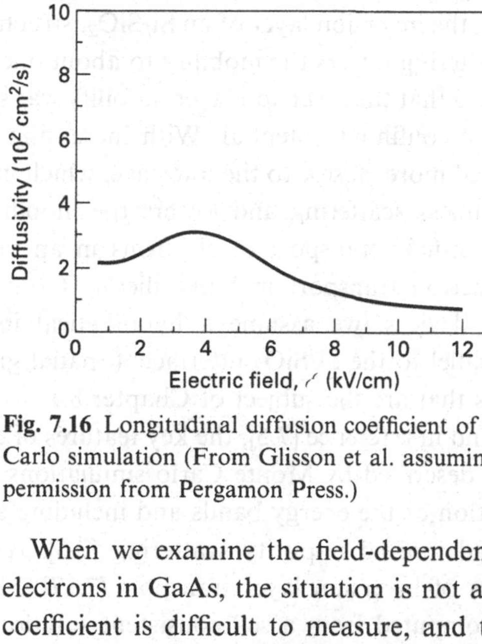

Fig. 7.16 Longitudinal diffusion coefficient of electrons in GaAs as computed by Monte Carlo simulation (From Glisson et al. assuming $N_D - 10^{16}\ \text{cm}^{-3}$ [7.13]. Reproduced with permission from Pergamon Press.)

When we examine the field-dependent, longitudinal diffusion coefficient for electrons in GaAs, the situation is not as clear. One reason is that the diffusion coefficient is difficult to measure, and the results are quite sensitive to circuit effects [7.12–7.13]. The Monte Carlo computations are also tricky, and conflicting results have been reported [7.13–7.15]. Some recent Monte Carlo results are displayed in Fig. 7.16 [7.13]. Below the critical field at which intervalley transfer occurs, the behavior is determined by how quickly μ_n falls with field and how fast u_{zz} rises. The net effect predicted by these simulations is that D_n is roughly constant. Above the critical field, however, the energy increases more slowly while the mobility plummets, so $D_n(\mathcal{E}_z)$ falls very rapidly with field. It is interesting to note that accurate simulations of impact ionization transit time devices (IMPATT's), which are especially sensitive to $D_n(\mathcal{E})$, result when the Monte Carlo simulated diffusion coefficient is used [7.13]. Because of the experimental difficulties associated with measuring $D_n(\mathcal{E})$, computer simulation using the Monte Carlo method may actually be preferable.

7.6 High-field transport of confined carriers

When carriers are confined so that they move in two-dimensions instead of three, new transport features appear. The confinement potential in the z-direction splits the conduction band into a system of subbands. For silicon, there are two sets of subbands; electrons in the two ellipsoids with their long axes in the z-direction respond to the confining potential with the longitudinal effective mass while the remaining four ellipsoids respond with the transverse effective mass. The new features of carrier transport are scattering between these subbands and scattering from surface roughness and charges in the oxide. In Section 4.9.1 we discussed low-field transport of electrons in the inversion layer of an Si/SiO_2, structure. We found that surface roughness scattering lowers the mobility to about one-half of its value in the bulk. We also found that the inversion layer mobility was strongly dependent on the strength of the confining potential. With increasing normal electric field, electrons are confined more closely to the interface, which increases the importance of surface roughness scattering and lowers the mobility. Our interest here is in examining high-field transport of electrons in an inversion layer and comparing it with electron transport in bulk silicon. To avoid the spatial gradients that occur in devices, we assume a hypothetical inversion layer which is homogeneous parallel to the Si/SiO_2 interface. (Spatial gradients introduce new transport features that are the subject of Chapter 8.)

As discussed in Section 7.3.1 and in reference [7.5], the key features of electron transport in bulk silicon can be described by Monte Carlo simulations using a nonspherical, ellipsoidal description of the energy bands and including acoustic phonon, intervalley phonon, and ionized impurity scattering [7.5]. A similar model describes the key features of inversion layer transport [7.17]. First, the system of subbands has to be computed from a self-consistent solution to the wave equation. Only a finite number of subbands can be treated in a numerical simulation. For high longitudinal electric fields, carriers gain substantial kinetic energy and 200 or more subbands may be occupied [7.17]. At low normal electric fields, the confining potential is weak. Subbands are closely spaced and it is still necessary to include on the order of 100 subbands [7.17].

The scattering rates of confined carriers differ from those in the bulk. Intra- and inter-subband scattering rates can be calculated as outlined in Section 2.11 using the same phonon energies and deformation potentials used for bulk silicon [7.17]. (Actually, we should not expect the phonon dispersion relation near an interface between two different materials to be the same as it is in the bulk of one of the materials. For some material systems, these interface modes have to be considered, but for the Si/SiO_2 system, the use of bulk phonons introduces little error [7.17]. Surface roughness scattering can be treated along the lines outlined in Section 2.12. Because the carrier densities in the inversion layer may be large, it is also important to treat carrier degeneracy. Scattering rates are suppressed by a $(1 - f)$ factor to account for the number of available states. Because of the high carrier density, screening of scattering potentials by the free carriers is also important. The use of a Debye length as discussed in Section 2.15 is extended to treat the response of a 2D, degenerate electron gas to rapidly varying potentials [7.17].

Figures 7.17a and 7.17b show Monte Carlo simulation results for electron transport in a homogeneous silicon inversion layer [7.17]. As discussed in Section 4.9.1, the low field mobility is reduced by almost a factor of two by surface roughness scattering. We see, however, that the saturated velocity of inversion layer electrons is identical to that of bulk electrons. As shown in Fig. 7.17b, strong normal electric fields reduce the low longitudinal electric field mobility, but they do not affect the magnitude of the saturated velocity. The Monte Carlo simulations and the experimental results indicate that carrier confinement and surface roughness scattering have a strong effect on the low field mobility but very little effect on the high-field transport of inversion layer

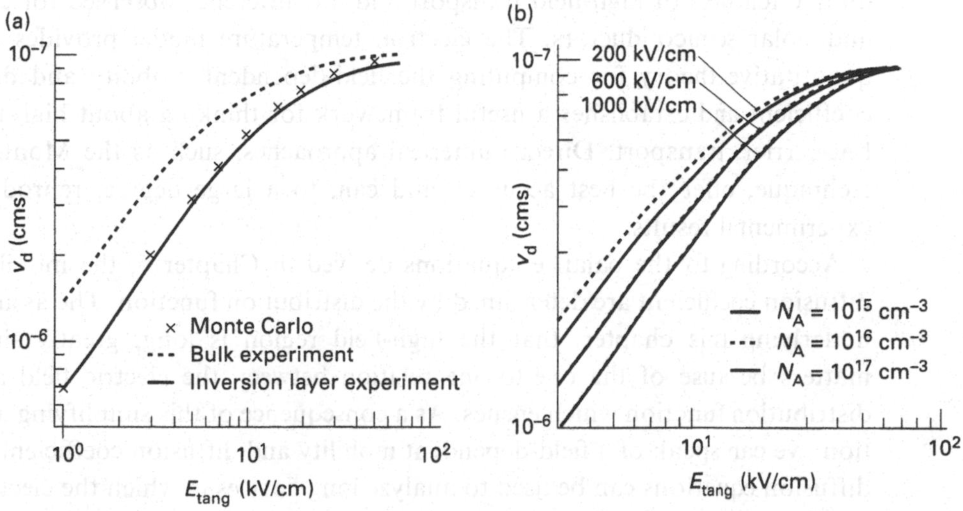

Fig. 7.17 Drift velocity vs. electric field parallel to the Si/SiO_2 interface for electron transport. (a) Monte Carlo simulations compared with experimental results for inversion layer and bulk electron transport. (b) Results for three different normal electric fields and three different doping densities. (From Jungemann et al. [7.17].) (Reproduced with permission of Pergamon Press.)

electrons (except for the increased critical field for velocity saturation caused by the reduced low-field mobility).

Although a relatively simple Monte Carlo model can account for the key features of inversion layer transport and provide good quantitative agreement too, unresolved issues remain. For example, the acoustic deformation potential is assumed to be isotropic and is taken as an adjustable parameter [7.17], but a more careful treatment shows that it is anisotropic [7.16]. A proper treatment of screening is difficult and surface optical modes which couple longitudinal optical modes and interface electromagnetic modes exist [7.16]. More careful calculations which do not include adjustable parameters over-estimate the mobility by about 20%, but they also predict a saturated velocity identical to the bulk [7.16]. Even here, however, the issue is still unsettled because some experimental results show a saturated velocity for inversion layer electrons that is as small as one-half the bulk value (see [7.16] for a fuller discussion of these issues).

7.7 Summary

Our objectives in this chapter have been to explain the general features of high-field transport in common semiconductors, to describe theoretical approaches for computing the high-field transport coefficients, and to examine some measured results for Si and GaAs. Simple physical arguments can explain the qualitative features of high-field transport and the difference observed for covalent and polar semiconductors. The electron temperature model provides a semi-quantitative theory for computing the field-dependent mobility and diffusion coefficient and establishes a useful framework for thinking about high-field, or hot-carrier, transport. Direct numerical approaches, such as the Monte Carlo technique, offer the best accuracy and can, to a large degree, reproduce the experimental results.

According to the balance equations derived in Chapter 5, the mobility and diffusion coefficient are determined by the distribution function. The assumption underlying this chapter, that the high-field region is long, greatly simplifies matters because of the one-to-one relation between the electric field and the distribution function which ensues. As a consequence of this simplifying assumption, we can speak of a field-dependent mobility and diffusion coefficient. Drift–diffusion equations can be used to analyze long devices in which the electric field changes slowly by determining $\mu(\mathcal{E})$ and $D(\mathcal{E})$ from the local electric field. But in short devices the situation is far more complicated because the fields may change too rapidly (in space or time) for the carriers to reach the steady-state, uniform field conditions assumed in this chapter. Transport in devices is the subject of the next chapter.

References and further reading

The measured characteristics of high-field transport in common semiconductors are surveyed in

7.1 Jacobini, C. and Reggiani, L. Bulk hot-electron properties of cubic semiconductors. *Advances in Physics*, **28**, 493–553, 1979.

7.2 Jacobini, C., Canali, C., Ottaviani, G. and Albergi Quaranta, A. A review of some charge transport properties of silicon. *Solid-State Electronics*, **20**, 77–89, 1977.

7.3 Ali Omar, M. and Reggiani, L. Drift and diffusion of charge carriers in silicon and their empirical relation to the electric field. *Solid-State Electronics*, **30**, 693–7, 1987.

7.4 Chang, C. S. and Fetterman, H. Electron drift velocity versus electric field in GaAs. *Solid-State Electronics*, **29**, 1295–6, 1986.

Jacoboni and Reggiani explain how Monte Carlo simulation is applied to covalent semiconductors in

7.5 Jacobini, C. and Reggiani, L. The Monte Carlo method for the solution of charge transport in semiconductors with application to covalent materials. *Reviews of Modern Physics*, **55**, 645–705, 1983.

And for a description of how the Monte Carlo technique is applied to high-field electron transport in GaAs, consult

7.6 Fawcett, W., Boardman, A. D. and Swain, S. Monte Carlo determination of electron transport properties in GaAs. *J. Physics and Chemistry of Solids*, **31**, 1963–90, 1970.

The electron temperature approach to high-field transport theory is discussed in

7.7 Seeger, K. *Semiconductor Physics*. 3rd edn., Springer-Verlag, New York, 1985.

7.8 Nag, B. R. *Electron Transport in Compound Semiconductors*. Springer-Verlag, New York, 1980.

For a classic treatment of high-field electron transport, consult

7.9 Conwell, E. M. High-field transport in semiconductors. In *Solid-State Physics*, Suppl. Vol. 9, Academic Press, New York, 1967.

And for a recent treatment, see

7.10 Hess, K. Advanced Theory of Semiconductor Devices. Prentice-Hall, Englewood Cliffs, NJ, 1988.

7.11 Hess, K. Phenomenological physics of hot carriers in semiconductors. In *Physics of Nonlinear Transport in Semiconductors*. Ed. by Ferry, D. K., Barker, J. and Jacoboni, C. Plenum, New York, 1980.

To review the status of measuring and simulating the field-dependent diffusion coefficient of electrons in GaAs, read

7.12 Ruch, J. G. and Kino, G. S. Transport properties of GaAs. *Physics Review*, **174**, 921–31, 1968.

7.13 Glisson, T. H., Sadler, R. A., Hauser, J. and Littlejohn, M. A. Circuit effects in time-of-flight diffusivity measurements. *Solid State Electronics*, **23**, 637, 1980.

7.14 Pozela, J. and Reklaitis, A. Diffusion coefficient of hot electrons in GaAs. *Solid State Communications*, **27**, 1073–77, 1978.

7.15 Glisson, T. H., Williams, C. K., Hauser, J. and Littlejohn, M. A. Transient response of electron transport in GaAs using the Monte Carlo method. In *VLSI Electronics: Microstructure Science*, Vol. 4, Chapter 3. Academic Press, New York, 1982.

Comprehensive full band Monte Carlo simulations of electron and hole transport in a variety of semiconductors are presented in

7.16 Fischetti, M. V. Monte Carlo simulation of transport in technologically significant semiconductors of the diamond and zinc-blende structures – Part I: homogeneous transport. *IEEE Transactions on Electron Devices*, **38**, 634–49, 1991.

Monte Carlo simulations of electron transport in the Si/SiO_2 inversion layer are presented in

7.17 Jungemann, Chr., Emunds, A. and Engl, W. L. Simulation of linear and nonlinear electron transport in homogeneous silicon inversion layers. *Solid-State Electronics*, **36**, 1529–40, 1993.

Problems

7.1 Equation (7.5) is often stated as

$$D_n = \frac{2}{3}\frac{u}{q}\mu_n.$$

Explain when this equation is valid.

7.2 Equation (7.15) is justified when

$$(m^* v_d)^2 \ll 2\mathbf{p} \cdot m^* v_d \qquad \text{(i)}$$

and when

$$2m^* \mathbf{p} \cdot v_d \ll p^2. \qquad \text{(ii)}$$

Apply these relations to a thermal average carrier, and show that both require that the average drift energy be a small component of the total energy.

7.3 Assume that the electron mobility in silicon is controlled by acoustic deformation potential scattering. Use the data presented in this chapter to deduce $T_e(\mathcal{E})$. Plot the electron temperature versus electric field for $0 < \mathcal{E} < 10^5$ V/cm.

7.4 Solve the Boltzmann transport equation in the relaxation time approximation for moderate strength fields and show that $\mu(\mathcal{E}) = \mu_0(1 + \beta\mathcal{E}^2)$. *Hint*: when evaluating $\partial f/\partial p_z$ approximate f by $f_0 + f_A^1$ where f_A^1 is the low-field solution we obtained in Chapter 3.

7.5 Compute the velocity versus field characteristic when acoustic deformation potential (ADP) scattering controls both energy and momentum relaxation.

(a) Use the results of homework problem 2.8 to show that

$$\langle\langle\frac{1}{\tau_E}\rangle\rangle = \langle\langle\frac{1}{\tau_m}\rangle\rangle\left(\frac{2m^* v_s^2}{k_B T_L}\right).$$

(b) Show the solution to the energy balance equation is

$$T_e = \frac{T_L}{2}\left[1 + \sqrt{\frac{4}{3}\left(\frac{\mu_o \mathcal{E}_z}{v_s}\right)^2}\,\right].$$

(c) Simplify the result for high-fields and show that the high-field mobility and drift velocity are:

$$\mu(\mathcal{E}_z) = \sqrt{\frac{\sqrt{3}\mu_o \upsilon_s}{\mathcal{E}_z}}$$

and

$$\upsilon_{dz} = \sqrt{\sqrt{3}\mu_o \upsilon_s \mathcal{E}_z}.$$

These conditions apply to lightly doped silicon or germanium at low temperatures where the energy loss is primarily by emission of acoustic phonons. The measured characteristics do display the predicted square root dependence on applied field [7.11].

7.6 Obtain an expression for the field-dependent mobility of silicon assuming that energy relaxation is accomplished by intervalley phonons and that momentum relaxation is controlled by ionized impurity scattering according to eq. (7.29). Examine your result for small electric fields. As the field increases from zero, does the mobility increase or decrease? Explain the result physically.

7.7 Use the Monte Carlo results displayed in Figs. 7.4a and 7.7a to answer these questions.

(a) Plot the drift energy divided by the kinetic energy versus electric field. Is the neglect of the drift energy valid for transport of electrons in silicon under uniform high electric fields?

(b) Repeat for electrons in GaAs.

7.8 Assume that a very high field is applied to silicon so that the average electron energy is much greater than a phonon energy. Obtain an expression for the mobility as a function of electron temperature assuming that equivalent intervalley scattering dominates. The fact that the result is just like eq. (7.25) which was derived for ADP scattering helps to explain why eq. (7.34) does such a good job of explaining the observed field dependence of the electron mobility for silicon.

7.9 A Monte Carlo simulation of electron transport in bulk ⟨100⟩ silicon generated the results shown in the table below. Use these results to answer the questions below.

$\mathcal{E}$ (V/cm)	υ_d (cm/s)	u (meV)	u_{zz} (meV)
1×10^3	1.32×10^6	41.5	13.8
3×10^3	3.06×10^6	44.8	16.1
1×10^4	6.36×10^6	61.6	24.7
3×10^4	9.06×10^6	117.3	49.0
1×10^5	1.02×10^7	339.0	137.0

(a) Deduce the field-dependent electron mobility versus electric field and compare the result with eq. (7.34).

(b) Use the expression $D_n = 2u_{zz}\mu_n/q$ to deduce the diffusion coefficient versus electric field. Compare it to the value you would expect from the conventional Einstein relation, $D_n = (k_B T_L/q)\mu_n$.

(c) Estimate the drift energy versus electric field. Proceed as follows. Estimate the average effective mass, $\langle m^* \rangle$ using $m^*(E) = m^*(0)(1 + \alpha E)$. Estimate $\langle m^* \rangle$ by using the average carrier energy printed in the output table (u) assuming that $\alpha = 0.5\,\text{eV}^{-1}$ and $m^*(0) = m_c^* = 0.26 m_0$. From the average effective mass, estimate the drift energy versus electric field. What fraction of the total carrier energy is drift energy?

(d) Using the average carrier energy and the estimated drift energy, estimate the electron temperature vs. electric field. Compare the result with eq. (7.33). Use an Einstein relation and the result of part (a) to estimate the diffusion coefficient versus electric field and compare the answer to the results of part (b).

(e) From the computed mobility versus electric field and electron temperature versus electric field, construct a table of mobility versus electron temperature. Compare your results to eq. (7.25).

(f) Define the mobility as $\mu_n = q\langle\hat{\tau}_{\mathrm{m}}\rangle/\langle m^*\rangle$, where $\langle\hat{\tau}_m\rangle$ is the average momentum relaxation time and $\langle m^*\rangle$ is the average effective mass as obtained in part (b). Estimate the momentum relaxation time versus electric field.

(g) The steady-state energy balance equation is $3nk_{\mathrm{B}}(T_{\mathrm{C}} - T_{\mathrm{L}})/(2\langle\hat{\tau}_{\mathrm{E}}\rangle) = J_{nz}\mathcal{E}_z$ where $\langle\hat{\tau}_{\mathrm{E}}\rangle$ is the average energy relaxation time. Use the Monte Carlo results to estimate the energy relaxation time versus electric field and compare it to the momentum relaxation time.

8 Carrier transport in devices

Carrier transport in semiconductor devices is complicated by the rapid spatial and temporal variations that often occur. For large devices, the low- and high-field transport theory developed in previous chapters is directly applicable. Such devices can be analyzed by drift–diffusion equations with field-dependent mobilities and diffusion coefficients. Transport in small devices, however, differs qualitatively from that in bulk semiconductors because the carrier distribution function is no longer determined by the local electric field. Since transport is nonlocal in both space and time, conventional drift–diffusion equations do not apply, but new possibilities for enhancing device performance arise.

In this chapter, we explain why the drift–diffusion equation loses validity for small devices and describe some important features of carrier transport in the presence of rapidly varying fields. The objective is to gain an intuitive understanding of carrier transport in modern devices such as small bipolar and field-effect transitors. To identify the kinds of transport problems that need to be addressed, we begin by describing a generic transistor. We then examine carrier transport under several specific situations that occur in modern devices and explain why the drift–diffusion equation often loses validity. Finally, we briefly examine device simulation to indicate how the transport equations are formulated for numerical solution, so that nonlocal transport can be simulated for realistic devices.

8.1 A generic view of transistors

A simple, conceptual model for a transistor consists of a carrier injector, a carrier collector, and a control region that meters the flow of carriers out of the source (Fig. 8.1). (For an introduction to semiconductor devices, consult Pierret [8.1], Singh [8.2], or Sze [8.3].) This simple model describes a bipolar transistor where the three regions are the emitter, base and collector, and the control voltage is the base-emitter voltage. It also describes a metal–oxide–semiconductor field-effect transistor (MOSFET) where the three regions are the source, channel, and drain, and the control voltage is the gate to source voltage [8.4]. (More precisely, the control region is the low-field portion of the channel near the source, and the collector is the high-field portion of the channel and the drain.)

Figure 8.1 also indicates the types of transport that often occur in modern semiconductor devices. Carriers injected from the source must first be transported over a barrier by diffusing against an electric field. Next, they drift and diffuse across the thin control region. Because this region (the base or the low-field portion of the channel) is so thin in modern transistors, carriers may experience few scattering events as they cross it. Finally, the carriers are collected by a thin, high-field region. What makes transport in devices interesting is the fact that these three regions are short so that bulk conditions are never achieved. As carriers travel from the source to the drain, a distance on the order of 100 nm in present-day transistors, they experience rapidly changing transport conditions. We will examine transport under conditions like these in the remainder of this chapter, but before we do, we explain why the simplest description of carrier

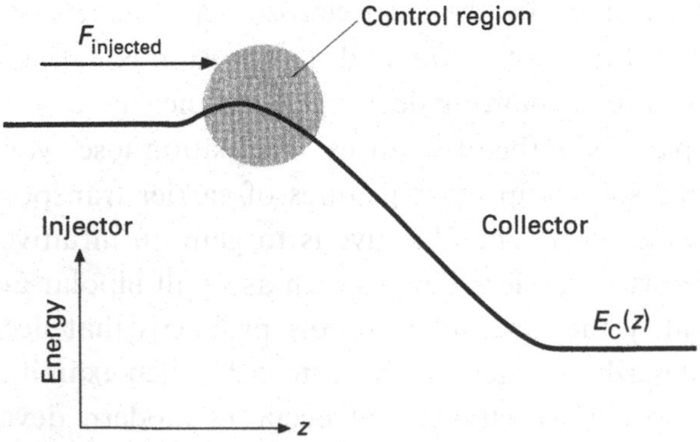

Fig. 8.1 A generic model for a transistor showing a carrier injector, a carrier collector, and a control region in between. The height of the potential barrier between the source and control region is set by a third terminal (e.g. the base or the gate). Normal, active region of operation for the transistor has been assumed.

transport, drift–diffusion, breaks down under conditions like those shown in Fig. 8.1.

8.2 Semiclassical transport and the drift-diffusion equation

We begin with the drift–diffusion equation because it has long served as the foundation of device analysis. To derive it, we assume that the energy-related tensor, $\overleftrightarrow{W}$, is diagonal, so that the momentum balance equation for electrons, eq. (5.32), can be written as

$$\hat{\tau}_m \frac{\partial J_{nz}}{\partial t} + J_{nz} = nq\mu_n \mathcal{E}_z + 2\mu_n \frac{dW_{zz}}{dz}, \tag{8.1a}$$

where

$$\mu_m \equiv \frac{q\hat{\tau}_m}{m^*} \tag{8.1b}$$

is the electron mobility, and

$$W_{zz} \equiv 1/2n\langle p_z v_z \rangle \equiv nu_{zz} \tag{8.1c}$$

is the average kinetic energy density associated with the degree of freedom along the $\hat{z}$-axis (u_{zz} is the average kinetic energy component per electron). To simplify the notation, an 'average' momentum relaxation time, $\hat{\tau}_m$, has been introduced. It is rigorously defined as $\hat{\tau}_m \equiv 1/\langle\langle 1/\tau_m(\mathbf{p})\rangle\rangle$, but in practice it is frequently estimated by Monte Carlo simulation. It is important to note that eq. (8.1) is essentially exact; it was derived directly from the Boltzmann Transport Equation (BTE) by integration. We did assume spherical energy bands and that $\overleftrightarrow{W}$ is diagonal, but these assumptions are readily removed (see Bandopadhyay et al., *IEEE Transactions on Electron Devices*, **34**, 392–399, 1987). Equation (8.1) reduces to a conventional drift–diffusion equation when: (1) the current density varies slowly on the scale of the momentum relaxation time, and (2) the kinetic energy component, u_{zz}, varies slowly with position. As device speeds increase and dimensions shrink, both assumptions are losing validity.

To make use of the drift–diffusion equation, the parameters μ_n and W_{zz} (or, equivalently, μ_n, D_n, and u_{zz}) must be specified. These parameters are directly related to the distribution function, which means that the BTE should first be solved. But if the BTE has been solved, then the state of the device is completely specified, and eq. (8.1) is not needed. Under low applied fields, however, the electron mobility becomes a *material-dependent* parameter which can be computed once and for all by invoking the relaxation time approximation or by more sophisticated numerical techniques. For low applied fields, the carriers are in equilibrium with the lattice, and $W_{zz} = nk_B T_L/2$. For high applied fields in bulk semiconductors, W_{zz} is constant, and there is a one-to-one mapping between the distribution function and the applied field. Consequently, for uniform, high fields in bulk semiconductors, a *field-dependent* mobility and diffusion coefficient can be defined. We can, therefore, describe transport in bulk semiconductors by a drift–diffusion equation.

Small devices pose special problems for drift–diffusion equations because the transport parameters become *device-dependent*. This occurs because the distribution function is no longer related to the local electric field but depends in a complicated way on the electric field throughout the device. Equation (8.1) continues to hold, but the transport parameters can vary widely from device to device, within a device, and can even change with the applied bias as the distribution function changes. One can still talk about mobilities and diffusion coefficients with the understanding that they are not the fundamental parameters they are in bulk semiconductors. In the next few sections, we'll examine transport under various conditions that can occur in modern devices. The objective is to develop physical insight into the nature of carrier transport in devices. In Section 8.9 we'll discuss device simulation whose objective is to provide accurate, quantitative predictions of device performance.

8.3 Transport across a barrier

Transport across a barrier occurs commonly in devices. For a metal–semiconductor junction, it determines the current–voltage characteristic. For a MOSFET, it occurs across the source–channel barrier and in a bipolar transistor, across the emitter–base junction. The problem is difficult because strong electric fields occur over short distances accompanied by strong gradients in the carrier concentration. Consider the simple *p–n* junction illustrated in Fig. 8.2. In equilibrium, the transition region is less than ≈ 100 nm wide, and the electric field within this region varies from over 10^5 V/cm at the junction to zero at the edge of the transition region. Across the same region, the carrier concentration may vary by nearly 20 orders of magnitude. In equilibrium, therefore, an enormous drift current is balanced by an equal and opposite diffusion current (Fig. 8.2a). Under forward bias, the electric field is reduced, and the net current consists of carriers diffusing over the potential barrier (Fig. 8.2b). For reverse bias conditions, the electric field increases, the net current consists of carriers drifting down the potential drop. Some very interesting effects, known as *velocity overshoot*, can occur for transport down a potential drop and are the subject of Section 8.6. Our interest in this section is to understand transport over the barrier, when carriers diffuse against the electric field. We examine the average carrier energy, or

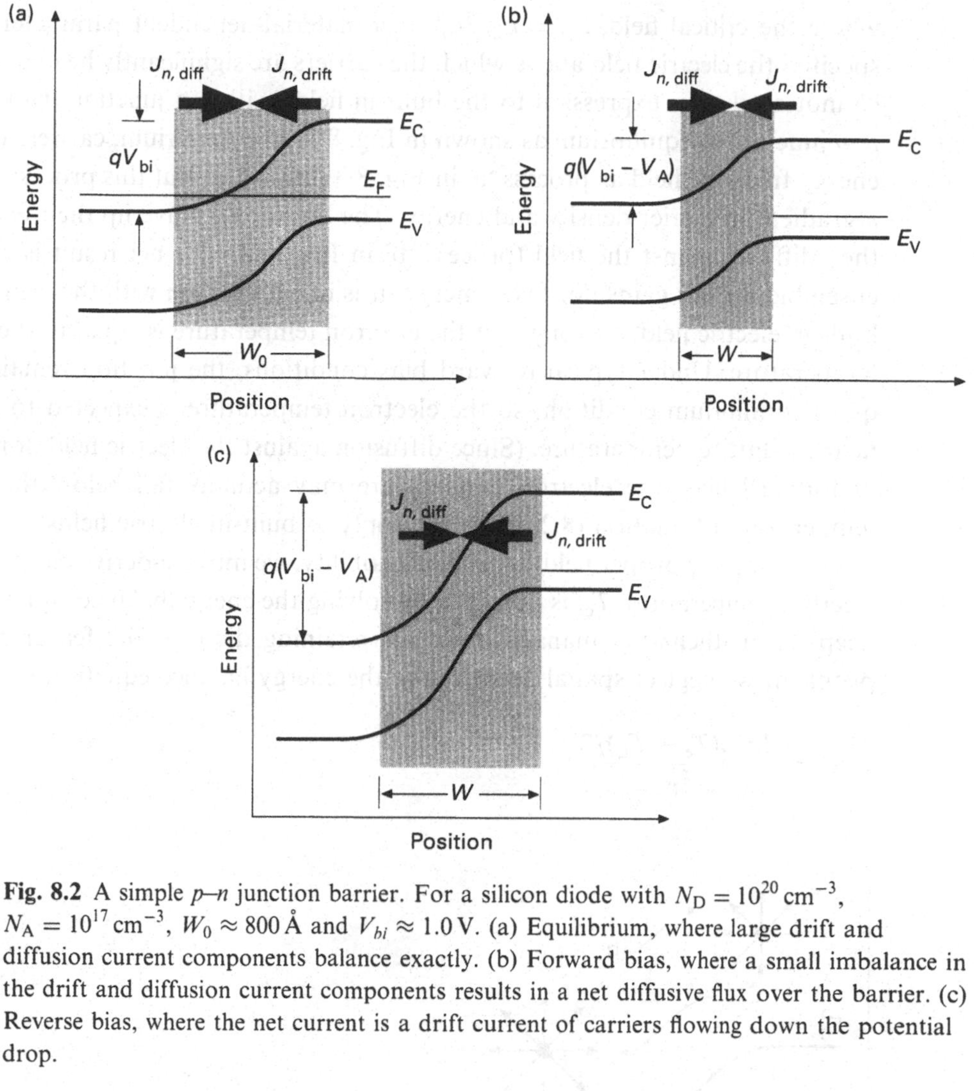

Fig. 8.2 A simple *p–n* junction barrier. For a silicon diode with $N_D = 10^{20}\ \text{cm}^{-3}$, $N_A = 10^{17}\ \text{cm}^{-3}$, $W_0 \approx 800\ \text{Å}$ and $V_{bi} \approx 1.0\ \text{V}$. (a) Equilibrium, where large drift and diffusion current components balance exactly. (b) Forward bias, where a small imbalance in the drift and diffusion current components results in a net diffusive flux over the barrier. (c) Reverse bias, where the net current is a drift current of carriers flowing down the potential drop.

temperature, within the transition region and then examine diffusion in the presence of a high concentration gradient.

8.3.1 The carrier temperature in the presence of built-in electric fields

We saw in Chapter 7 that for bulk semiconductors, electric fields heat the carriers. For electron transport in silicon, the carrier temperature was related to the electric field by

$$\frac{T_e}{T_L} = 1 + \left(\frac{\mathcal{E}_z}{\mathcal{E}_{cr}}\right)^2, \tag{8.2}$$

where the critical field, $\mathcal{E}_{cr} = \upsilon_{sat}/\mu_n^o$, is a material-dependent parameter which specifies the electric field above which the carriers are significantly heated. But we cannot apply this expression to the built-in fields within a junction. Consider a *p–n* junction in equilibrium as shown in Fig. 8.3. In equilibrium, carriers do gain energy from the field as process 'a' in Fig. 8.3 illustrates. But this process sets up a gradient in carrier density and energy. The hot carriers give up their energy as they diffuse against the field (process 'b' in Fig. 8.3). The net result is that the ensemble neither gains nor loses energy; it is in equilibrium with the lattice. The built-in electric field is strong, but the electron temperature is equal to the lattice temperature. Under typical forward bias conditions, the junction remains in a quasi-equilibrium condition, so the electron temperature is expected to remain near the lattice temperature. (Since diffusion against the electric field dominates in forward bias, the electron temperature may actually fall below the lattice temperature.) Equation (8.2) does not apply to built-in electric fields.

To obtain the proper field-dependent mobility, we must re-derive eq. (8.2). The electron temperature, T_e, is obtained by solving the energy balance equation. To keep the mathematics manageable while retaining the essential features of the problem, we neglect spatial gradients in the energy balance equation to write

$$J_{nz}\mathcal{E}_z = \frac{3nk_B(T_e - T_L)/2}{\hat{\tau}_E} \tag{8.3}$$

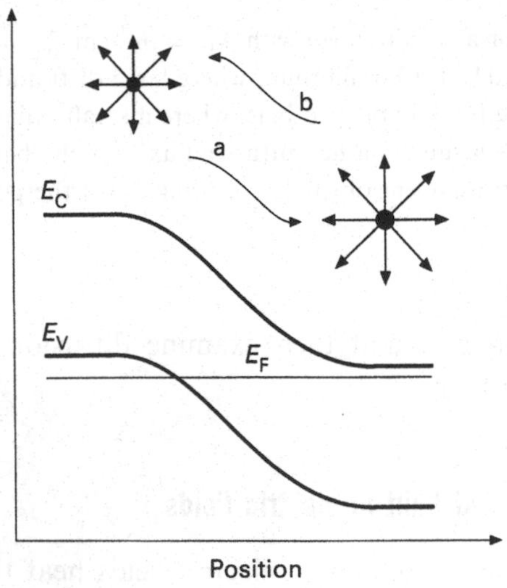

Fig. 8.3 Illustration of energy balance in the depletion region of a *p–n* junction in equilibrium. Process 'a' depicts the gain in energy from the field, and process 'b' the loss in energy by diffusion against the field. In equilibrium, the two processes balance. Under reverse bias, process 'a' dominates, and under forward bias process 'b' dominates. The length of the arrows represents the kinetic energy of carriers.

After solving (8.3) for T_e, we find a result like eq. (8.2) except that

$$\mathcal{E}_{cr}^2 \equiv \frac{3}{2}\frac{nk_B T_L \mathcal{E}_z}{\hat{\tau}_E J_{nz}}. \tag{8.4}$$

Note from the energy balance relation that the important quantity is not the electric field but the product of field and current, which is the input power. In equilibrium, the built-in electric field may be high, but $J_{nz} = 0$, so no power is input from the field, the critical field approaches infinity, and $T_e = T_L$. Only for uniform semiconductors, in which $J_{nz} = nq\mu_n\mathcal{E}_z$, does eq. (8.4) reduce to a material dependent parameter. In general, the critical field depends on the current flow.

Under reverse bias, J_{nz} and $\mathcal{E}_z$ are less than zero, so $\mathcal{E}_{cr}^2$ is a positive number, and eq. (8.2) shows that the electron temperature increases. This situation describes the collection of electrons from the *p*-layer which would occur if the *p–n* junction were the base–collector junction of a bipolar transistor. As the carriers are swept through the transition region, their energy increases. Since the carrier scattering rate increases with carrier energy, we expect that the mobility will decrease under these conditions. If the transition region is wide, then the electric field varies slowly with position, and the use of a field-dependent mobility as discussed in Chapter 7 is justified. On the other hand, if the transition region is narrow, the electric field varies rapidly with position, and nonlocal transport effects such as velocity overshoot occur. Such effects will be discussed in Section 8.6.

When the junction is forward biased, $J_{nz} > 0$ but $\mathcal{E}_z < 0$, so eq. (8.4) shows that $\mathcal{E}_{cr}^2$ is negative. The result is that $T_e < T_L$; electrons are cooled by the electric field as they diffuse against it. Typical junctions frequently operate under near-equilibrium condition for typical forward biases, so the carrier temperature is close to T_L. The cooling can be significant under some conditions, however, and because the scattering rate decreases as the electron temperature decreases, the electron mobility could actually increase above its thermal equilibrium value. The electron mobility, however, is related to the carrier distribution function through eq. (8.1b). In general, the mobility will be related to all moments of the distribution function, not only to the second moment (the electron temperature). So it is not clear whether the mobility in a forward-biased junction is greater or less than the near-equilibrium mobility.

8.3.2 Diffusion in strong concentration gradients

Very strong concentration gradients can also occur within a potential barrier. Simple, closed-form treatments are difficult to obtain because of the coexistence of large gradients in both the potential and the carrier concentration. To gain some insight into diffusion in a strong concentration gradient, we consider a simpler problem, diffusion in the absence of an electric field. Fick's law gives the electron current as

$$J_{nz} = qD_n \frac{dn}{dz} = (-q)n(z)\upsilon_{DIFF}(z), \tag{8.5a}$$

where the diffusion velocity is

$$\upsilon_{DIFF}(z) = -D_n\left(\frac{1}{n(z)}\frac{dn}{dz}\right). \tag{8.5b}$$

Since diffusion consists of the random thermal motion of the carriers, the diffusion velocity should not exceed the thermal velocity, but eq. (8.5b) gives a diffusion velocity that increases without limit as the concentration gradient increases.

To compute the current properly, we begin with eq. (8.1a) in steady-state, which becomes

$$J_{nz} = qD_n \frac{dn}{dz} + 2n\mu_n \frac{du_{zz}}{dz} \tag{8.6}$$

in a quasi-neutral base. Recall that $D_n = 2u_{zz}\mu_n/q$ and that u_{zz} comprises both thermal and drift energy components, i.e.

$$u_{zz} = \frac{1}{2}k_B T_e + \frac{1}{2}m^* \upsilon_{dz}^2. \tag{8.7}$$

If the base is thin, then we can assume

$$u_{zz} = \frac{dT_e}{dz} \simeq 0. \tag{8.8}$$

When no recombination occurs, $\nabla \cdot J_{nz} = 0$, and we find

$$\frac{du_{zz}}{dz} = -\frac{m^* J_{nz}^2}{q^2 n^3}\frac{dn}{dz}, \tag{8.9}$$

which can be inserted in eq. (8.6) to obtain

$$J_{nz}^2 + \frac{(nq)^2}{\mu_n m^* dn/dz} J_{nz} - \frac{(qn)^2 k_B T_e}{m^*} = 0. \tag{8.10}$$

Equation (8.10) is a quadratic equation for J_{nz} that can be solved for

$$J_{nz} = \frac{-(nq)^2}{2\mu_n m^* dn/dz}\left[1 - \sqrt{1 + 8m^*(\mu_n dn/dz)^2 k_B T_e/(qn)^2}\right], \tag{8.11}$$

which is the proper current equation to apply in a field-free region with a concentration gradient.

When the concentration gradient is gentle, the second term under the square root in eq. (8.11) is small, so the square root can be expanded to find

$$J_{nz} = qD_n \mathrm{d}n/\mathrm{d}z. \tag{8.12a}$$

As expected, the result is just Fick's law. For strong concentration gradients, however, eq. (8.11) becomes

$$J_{nz} = qn\sqrt{2k_B T_e/m^*}. \tag{8.12b}$$

Since $\sqrt{2k_B T_e/m^*}$ is approximately the thermal velocity, eq. (8.12b) states that carriers cannot diffuse faster than the thermal velocity.

From eq. (8.11), we observe that Fick's law is valid when

$$\frac{8\mu_n^2 m^* k_B T_e}{q^2}\left(\frac{\mathrm{d}n/\mathrm{d}z}{n}\right)^2 \ll 1,$$

which can be re-arranged as

$$\mathrm{d}n/\mathrm{d}z \ll \frac{n}{\hat{\tau}_m\sqrt{k_B T_e/m^*}},$$

or

$$\mathrm{d}n/\mathrm{d}z \ll \frac{n}{\lambda}. \tag{8.13}$$

The parameter, λ, is the mean-free-path between scattering events. According to eq. (8.13), Fick's law holds when the carrier concentration changes slowly over a distance equal to a mean-free-path.

From eq. (8.11), we obtain an expression for the diffusion velocity in the presence of a general concentration gradient as

$$\upsilon_{\mathrm{DIFF}} = -D_{\mathrm{eff}}\left(\frac{1}{n}\frac{\mathrm{d}n}{\mathrm{d}z}\right), \tag{8.14a}$$

where

$$D_{\mathrm{eff}} = \left\{\frac{(-q)n^2}{2\mu_n m^*(\mathrm{d}n/\mathrm{d}z)^2}\left[1-\sqrt{1+8m^*(\mu_n \mathrm{d}n/\mathrm{d}z)^2 k_B T_e/(qn)^2}\right]\right\}. \tag{8.14b}$$

The effective diffusion coefficient is simply the near-equilibrium diffusion coefficient, D_n, when the concentration gradient is gentle, but it is reduced for high concentration gradients so that unphysical carrier velocities do not occur. A plot of the diffusion velocity versus the driving force for diffusion, Fig. 8.4, looks much like a velocity versus field plot for a bulk semiconductor. For gentle concentration gradients, the diffusion velocity is proportional to the driving force, but for strong concentration gradients, the diffusion velocity saturates at the thermal velocity.

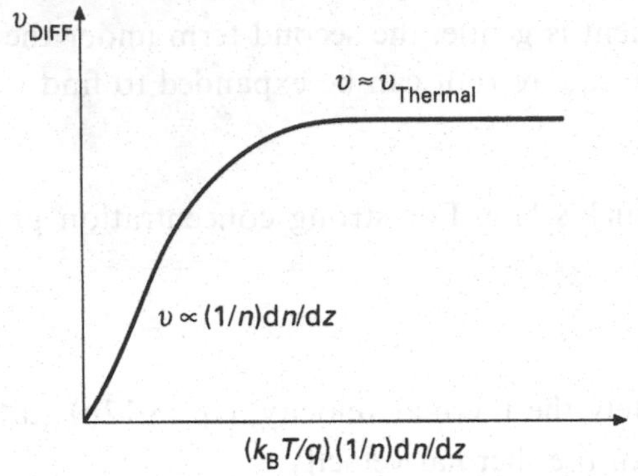

Fig. 8.4 The diffusion velocity versus the effective driving field for diffusion, $\mathcal{E}_{\rm effective} = (k_B T/q)(1/n){\rm d}n/{\rm d}z$.

8.3.3 Discussion of barrier transport

We can now apply the concepts developed in the previous two sub-sections to the barrier transport problem. Let's separate the current into drift and diffusion components (although the two are so strongly coupled in the barrier that it's not clear that this separation is meaningful). Since the concentration gradient is strong, we expect that diffusion should be described by an effective diffusion coefficient that is smaller than the near-equilibrium value. We also know that the electron temperature is near the lattice temperature under forward bias conditions, so if the Einstein relation holds, we conclude that the mobility is also reduced. On the other hand, since the electron temperature is near the lattice temperature (or lower), scattering should not increase, which leads to the conclusion that the mobility is near its equilibrium value (or higher).

Barrier transport is an important feature of transport in most devices, but our discussion shows that several uncertainties remain. In many cases, the junction operates in a quasi-equilibrium mode, so the results are nearly independent of the mobility and depend primarily on the boundary conditions assumed. For a metal–semiconductor junction, however, the results do depend on the transport model assumed. For low barriers, the proper current is obtained by integrating a drift–diffusion equation with near-equilibrium mobility and diffusion coefficient, but for high barriers, the proper current is obtained by thermionic emission, which limits the current to the thermal velocity [8.3]. One can obtain the proper results for any barrier, however, if one assumes a reduced diffusion coefficient and mobility related by the equilibrium Einstein relation [Lundstrom, M. S. and Tanaka, S. I. *Journal of Applied Physics*, **66**, 962–964, 1995]. This suggests that the primary effect is diffusion over a barrier. High barriers lead to high concentration gradients and therefore to a reduced effective diffusion coefficient (and a corresponding reduced mobility through the Einstein relation).

8.4 Diffusion across a thin base

In a device such as a bipolar transistor, transport is controlled by the diffusion of carriers across a thin, quasi-neutral region. Consider a model problem in which carriers are injected at the left side of a thin, neutral region and collected at the right side. We will assume that the collector is a perfect absorber of particles; imperfect collectors will be treated in Section 8.8. We consider the two different injection conditions shown in Fig. 8.5. In the first case, thermal equilibrium carriers are injected and in the second case, hot carriers, Our objective is to understand how transport changes as the width of the neutral region varies from $W \gg \lambda$ to $W \ll \lambda$, where λ is the mean-free-path for scattering.

8.4.1 Diffusion: near-equilibrium injection

Our model problem, sketched in Fig. 8.6, assumes a thermal flux injected from the left and perfect absorption at the right. The essential features of the problem can be illustrated by Fick's law,

$$F = -D\mathrm{d}n/\mathrm{d}z, \tag{8.15}$$

where F is the steady-state flux. A common approach assumes that the perfect collector forces $n(W) \approx 0$, so

$$F = \left(\frac{D}{W}\right)n(0) = \upsilon_{\mathrm{DIFF}}n(0). \tag{8.16}$$

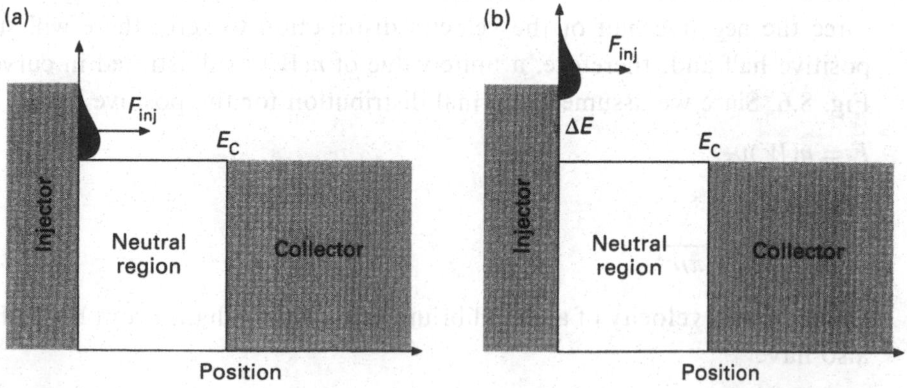

Fig. 8.5 Illustration of: (a) Near-equilibrium injection into a neutral region and (b) Energetic (hot) carrier injection.

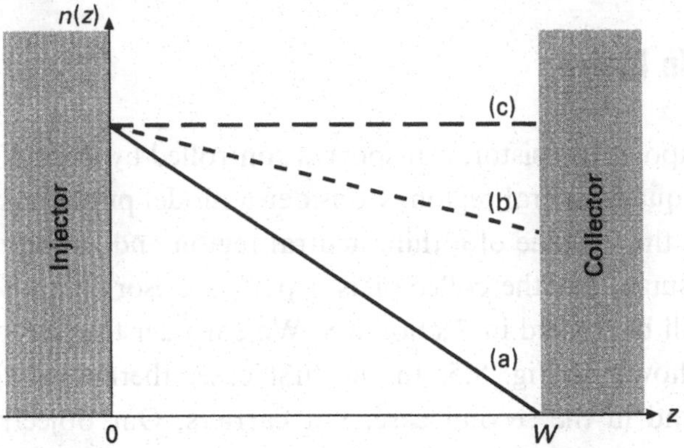

Fig. 8.6 Illustration of diffusion across a thin region. (a) Solution assuming $n(W) \approx 0$; (b) solution assuming a more realistic boundary condition of $n(W) = F_{\text{inj}}/\upsilon_T$; and (c) solution assuming ballistic transport.

This approach is accurate when W is large, but as W shrinks, the diffusion velocity, (D/W), increases without bound. Since diffusion is a result of random, thermal motion, υ_{DIFF} should be limited to υ_T, the thermal velocity.

Equation (8.16) holds only when

$$(D/W) \ll \upsilon_T. \tag{8.17}$$

The diffusion coefficient is proportional to $\upsilon_T\lambda$, where λ is the mean-free-path for scattering. Equation (8.6), therefore, holds when

$$W \gg \lambda; \tag{8.18}$$

the region must be several mean-free-paths wide.

Fick's Law can be used for a thin region *if* we are careful about the boundary conditions. Problems arise when we force $n(W) = 0$, which is non-physical and produces artificially steep concentration gradients. A perfect collector can only force the negative half of the velocity distribution to zero; there will still be a positive half and, therefore, a finite value of $n(W)$ as illustrated in curve (b) of Fig. 8.6. Since we assume a thermal distribution for the positive half,

$$F = n(W)\upsilon_T, \tag{8.19}$$

where

$$\upsilon_T = \sqrt{2k_BT/\pi m^*} \tag{8.20}$$

is the thermal velocity of an equilibrium hemi-Maxwellian. From Fick's Law, we also have

$$F = \left(\frac{D}{W}\right)[n(0) - n(W)]. \tag{8.21}$$

From eqs. (8.19) and (8.21) we find

$$F = \left(\frac{D}{W}\right)\left[\frac{n(0)}{1+(D/W)/\upsilon_T}\right], \tag{8.22a}$$

which should be compared to eq. (8.16). We also find that

$$n(W) = \left[\frac{n(0)}{1+\upsilon_T/(D/W)}\right]. \tag{8.22b}$$

In the thick region limit, $(D/W) \ll \upsilon_T$, and eq. (8.22a) reduces to eq. (8.16). In the thin region limit, $(D/W) \gg \upsilon_T$, and eq. (8.22a) reduces to

$$F = \upsilon_T n(0), \tag{8.23}$$

which is the ballistic limit flux. In this limit, $n(W)$ approaches $n(0)$. The concentration gradient can be expressed as

$$\frac{dn}{dz} = \left[\frac{n(0)}{(D/\upsilon_T)+W}\right] \approx \frac{n(0)}{\lambda+W} \tag{8.24}$$

which shows that the criterion, eq. (8.13), is satisfied so Fick's Law can be used. The steep concentration gradients that invalidate Fick's Law do not occur in diffusion across a thin, neutral region.

8.4.2 Diffusion: off-equilibrium injection

Transport across a thin, neutral region is more complex when carriers are injected energetically, as shown in Fig. 8.5b. We consider a specific example, electrons injected from an InP emitter into an $In_{0.53}Ga_{0.47}As$ base (Dodd and Lundstrom, *Applied Physics Letters*, **61**, 465, 1992.) The injection energy is $\approx 10\,k_BT$, which corresponds to an injection velocity of $\approx 9 \times 10^7$ cm/s. If there were no scattering, the corresponding transit time across a 300 Å region would be ≈ 0.03 ps. Monte Carlo simulations of electron transport for this problem show that most electrons cross the region with no, or only a few, scattering events. This suggests that transport is nearly ballistic. The steady-state, average transit time, however, is about 0.1 ps, which is over three times the ballistic transit time. It also scales as W^2. These results suggest that transport is diffusive.

Figure 8.7 shows the computed, steady-state velocity distribution in the middle of the neutral region. The ballistic peak due to energetic electrons is observed along with a Maxwellian component caused by the scattered carriers. From the areas under the two distribution components, we fund that $\approx 80\%$ of the carriers are in the Maxwellian distribution even though $\approx 80\%$ of the injected electrons cross with three or fewer scattering events. The large Maxwellian component

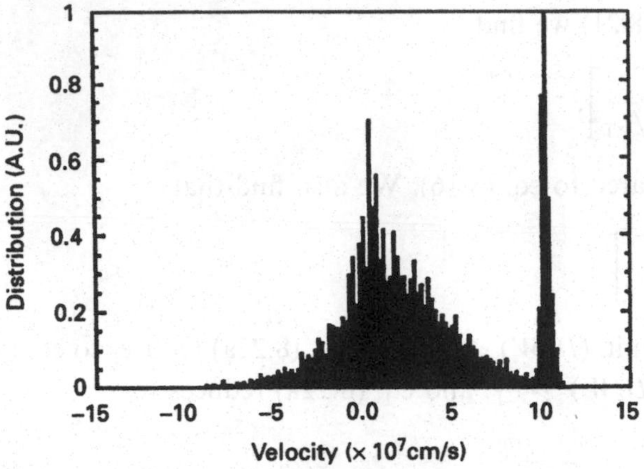

Fig. 8.7 Steady-state velocity distribution for electrons in the center of a 300 Å wide p^+n GaAs base. The distribution function is plotted versus longitudinal velocity. (From Dodd, and • Lundstrom, M. *Applied Physics Letters*, **61**, 465, 1992.)

explains why the average transit time is so much longer than the ballistic time, but why does it occur? It turns out that under steady-state conditions, relatively small amounts of scattering can produce relatively large populations of scattered carriers. Ballistic electrons cross the region very quickly, but those that suffer a wide angle scattering event spend a much longer time in the region as they diffuse out. While the scattered carriers diffuse out, many more are being injected, so the steady-state population of carriers builds up.

To estimate the steady-state population of scattered carriers, we write the injected flux as $F_{inj} = n_0 \upsilon_{inj}$. The ballistic flux that exists is $(1 - \xi) n_0 \upsilon_{\mathrm{inj}}$, where ξ is the probability of a large angle scattering event. Those that do scatter must diffuse out, so the flux of scattered carriers that exists is $n_{\mathrm{MW}}(D/W)$. Equating injected and exiting fluxes,

$$n_0 \upsilon_{inj} = (1 - \xi) n_0 \upsilon_{inj} + n_{\mathrm{MW}}(D/W)$$

we find

$$\frac{n_{\mathrm{MW}}}{n_0} = \xi \times \left(\frac{\upsilon_{inj}}{(D/W)} \right). \tag{8.25}$$

The fraction of carriers in the scattered population is not simply the fraction that scatters but is multiplied by the ratio of ballistic to diffusive velocities. The diffusion coefficient is enhanced because the scattered carriers are hot, but the ratio of velocities is still large so high steady-state populations of scattered carriers results.

The important point in this example is that small amounts of scattering can produce large steady-state populations of scattered carriers which slow down devices. Device designers frequently build an electric field into the region so that carriers unlucky enough to scatter are swept out before their population can build up.

8.5 Ballistic transport

As devices continue to shrink in size, they could eventually become so small that carriers might traverse the active region without scattering. Under such conditions, they would move ballistically, and Newton's laws could be used to compute their trajectories. One possibility is that carriers could ballistically cross a short, high-field region like the collector in Fig. 8.1. Another possibility is that energetic carriers could ballistically cross a neutral region, as in Fig. 8.5b. A third possibility, which has been discussed in Section 8.4, is that near-equilibrium carriers could ballistically cross a thin, neutral region. In this section, we examine ballistic transport across these kinds of structures, but we should stress that the contacts establish boundary conditions which often control device performance.

8.5.1 Ballistic transport in model structures

If near-equilibrium electrons are injected into a high-field region with nearly zero velocity and move across it without scattering, then the average velocity versus position within the channel is simply

$$v_{dz}(z) = \sqrt{2qV(z)/m^*}, \tag{8.26}$$

where $V(z)$ is the electrostatic potential. The electron current density is related to the carrier velocity by

$$J_{nz} = (-q)n(z)v_{dz}(z). \tag{8.27}$$

Because the current density may be quite high, the injected electrons may perturb the electric field. Assuming that the density of injected electrons is much greater than the background doping density, Poisson's equation becomes

$$\frac{d\mathcal{E}_z(z)}{dz} = \frac{(-q)n}{\kappa_S\varepsilon_0} \tag{8.28}$$

which, using eq. (8.27), can be expressed as

$$\frac{d^2V(z)}{dz^2} = \left[\frac{J_{nz}\sqrt{m^*/2q}}{\kappa_S\varepsilon_0}\right]V(z)^{-1/2}. \tag{8.29}$$

Equation (8.29) can be integrated from 0 to L to obtain

$$J_{nz} = \frac{4}{9}\frac{\kappa_s\varepsilon_0}{L^2}\sqrt{\frac{2q}{m^*}}V_A^{3/2}, \tag{8.30}$$

where V_A is the voltage across the device. (To obtain this result, we assumed that the electric field was zero at $z = 0$.) Equation (8.30) describes the J–V_A relation *if* the electrons move ballistically across the region. It is an old result known as *Child's law*, which has long been used to describe vacuum tubes.

Child's law should be contrasted with the result obtained by assuming that transport is collision-dominated for which electrons drift in the electric field according to

$$\upsilon_{dz}(z) = \mu_n \mathcal{E}_z. \tag{8.31}$$

When the derivation is repeated for this assumption, we find

$$J_{nz} = \frac{9}{8}\frac{\kappa_S\varepsilon_0}{L^3}\mu_n V_A^2, \tag{8.32}$$

which is known as the *Mott-Gurney law*.

The question of what mobility to use in eq. (8.31) is not so clear (the low-field mobility, the field-dependent mobility, or something else?). We shall return to this question in Section 8.6. It is interesting to observe, however, that the two assumptions, collision-free transport and collision-dominated transport, produce characteristic dependencies of current voltage that are not too different. The effects of the contacts, which may influence or even dominate device performance, also tend to obscure the transport-related effects, so the measured current versus voltage characteristic provides no clear evidence of ballistic transport within such structures. The message is that to exploit the speed advantages of ballistic transport, devices must be carefully designed to minimize space–charge and contact effects.

Consider next the energetic injection of electrons into a neutral region as sketched in Fig. 8.5b. (Such conditions may occur in a heterojunction bipolar transistor.) When electrons are injected into the base, the potential energy associated with the change in conduction band, ΔE_C, is converted to kinetic energy according to

$$\upsilon_{dz}(z) = \sqrt{2\Delta E_C/m^*}. \tag{8.33}$$

(This result assumes that electrons are injected with near zero velocity and that they traverse the base without scattering.) The heterojunction step acts like an electric field impulse which accelerates carriers quickly. With modern epitaxial technology, the conduction band discontinuity can be engineered to be a few tenths of an electron volt high in material systems such as the $Al_xGa_{1-x}As$ ternary. Equation (8.33) predicts very high electron velocities which would substantially lower base transit time and significantly lower the stored electron charge in the base.

8.5.2 Ballistic transport in real devices

To determine whether ballistic transport can occur in real devices, we need to examine the scattering rates in semiconductors. A rough estimate of the mean free path of electrons can be obtained from eq. (8.1b) using the measured mobilities of electrons in high-purity silicon and gallium arsenide. At room temperature, we find a mean free path of roughly 100 nm for both Si and GaAs, so the active region of the device would have to be much shorter than 100 nm to achieve ballistic transport. At liquid nitrogen temperature, however, the mobility of electrons in pure GaAs can be extremely high ($\approx 150\,000\,\mathrm{cm}^2/V\,\mathrm{s}$), and the corresponding mean free path is about 1000 nm.

Before we conclude that ballistic transport is the rule in small GaAs devices operating at 77 K, we should examine the scattering rate versus energy for electrons in GaAs. From Fig. 2.25b, the reason for the especially high low-temperature mobility of GaAs is apparent. At 77 K the thermal energy is only about 10 meV, which is below the longitudinal optical phonon energy of about 32 meV. As a consequence, electrons in pure GaAs at low temperatures can scatter only by absorbing optical phonons, and Fig. 2.25b shows that the time between collisions is on the order of one picosecond. But electrons ballistically crossing a high-field region gain an energy equal to the electrostatic potential. If the applied bias exceeds 0.032 V, the electrons will have sufficient energy to scatter by emitting optical phonons, and Fig. 2.25b shows that the average time between collisions will decrease by almost an order of magnitude. But if the bias is restricted to 0.032 V, eq. (8.26) shows that the maximum velocity will be only $\simeq 3 \times 10^7$ cm /s. Since the 77 K saturated velocity for electrons in bulk GaAs under a high applied bias is roughly the same, such a ballistic device offers little speed advantage. We conclude that ballistic transport should not be expected unless the applied voltage is unreasonably low.

For an applied bias (or ΔE_C) greater than 0.032 V but less than about 0.3 V, the scattering rate for electrons in pure GaAs is nearly constant at about 10^{13} collisions per second (refer again to Fig. 2.25b). If we set ΔE_C to 0.3 eV, then eq. (8.33) predicts a velocity of about 10^8 cm/sec. (Note that m^* is actually a function of energy, which is important because these hot electrons are far above the band minimum.) The limit to the achievable velocities is not set by ΔE_C as implied by eq. (8.33) but by the minimum curvature of $E(\mathbf{p})$. For GaAs, this band-structure-limited velocity is about 10^8 cm/s. Since the carriers are moving at $\approx 10^8$ cm/s and travel for $\approx 10^{-13}$ s before scattering, they can travel about 1000 Å between collisions. For heterojunction bipolar transistors with a base thickness much less than 1000 Å, a sizable fraction of carriers might be expected to traverse the base without scattering.

Although ballistic transport has been observed in structures like that of Fig. 8.5b, the number of electrons that traverse the structure without scattering is typically quite small [8.5, 8.6, 8.7]. The reason is that the base regions need to be heavily doped which results in a high rate of carrier–plasmon scattering, so the mean-free-path is much shorter than the estimate above which assumed pure GaAs. At very low temperatures and for very low applied biases, however, mean-free-paths can be quite long. Under such conditions, devices can even display interference effects due to the wave nature of electrons. By exploiting the electron's wave properties, electronic devices analogous to optical or microwave devices are a possibility [8.8, 8.9].

8.6 Velocity overshoot

For most devices operating near room temperature under modest bias, transport is dominated by scattering. Between the frequent collisions, electrons move according to

$$m^* \frac{\mathrm{d}\upsilon_{dz}}{\mathrm{d}t} = (-q)\mathcal{E}_z, \tag{8.34a}$$

but collisions reduce the average velocity just as friction opposes the acceleration of a mass. A simple way to treat collisions is to add a term to eq. (8.34a) so that

$$m^* \frac{\mathrm{d}\upsilon_{dz}}{\mathrm{d}t} = (-q)\mathcal{E}_z - \frac{m^*\upsilon_{dz}}{\hat{\tau}_m}, \tag{8.34b}$$

which accounts for the 'friction' introduced by scattering. Equation (8.34b) is a simple momentum balance equation for an average carrier. By comparing with eq. (5.27b), the proper momentum balance equation derived from the BTE, we observe that eq. (8.34b) neglects diffusion. It works well in strong, uniform electric fields where diffusion effects are small but must be used with caution in other situations.

For a uniform electric field applied at $t = 0$, the solution to eq. (8.34b) is

$$\upsilon_{dz}(t) = (q\hat{\tau}_m/m^*)\mathcal{E}_z(\mathrm{e}^{-t/\hat{\tau}_m} - 1). \tag{8.35}$$

This result, which is plotted in Fig. 8.8a, shows that steady-state is achieved in a time on the order of a momentum relaxation time. While steady-state is being approached, carriers travel a distance

$$d = \int_0^{\hat{\tau}_m} \upsilon_{dz}(t)\mathrm{d}t = \frac{(-q)\hat{\tau}_m^2}{m^*}\mathcal{E}_z e^{-1}. \tag{8.36}$$

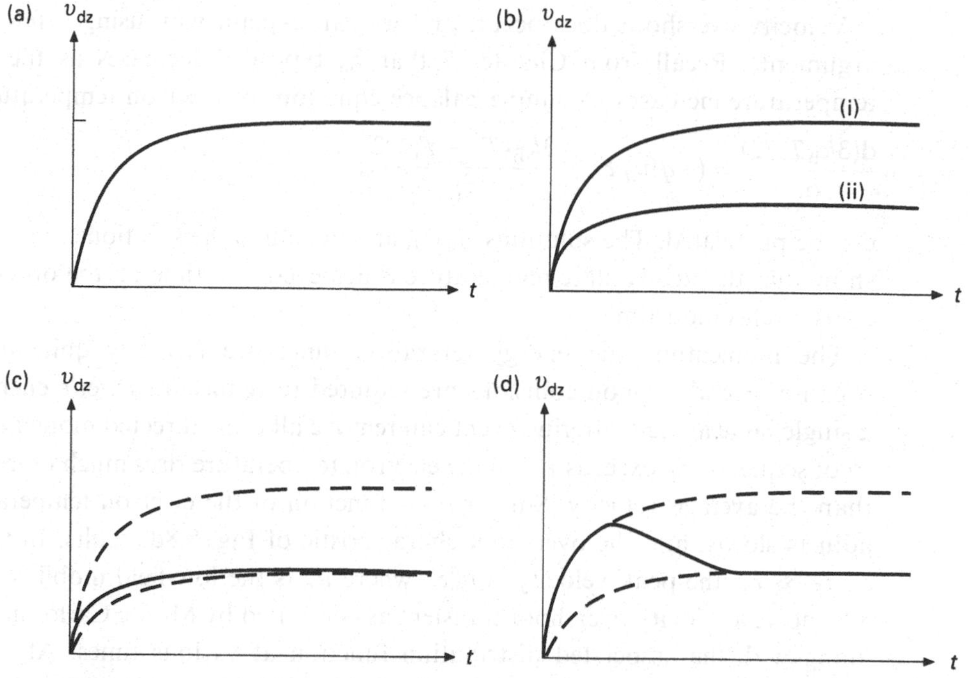

Fig. 8.8 Carrier velocity versus time assuming several different momentum relaxation times. (a) A constant $\hat{\tau}_m$. (b) Two different, constant $\hat{\tau}_m$'s; case (i) assumes the low-field ensemble relaxation time, and case (ii) the ensemble relaxation time characteristic of uniform, high fields. (c) A variable $\hat{\tau}_m$; the plot shows the characteristic expected when $\hat{\tau}_m$ adjusts rapidly from low- to high-field conditions. (d) A variable $\hat{\tau}_m$; the plot shows the characteristic expected when $\hat{\tau}_m$ adjusts slowly from low- to high-field conditions.

If we estimate $\hat{\tau}_m$ from the measured low-field mobility, then for $\mathcal{E}_z = -10^4$ V/cm we find $d \approx 200$ Å for electrons in Si and $d \approx 1500$ Å for electrons in GaAs. In small GaAs devices, the active regions are not much bigger, so the steady-state velocity is never achieved. Transport under such conditions is known as *transient, nonstationary, off-equilibrium, or nonlocal transport.*

Because the scattering rate generally increases with energy, $\hat{\tau}_m$ is not a constant but, rather, decreases as the carriers are accelerated. In Fig. 8.8b the $v_{dz}(t)$ characteristic is plotted for two different momentum relaxation times; the first corresponds to a low applied field and the second to a high field in a bulk semiconductor. Because $\hat{\tau}_m$ varies during the transient, the actual characteristic should lie between the limits displayed. If $\hat{\tau}_m$ adjusts very rapidly, then the characteristic displayed in Fig. 8.8c results. On the other hand, if $\hat{\tau}_m$ adjusts slowly, then the characteristic shown in Fig. 8.8d is obtained. For this second case, the transient velocity overshoots its steady-state value. *Velocity overshoot* can improve the performance of small devices because the average carrier velocity can exceed the bulk limit.

Velocity overshoot does occur, and we can explain why using a few simple arguments. Recall from Chapter 7 that $\hat{\tau}_m$ typically decreases as the carrier temperature increases. A simple balance equation for electron temperature,

$$\frac{d(3k_B T_e/2)}{dt} = (-q)\upsilon_{dz}\mathcal{E}_z - \frac{3k_B(T_e - T_L)/2}{\hat{\tau}_E}, \tag{8.37}$$

can be postulated. The solutions, $T_e(t)$, are much like the solutions to (8.34) and show that the steady-state temperature is achieved in a time on the order of the energy relaxation time.

The momentum and energy relaxation times are typically quite different because several phonon collisions are required to reduce a carrier's energy, but a single large angle scattering event can remove all of its directed momentum. As a consequence, $\hat{\tau}_E$ exceeds $\hat{\tau}_m$, so the electron temperature rises much more slowly than the average velocity. Since $\hat{\tau}_m$ is a function of the electron temperature, it adjusts slowly, and the overshoot characteristic of Fig. 8.8d results. In the limit of $\hat{\tau}_E \gg \hat{\tau}_m$, the peak velocity is $\mu_n^o\mathcal{E}_z$ where μ_n^o is the low-field mobility. Figure 8.9 shows a velocity overshoot transient as computed by Monte Carlo simulation along with the associated distribution function at various times. At first, the electric field simply displaces the distribution function with little change in shape. Although the drift velocity is high, the temperature is not, so the mobility remains high. Later on, collisions broaden the distribution, the electron temperature increases, $\hat{\tau}_m$ decreases, so the drift velocity drops.

8.7 Transport in rapidly varying electric fields

We have seen that when an electric field is quickly switched on, velocity overshoot occurs. For electrons in silicon, velocity overshoot occurs because $\hat{\tau}_E > \hat{\tau}_m$. For electrons in GaAs, intervalley transfer gives rise to even stronger velocity overshoot. When the electric field varies in time, complex effects can occur. Velocity overshoot also occurs under steady-state conditions. In this section, we examine some of the nonlocal transport effects than can occur in devices.

8.7.1 Transport in time-varying electric fields

The velocity transient of electrons in GaAs subjected to the electric field pulse displayed in Fig. 8.10a illustrates the rich variety of effects possible. The average electron velocity versus time and the average energy versus time are displayed in Figs 8.10b and 8.10c. (All results were obtained by Monte Carlo simulation [8.10].) For $t < 2$ ps, electrons simply drift in a modest electric field. At $t = 2$ ps, a high field is applied, and velocity overshoot occurs. Note that velocity

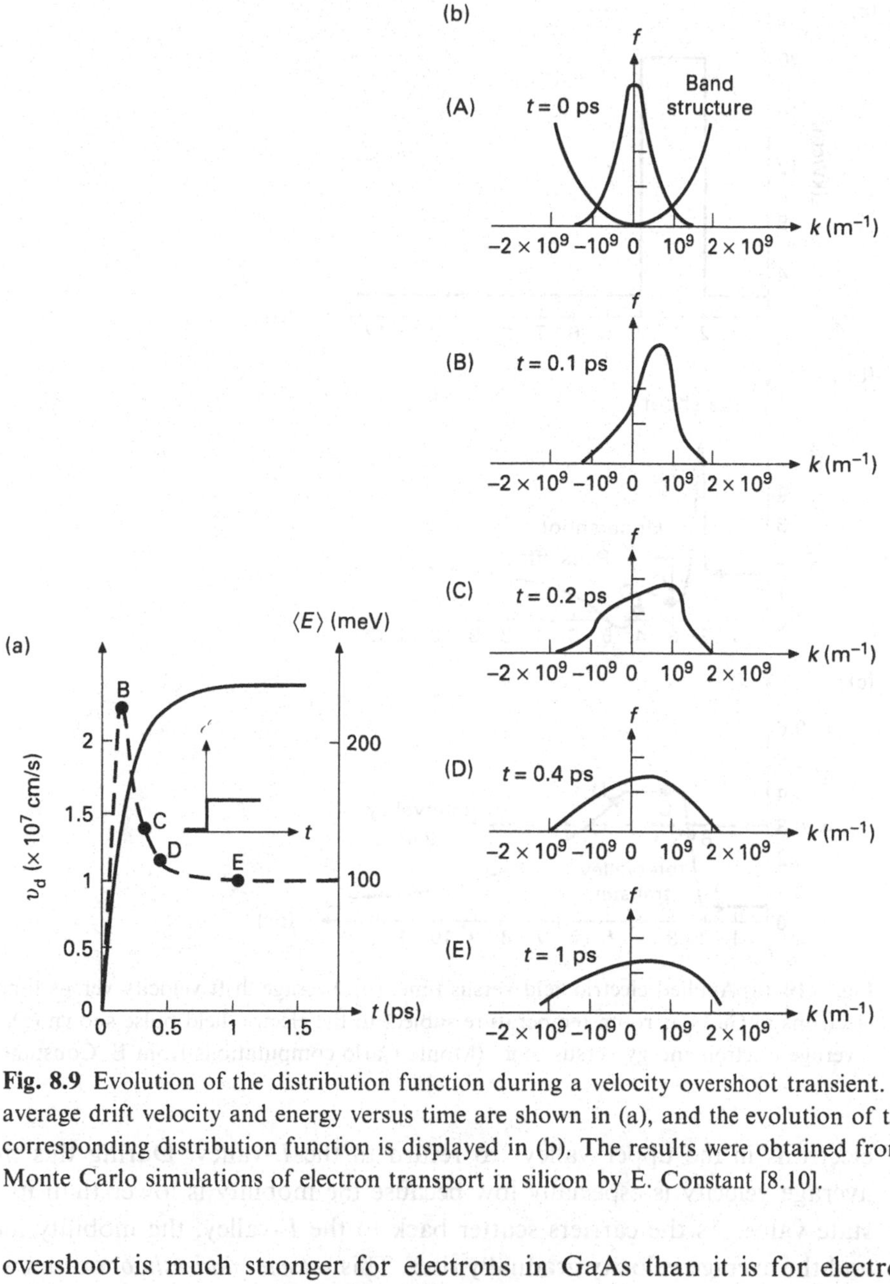

Fig. 8.9 Evolution of the distribution function during a velocity overshoot transient. The average drift velocity and energy versus time are shown in (a), and the evolution of the corresponding distribution function is displayed in (b). The results were obtained from Monte Carlo simulations of electron transport in silicon by E. Constant [8.10].

overshoot is much stronger for electrons in GaAs than it is for electrons in silicon.

The velocity versus time transient displayed in Fig. 8.10b can be understood by referring to the corresponding energy versus time transient displayed in Fig. 8.10c. At time B the average electron is still low, so little intervalley scattering occurs, and the electron velocity is high. By time C, the electron energy is very high, so the intervalley scattering rate is high and the average velocity has plummeted. At $t = 4$ ps, the high field is removed, but it takes some time for the

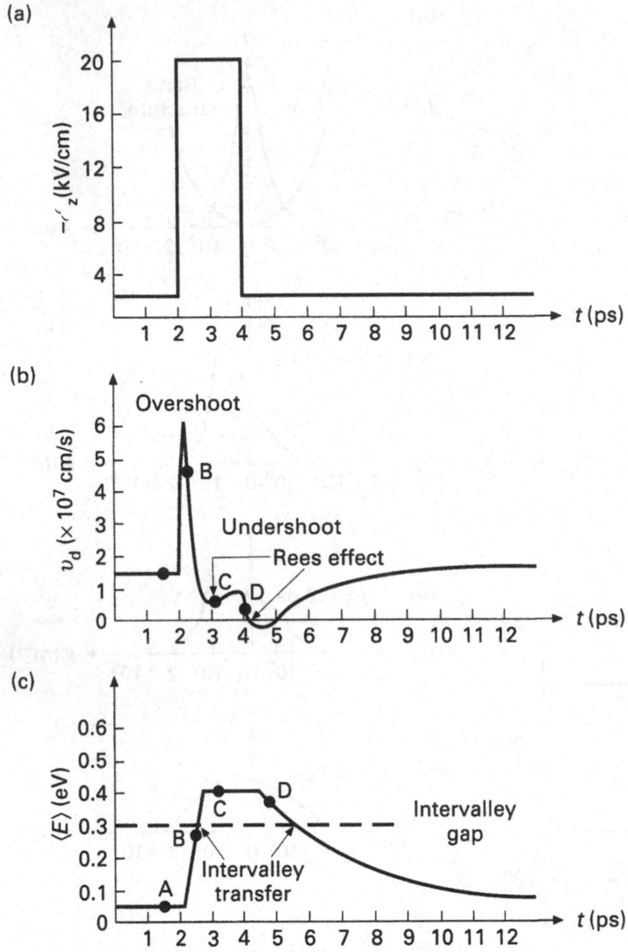

Fig. 8.10 (a) Applied electric field versus time. (b) Average drift velocity versus time for electrons in GaAs at room temperature subject to the electric field pulse shown (a). (c) Average electron energy versus time. (Monte Carlo computations from E. Constant [8.10].)

electrons in the upper valleys to return to the Γ-valley. During this time, the average velocity is especially low because the mobility is lower than its steady-state value. As the carriers scatter back to the Γ-valley, the mobility increases, and the average velocity gradually rises. This *velocity undershoot* occurs when the average carrier energy exceeds its steady-state value.

The dips in the velocity transient just before time C, and just after time D require some explanation. Both dips occur just after intervalley transfer begins; the second dip is so severe that the velocity actually becomes negative. This *Rees effect* can be explained by examining Fig. 8.11, which shows that carriers which scatter back to the Γ-valley may end up with either positive or negative momentum. Those with positive momentum may gain energy from the field and scatter

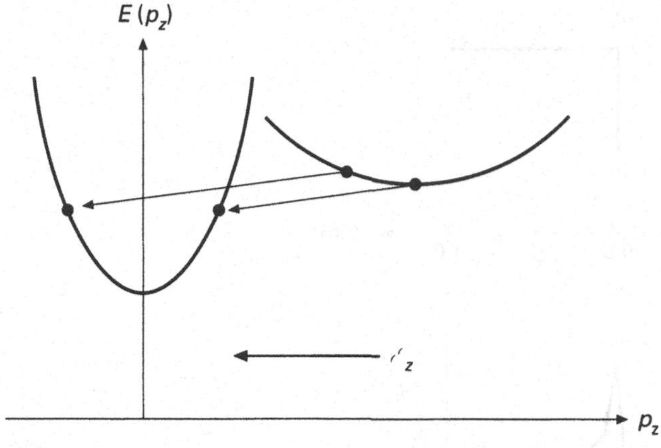

Fig. 8.11 Illustration of why electrons scattered from the upper to central valley with negative momentum remain in the central valley longer than carriers which enter with positive momentum. This effect explains the two dips in the velocity transient of Fig. 8.10b. (From E. Constant [8.10].)

out very quickly, but carriers with negative momentum must undergo a very long flight before they can scatter out. As a consequence, electrons which enter the Γ-valley with negative momentum remain in the valley longer and lower the average velocity.

8.7.2 Steady-state transport in spatially varying electric fields

Electric fields within devices may vary both with time and position. In Section 8.6, we described how carriers respond to a time-varying, but spatially uniform field. In this section, we describe how carriers respond to a spatially varying field under steady-state conditions in time. The simplest case to consider is that of electrons being injected from a low-field region into a long, high-field region. Eventually, the carriers simply drift at a velocity equal to the product of the high-field mobility times the electric field, but a transient occurs first. Spatial transients for electrons in Si and GaAs are displayed in Fig. 8.12, which shows steady-state velocity overshoot in space. Note that overshoot is more pronounced for electrons in GaAs and that its extent is an appreciable fraction of one micron for GaAs, but only a few hundred angstroms for Si.

It might be expected that temporal transients could be converted to spatial transients according to

$$z = \int_0^t v_{dz}(t')\mathrm{d}t'. \tag{8.38}$$

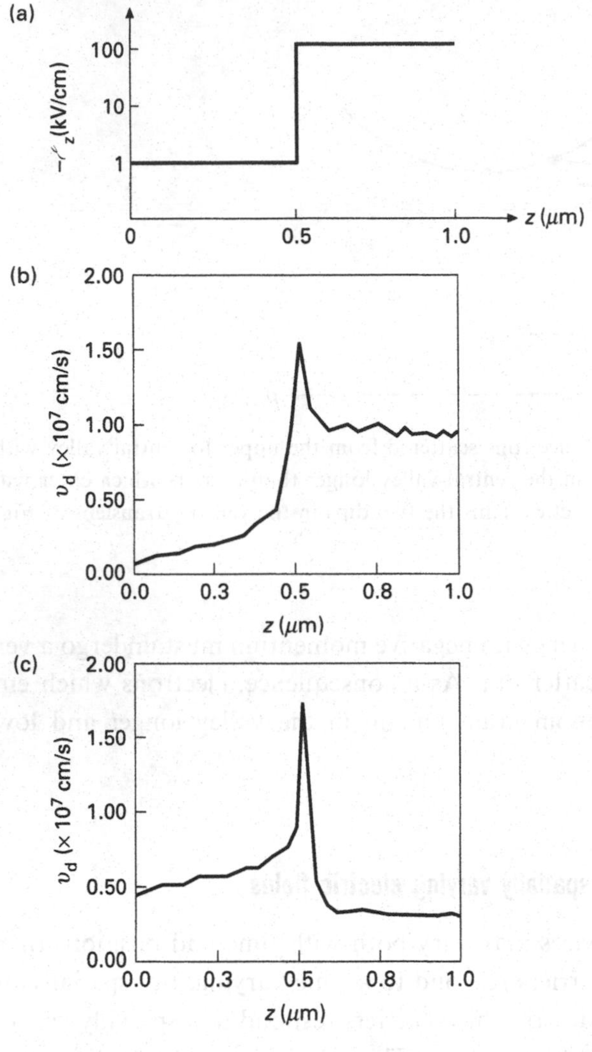

Fig. 8.12 (a) Device structure with a low-field to high-field step. (b) Steady-state velocity versus position for electrons in silicon subjected to the electric field profile shown in (a). (c) Steady-state velocity versus position for electrons in GaAs subjected to the electric field profile shown in (a). (Monte Carlo computations (b and c) courtesy of M. A. Stettler and A. Das.)

According to this prescription, the only difference between temporal and spatial transients is a re-definition of the horizontal axis. This prescription cannot be correct, however, because if the horizontal axis in Fig. 8.12 is time, then the velocity profile would be non-causal; it would increase before the electric field increased. To understand the difference between temporal and spatial transients, compare the response of electrons to a field pulse in time with the steady-state response to a field pulse in space. The results of a Monte

Carlo simulation are shown in Fig. 8.13 [8.10]. For the spatial pulse, the steady-state, average carrier velocity versus position is plotted in Fig. 8.13b, and for the temporal transient the average carrier velocity is plotted on the same scale by using (8.38) to convert the time axis to distance. Note the distinct difference between the two plots; the temporal transient achieves much higher velocities which persist much longer. The differences between temporal and spatial transients are a result of *ensemble effects*, or more simply, diffusion. The flow of momentum density into or out from the ensemble of carriers is described by the term in eq. (5.27b) involving the spatial gradient. It takes some time for the ensemble to build-up, so diffusion is most pronounced under steady-state conditions. Equation (8.34b) describes temporal transients rather well, but we cannot use eq. (8.38) to convert those results to the steady-state case where diffusion is strong.

Figure 8.13b also shows an *anticipatory effect*, which is the increase in the steady-state, average velocity prior to the electric field pulse. This effect can be explained from the plots of average carrier velocity and density versus position displayed in Figs. 8.13b and 8.13c. Because recombination–generation is assumed to be absent, the steady-state electron current is constant which means that the product of electron velocity and density must be constant. The strong overshoot in velocity is accompanied by a dip in the carrier density which leads to strong diffusion effects. The spatially varying energy also produces diffusion currents via the second term in (8.1a). The diffusion flux increases the average velocity before the field step and decreases it after the step.

We now have a phenomenological understanding of why spatial transients 'anticipate' the change in electric field and why they decay so rapidly after the pulse. To develop a microscopic understanding, we need to examine the distribution function versus position as plotted in Fig. 8.13d. Well before the field pulse and just after it (locations A and D) the distribution function is approximately Maxwellian. As the field pulse is approached, however, the distribution becomes increasingly non-Maxwellian. At location B, which is just before the pulse, the distribution function contains virtually no negative-velocity electrons. This distortion, which anticipates the field pulse, occurs because the negative-going electrons must have come from the right, but the strong field pulse prevents electrons with negative velocities from reaching that portion of the structure to the left of the pulse. Within the pulse, at location C, the distribution function displays a strong 'ballistic peak' of electrons that have been accelerated to very high energies within the Γ valley. Just after the pulse, at location D, the distribution function picks up a large component with negative velocities which sharply lowers the ensemble velocity. The region to the right of the pulse acts as a source of negative velocity

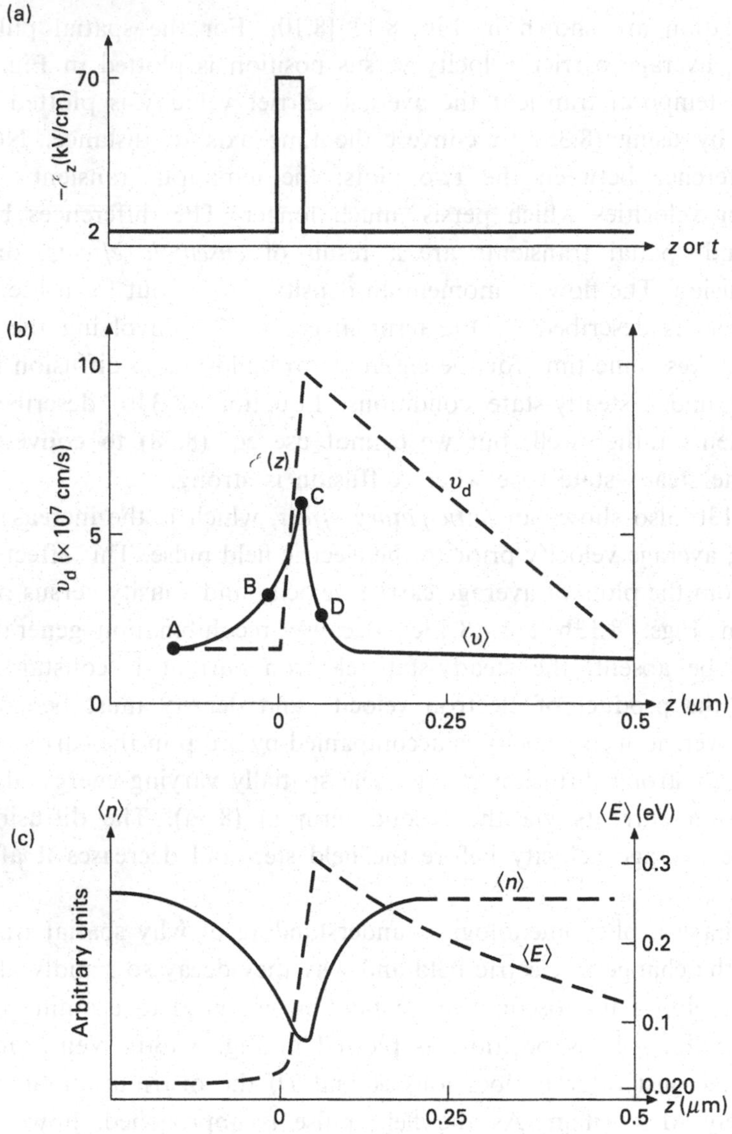

Fig. 8.13 (a) Applied electric field pulse in time and in space. (b) Average velocity versus position for a pulse applied in space (solid line) and in time (dashed line). The dashed line was obtained from the temporal transient according to eq. (8.38). The solid line is the steady-state velocity for the spatial pulse. (c) Steady-state carrier density (solid line) and energy density (dashed line) for the spatial electric field pulse. (From E. Constant, [8.10].) (d) Steady-state electron distribution function versus v_z at selected locations within the device structure with a spatial pulse. The letters refer to the various locations specified in (b). Electron transport in GaAs is assumed. (Monte Carlo computations courtesy of M. A. Stettler and A. Das.)

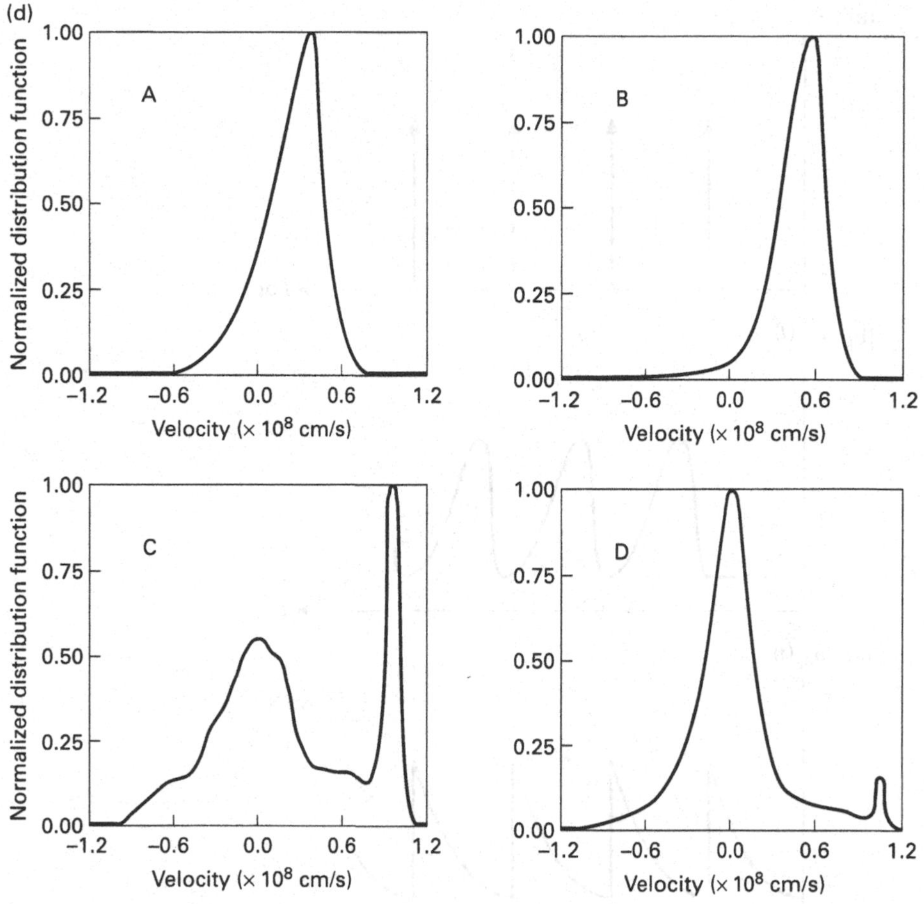

Fig. 8.13 *(continued)*

electrons, which are produced by scattering. The electric field pulse prevents them from flowing further to the left, but where the field pulse terminates, a large negative velocity component of the distribution function appears and lowers the velocity abruptly.

Diffusion accounts for the sharp differences between the average velocity resulting from a temporal transient and the steady-state velocity produced by a spatial transient. The differences can be profound, as illustrated in Fig. 8.14. For this example, a series of field impulses is applied – first in time, then in space. The resulting average velocity versus time plotted in Fig. 8.14b is easy to understand as a series of velocity overshoot transients. The steady-state velocity versus position for the impulses applied in space, however, is sharply different. Figure 8.14c shows that the steady-state velocity increases in anticipation of the field impulse and then plummets abruptly after the impulse. The

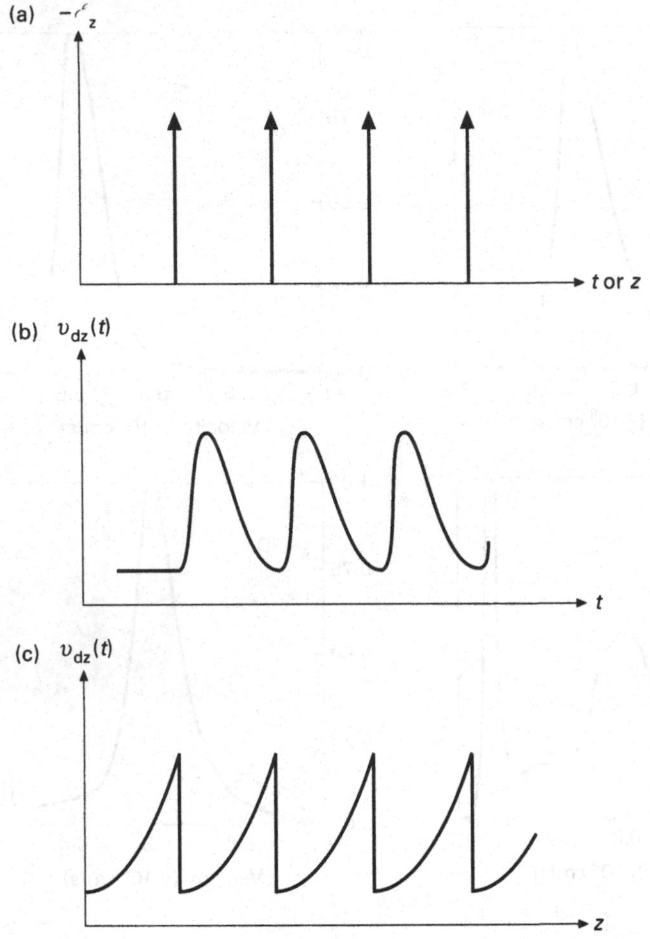

Fig. 8.14 (a) A series of electric field impulses versus time or distance. (b) Sketch of the expected average velocity versus time for a series of electric field impulses applied in time. (c) Sketch of the expected average velocity versus position for a series of electric field pulses applied in space.

velocity versus time and velocity versus position plots are 180° out of phase. The steady-state velocity versus position characteristic is a result of diffusion. The high velocity at the impulse implies that the steady-state carrier density is low. The resulting diffusion fluxes increase the average velocity to the left of the impulse and decrease it to the right of the impulse. The moral is that for devices in which the electric field varies rapidly in both time and space, diffusion effects are pronounced. Equation (8.34b) cannot describe the momentum versus time in most devices; eq. (5.27b), which was derived from the BTE, must be used.

8.8 Carrier collection by a high-field region

The collection of carriers by a high-field region is a common feature in devices. A model collector is sketched in Fig. 8.15a. Carriers may be injected into the active region from the left (in the case of a transistor) or photogenerated in the case of a detector. Any carrier incident on the high-field collector should be swept across and out the right contact. Transport across the collector is a complex problem in off-equilibrium transport. As illustrated in Fig. 8.15b, strong velocity overshoot can occur, and detailed numerical treatments are necessary, especially if the self-consistent electric field is to be treated.

As illustrated in Fig. 8.15a, we characterize a collector by the fraction, R_C, of the flux incident on the collector that backscatters within the collector and returns to the active region. A fraction $(1 - R_C)$ crosses the collector and emerges from the right contact. For a perfect collector, $R_C = 0$. It turns out that we can estimate R_C rather easily, unless the carriers being collected perturb the electric field in which case a full solution to the self-consistent transport problem is necessary. Characterizing transport by reflection coefficients is the subject of Chapter 9 where the ideas in this section are more thoroughly developed.

Consider first a simpler problem. If carriers are injected into the left face of a neutral region of length L, the fraction that backscatters and re-emerges from the left face is [8.9]

$$R_0 = \frac{L}{L+\lambda}, \tag{8.39a}$$

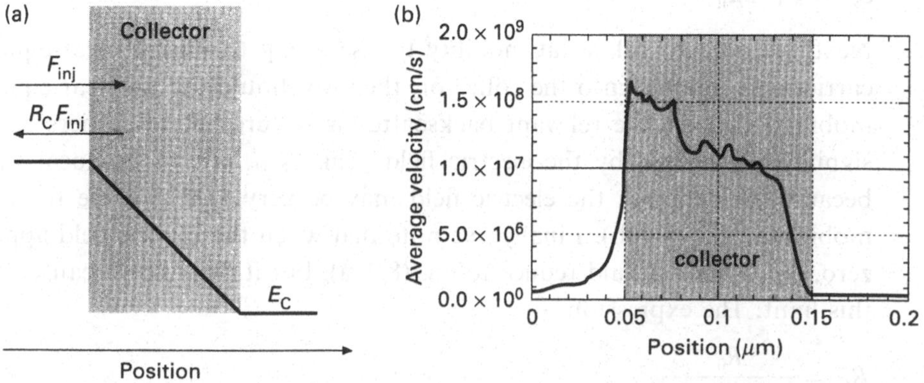

Fig. 8.15 (a) A model electron collector and (b) the velocity versus position profile for the model collector assuming electron transport in silicon.

where λ is the mean-free-path for carrier scattering. (Equation (8.39a) is derived in Section 9.4.1) The problem is more complex, however, when an electric field exists.

Detailed simulations show that when carriers are injected at the top of a potential drop, only those that backscatter very near the top can re-emerge and exit the region. One reason is that scattering by phonon emission lowers their energy, so they can not get back over the barrier and out, but the same effect occurs for elastic scattering too. The reason is that only a small fraction backscatter directly normal to the barrier and have

$$1/2m^* \upsilon_z^2 > \Delta E,$$

where ΔE is the potential drop experienced by carriers before they scatter. Detailed simulations show that if carriers travel more than $\approx 1 - 2k_B T/q$ down a potential drop, they are unlikely to re-emerge, even if they do backscatter.

We can use these ideas to develop a simple estimate for R_C. When an electric field is present, the relevant distance is not the length of the collector, L, but rather, a length, ℓ, over which the first $k_B T/q$ of potential drop occurs. Equation (8.39a) must be modified to

$$R = \frac{\ell}{\ell + \lambda}, \tag{8.39b}$$

where $\ell = (k_B T/q)/|\mathcal{E}_z|$ with $|\mathcal{E}_z|$ assumed to be constant. We can estimate the mean-free-path, λ, from the mobility. Recall that $D = \lambda \upsilon_T/2$. Using the Einstein relation, we can relate λ to the mobility. Equation (8.39b) then becomes

$$R_C = \frac{1}{1 + |\mathcal{E}_z|/\mathcal{E}_{cr}}, \tag{8.40a}$$

where

$$\mathcal{E}_{cr} = \upsilon_T/2\mu_n. \tag{8.40b}$$

Next, we need to ask what mobility to use in eq. (8.40b). If near-equilibrium carriers are injected into the collector, then we should use the near-equilibrium mobility, because the relevant backscattering occurs before carriers have been significantly heated by the electric field. This is a case of nonlocal transport because even though the electric field may be very high, we use the low-field mobility in eq. (8.40b). Finally, we note that when the electric field approaches zero, eq. (8.40a) should reduce to eq. (8.39a), but it does not because $\ell \gg L$ in this limit. The expression

$$R_C = \frac{R_0}{1 + |\mathcal{E}_z|/\mathcal{E}_{cr}}, \tag{8.40c}$$

has the correct low-field limit.

Figure 8.16 is a plot of R_C versus electric field in our model collector (electron transport in pure silicon is assumed). The solid line is the result of a Monte Carlo simulation. As predicted by eq. (8.40), R_C decreases as the electric field increases. In fact, eq. (8.40c) provides a rather good estimate for R_C. In the limit of a very high electric field, R_C is only a few percent. In that case, the average velocity at the beginning of the collector should just be υ_T, the injected, thermal velocity. For electrons in silicon, $\upsilon_T = 1 \times 10^7$ cm/s, which is very close to the velocity at the beginning of the collector as shown in Fig. 8.15b.

In summary, under steady-state conditions, a carrier collector can be characterized by its backscattering coefficient, R_C, which is zero for a perfect collector. In a high-field collector, this parameter is determined by the low-field mobility and by the electric field, or potential drop, within the first mean-free-path of the collector. The design of efficient collectors, therefore, focuses on a thin region at the beginning of the collector.

8.9 Device simulation

In previous sections we described the qualitative features of carrier transport in devices, but device analysis, design, and optimization require accurate, quantitative predictions of device performance. To simulate a device, we solve a transport equation (to describe the response of carriers to the fields) and Poisson's equation (because as they move, they perturb the field). The simulation process is summarized in Fig. 8.17.

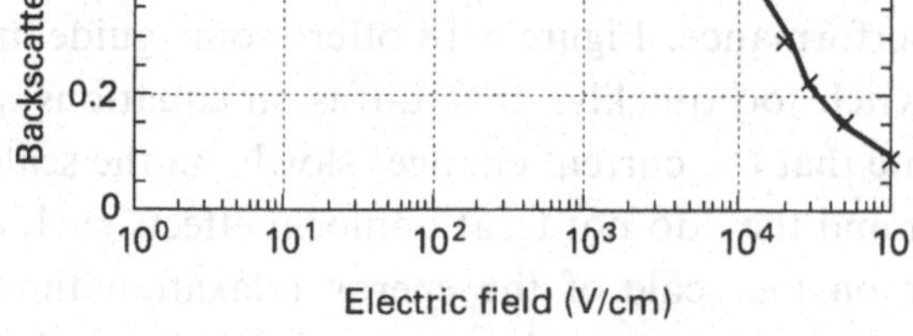

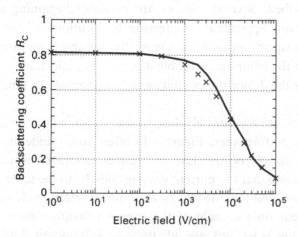

Fig. 8.16 The backscattering coefficient versus electric field for the model collector. Solid line, Monte Carlo simulation, points, the estimate of eq. (8.40c). The mean-free-path, l, was selected so that eq. (8.39a) matched the Monte Carlo results at low electric fields.

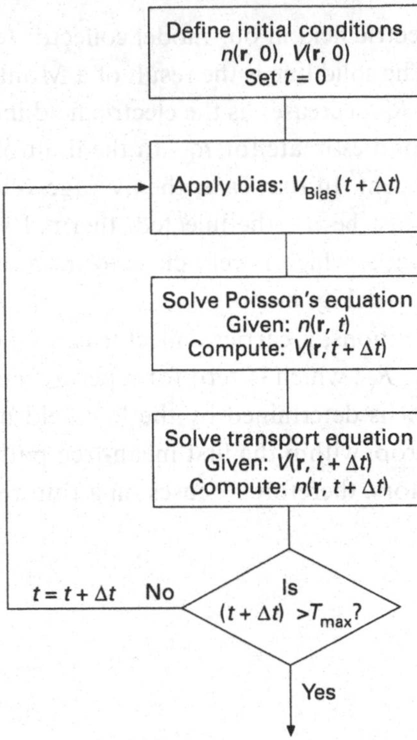

Fig. 8.17 The basic algorithm for time-dependent device simulation. (A unipolar device is assumed, so hole transport is neglected.)

The accuracy of a device simulation is often determined by how accurately carrier transport is described. Several options are available beginning at the conventional, drift–diffusion approach, progressing to solutions of the BTE and then to quantum transport approaches. Generally, the more sophisticated the approach, the heavier the computational burden, so it is important to select an approach adequate for the device under study and to appreciate its limits and range of validity.

The simulation approach must accurately describe transport in the portions of the device that control its performance. Figure 8.18 offers some guidelines. For large devices that do not switch too quickly, drift–diffusion equations are adequate. Such equations assume that the current changes slowly on the scale of the momentum relaxation time, and they do not treat nonlocal effects such as velocity overshoot which occur on the scale of the energy relaxation time. On a spatial scale, drift–diffusion equations are adequate to sub-micron dimension for Si devices but not below about 1 μm for hot electrons in GaAs devices. (The question of exactly when a drift–diffusion equation can be trusted has

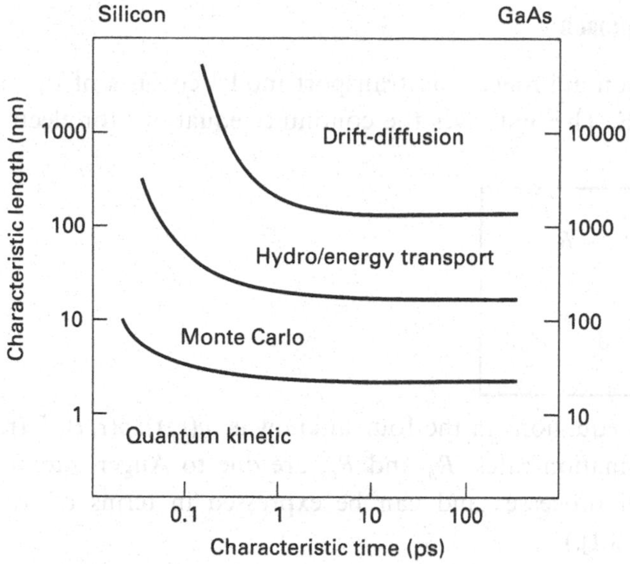

Fig. 8.18 Illustration of the regions of validity for several device simulation approaches. (The boundaries should be viewed as rough guidelines only.)

still not been clearly answered, but it has proven to be useful for far smaller devices than originally expected.)

Some of the nonlocal effects missed by drift–diffusion equations can be captured by momentum–energy balance equations. This approach, however, requires a large number of simplifying assumptions about the shape of the distribution function and about the ensemble relaxation times. For very small devices, in which these assumptions become suspect, direct solutions of the BTE, using techniques such as Monte Carlo simulation, may be necessary. The BTE loses validity under very short time scales, where collisions can no longer be treated as instantaneous, and when the critical regions of the device approaches the carrier's wavelength. Quantum transport approaches may be necessary to treat devices on this scale.

For the remainder of this section, some common device simulation techniques will be surveyed with an emphasis on formulating and understanding each approach, not on numerical solution techniques (for which the reader is referred to the chapter references). We remind the reader again that the region of validity for each transport model still has not been clearly identified. Those who use such simulations must be able to ascertain from their understanding of transport physics and the specific device being examined whether to trust the simulation or not.

8.9.1 The drift-diffusion approach

For the drift–diffusion approach, the transport model consists of the first two moments of the BTE. The first gives the continuity equations for electrons and holes,

$$\frac{\partial n}{\partial t} = \frac{1}{q}\nabla \cdot \mathbf{J}_n - R_n \tag{8.41a}$$

and

$$\frac{\partial p}{\partial t} = -\frac{1}{q}\nabla \cdot \mathbf{J}_p - R_p, \tag{8.41b}$$

which comprise two equations in the four unknowns, $n(\mathbf{r}, \mathbf{t}), p(\mathbf{r}, t), \mathbf{J}_n(\mathbf{r}, t)$, and $\mathbf{J}_p(\mathbf{r}, t)$. The recombination rates, R_n and R_p, are due to Auger, radiative, and Shockley–Read–Hall processes and can be expressed in terms of $n(\mathbf{r}, t)$ and $p(\mathbf{r}, t)$. (See Pierret, [8.1].)

The two additional equations needed to specify the problem are the current flow equations. The simplest flow equations are the drift–diffusion equations,

$$\mathbf{J}_n = -nq\mu_p \nabla V + qD_n \nabla n \tag{8.42a}$$

and

$$\mathbf{J}_p = -pq\mu_p \nabla V - qD_n \nabla p \tag{8.42b}$$

which express the current densities in terms of $n(\mathbf{r}, t)$ and $p(\mathbf{r}, t)$. When Eqs. (8.42) are inserted in eq. (8.41), a system of two nonlinear equations in two unknowns results. Along with the boundary conditions, these equations completely specify the transport problem for drift–diffusion-based device simulation. The complete simulation couples the solution of the transport problem to the self-consistent electrostatic potential by solving Poisson's equation too. For drift–diffusion transport, three, coupled, nonlinear partial differential equations must be solved to simulate a device.

Application of the technique to a two-dimensional, Si-MOSFET is illustrated in Fig. 8.19. The vertical boundaries of the device are assumed to be lines of symmetry. Ideal contacts are assumed; they maintain the carrier densities at their equilibrium values. Along such contacts, the electrostatic potential is its value in equilibrium, shifted by the applied bias. Along noncontacted boundaries, the normal components of the currents are set to the surface recombination rate and the normal component of the displacement field to the surface charge.

For numerical solution, the device is subdivided by a mesh as illustrated in Fig. 8.19b. To obtain n and p at each of the N vertices, the defining equations are applied to each element which results in a system of $3N$ nonlinear equations for the $3N$ unknowns. The unknowns at node, (i, j) are $n(i, j), p(i, j)$, and $V(i, j)$. Numerical techniques for solving such equations are highly refined and are dis-

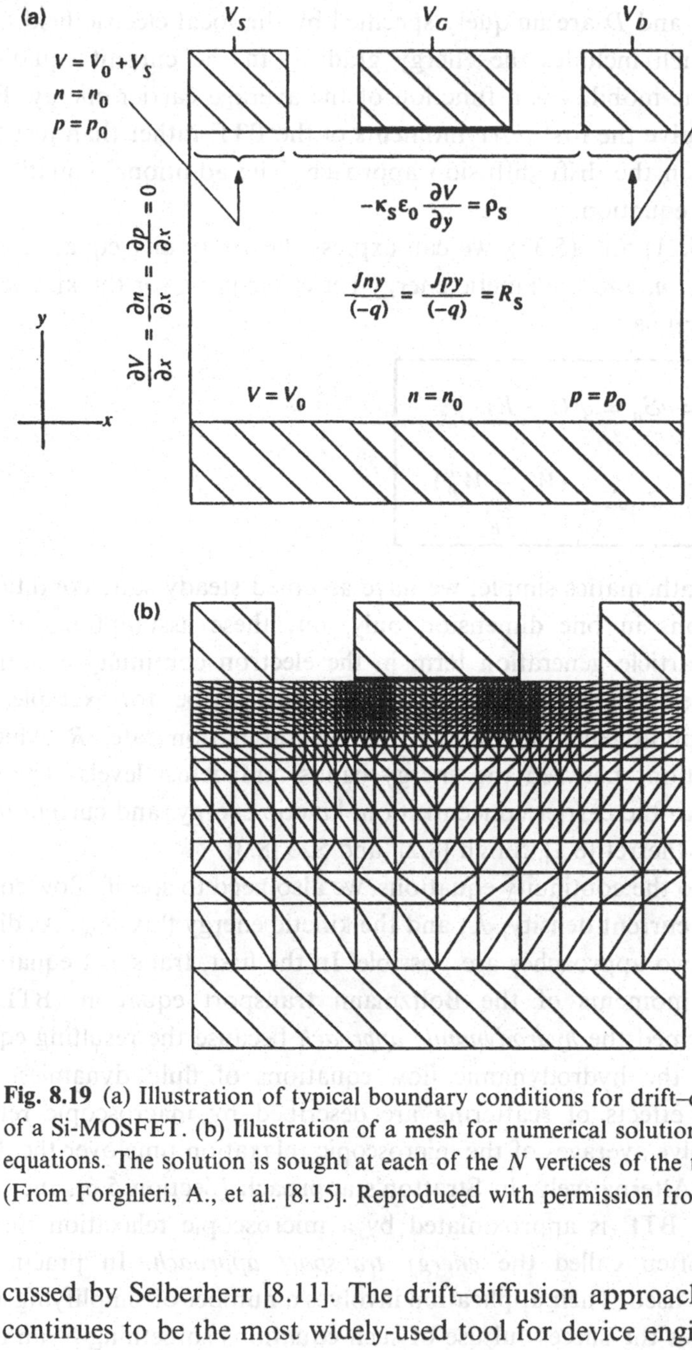

Fig. 8.19 (a) Illustration of typical boundary conditions for drift–diffusion-based simulation of a Si-MOSFET. (b) Illustration of a mesh for numerical solution of the drift–diffusion equations. The solution is sought at each of the N vertices of the triangular elements. (From Forghieri, A., et al. [8.15]. Reproduced with permission from IEEE.)

cussed by Selberherr [8.11]. The drift–diffusion approach to device simulation continues to be the most widely-used tool for device engineering.

8.9.2 The hydrodynamic/energy transport approach

The drift–diffusion equation is a rough approximation to the momentum balance equation. It ignores the spatial variation of the average carrier energy and assumes that μ and D are uniquely specified by the local electric field. A more detailed approach includes the energy gradient in the current equation and assumes that the mobility is a function of the average carrier energy. For this approach, we solve the first *three* moments of the BTE rather than just the first *two* as is done in the drift–diffusion approach. The additional equation is the energy balance equation.

From eqs. (5.21) and (5.37), we can express the continuity equations for the electron density, n, and the kinetic energy per electron, u, (or the kinetic energy density, $W = nu$) as

$$\frac{\mathrm{d}J_{nz}}{\mathrm{d}z} = q\mathcal{S}_n = q(G - R), \tag{8.43a}$$

and

$$\frac{\mathrm{d}F_{Wz}}{\mathrm{d}z} = J_{nz}\mathcal{E}_z - \frac{(W_n - W_n^o)}{\hat{\tau}_E} \tag{8.43b}$$

To keep the mathematics simple, we have assumed steady-state conditions and spatial variations in one dimension only, but these assumptions are easily relaxed. The particle generation term in the electron continuity equation, $\mathcal{S}_n$, consists of a net generation rate, G, which may describe, for example, impact ionization or photogeneration and a net recombination rate, R, which may describe transitions between the energy bands and defect levels. These terms can be related to the carrier concentration, kinetic energy, and current densities as discussed in Pierret [8.1], Singh [8.2], and Sze [8.3].

In addition to the continuity equations, we also need to specify flow equations for the electric current density, J_n, and the kinetic energy flux, F_W. As discussed in Chapter 5, two approaches are possible. In the first, transport equations are derived from moments of the Boltzmann transport equation (BTE). This approach is termed the *hydrodynamic approach* because the resulting equations are similar to the hydrodynamic flow equations of fluid dynamics. In this approach, the effects of scattering are described by macroscopic relaxation times that involve averages of the microscopic relaxation time over the distribution function. Alternatively, in Stratton's approach (Section 5.6), the collision integral in the BTE is approximated by a microscopic relaxation time. This approach is often called the *energy transport approach.* In practice, both approaches are used. Their application involves a number of simplifying assumptions that makes the successful use of such equations something of an art. The final accuracy can only be determined by comparing the results to those of a more rigorous simulation. Fortunately, the form of the final equations is similar for the two approaches, and good results have been obtained for both approaches. Since the two approaches are so similar, we refer to them as the hydrodynamic/energy transport approach.

We first present the hydrodynamic flow equations. From eqs. (5.32) and (5.38), we can write the steady-state flow equations for spatial variation in one dimension as

$$J_{nz} = nq\mu_n\mathcal{E}_z + 2\mu_n\frac{\mathrm{d}}{\mathrm{d}z}(nu_{zz}) \tag{8.44a}$$

and

$$F_{Wz} = -W\mu_E\mathcal{E}_z - 2\mu_E\frac{\mathrm{d}}{\mathrm{d}z}(nR_{zz}), \tag{8.44b}$$

where μ_E is related to the energy relaxation time by

$$\mu_E = \frac{q\hat{\tau}_E}{m^*} \tag{8.45a}$$

and R_{zz} is related to the fourth moment of the BTE, X_{zz}, [as defined in eq. (5.39)], by

$$R_{zz} = \frac{m^* X_{zz}}{qn}. \tag{8.45b}$$

(We have assumed a constant effective mass, but that assumption can be eliminated.)

Flow equations can also be developed by Stratton's approach (the so-called energy transport model). From eqs. (5.86) and (5.91), we find

$$J_{nz} = nq\mu_n\mathcal{E}_z + 2\mu_n\frac{\mathrm{d}}{\mathrm{d}z}(nu/3) \tag{8.46a}$$

and

$$F_{Wz} = C_e n\mu_n k_B T_e\mathcal{E}_z - C_e\mu_n\frac{\mathrm{d}}{\mathrm{d}z}(nk_B^2T_e^2/q), \tag{8.46b}$$

where $C_e = (5/2 - s)$ depends on the power law, s, for scattering.

To complete the specification of the hydrodynamic/energy transport model, we need to close the equations and specify the relaxation times. We desire a set of equations in two unknowns, the electron density, n, and the kinetic energy per electron, u, which means that u_{zz} and R_{zz} in eqs. (8.45) must be expressed in terms of these parameters. The simplest closure approach is to neglect the drift energy and assume equipartition of energy to write

$$u_{zz} = W/3n. \tag{8.47a}$$

A better assumption is to assume that the component of the kinetic energy associated with random, thermal motion is equally distributed among the three degrees of freedom and to include the drift energy. The result is

$$nu_{zz} = \frac{1}{3}\left(W - \frac{1}{2}m^* J_{nz}^2/nq^2\right) + \frac{1}{2}m^* J_{nz}^2/nq^2 = \frac{1}{3}\left(W + m^* J_{nz}^2/nq^2\right), \tag{8.47b}$$

which is much better, but note that the current equation, eq. (8.44a) is no longer an explicit expression for the current because J_{nz} appears on both sides of the equation.

The tensor component, R_{zz}, can be expressed as (see homework problem 5.6),

$$R_{zz} = \frac{10}{9}\frac{W^2}{q}, \tag{8.48}$$

if we assume a displaced Maxwellian distribution and ignore the drift energy. More refined treatments are possible [8.12]. Finally, we need to express the relaxation times, $\hat{\tau}_m$ and $\hat{\tau}_E$, as functions of n and u. After doing so, which we discuss in Section 8.9.3, the problem specification is complete. By using eqs, (8.44) in eqs. (8.43), we obtain two coupled partial differential equations for the two unknowns, $n(z)$ and $u(z)$. These equations are then solved self-consistently with Poisson's equations using techniques much like those for solving the drift–diffusion equations, but instead of obtaining $n(z)$ and $V(z)$ throughout the device, we also obtain $u(z)$.

The energy transport equations (Stratton's approach) can also be used for the flow equations. Instead of eqs. (8.44), we use eqs. (8.46). Note that there is no closure problem for these equations because we assumed a displaced Maxwellian when deriving them, so they are expressed in terms of the unknowns, $n(z)$ and $u(z)$. Note also that the flow equations for the two approaches are very similar. In fact, if we ignore the drift energy in the hydrodynamic equations and approximate u_{zz} by eq. (8.47a), then the hydrodynamic flow equation, eq. (8.44a), is identical to the energy transport flow equation, eq. (8.46a). In the energy transport approach,

$$W = \frac{3}{2}k_B T_e \tag{8.49a}$$

$$\mu_E = \frac{2}{3}C_e \mu_n \tag{8.49b}$$

and

$$R_{zz} = \frac{2}{3}\frac{W^2}{q}. \tag{8.49c}$$

We can, therefore, think of the energy transport flow equations as the hydrodynamic equations with a specific set of closure relations. The successful use of these equations may require a careful examination of the closure relations (in the hydrodynamic approach) or the assumed shape of the distribution (energy transport approach). One of the key issues is determining the relaxation times, $\hat{\tau}_m$ and $\hat{\tau}_E$, or equivalently the mobilities, μ_n and μ_E.

Hydrodynamic/energy balance simulations are much like drift–diffusion simulations – with the addition of equations for electron and hole energy balance. Instead of $3N$ equations, we have $5N$ for the five unknowns, V, n, p, T_e, and T_h, at each of the N vertices. Figure 8.20 shows some results of simulating an n-channel Si-MOSFET using the momentum/energy balance approach [8.15]. The large applied bias at the drain and the positive gate voltage are apparent in Fig. 8.20a which is a plot of the electrostatic potential within the device. The simulated electron concentration profile plotted in Fig. 8.20b shows the high electron concentration expected in the channel, but it also shows that the channel electrons are not well-confined near the drain. Hot electron effects are illustrated by Fig. 8.20c, which is a plot of the electron temperature within the transistor. Note that the electron temperature is very high, in excess of 3000 K, near the drain end of the channel where the electric field is high. The high-temperature ridge along the drain-substrate junction occurs because the reverse-biased collector-substrate junction collects electrons from the substrate which are then accelerated as they traverse the junction. For the particular example considered, the channel length was rather long (1 μm), so the differences between the drift–diffusion and momentum–energy balance approaches are quite small. For deep sub-micron MOSFETs, however, the differences can be substantial.

8.9.3 Transport parameters in hydrodynamic/energy transport models

A key issue in using hydrodynamic or energy transport models lies in expressing the mobilities, $\mu_{\rm n}$ and $\mu_{\rm E}$, or alternatively the relaxation times, $\hat{\tau}_{\rm m}$ and $\hat{\tau}_{\rm E}$, in terms of the unknowns, $n(z)$ and $u(z)$. A common approximation involves determining these expressions under steady-state, spatially uniform conditions and then *assuming* that they apply to the transient, nonuniform conditions within a device. The steady-state results may be measured or they may be obtained by Monte Carlo simulation.

Determining transport parameters from measured characteristics

We outline a simple approach that can be used to obtain transport parameters from measured results [8.14]. Recall that D and μ are related by

$$\frac{D}{\mu} = \frac{k_{\rm B} T_e}{q} \tag{8.50}$$

for bulk silicon, where the drift energy can be ignored, and that for electrons in silicon, the measured diffusion coefficient is roughly independent of field. If we evaluate eq. (8.50) at low applied fields, where $\mu_n = \mu_n^o$ and $T_e = T_{\rm L}$ and at high applied fields and then divide the two results, we find

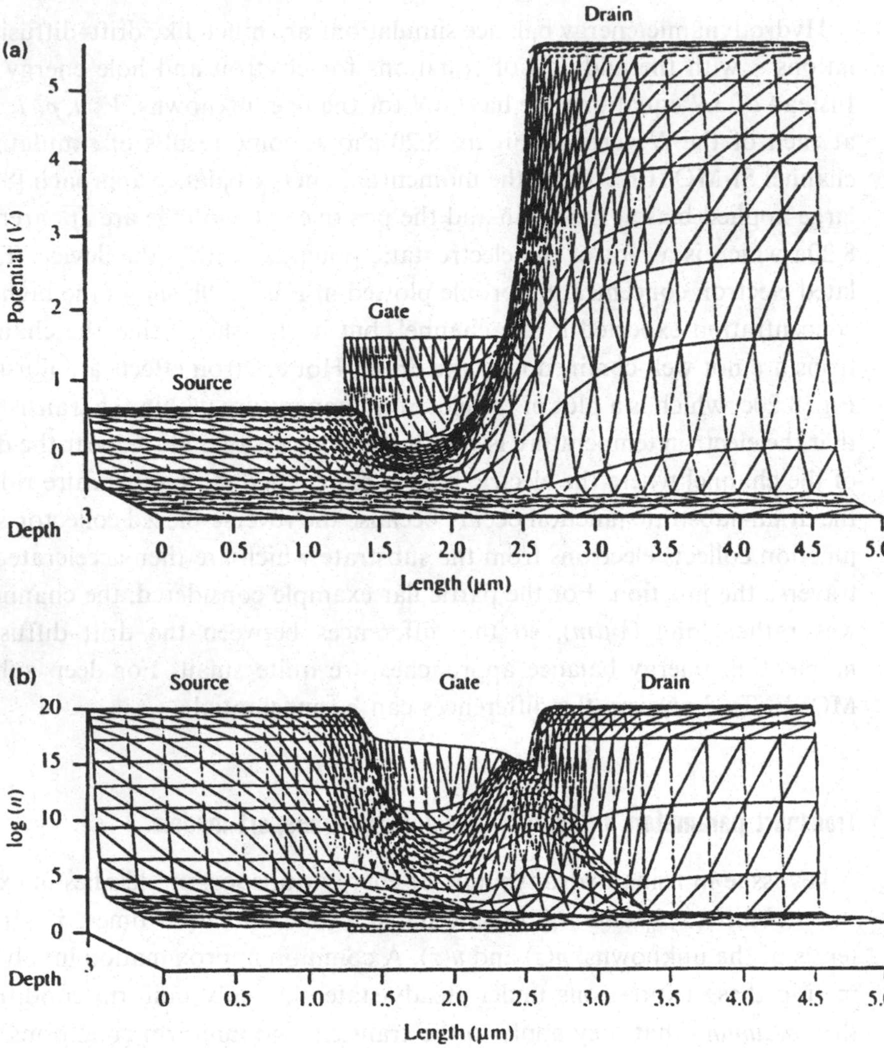

Fig. 8.20 Examples of hydrodynamic device simulation. (a) Plot of the electrostatic potential within a simulated MOSFET under a bias of $V_{DS} = 5\,V$ and $V_{GS} = 1\,V$. (b) Plot of the electron concentration versus position. (c) Plot of the electron temperature versus position. Computed by the momentum–energy balance equation approach by Forghieri, A., et al. [8.15]. (Reproduced with permission from IEEE.)

$$\mu_n(T_e) = \mu_n^o \frac{T_L}{T_e}. \tag{8.51}$$

Equation (8.51) expresses the electron mobility in terms of the unknown electron temperature, T_e.

The energy relaxation time is determined by energy balance under steady-state, spatially uniform conditions. Under such conditions, eq. (8.34b) becomes

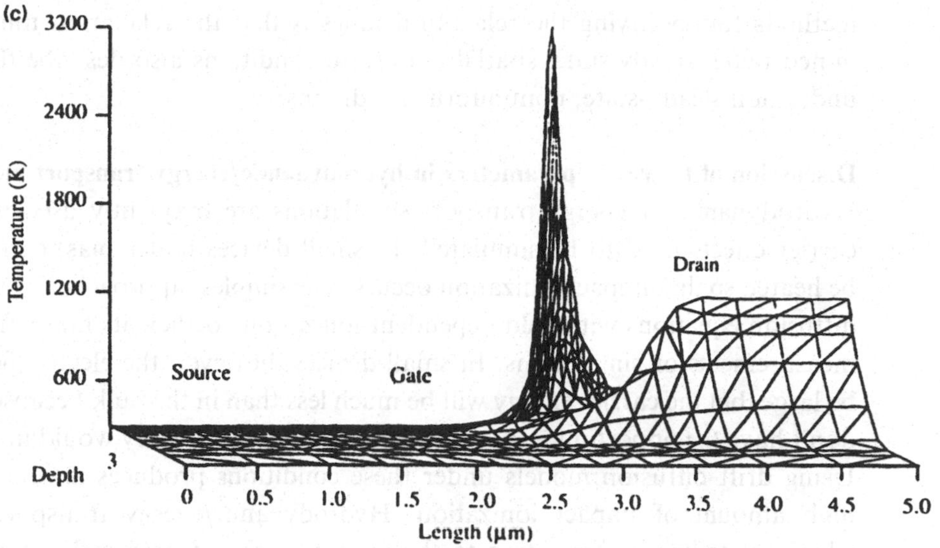

Fig. 8.20 (*continued*)

$$q\mu_n(\mathcal{E}_z)\mathcal{E}_z^2 = \frac{m_n^* \upsilon_{dz}^2/2 + 3k_B(T_e - T_L)/2}{\hat{\tau}_E}. \tag{8.52}$$

The measured high-field mobility of electrons in bulk silicon is empirically described by

$$\mu_n(\mathcal{E}_z) = \frac{\mu_n^o}{\sqrt{1 + \mathcal{E}_z^2/\mathcal{E}_{cr}^2}}, \tag{8.53}$$

where $\mathcal{E}_{cr}$, the critical field, is a material constant. From eq. (8.51)–(8.53), we find

$$\hat{\tau}_E(T_e) = \frac{3k_B}{2q\mu_n^o\mathcal{E}_{cr}^2}\frac{T_e T_L}{T_e + T_L} + \frac{m_n^*\mu_n^o}{2q}\frac{T_L}{T_e}. \tag{8.54}$$

This very simple approach has been used with some success for transport in silicon [8.14], but as devices shrink, more sophisticated approaches become necessary [8.12].

Determining transport parameters by Monte Carlo simulation

An alternative method for specifying the mobility (or momentum relaxation time) and the energy relaxation time is to make direct use of Monte Carlo simulation. Simulations of carrier transport in uniform, bulk semiconductors are performed at a variety of electric fields. Since the average carrier temperature and drift velocity are obtained from the simulation, a table of $\mu(T) = \upsilon_{dz}/\mathcal{E}_z$ is readily constructed. Similarly, from the average carrier energy and the input power, a table of $\hat{\tau}_E(T)$ can be constructed. A key assumption underlying both methods for specifying the relaxation times is that the relaxation times determined under steady-state, spatially uniform conditions also describe the device under non steady-state, nonuniform conditions.

Discussion of transport parameters in hydrodynamic/energy transport models
Hydrodynamic or energy transport simulations are frequently used when hot carrier effects need to be simulated. In small devices under bias, carriers may be heated so that impact ionization occurs. The simplest approach is to use drift–diffusion equations with field dependent ionization coefficients taken from bulk measurements or simulations. In small devices, however, the electric fields may be large, but the carrier energy will be much less than in the bulk because carriers don't have the opportunity to gain the amount of energy they would in the bulk. Using drift–diffusion models under these conditions produces an unphysically high amount of impact ionization. Hydrodynamic/energy transport models which relate impact ionization to the average carrier energy rather than to the local electric field are much better for examining such effects in small devices.

Hydrodynamic/energy transport models are also used to treat velocity overshoot, which can increase the current and the high frequency performance of a device. For accurate predictions, great care in specifying transport parameters is essential. We know, for example, that the ensemble relaxation times are averages over the carrier distribution, but most approaches relate these parameters to the second moment (the kinetic energy) alone. Consider a device like that shown in Fig. 8.1. Carriers enter the collector with little kinetic energy, gain energy as they cross the collector, then lose energy at the end of the collector. If we pick a specific kinetic energy, say $10k_BT$, then it will occur both near the beginning of the collector and near the end. Most approaches, which express the mobility as a function of the kinetic energy alone, would predict the same mobility at the two locations, but careful examination shows that it can be quite different. Tang et al. [8.12] and Stettler et al. [8.13] discuss these issues, which are becoming increasing important as devices continue to shrink. The moral is that one who uses hydrodynamic/energy transport simulations should be aware of the simplifying assumptions used. At times, it is necessary to test these models by comparing them to more rigorous approaches, such as Monte Carlo simulation.

8.9.4 The Monte Carlo approach

Drift–diffusion and hydrodynamic/energy transport approaches are based on simplification of the more general Boltzmann transport equation, but for very small devices, the BTE itself should be solved. The Monte Carlo method, a numerical technique for solving the BTE, is becoming an important tool for advanced device simulation. For this approach, the transport component of the simulation algorithm (as illustrated in Fig. 8.17) consists of moving each particle within the device for a time, Δt by the Monte Carlo method.

The simulation procedure is basically the Monte Carlo method described in Chapter 6, Section 5. It begins by specifying the initial conditions using the results of a previous simulation, or by employing a simpler method such as the drift–diffusion approach. The device is sub-divided into small elements, and each element is populated with electrons consistent with the known initial conditions. Because it is not generally possible to simulate each electron, a small sample of typically several thousand electrons is employed. The charge of each of these super-electrons is weighted to ensure overall space–charge neutrality. After applying the bias and solving Poisson's equation, each particle is moved by Monte Carlo methods for a time, Δt. Poisson's equation is re-solved for the new distribution of super-electrons, and the process is repeated until steady-state conditions are achieved.

Application of Monte Carlo simulation to advanced device simulation is discussed in Venturi et al. [8.16] and Fischetti and Laux [8.17]. A key issue is the coupling of Poisson's equation to the transport model. Because of the statistical noise associated with Monte Carlo simulation, charge density fluctuations occur and can be amplified by Poisson's equation thereby preventing convergence. Techniques to suppress the fluctuations, or to minimize their effects, are essential [8.16, 8.17]. The heavy computational burden of the method also makes it imperative to carefully structure the program for maximum efficiency [8.14, 8.17].

Figure 8.21 is a glimpse at some Monte Carlo simulation results. As Fig. 8.21a shows, the Monte Carlo computed average velocity of electrons in the channel of a deep sub-micron, *n*-channel MOSFET is distinctly higher than that obtained from a drift–diffusion simulation. This effect, a manifestation of velocity overshoot, can enhance the performance of very small transistors. Figure 8.21b compares the steady-state output characteristics from Monte Carlo and drift–diffusion simulations. The higher drain current obtained by Monte Carlo simulation is a consequence of the high average velocity of electrons in the channel. The reduced transit time also affects the transient characteristics; Monte Carlo simulation predicts a gain-bandwidth product 70% higher than the drift–diffusion prediction for this particular device.

8.10 Summary

Semiconductor devices display a rich variety of transport effects, many of which were illustrated in this chapter. To analyse or design a device, we need to select an appropriate conceptual framework in which to think about the device. To

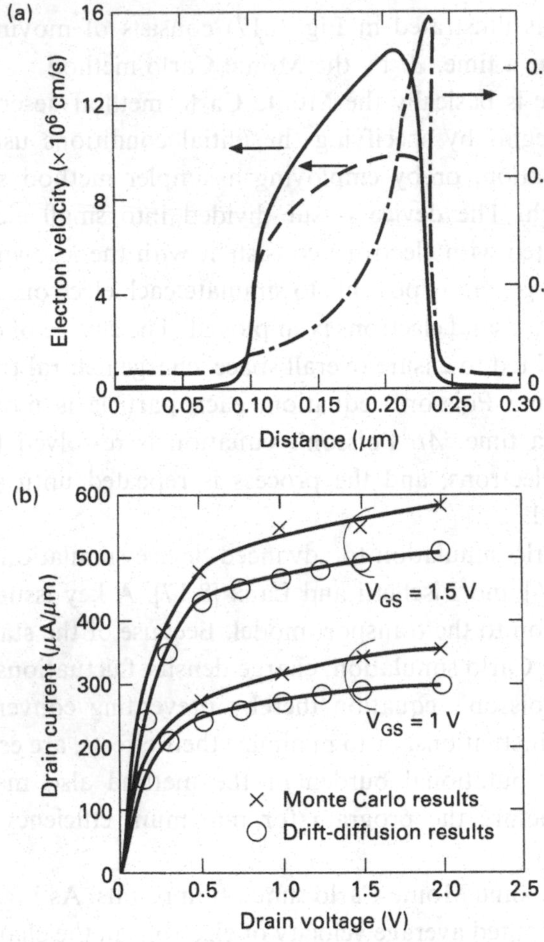

Fig. 8.21 Example Monte Carlo simulations of a silicon MOSFET. (a) Average electron velocity and energy versus position for electrons in the channel of a 0.15 μm channel length, Si-MOSFET. The dashed line for velocity was obtained by the drift–diffusion approach. (b) Comparison of the simulated output characteristics of a 0.15 μm Si-MOSFET. The × denotes Monte Carlo results, and the O denotes the drift–diffusion results. [From Venturi F., et al. [8.16]. Reproduced with permission from IEEE.)

simulate a device, we need to translate this conceptual framework into a mathematical formulation. The appropriate framework is determined by the size of the device and relevant time scale as summarized in Fig. 8.18. When the potentials in the device vary on the scale of the lattice spacing, then a fully quantum mechanical treatment is required. The wave equation, (1.1), should be solved with the applied, built-in crystal, and scattering potentials all included. If the device size is larger, but still comparable to the wavelength of the carriers, then the rapidly varying crystal potential may be factored out and an effective mass equation solved. If the potential varies slowly on the scale of the carrier's wavelength, then quantum mechanical reflections do not occur. When the device size is also large compared to a mean free path, then the collisions randomize the phase and the various electron waves don't interfere. For such devices, a classical approach which treats carriers as particles is appropriate. The most accurate treatment consists of solving the Boltzmann transport equation directly. For this approach, the quantum mechanics appears in the $E(\mathbf{p})$ relation which is used to evaluate the carrier velocity and in the scattering rates. A somewhat less rigorous treatment consists of solving the carrier, momentum, and energy balance equations derived from the BTE. Finally, when the device is large and its response slow, then the balance equations reduce to drift–diffusion equations with mobility and diffusion coefficients determined by the local electric field.

Most silicon transistors can still be described by drift–diffusion equations, although advanced silicon devices are now often simulated with the balance equation or Monte Carlo approaches. Present-day GaAs devices are generally not well-described by the conventional drift–diffusion approach. They require the balance equations, or Monte Carlo solutions to the BTE. Quantum mechanical effects are becoming increasingly important in devices. Quantum confinement affects inversion layer carriers in silicon MOSFETs, and the resonant tunneling diode uses quantum effects to produce negative differential resistance. For most of this text, our focus has been on semiclassical transport for which the wave nature of electrons is not a critical factor. In the next chapter, we generalize these ideas to show how quantum transport is treated.

References and further reading

For an introduction to semiconductor devices, consult

8.1 Pierret, R. F. *Semiconductor Device Fundamentals*. Addison-Wesley Publishing Company, Reading, MA, 1996.

8.2 Singh, J. *Semiconductor Devices*. McGraw-Hill, Inc., New York, 1994.

8.3 Sze, S. M. *Physics of Semiconductor Devices*, 2nd edn. John Wiley and Sons, New York, 1981.

An interesting analogy between the MOSFET and the bipolar transistors is described by Johnson in

8.4 Johnson, E. O. The insulated-gate field-effect transistor – a bipolar transistor in disguise. *RCA Review*, Vol. 34, pp. 80–94, 1973.

Experiments to characterize ballistic transport in devices like that shown in Fig. 8.1b are reported in

8.5 Hayes, J. R., Levi, A. F. J., Gossard, A. C., Hutchinson, A. L. and English, J. H. Base transport dynamics in a heterojunction bipolar transistor. *Applied Physics Letters*, **49**, 1481–3, 1986.

8.6 Levi, A. F. J. and Yafet, Y. Nonequilibrium electron transport in bipolar devices. *Applied Physics Letters*, **51**, 42–4, 1987.

8.7 Hayes, J. R., Levi, A. J. F. and Weigmann, W. Dynamics of electron cooling in GaAs. *Applied Physics Letters*, **48**, 1365–7, 1986.

For a description of the quantum mechanical effects which can occur when electrons move ballistically in ultra-small structures, consult

8.8 Capasso, F. (ed.) *Physics of Quantum Devices*. Springer Series in Electronics and Photonics, Vol. 28. Springer-Verlag, New York, 1989.

Quantum effects are also described by Datta in

8.9 Datta, S, *Electronic Transport in Mesoscopic Systems*. Cambridge University Press, Cambridge, UK, 1995.

Several of the examples in this chapter were taken from

8.10 Constant, E. Non-steady-state carrier transport in semiconductors in perspective with semiconductor devices. In *Hot-Electron Transport in Semiconductors*, Ed. by L. Reggiani, Vol. 58 of Topics in Applied Physics, Chapter 7. Springer-Verlag, New York, 1985.

Numerical techniques for solving the device equations are discussed in

8.11 Selberherr, S. *Analysis and Simulation of Semiconductor Devices*. Springer-Verlag, New York, 1984.

The issues in formulating macroscopic transport equations, closing the hierarchy of balance equations, and determining transport parameters are discussed in

8.12 Tang, T.-W., Ramaswamy, S. and Nam, J. An improved hydrodynamic transport model for silicon. *IEEE Transactions on Electron Devices*, **40**, 1469–77, 1993.

8.13 Stettler, M. A., Alam, M. A. and Lundstrom, M. S. A critical examination of the assumptions underlying macroscopic transport equations for silicon devices. *IEEE Transactions on Electron Devices*, **40**, 733–9, 1993.

For a discussion of how the momentum and energy balance equations are formulated and solved, consult

8.14 Baccarani, G. and Wordemann, M. R. An investigation of steady-state velocity overshoot in silicon. *Solid-State Electronics*, **28**, 407–16, 1985.

8.15 Forghieri, A., Guerrieri, R., Ciampolini, P., Gnuidi, A., Rudan, M. and Baccarani, G. A new discretization strategy of the semiconductor equations comprising momentum and energy balance. *IEEE Transactions on Computer-Aided Design*, **7**, 231–42, 1988.

Application of the Monte Carlo technique to semiconductor device simulation is described in

8.16 Venturi, F., Smith, R. K., Sangiorgi, E. C., Pinto, M. R. and Ricco, B. A general purpose device simulator coupling Poisson and Monte Carlo transport with application to deep submicron MOSFET's. *IEEE Transactions Computer Aided Design*, **8**, 360–9, 1989.

8.17 Fischetti, M. V. and Laux, S. E. Monte Carlo analysis of electron transport in small semiconductor devices including band-structure and space-charge effects. *Physical Review B*, **38**, 9721–45, 1988.

Problems

8.1 Assume that electrons are injected into GaAs from a heterojunction launching ramp like that shown in Fig. 8.5b. If $p_x = p_y = 0$ and p_z is initially very small, then

(a) Derive an expression which relates the electron velocity, v_d, to the energy step, ΔE_C. Include the effects of conduction band nonparabolicity as described by

$$\frac{p^2}{2m^*(0)} = E(1 + \alpha E).$$

(b) Plot v_{dz} versus ΔE_C for $0.0 < \Delta E_C < 0.5\,\mathrm{eV}$. Compare the results to eq. (8.33) for parabolic energy bands.

(c) Define an energy-dependent effective mass so that the result obtained in part (a) has the form of eq. (8.33). Plot $m^*(E)$ for $0 < E < 0.5\,\mathrm{eV}$.

8.2 Assume ballistic transport across the base of a heterojunction bipolar transistor. Compute the base transit time assuming a heterojunction launching ramp with $\Delta E_C = 0.3\,\mathrm{eV}$. Compare the result with the standard expression assuming that electrons diffuse across the base. Assume $m^* = 0.067\,m_0$, $\mu_n = 1000\,\mathrm{cm^2/V\,s}$, and $W_B = 750\,\text{Å}$.

8.3 Examine the 'diffusion velocity' versus 'diffusion field' characteristic by answering the following:

(a) Show that an effective field which derives diffusion can be defined as

$$\mathcal{E}_{\mathrm{effective}} = \frac{k_B T_e}{q}\frac{\nabla n}{n}.$$

Hint: examine the gradient of the quasi-Fermi level.

(b) Derive an expression for the average velocity versus effective field [assume that $\mathcal{E} = 0$ and use eq. (8.11)].

(c) Show that for 'low-field' diffusion (i.e. when the concentration gradient is gentle)

$$v_d = -\mu_n \mathcal{E}_{\mathrm{effective}}.$$

(d) Plot the average velocity versus effective field characteristic and show that the velocity saturates when

$$\mu_n \mathcal{E}_{\mathrm{effective}} \gg \sqrt{k_B T_e / 2m^*}.$$

8.4 Derive eq. (8.32), the Mott–Gurney law.

8.5 Read the paper: Investigation of transient electronic transport in GaAs following high energy injection, by Tang and Hess in *IEEE Transactions on Electron Devices*, **ED-29**, 1906–1911. This paper presents Monte Carlo simulations of electron injection across a launching ramp similar to the one illustrated in Fig. 8.5b.

(a) Assume ballistic transport and compute the average velocity for launching ramps of 0.04, 0.07, 0.11, 0.16, 0.21, 0.27 and 0.34 eV.

(b) Compare the answers to part (a) with the Monte Carlo simulation results presented by Tang and Hess in Fig. 1a of their paper. Is ballistic transport a reasonable assumption for such a GaAs device? Explain why the result for the 0.34 eV launching ramp is so different from the rest.

8.6 Use the approach summarized by eqs. (8.50)–(8.54) to answer the following.

(a) Construct a plot of electron temperature versus electric field. Assume electrons in Si, a spatially uniform field, and plot the result from 100 to 100 000 V/cm.

(b) Plot the mobility and the energy relaxation time versus electron temperature for T_e from 100 to 10 000 K.

8.7 Use the Monte Carlo simulation results presented in Fig. 7.4a to construct a plot of electron mobility versus electron temperature. How do the results obtained from Monte Carlo simulation compare with those obtained in problem 8.6b above?

8.8 Derive eq. (8.54).

8.9 Using the effect summarized in Fig. 8.11, explain why the distribution functions for high-field electron transport in Si, as illustrated in Fig. 7.6, are skewed towards negative velocities.

9 Transport in mesoscopic structures

9.1 The mesoscopic regime

Classical physics describes the everyday, macroscopic, world, but quantum mechanics describes the microscopic world of atoms and molecules. Traditionally, semiconductor devices could be thought of as macroscopic objects describable by semiclasical concepts (i.e. we describe particle dynamics by equations like Newton's law of motion generalized to include the concept of bandstructure). It is now possible, however, to produce devices and structures for which these semiclassical concepts break down. Such devices are still larger than the atomic or molecular scale, but they are smaller than some critical length scales above which traditional transport theories apply. The size scale between the microscopic and macroscopic regimes is known as the *mesoscopic* regime. Our objectives in this chapter are to describe some key approaches and important results concerning transport at the mesoscopic scale.

The mean-free-path for scattering is an important length scale for carrier transport. One can show (see homework problem 9.1) that the mean-free-path, λ, is related to the diffusion coefficient, D, by $D \approx \lambda v_R$, where $v_R \approx \sqrt{k_B T_L / 2\pi m^*}$ is the so-called Richardson velocity. For room-temperature electrons in pure silicon, $\lambda \approx 700$ Å, while for GaAs, $\lambda \approx 200$ Å. (At $T = 77$ K, the mean-free-path for electrons in pure GaAs increases to $\approx 1\ \mu$m.) Another characteristic length associated with scattering is the energy relaxation length for hot electrons, which is on the order of a few hundred Angstroms for Si and a few thousand Angstroms for GaAs. Semiconductor devices with critical regions of this length scale are readily produced. At these length scales, effects such as quasi-ballistic transport and velocity overshoot come into play.

A second important length scale is the *coherence* length for the electron wave. Collisions can impart a random phase to the electron, thereby washing out quantum interference effects. For macroscopic devices, phase randomizing scattering dominates, so quantum interference effects can be neglected. For very small devices, however, one may need to use a wave approach to electron transport. It might be thought that the phase randomization length would be closely related to the mean-free-path for scattering, but this is not always the case. Elastic scattering from ionized impurities, for example, imparts no random component to the phase, so interference effects can still occur. Phonon scattering is phase randomizing, so quantum interference effects are typically difficult to observe at room temperature. At low applied biases, only phonon absorption occurs. At very low temperatures, few phonons are present. Under these conditions, the major source of phase randomizing scattering is electron–electron scattering. The electron systems used for these studies tend to be degenerate at low temperatures, so final state filling greatly suppresses scattering by the $(1-f)$ factor. The result is that quite long phase randomization lengths can be obtained at temperatures of a few Kelvin and below. Under these conditions, it is essential to treat the wave nature of electrons when describing their transport.

Our purpose in this chapter is to provide an introduction to carrier transport in the mesoscopic regime. Mesoscopic transport is a term usually applied to quantum transport effects. For a thorough introduction to this field, consult Datta's book, *Electronic Transport in Mesoscopic Systems* [9.1]. In this chapter, we will also discuss semiclassical transport in structures on the order of a mean-free-path in length. Since semiclassical transport is more familiar, we will begin there and introduce a general conceptual picture that will later be extended to treat quantum transport.

Figure 9.1 shows the conceptual view that we will use to describe transport in small structures and devices. The device is a structure connected to contacts that are assumed to be in thermodynamic equilibrium. Each contact injects a known

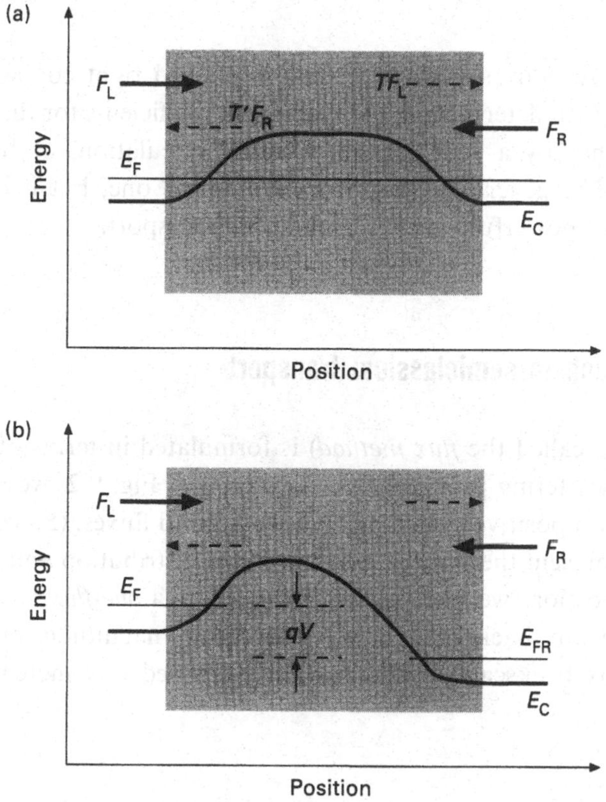

Fig. 9.1 A general, conceptual picture of a semiconductor device. The contacts are assumed to be reservoirs of thermal equilibrium carriers. Each contact injects a flux of carriers into the device itself (F_L and F_R in this case). A fraction, T, of the flux injected from the left transmits across the device and a fraction T' of the flux injected from the right transmits to the left. (a) Equilibrium and (b) under bias. The net flux is $F = TF_L - T'F_R$.

flux of electrons into the device. If the contacts are heavily doped semiconductors with n_0 carriers, then each contact injects a flux of

$$F_{inj} = \left(\frac{n_0}{2}\right)\upsilon_T \tag{9.1a}$$

where

$$\upsilon_T = \sqrt{\frac{2k_B T_L}{\pi m^*}} \tag{9.1b}$$

is the average velocity of nondegenerate, thermal electrons crossing a plane (recall homework problem 3.5).

The device itself is described by its transmission coefficients, T and T', and the net flux through the device is

$$F_{\text{net}} = TF_{\text{L}} - T'F_{\text{R}'} \tag{9.1c}$$

where F_{L} and F_{R} are the fluxes injected from the left and right contacts. The central problem. then, is to determine the transmission coefficients for the device. This can be done either by a semiclassical classical calculation, or by using quantum mechanics. The general picture is a very simple one, but it has also turned out to be a very powerful way to think about transport.

9.2 The scattering approach to semiclassical transport

Scattering theory (also called the *flux method*) is formulated in terms of carrier fluxes and their backscattering probabilities. As shown in Fig. 9.2, we separate the flux distribution into positively- and negatively-directed fluxes. (Since we use one average flux to represent the positive portion of the distribution and another one for the negative portion, we refer to this approach as a *one-flux method.*) If the positively-directed flux backscatters, it is reduced in magnitude, but if the negatively-directed flux backscatters, the positively-directed flux increases. We can, therefore, write

$$\frac{\mathrm{d}a}{\mathrm{d}z} = -\xi a + \xi' b \tag{9.2a}$$

$$\frac{\mathrm{d}b}{\mathrm{d}z} = -\xi a + \xi' b, \tag{9.2b}$$

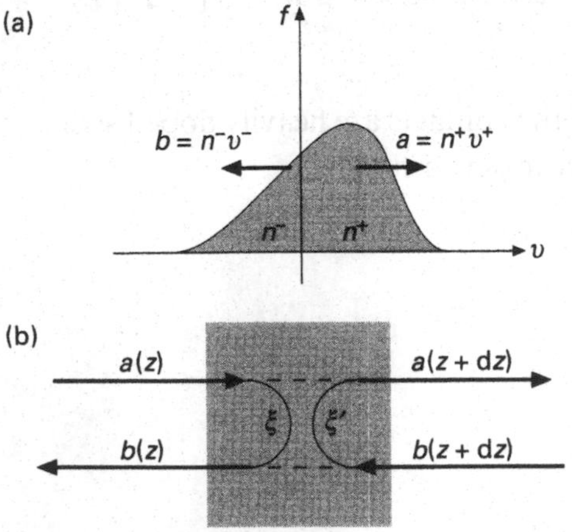

Fig. 9.2 Illustration of the scattering approach to carrier transport (which is also known as the flux method). (a) The carrier distribution is separated into positively and negatively directed components. (b) Carrier backscattering from a region of length, *dz*, changes the magnitude of the two fluxes.

where ξ and ξ' are the backscattering probabilities per unit length for the right- and the left-directed fluxes respectively. (We assume steady-state conditions, so $a(z)$ and $b(z)$ are the position-dependent, steady-state, right- and left-directed fluxes. We also neglect recombination–generation processes. See [9.3–9.5] for discussion about treating these effects.)

If we examine a semiconductor slab with a finite thickness, Δz, then as shown in Fig. 9.3, there is a right-directed flux incident on the left face of the slab and a left-directed flux incident on the right face. The problem is to determine the two fluxes, $b(z)$ and $a(z+\Delta z)$ that emerge from the slab, which is readily accomplished by integrating eqs. (9.2a) and (9.2b) across the slab (see Appendices A and B of [9.3]). The result can be expressed in terms of the *scattering matrix* for the slab, which relates the two fluxes emerging from the slab to the two fluxes incident on the slab by

$$\begin{pmatrix} a(z+\Delta z) \\ b(z) \end{pmatrix} = \begin{bmatrix} T & 1-T' \\ 1-T & T' \end{bmatrix} \begin{pmatrix} a(z) \\ b(z+\Delta z) \end{pmatrix}, \tag{9.3}$$

where T and T' represent the fraction of the steady-state right- and left-directed fluxes that transmit across the slab. The column sum of one is a statement of conservation of flux (recall that we have neglected recombination and generation within the slab). The elements of the scattering matrices used to describe particle transport are real numbers between zero and one (in contrast to the scattering matrices used for electromagnetic problems for which the elements are complex numbers).

Scattering theory can be applied to semiconductor devices in two different ways. One way is to simply integrate eqs. (9.2a) and (9.2b) across the device. Another way is to divide the device into a set of thin slabs connected so that the output fluxes from one slab provide the input fluxes to its neighboring slabs. In both cases, the boundary conditions are the two fluxes injected from the contacts. The central problem is to specify the backscattering probabilities per unit length, ξ and ξ', or the transmission coefficients of the finite thickness slabs.

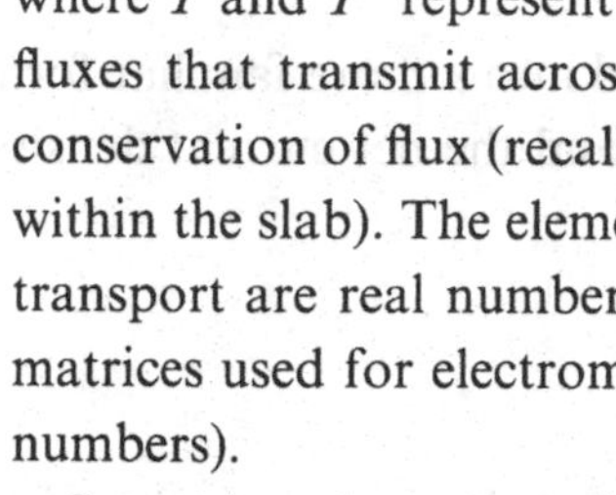

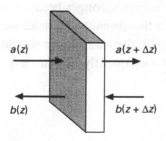

Fig. 9.3 Carrier fluxes incident upon and backscattering from a slab of finite thickness.

9.2.1 Scattering theory and the drift-diffusion equation

Scattering theory is an alternative to the conventional transport approaches discussed in earlier chapters of this text. There is, however, a close connection between scattering theory and the conventional approaches. Conventional approaches are formulated in terms of total quantities such as carrier density and the net current. To keep things simple, assume that each half of the flux distribution is described by a thermal equilibrium hemi-Maxwellian. The average velocity of each half is

$$v^+ = v^- = v_T = \sqrt{2k_B T_L/\pi m^*}. \tag{9.4}$$

The carrier density is related to the fluxes by

$$n(z) = \frac{a(z) + b(z)}{v_T}. \tag{9.5a}$$

Similarly, the net electron current is

$$J_n(z) = (-q)[a(z) - b(z)]. \tag{9.5b}$$

Equations (9.2a) and (9.2b) can be manipulated into a more familiar form which expresses them in terms of the carrier density and the net current density. First, by subtracting the two equations, we find

$$\frac{dJ}{dz} = 0, \tag{9.6}$$

which is recognized as the continuity equation for electrons (recall again that we have neglected recombination–generation processes and have assumed steady-state conditions). Next, by adding the two equations, we find

$$J_{nz}(z) = (-q)\left(\frac{\xi' - \xi}{\xi + \xi'}\right)v_T n(z) + q\frac{v_T}{(\xi + \xi')}\frac{dn}{dz}. \tag{9.7}$$

Equation (9.7) has a term proportional to the carrier density and one proportional to the gradient of the carrier density, which suggests that it can be written in drift–diffusion form. In fact, the suggestion is even stronger because the term $(\xi' - \xi)$ should be related to the electric field. In the absence of a field, we expect that $\xi = \xi'$, and in the presence of an electric field, their difference should be (to first order) proportional to the electric field. Consequently, we can express eq. (9.7) as

$$J_{nz}(z) = nq\mu_n \mathcal{E}_z + D_n\frac{dn}{dz}, \tag{9.8a}$$

where

$$\mu_n \equiv \left(\frac{\xi - \xi'}{\xi + \xi'}\right)\left(\frac{\upsilon_T}{\mathcal{E}_z}\right) \tag{9.8b}$$

and

$$D_n = \frac{\upsilon_T}{(\xi + \xi')}. \tag{9.8c}$$

As we will discuss in Section 9.4, the backscattering probability per unit length is simply one over the carrier mean free path, so eq. (9.8c) just states that the diffusion coefficient is proportional to the product of the mean-free-path and the thermal velocity, a well-known fact [see homework problem (9.1)].

These results show that there is a close connection between the scattering equations, eqs. (9.2a) and (9.2b), and the conventional current continuity and drift–diffusion equation, eqs (9.6) and (9.8a). When eq. (9.8a) is inserted into the continuity equation, eq. (9.6), we obtain a second order differential equation for the carrier density, $n(z)$. The process has simply converted the system of two first order differential equations, eqs. (9.2a) and (9.2b), into a second order equation for $n(z)$.

9.3 Working with scattering matrices

To analyze a device in terms of interconnected scattering matrices, we need to learn how to manipulate scattering matrices. Figure 9.4 shows two interconnected scattering matrices. If we are only interested in the fluxes emerging from this set of two scattering matrices, b_0 and a_2, then we can replace the two scattering matrices by a single, composite scattering matrix. The procedure is more involved than simple matrix multiplication, but it is readily derived (see homework problem 9.2). If the two scattering matrices have elements, t_1, t_1', t_2, and t_2', then the elements of the composite scattering matrix are:

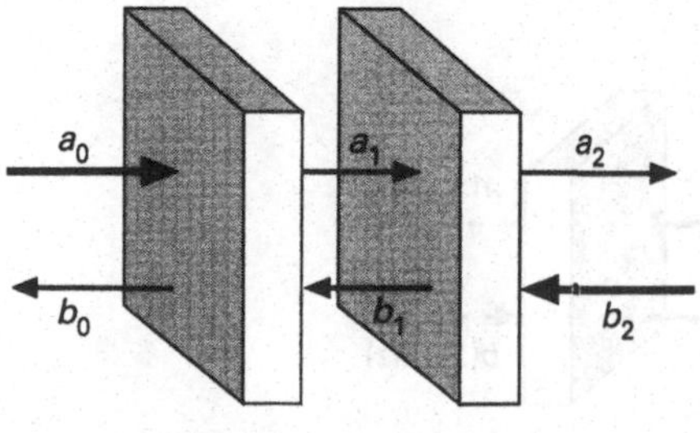

Fig. 9.4 Illustration of how two scattering matrices are cascaded to produce a single, composite scattering matrix. For the composite scattering matrix, indicated by the large gray box, the incident fluxes are a_0 and b_2 and the emerging fluxes are b_0 and a_2.

$$t_{21} = t_1[1 - r_2' r_2]^{-1} t_2 \tag{9.9a}$$

$$r_{21} = r_1 + t_1' r_2 [1 - r_1' r_2]^{-1} t_1 \tag{9.9b}$$

$$r_{21}' = r_2' + t_2 [1 - r_1' r_2]^{-1} r_1' t_2' \tag{9.9c}$$

$$t_{21}' = t_1' [1 - r_1' r_2]^{-1} t_2', \tag{9.9d}$$

where $r_1 = 1 - t_1$, etc. The cascading rules are exactly the same ones that are used for microwave analysis. Note that by flux conservation, $r_{12} = 1 - t_{12}$, and $r_{12}' = 1 - t_{12}'$.

Equations (9.9) describe the multiple reflection processes that occur as a flux injected from the left or right, transmits across the first slab then backscatters and reflects from the interiors of the two slabs infinitely many times. To analyze a device, it is divided into a finite number of scattering matrices which are cascaded two at a time until the entire device is described by a single, composite scattering matrix. The two fluxes that emerge into the contacts can then be evaluated from the known fluxes injected from the two contacts and the composite scattering matrix. Once all of the fluxes are known, the current through the device can be obtained by subtracting the right- and left-directed fluxes.

It is sometimes preferable to work with transmission matrices rather than with scattering matrices. As illustrated in Fig. 9.5, a transmission matrix relates the two fluxes at the left of the slab to the two fluxes at the right. The elements of the transmission matrix are readily determined from those of a given scattering matrix (see homework problem 9.3). Cascading two transmission matrices together is accomplished by matrix multiplication, which is much simpler than the process for cascading scattering matrices. Using transmission matrices, one could begin at the left slab of the device and evaluate the fluxes throughout the device by matrix multiplication from left to right. There are two difficulties. First, the two fluxes at the left side of the device are not known. We only know that the left contact injects a known flux into the device, we don't know

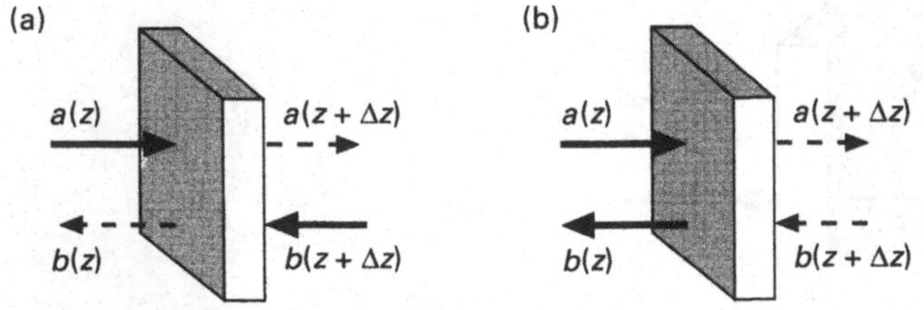

Fig. 9.5 Illustration of the difference between (a) a scattering matrix and (b) a transmission matrix. The heavy arrows indicate the given fluxes and the dashed lines are the fluxes determined by the scattering matrix or by the transmission matrix.

the flux that emerges until the problem has been solved. There are ways around this (e.g. iterative solutions), but the second problem is that small errors can grow exponentially when transmission matrices are multiplied.

9.4 One-flux scattering matrices

In this section, we derive some simple scattering matrices that describe transport under conditions that commonly occur in devices. In the following section, we'll apply these scattering matrices to devices.

9.4.1 Transport across a field-free slab

If there is no electric field present, the scattering matrix is symmetrical, so $T = T' = T_0$ and $\xi = \xi' = \xi_0$. Recall that ξ represents the probability per unit length of backscattering. In the absence of an electric field, this is just $1/\lambda$, where λ is the mean-free-path for backscattering. If the slab is thin, the probability that a carrier backscatters while crossing the slab is just $\mathrm{d}z/\lambda$. The scattering matrix for a slab of thickness dz is, therefore,

$$[\mathrm{d}S] = \begin{bmatrix} \mathrm{d}t & \mathrm{d}r \\ \mathrm{d}r & \mathrm{d}t \end{bmatrix} = \begin{bmatrix} 1-\mathrm{d}z/\lambda & \mathrm{d}z/\lambda \\ \mathrm{d}z/\lambda & 1-\mathrm{d}z/\lambda \end{bmatrix}. \tag{9.10}$$

Because of the multiple scattering events that occur, it is more difficult to deduce the scattering matrix for a slab of finite thickness. Within the slab, we have

$$\frac{\mathrm{d}a}{\mathrm{d}z} = -\frac{a}{\lambda} + \frac{b}{\lambda}. \tag{9.11}$$

Since flux is continuous, $F = a - b$ is constant, and eq. (9.11) can be written as

$$\frac{\mathrm{d}a}{\mathrm{d}z} = -F/\lambda \tag{9.12}$$

and integrated to obtain

$$a(z) = a(0) - (F/\lambda)z. \tag{9.13}$$

Here, zero is at the left face of the slab. If we again use current continuity,

$$a(z) = a(0) - [a(z) - b(z)](z/\lambda), \tag{9.14}$$

and evaluate the expression and at $z = L$, the end of the slab, we find

$$a(L) = \left(\frac{\lambda}{L+\lambda}\right)a(0) + \left(\frac{L}{L+\lambda}\right)b(L) = T_0 a(0) + R_0' b(L). \tag{9.15}$$

We conclude that in the absence of an electric field, the transmission coefficient of a slab of thickness L is

$$\boxed{T_0 = 1 - R_0 = \frac{\lambda}{L+\lambda} = T_0'} \,. \tag{9.16}$$

As expected, T_0 goes to zero for a long slab and approaches one for a thin slab.

Transport across a thin base

Figure 9.6 illustrates a common situation which represents, for example, the base of a bipolar transistor. Electrons are injected from the left (from the emitter) and collected at the right. The collector is modeled as an ideal, absorbing contact, which means that any carrier incident upon it is absorbed (collected) and none backscatters into the device. A real collector, such as the one in a bipolar transistor, is usually a good approximation to the ideal, absorbing collector.

Since $b_{\mathrm{L}} = 0$, the net flux at the right of the slab (and at the left by current continuity) is given by

$$F = T_0 a_0. \tag{9.17}$$

The carrier density at the left is

$$n(0) = \frac{(a_0 + R_0 a_0)}{\upsilon_{\mathrm{T}}} = \frac{a_0}{\upsilon_T}(1 + R_0). \tag{9.18}$$

From eqs. (9.16)–(9.18), we find

$$F = (\lambda \upsilon_{\mathrm{T}}/2)\frac{n(0)}{L+\lambda/2}, \tag{9.19}$$

which applies for an arbitrary thickness, L, from much thinner than the mean-free-path to much thicker. When the slab is much thicker than a mean-free-path, the flux is also described by a diffusion equation and can be written as

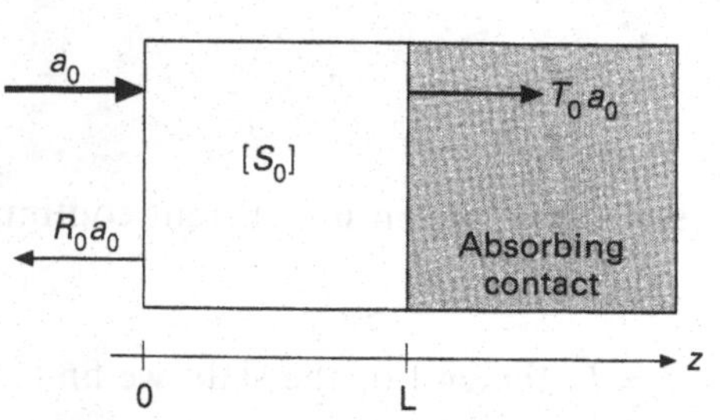

Fig. 9.6 Carrier transport across a semiconductor slab. Carriers are injected at the left and collected by a perfectly absorbing contact at the right.

$$F = D_n \frac{n(0)}{L}. \tag{9.20}$$

By equating these two expressions for the carrier flux in the limit $L \gg \lambda$, we find that the diffusion coefficient is related to the mean-free-path for backscattering by

$$D_n = \lambda \upsilon_R, \tag{9.21}$$

where $\upsilon_R = \upsilon_T/2$ is the so-called *Richardson velocity*.

Equation (9.21) relates the mean-free-path for backscattering to a more familiar quantity, the bulk diffusion coefficient, but eq. (9.19) applies from the ballistic to the diffusive limit. When using a diffusion equation, one often assumes carrier densities at the two boundaries, $n(0)$ and $n(L)$, and then computes the net flux. The physically appropriate boundary conditions, however, are the fluxes injected at the two ends of the slab, $a(0)$ and $b(L)$ [for the perfect absorber at the right, $b(L) = 0$]. The response of the slab is determined by its transmission coefficient,

$$T_0 = \frac{1}{1 + \upsilon_R/(D_n/L)}, \tag{9.22}$$

and the net flux is given by eq. (9.17). The carrier densities that result from the injected flux can be written as

$$n(0) = \frac{F}{\upsilon_T}\left(\frac{2 - T_0}{T_0}\right) \tag{9.23a}$$

and

$$n(L) = \frac{F}{\upsilon_T}, \tag{9.23b}$$

and it is readily shown that the carrier density varies linearly within the slab. In the limit of a thick slab, $(D_n/L) \ll \upsilon_R$, and one can shows that the net flux is given by eq. (9.20) and that $n(L)$ approaches zero, the expected answers for diffusive transport. On the other hand, for a thin slab, $(D_n/L) \gg \upsilon_R$, and $n(0)$ and $n(L)$ approach F/υ_T. In this ballistic limit the carrier density approaches a constant. The average transit time for carriers to cross the slab can also be evaluated:

$$\tau \equiv \frac{\int_0^L n(z)\mathrm{d}z}{F} = \frac{L^2}{2D_n} + \frac{L}{\upsilon_T}, \tag{9.24}$$

which has the expected limits for the diffusive and ballistic cases.

Figure 9.7 summarizes the results and shows carrier profiles in the diffusive and ballistic limits. This simple version of scattering theory accurately describes transport as L varies from much less than to much greater than a mean-free-path. Comparisons to solutions of the Boltzmann equation show that the errors

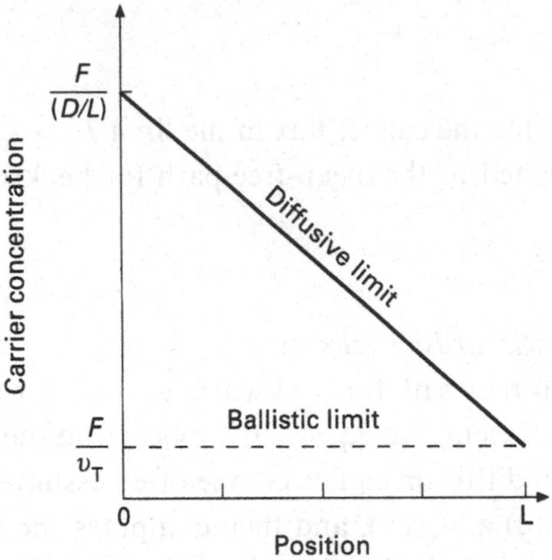

Fig. 9.7 Carrier profiles across a thin slab for a fixed net carrier flux, F, and a perfectly absorbing contact at the right. Both the ballistic and diffusive limits are shown.

are typically only a few percent. It may be surprising, but one can also derive the same equations by solving the diffusion equation if one is careful not to impose physically incorrect boundary conditions (see homework problem 9.4). The reason is that, as we showed in Section 9.2, the drift–diffusion equation is equivalent to the flux equations.

9.4.2 Transport in the presence of an electric field

When an electric field is present, the scattering matrix is not symmetrical and $\xi \neq \xi'$ and $T \neq T'$. One can determine T and T' by integrating eqs. (9.2a) and (9.2b) across a slab as discussed in Appendices A and B of [9.3]. In this section, we make some approximations in order to simplify the mathematics. Beginning with eq. (9.2a) and using current continuity, $F = a - b$, we can generalize eq. (9.12) to

$$\frac{\mathrm{d}a}{\mathrm{d}z} + (\xi - \xi')a = -F\xi', \tag{9.25}$$

which can be solved for

$$a(z) = -\frac{\xi'}{(\xi - \xi')}F + A\mathrm{e}^{-(\xi-\xi')z}, \tag{9.26}$$

where A is an integration constant. If $\mathcal{E} > 0$, then $\xi > \xi'$. According to eq. (9.26),

$$A = a(0) + \frac{\xi'}{(\xi - \xi')}F, \tag{9.27}$$

which, when used in eq. (9.26) gives

$$a(L) = \frac{-\xi'}{(\xi - \xi')} F + \left[a(0) + \frac{\xi'}{(\xi - \xi')} F \right] e^{-(\xi - \xi')L}. \tag{9.28}$$

Finally, using $F = a(L) - b(L)$, we can express eq. (9.28) as

$$a(L) = \left\{ \frac{e^{-(\xi-\xi')L}}{\left[1 + \frac{\xi'}{(\xi-\xi')}\left(1 - e^{-(\xi-\xi')L}\right)\right]} \right\} a(0) + \left\{ \frac{\frac{\xi'}{(\xi-\xi')}\left(1 - e^{-(\xi-\xi')L}\right)}{\left[1 + \frac{\xi'}{(\xi-\xi')}\left(1 - e^{-(\xi-\xi')L}\right)\right]} \right\} b(L). \tag{9.29}$$

The transmission and reflection coefficients for a slab of length L with a nonzero electric field are read directly from eq. (9.29) as

$$T = \frac{(\xi - \xi')}{(\xi e^{(\xi-\xi')L} - \xi')} \tag{9.30a}$$

$$R = 1 - T \tag{9.30b}$$

$$T' = \frac{(\xi' - \xi)}{(\xi' e^{(\xi'-\xi)L} - \xi)} \tag{9.30c}$$

$$R' = 1 - T'. \tag{9.30d}$$

Finally, we need to specify the parameters, ξ and ξ'. The easiest way to do so is to recall that the near-equilibrium flux equations are formally identical to near-equilibrium drift–diffusion equations. According to eqs. (9.8b) and (9.8c),

$$\frac{D_n}{\mu_n} = \frac{\mathcal{E}_z}{\xi - \xi'}. \tag{9.31}$$

If we assume the Einstein relation, we find

$$(\xi - \xi') = \frac{\mathcal{E}_z}{(k_B T_L / q)}. \tag{9.32}$$

Having specified the difference of the backscattering parameters, we now need to specify one of them. There are a couple of plausible ways to do this. One is to assume that for small electric fields,

$$\xi = \xi_0 + \xi_1/2 \tag{9.33a}$$

and

$$\xi' = \xi_0 - \xi_1/2, \tag{9.33b}$$

where $\xi_1 = \xi - \xi'$. Another way to specify the backscattering probabilities, which also works for large electric fields, is to assume that the flux traveling down the potential drop is unaffected by the electric field and all of the effect is felt by the flux traveling against the electric field. So,

$$\xi = \xi_0 \text{ and } \xi' = \frac{|\mathcal{E}_z|}{(k_B T_L/q)} + \xi_0 \quad \text{for } \mathcal{E}_z < 0 \tag{9.34a}$$

and

$$\xi' = \xi_0 \text{ and } \xi = \frac{|\mathcal{E}_z|}{(k_B T_L/q)} + \xi_0 \quad \text{for } \mathcal{E}_z > 0. \tag{9.34b}$$

At this point, we should check to be sure that the final result is consistent with our expectations. For an infinitely long slab, any flux incident on a finite slab imbedded inside it will be the same as the corresponding flux that emerges (see Fig. 9.8). We conclude, therefore, that

$$[S]\begin{pmatrix} a \\ b \end{pmatrix} = \begin{pmatrix} a \\ b \end{pmatrix}. \tag{9.35}$$

The bulk solution corresponds to the eigenvector of the scattering matrix associated with an eigenvalue of one. This eigenvector is

$$\begin{pmatrix} a \\ b \end{pmatrix} = \begin{pmatrix} R'/R \\ 1 \end{pmatrix} b, \tag{9.36}$$

from which we can find the average velocity as

$$\langle v_z \rangle = \frac{F}{n} = \left(\frac{a-b}{a+b}\right) v_T = \left(\frac{R'-R}{R'+R}\right) v_T. \tag{9.37}$$

Using eqs. (9.30) and (9.33) with $L \to \infty$, we find

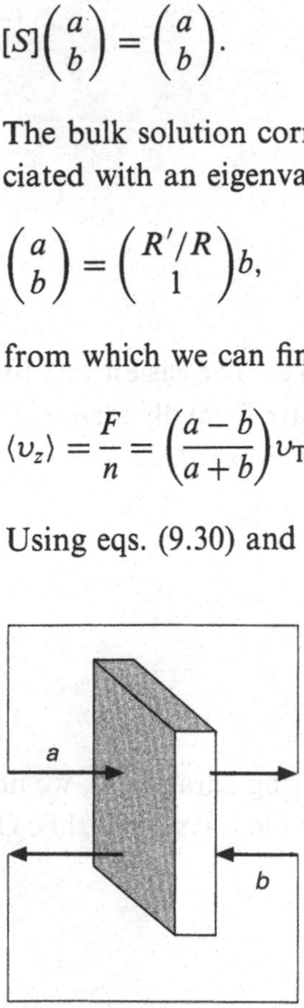

Fig. 9.8 Illustration of the fluxes in an infinite, bulk semiconductor.

$$\langle v_z \rangle = -\frac{\xi_1/\xi_0}{2} v_T. \tag{9.38}$$

Using $\xi_0 = 1/\lambda$ and $\xi_1 = \mathcal{E}_z/(k_B T_L/q)$, we find

$$\langle v_z \rangle = -\frac{\lambda v_R}{(k_B T_L/q)} \mathcal{E}_z = -\frac{D_n}{(k_B T_L/q)} \mathcal{E}_z = -\mu_n \mathcal{E}_z, \tag{9.39}$$

which is exactly what we expected.

9.4.3 Transport over a barrier

Figure 9.9 shows the energy band diagram of a barrier with fluxes injected from the left and right. We can use the methods of the previous section and integrate the flux equations across the barrier. We expect that the transmission coefficient, T, from the left to right will be small and depend exponentially on the height of the barrier while the transmission coefficient for the flux injected from the right, T', will be large and not depend sensitively on the barrier height. We can develop the scattering matrix for a barrier directly from the results of the previous section, but first we develop the results by physical reasoning.

If we assume ballistic transport within the barrier, the scattering matrix is simply

$$[S] = \begin{bmatrix} e^{-\Delta E/k_B T} & 0 \\ (1 - e^{-\Delta E/k_B T}) & 1 \end{bmatrix}. \tag{9.40}$$

Equation (9.40) is based on thermionic emission and works adequately when the net flux is against the barrier. When the carrier flow is predominately down the potential barrier, then the junction is a carrier collector. If T' is 1.0, as assumed

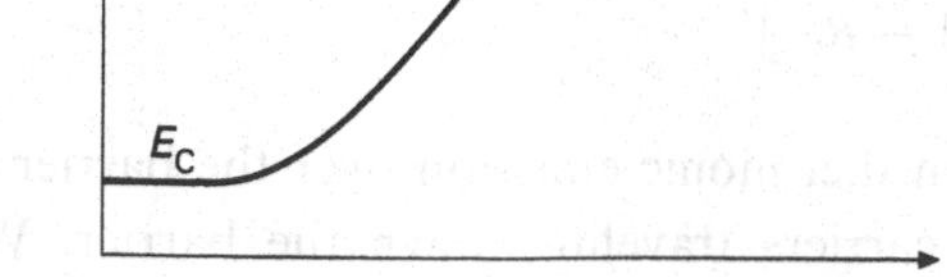

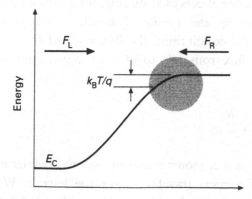

Fig. 9.9 The carrier fluxes for an energy barrier. The critical distance, l, corresponding to a potential drop of $k_B T_L/q$ is also shown.

in eq. (9.40), then the collector is an ideal absorbing collector, but there will be some backscattering from a real collector.

Consider a flux of carriers injected from the top of a barrier of length L, as shown in Fig. 9.9. If there were no scattering, $T' = 1$. If the electric field in the barrier were very small, then we could calculate T' in the presence of scattering using the approach of Section 9.4.2, but the electric field is typically very large in collectors. Recall that when there is no electric field at all, then R' is (Section 9.4.1)

$$R' = \frac{L}{L+\lambda}. \tag{9.41}$$

Monte Carlo simulations show that if a carrier penetrates only a little way into the collector, then even if it backscatters, it has little chance of making it over the barrier and back out, so the length, L, of the collector is not relevant. Since carriers enter with a kinetic energy of about $k_B T_L$, the critical distance is the distance over which the first $k_B T_L/q$ potential drop occurs, as illustrated in Fig. 9.9. We can, therefore, estimate the backscattering coefficient of the collector by

$$R_C \approx \frac{\ell}{\ell+\lambda} \approx \frac{\ell}{l+(\mu_n k_B T/q\upsilon_R)}. \tag{9.42}$$

Note that the appropriate mobility to use is the near-equilibrium mobility (if near-equilibrium carriers are injected into the collector) because the backscattering that contributes to R_C occurs before the carriers have acquired much kinetic energy. So the backscattering coefficient of a collector can be readily estimated from the low-field mobility, even if the electric field within the collector is high and strong off-equilibrium transport effects occur.

To modify eq. (9.40) to include the effects of scattering, we must do more than simply replace the 0 and 1 with R_C and $1 - R_C$. The reason is that detailed balance would not be satisfied (in equilibrium, the flux injected from the left to right should exactly equal the flux from right to left). The appropriate scattering matrix to use is

$$[S] = \begin{bmatrix} (1-R_C)e^{-\Delta E/k_B T} & R_C \\ [1-(1-R_C)e^{-\Delta E/k_B T}] & 1-R_C \end{bmatrix}, \tag{9.43}$$

where the exponential arises from thermionic emission over the barrier and R_C describes the backscattering of carriers traveling down the barrier. We have developed eq. (9.43) by physical reasoning, but we could have obtained it directly from eq. (9.30) using $(\xi - \xi')L = \Delta E/k_B T$.

9.5 One-flux treatment of devices

Having shown how scattering matrices are defined, we can now demonstrate how they are used to describe transport in devices. Only two short examples will be considered, a *p–n* diode and a MOSFET. The objective is not an exhaustive analysis of these devices, but merely to illustrate the approach.

9.5.1 Scattering matrix analysis of a *p-n* diode

Figure 9.10 shows an np^+ diode and how electron transport can be analyzed by dividing it into two scattering matrices. If we describe the barrier by a scattering matrix, $[S_J]$, and the quasi-neutral *p*-type base by a scattering matrix, $[S_B]$, then the *p–n* diode is described by

$$[S_{PN}] = [S_J] \otimes [S_B], \tag{9.44}$$

where $\otimes$ stands for the scattering matrix cascading procedure described by eqs. (9.9). The scattering matrix for the *p–n* diode is found to be

$$[S_{PN}] = \begin{bmatrix} T_{PN} & 1 - T'_{PN} \\ 1 - T_{PN} & T'_{PN} \end{bmatrix}, \tag{9.45a}$$

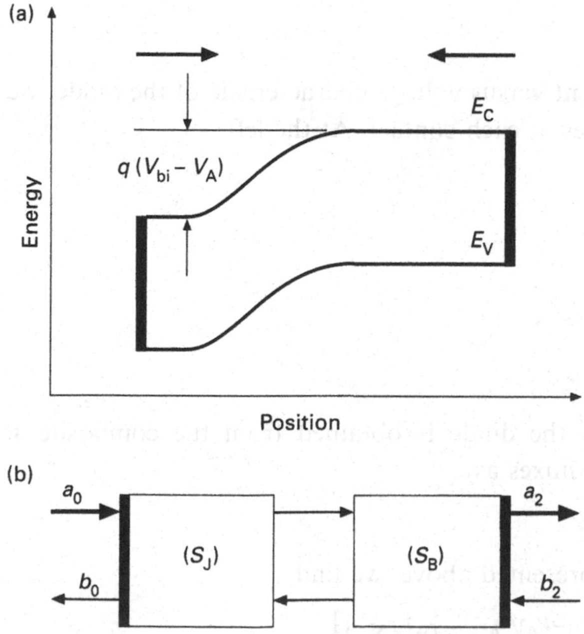

Fig. 9.10 Illustration of a *p–n* diode and how it is described in terms of two interconnected scattering matrices.

where

$$T_{\rm PN} = \frac{T_{\rm J}T_{\rm B}}{(1 - R_{\rm J}R_{\rm B})} \tag{9.45b}$$

and

$$T'_{\rm PN} = \frac{T'_{\rm J}T'_{\rm B}}{(1 - R_{\rm J}R_{\rm B})}. \tag{9.45c}$$

From Section 9.4.1, we have

$$T_{\rm B} = T'_{\rm B} = \frac{\lambda}{\lambda + W_{\rm B}} = \frac{1}{1 + \upsilon_{\rm R}/(D_n/W_{\rm B})}. \tag{9.46}$$

For the barrier, it is simplest to use the thermionic emission scattering matrix of Section 9.4.3,

$$T_{\rm J} = {\rm e}^{-q(V_{\rm bi} - V_{\rm A})/k_{\rm B}T} \tag{9.47a}$$

and

$$T'_{\rm J} = 1. \tag{9.47b}$$

By using these expressions, we find

$$T_{\rm PN} = T_{\rm J}T_{\rm B} = T_{\rm B}{\rm e}^{-q(V_{\rm bi} - V_A)/k_{\rm B}T} \tag{9.48a}$$

and

$$T'_{\rm PN} = T_{\rm B}. \tag{9.48b}$$

To compute the current versus voltage characteristic of the diode, we need to specify the injected fluxes at each contact. At the left,

$$a_0 = \left(\frac{N_{\rm D}}{2}\right)\upsilon_{\rm T} \tag{9.49a}$$

and at the right,

$$b_2 = \left(\frac{n_i^2}{2N_{\rm A}}\right)\upsilon_{\rm T}. \tag{9.49b}$$

The current through the diode is obtained from the composite scattering matrix and the injected fluxes as

$$J_{\rm D} = q(T_{\rm PN}a_0 - T'_{\rm PN}b_2). \tag{9.50}$$

Using the expressions presented above, we find

$$J_{\rm D} = q\left(\frac{\upsilon_{\rm T}}{2}\right)T_{\rm B}\left[N_{\rm D}{\rm e}^{-q(V_{\rm bi} - V_{\rm A})/k_{\rm B}T} - (n_i^2/N_{\rm A})\right], \tag{9.51a}$$

which can be expressed as

$$J_{\mathrm{D}} = q\left(\frac{n_i^2}{N_{\mathrm{A}}}\right)\upsilon_{\mathrm{R}} T_{\mathrm{B}}(\mathrm{e}^{qV_{\mathrm{A}}/k_{\mathrm{B}}T} - 1), \tag{9.51b}$$

or as

$$J_{\mathrm{D}} = q\left(\frac{n_i^2}{N_{\mathrm{A}}}\right)\left[\frac{(D_n/W_{\mathrm{B}})}{1 + \frac{(D_n/W_{\mathrm{B}})}{\upsilon_{\mathrm{R}}}}\right](\mathrm{e}^{qV_{\mathrm{A}}/k_{\mathrm{B}}T} - 1). \tag{9.51c}$$

Equation (9.51c) is the conventional result for the current versus voltage characteristic of a *p–n* diode – except that it is reduced by a factor of $[1 + \frac{(D_n/W_{\mathrm{B}})}{\upsilon_{\mathrm{R}}}]^{-1}$. In the limit that the base is wide, $(D_n/W_{\mathrm{B}}) \ll \upsilon_{\mathrm{R}}$, and eq. (9.51c) reduces to the conventional result. In the limit that the base is thin, $T_{\mathrm{B}} \to 1$, we find

$$J_{\mathrm{D}} = q\left(\frac{n_i^2}{N_{\mathrm{A}}}\right)\upsilon_{\mathrm{R}}(\mathrm{e}^{qV_{\mathrm{A}}/k_{\mathrm{B}}T} - 1), \tag{9.52}$$

which gives the current in a *p–n* diode with a ballistic quasi-neutral base.

The scattering analysis of the *p–n* diode produces simple results that apply over a wider range than the conventional treatment. The conventional approach does not treat transport across the barrier itself but, rather, imposes a boundary condition,

$$\Delta n(0) = n(0) - n_o(0) = \left(\frac{n_i^2}{N_{\mathrm{A}}}\right)(\mathrm{e}^{qV_{\mathrm{A}}/k_{\mathrm{B}}T} - 1) \tag{9.53}$$

at the beginning of the quasi-neutral base. We can check the validity of eq. (9.53), which is known as the *Law of the Junction*, by using our scattering theory results. The carrier density at the beginning of the base is, therefore,

$$n(0) = \frac{(T_{\mathrm{J}}a_0 + T_{\mathrm{J}}a_0R_{\mathrm{B}} + T_{\mathrm{B}}b_2)}{\upsilon_{\mathrm{T}}} = T_{\mathrm{J}}(1 + R_{\mathrm{B}})\frac{a_0}{\upsilon_{\mathrm{T}}} + (1 - R_{\mathrm{B}})\frac{b_2}{\upsilon_{\mathrm{T}}}. \tag{9.54a}$$

Using the expression for T_{J} that we developed earlier, eq. (9.54a) becomes

$$\Delta n(0) = \left(\frac{n_i^2}{N_{\mathrm{A}}}\right)(\mathrm{e}^{qV_{\mathrm{A}}/k_{\mathrm{B}}T} - 1) \times \left[\frac{1 + R_{\mathrm{B}}}{2}\right], \tag{9.54b}$$

which is the Law of the Junction, eq. (9.53), multiplied by a correction factor. For a thick base, $R_{\mathrm{B}} \to 1$, and eq. (9.54b) approaches the Law of the Junction, but for a thin base, $R_{\mathrm{B}} \to 0$, and eq. (9.54b) is a factor of two smaller than the value given by the Law of the Junction.

The *p–n* diode example shows how scattering theory is applied to devices. For a more complete treatment of this particular problem, including the effects of scattering in the barrier, see Tanaka and Lundstrom [9.5].

9.5.2 Scattering matrix analysis of a MOSFET

For low drain biases, a MOSFET acts as a gate voltage-controlled resistor with the drain current given by

$$I_{\mathrm{Dlin}} = \mu_{\mathrm{eff}} C_{\mathrm{ox}}(V_{\mathrm{GS}} - V_{\mathrm{T}})\left(\frac{W}{L}\right)V_{\mathrm{DS}} \qquad V_{\mathrm{DS}} \ll V_{\mathrm{Dsat}}, \tag{9.55a}$$

while for large drain bias it acts as a current source with the drain current given by

$$I_{\mathrm{Dsat}} = WC_{\mathrm{ox}}\langle \upsilon(0)\rangle(V_{\mathrm{GS}} - V_{\mathrm{T}}) \qquad V_{\mathrm{DS}} \geq V_{\mathrm{Dsat}}, \tag{9.55b}$$

where $\langle \upsilon(0)\rangle$ is the average velocity at the source end of the channel. As MOSFET channel lengths approach zero, questions about the limiting performance of the device become important. According to eq. (9.55a), the channel resistance is proportional to L, so it should approach zero as $L \to 0$. It is not clear what value $\langle \upsilon(0)\rangle$ will approach. For some time, it was thought that because short channels lead to high electric fields, the saturation velocity would be the limit, but it is now understood that velocity overshoot occurs in short, high-field regions, which has led to the belief that $\langle \upsilon(0)\rangle$ can exceed υ_{sat}. Questions like these are readily addressed by a simple scattering theory of the MOSFET.

Figure 9.11 shows the energy band diagram of a metal–oxide–semiconductor field-effect transistor (MOSFET). Since electrons are injected out of the source and drain and into the channel, we could describe the MOSFET by three interconnected scattering matrices. However, we can simplify the problem by making an assumption similar to the Law of the Junction for the diode. A quasi-equilibrium assumption that the electron concentrations at the drain and source ends are set by the gate voltage as they are in equilibrium results in

$$qn_s(0) = C_{\mathrm{ox}}(V_{\mathrm{GS}} - V_{\mathrm{T}}) \qquad (V_{\mathrm{GS}} > V_{\mathrm{T}}) \tag{9.56a}$$

and

$$qn_{\mathrm{S}}(L) = C_{\mathrm{ox}}(V_{\mathrm{GS}} - V_{\mathrm{DS}} - V_{\mathrm{T}}) \qquad (V_{\mathrm{GS}} > V_{\mathrm{DS}} + V_{\mathrm{T}}) \tag{9.56b}$$

where V_{T} is the threshold voltage. (It should be understood that n_{S} is zero if the terms in the parentheses in eqns. (9.56) are less than zero.) The use of these quasi-equilibrium assumptions in the channel of a MOSFET where the conditions may be far from equilibrium is known as the *Gradual Channel Approximation*. In

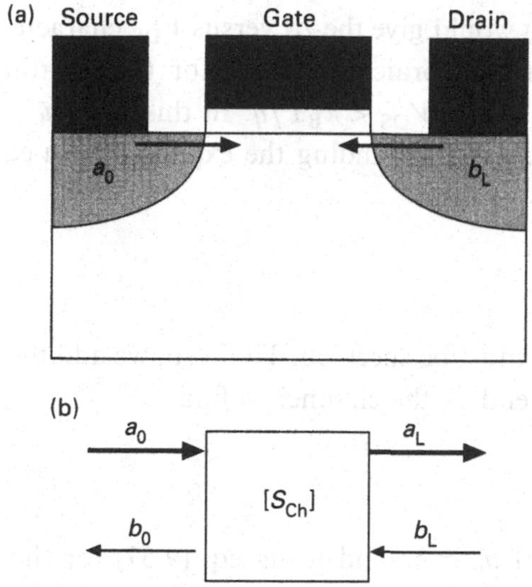

Fig. 9.11 Illustration of a MOSFET (a) and how it is described in terms of a channel scattering matrix (b).

contrast to the first example, we don't view the injected fluxes as being fixed, rather the MOS electrostatics fixes the carrier densities.

Before we can evaluate the I_{DS} versus V_{DS} characteristic, the channel reflection coefficient must be defined. The channel is a collector from source to drain and a barrier from drain to source, so the channel reflection coefficient can be obtained from eq. (9.43) as

$$[S_{\mathrm{Ch}}] = \begin{bmatrix} (1-R_{\mathrm{C}}) & [1-(1-R_{\mathrm{C}})\mathrm{e}^{-qV_{\mathrm{DS}}/k_{\mathrm{B}}T}] \\ R_{\mathrm{C}} & (1-R_{\mathrm{C}})\mathrm{e}^{-qV_{\mathrm{DS}}/k_{\mathrm{B}}T} \end{bmatrix}, \tag{9.57}$$

where R_{C} is given by eq. (9.42) or by eq. (9.41).

Having defined the scattering matrix, we can write the drain current as

$$I_{\mathrm{DS}} = q(a_0 T_{\mathrm{Ch}} - b_{\mathrm{L}} T'_{\mathrm{Ch}})W, \tag{9.58}$$

where

$$T_{\mathrm{Ch}} = 1 - R_{\mathrm{C}}$$

and

$$T'_{\mathrm{Ch}} = (1-R_{\mathrm{C}})\mathrm{e}^{-qV_{\mathrm{DS}}/k_{\mathrm{B}}T}.$$

With the use of these results, eq. (9.58) becomes

$$I_{\mathrm{D}} = qW(1-R_{\mathrm{C}})[a_0 - b_{\mathrm{L}}\mathrm{e}^{-qV_{\mathrm{DS}}/k_{\mathrm{B}}T}]. \tag{9.59}$$

A general solution to eq. (9.59) would give the I_D versus V_{DS} characteristic for the MOSFET. The solutions are algebraically simple for two limiting cases. Consider first the ohmic region, where $V_{DS} < k_B T/q$. In this case, $a_0 \approx b_L$ and $R_C \approx R_{C0}$ as given by eq. (9.41). After expanding the exponential in eq. (9.59), we find

$$I_{\text{Dlin}} = qW(1 - R_{C0})b_L\left(\frac{qV_{DS}}{k_B T}\right). \tag{9.60}$$

To proceed, the injected flux, b_L, must be specified. To do so, we add the positive and negative fluxes at the drain end of the channel to find

$$n_S(L) = \frac{b_L(1 + R'_{Ch}) + T_{Ch}a_0}{\upsilon_T}. \tag{9.61a}$$

With the low bias assumption of $a_0 \approx b_L$ and using eq. (9.57) for the channel scattering matrix, eq. (9.61a) becomes

$$b_L = n_S(L)\upsilon_R = \frac{C_{ox}}{q}(V_{GS} - V_T)\upsilon_R. \tag{9.61b}$$

From eq. (9.41), the backscattering coefficient at the beginning of the channel is

$$1 - R_{C0} = \frac{\lambda}{L + \lambda}.$$

Using these results in eq. (9.60), we find

$$I_{\text{Dlin}} = \frac{WC_{ox}}{L + \lambda}\left(\frac{\lambda \upsilon_R}{k_B T/q}\right)(V_{GS} - V_T)V_{DS}$$

or

$$I_{\text{Dlin}} = \mu_{\text{eff}} C_{ox}(V_{GS} - V_T)\frac{W}{L + \lambda}V_{DS} \qquad V_{DS} < k_B T/q. \tag{9.62}$$

Equation (9.62) is a generalization of the standard result, eq. (9.55a). It shows that as the channel length approaches zero, the current approaches a finite value. Alternatively, it says that the drain-to-source resistance has a fundamental lower limit. This resistance is analogous to the fundamental quantum contact resistance of $h/2q^2$ that is observed in mesoscopic devices at low temperatures (see Section 9.9). The effect is important when the channel length is on the order of a mean-free-path for scattering.

The second case of interest is for $V_{DS} \gg k_B T/q$ for which eq. (9.59) simplifies to

$$I_D \approx qWT_{Ch}a_0. \tag{9.63}$$

Since the flux injected at the drain end of the channel cannot transmit across when the drain voltage is high,

$$n_S(0) = \frac{(a_0 + R_C a_0)}{\upsilon_T} \tag{9.64a}$$

from which we find

$$a_0 = \frac{C_{ox}(V_{GS} - V_T)\upsilon_T}{q(1 + R_C)}. \tag{9.64b}$$

By using eq. (9.64b) in eq. (9.63), we finally obtain

$$I_D \approx WC_{ox}\langle\upsilon(0)\rangle(V_{GS} - V_T) \qquad V_{DS} \gg k_B T/q \tag{9.65a}$$

$$\langle\upsilon(0)\rangle = \left(\frac{1 - R_C}{1 + R_C}\right)\upsilon_T. \tag{9.65b}$$

There is a difficulty in using eqs. (9.65) because the channel reflection coefficient, R_C, depends on the potential drop within the channel as given by eq. (9.42). To estimate the potential drop, a self-consistent treatment is required. Nevertheless, eq. (9.65a) does clearly specify the limiting current through the device; it occurs when R_C approaches zero and the channel becomes ballistic. In this limit, $\langle\upsilon(0)\rangle \to \upsilon_T$. The thermal injection velocity sets the upper limit for the drain current; its value may exceed υ_{sat}. Velocity overshoot may occur within the channel, but it does not directly control the drain current. It can, however, influence the self-consistent electric field in the channel and therefore affect the drain current by changing the critical distance, ℓ in eq. (9.42).

9.6 Discussion: semiclassical transport

The version of scattering theory that we have described deals with the average fluxes traveling in the positive and negative directions and with average mean-free-paths for scattering. Equations (9.2) were written down by simple, physical reasoning, so the question of how scattering theory is related to the Boltzmann equation arises. Equations (9.2) can be derived from the Boltzmann equation by taking moments, but instead of summing over all momentum states, as in Chapter 5, we only sum over the positive or negative states. For example,

$$a(z) = \frac{1}{\Omega}\sum_{\mathbf{p}, p_z > 0} \upsilon_z f(z, \mathbf{p}). \tag{9.66a}$$

Alternatively, eqs. (9.2) can be obtained from an integral solution to the Boltzmann equation as discussed in Alam et al. [9.7].

Instead of deriving an equation for the average carrier flux from the Boltzmann equation, we could use scattering theory to solve the Boltzmann equation. Instead of writing the average flux as a scalar, we discretize momentum space into M orthogonal basis functions (the simplest being cubic bins representing a volume $dp_x dp_y dp_z$ of momentum space) and write the flux as an $M \times 1$ vector,

$$|a(z)\rangle = \begin{bmatrix} a_1(z) \\ a_2(z) \\ \vdots \\ a_M(z) \end{bmatrix}, \tag{9.66b}$$

where each element represents the flux carrier by a discrete 'mode' in momentum space. The vector, $|a(z)\rangle$ is a discrete representation of the flux distribution in momentum space. The generalization of one-flux scattering theory to a rigorous treatment of semiclassical transport leads directly to the Boltzmann transport equation.

To pursue these ideas further would carry us too far astray. Our objective here has been merely to convey the spirit of scattering theory and to show that even in a simple form it presents a useful way to think about transport in small devices. Those interested in exploring the connections to the Boltzmann equation should consult Alam et al. [9.7].

9.7 The tunneling approach to quantum transport

We now consider a different class of devices, those for which transport is controlled by quantum mechanics rather than by Newton's laws and scattering. Quantum transport can still be analyzed in terms of transmission coefficients. Our approach will follow that of Datta quite closely. The reader is referred to Datta [9.1, 9.2] for additional details.

Figure 9.12 is an illustration of a 'quantum device'. For such devices, the transmission coefficients depend strongly on the energy of the injected carrier. (It is easier for high energy carriers to tunnel through the barrier, and for the double barrier structure shown in Fig. 9.12, resonances occur which produce very high transmission at critical energies.) We should, therefore, generalize eq. (9.1c) to resolve the incident fluxes and transmission coefficients in energy (or momentum) space. To begin, we assume spatial variation in only one direction and write the flux from the left contact as

$$F_{\mathrm{L}} = \frac{1}{\Omega}\sum_{\mathbf{p}} v_z f_{\mathrm{L}}(E - E_{\mathrm{FL}}) = \frac{1}{A}\sum_{p_t}\frac{1}{L}\sum_{p_z} v_z f_{\mathrm{L}}(E - E_{\mathrm{FL}}), \tag{9.67}$$

where f_{L} is Fermi function in the left contact, and E_{FL} is the Fermi level there. The sum over momentum states has been separated into a sum over the transverse momentum states, p_t, (in the x–y plane) and the longitudinal states, p_z. The

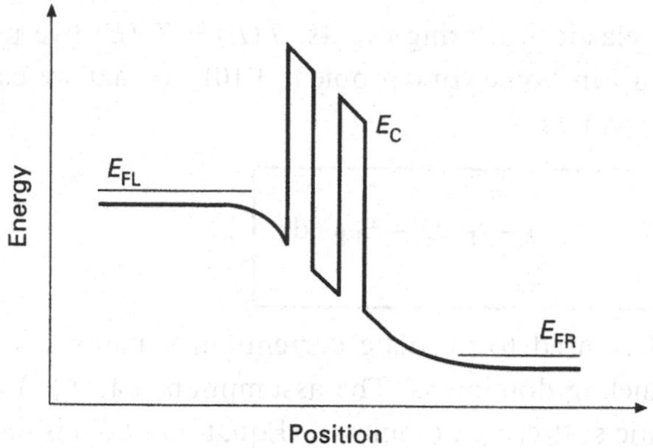

Fig. 9.12 A general picture of a device for which quantum transport dominates. The region within this device requires a quantum mechanical evaluation of the transmission coefficient. The particular device shown here is a double barrier device which shows resonance at critical energies.

sum over p_z can be converted to an integral over energy (see homework problem 9.6) to write eq. (9.67) as

$$F_L = \frac{2}{h}\frac{1}{A}\sum_n \int f_L(E - E_{FL})dE, \tag{9.68}$$

where the transverse momentum states have been labeled by the index, n. To define the current that transmits from the left contact to the right, we define $\tau_{mn}(E)$ as the fraction of electrons incident from transverse momentum state n in the left contact that transmit across to transverse momentum state m in the right contact. We can then write the electric current as

$$I_{L\to R} = \frac{-2q}{h}\sum_{n,m}\int \tau_{mn}(E) f_L(E - E_{FL})dE = \frac{-2q}{h}\sum_n \int T_n(E) f_L(E - E_{FL})dE, \tag{9.69}$$

where

$$T_n(E) \equiv \sum_m \tau_{mn}(E) \tag{9.70}$$

is the probability that an electron injected in transverse momentum state, n, from the left contact transmits across the device into any of the m transverse states in the right contact. Following the same procedure for the current injected from the right contact, we obtain the net current as

$$I = (I_{L\to R} - I_{R\to L}) = \frac{-2q}{h}\sum_n \int [T_n(E) f_L(E - E_{FL}) - T'_n(E) f_R(E - E_{FR})]dE. \tag{9.71}$$

Finally, we note that for elastic scattering events, $T(E) = T'(E)$ (we proved a similar property for $S(\mathbf{p}, \mathbf{p}')$ in homework problem 3.10), so that we can write our final result for the current as

$$\boxed{I = \frac{-2q}{h} \sum_n \int T_n(E)[f_L(E - E_{FL}) - f_R(E - E_{FR})]\mathrm{d}E} \quad . \tag{9.72}$$

Equation (9.72) is widely used to evaluate currents in situations for which quantum mechanical tunneling dominates. The assumption that $T(E) = T'(E)$ can be justified when elastic scattering dominates. (Equation (9.71) is more general, but the evaluation of T and T' in the presence of quantum mechanical tunneling and inelastic scattering is quite difficult.) When the cross-sectional area in the x–y plane is large, then the sum over transverse momentum states can be converted to an integral and evaluated analytically (see homework problem 9.7). For structures with small lateral dimensions that we will discuss in Section 9.9, there are few transverse modes, so the contributions of each can be added. Finally, one may ask why there is no $[1 - f_R]$ factor in the first term of eq. (9.72) or a $[1 - f_L]$ factor in the second term. It would appear that these filling factors should be present to account for the fact that the final state to which the electron is being injected may be filled. It is easy to show, however, that these factors cancel out, when the scattering is elastic, as eq. (9.72) assumes (see homework problem 9.8). More generally, however, the question arises as to whether they should appear in eq. (9.71). It turns out that they should not appear when T and T' are appreciable because in this case, the states on the left and right of the device are not isolated, but rather a single *scattering state* that includes incident, transmitted, and reflected components. See Datta [9.1] for a discussion of this subtle point.

According to eqs. (9.71) and (9.72), devices controlled by quantum, mechanical transport can be analyzed in much the same manner as classical devices, we just need to calculate the transmission coefficients properly.

9.8 Calculating transmission coefficients

Figure 9.13 shows a simple potential energy step. Classical particles would transmit across with a probability of unity if their energy were higher than the step and would not transmit at all if their energy were lower. From a quantum mechanical perspective, we expect reflections to occur and to modify the result. To calculate the transmission coefficient, we begin with the time-independent wave equation,

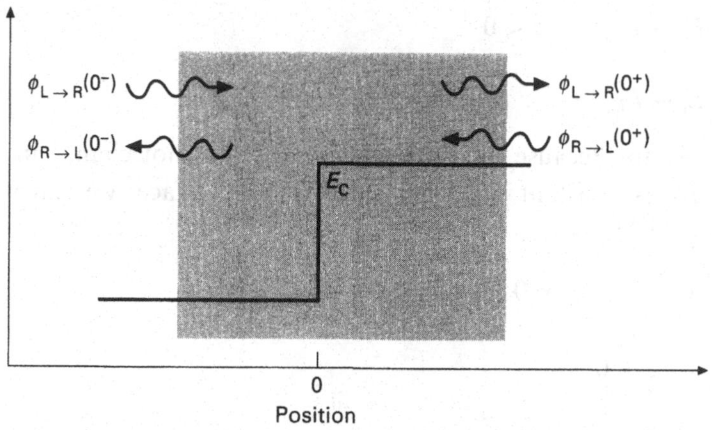

Fig. 9.13 Illustration of a potential step and the incident and emerging waves used to define a scattering matrix.

$$-\left(\frac{\hbar^2}{2m^*}\right)\nabla^2\psi(\mathbf{r}) + E_{C0}(\mathbf{r})\psi(\mathbf{r}) = E\psi(\mathbf{r}), \tag{9.73}$$

where $E_{C0}(\mathbf{r})$ is the bottom of the conduction band as determined by the self-consistent electrostatic potential and any compositional variations if the device is a heterostructure. If we assume spatial variations in the z-direction alone, then writing the solution as $\psi(\mathbf{r}) = \chi(x, y)\phi(z)$ leads to an equation for $\phi(z)$,

$$-\left(\frac{\hbar^2}{2m^*}\right)\frac{d^2\phi(z)}{dz^2} + E_{C0}(z)\phi(z) = (E - E_t)\phi(z), \tag{9.74}$$

where E_t represent the transverse energy (see homework problem 9.9). Equation (9.74) can be written more compactly as

$$\frac{d^2\phi(z)}{dz^2} + \frac{2m^*}{\hbar^2}\varepsilon(z)\phi(z) = 0, \tag{9.75}$$

where

$$\varepsilon(z) = E - E_t - E_{C0}(z). \tag{9.76}$$

In solving problems with the wave equation, note that E, the electron's energy, is a constant, because we assume that no inelastic scattering events take place. Because we are solving a wave equation, interference effects can occur. (Phonon scattering would impart random phase changes to the electron, which would tend to wash out interference effects.)

Consider now our potential step problem. The conduction band is piecewise constant, so we have

$$\varepsilon = \varepsilon_1 = E - E_t - E_{C1} \qquad z < 0 \tag{9.77a}$$

$$\varepsilon = \varepsilon_2 = E - E_t - E_{C2}, \qquad z > 0 \tag{9.77b}$$

where E_t is constant because the potential energy does not change in the x–y plane. Because ε is constant on either side of the interface, we can write the solutions to eq. (9.75) as

$$\phi(z) = e^{ik_1 z} + re^{-ik_1 z} \qquad z < 0 \tag{9.78a}$$

$$\phi(z) = te^{ik_2 z}, \qquad z > 0 \tag{9.78b}$$

where

$$k_1 = \sqrt{2m^* \varepsilon_1}/\hbar \tag{9.79a}$$

$$k_2 = \sqrt{2m^* \varepsilon_2}/\hbar. \tag{9.79b}$$

To solve for the amplitudes, r and t, we impose the two boundary conditions that ϕ and $d\phi/dz$ are continuous at the interface to find

$$1 + r = t \tag{9.80a}$$

$$k_1(1 - r) = tk_2, \tag{9.80b}$$

which can be solved for

$$r = \frac{k_1 - k_2}{k_1 + k_2} \tag{9.81a}$$

$$t = \frac{2k_1}{k_1 + k_2}. \tag{9.81b}$$

The procedure for evaluating r and t can be used to evaluate r' and t' in the reverse problem; a wave traveling in the $-z$-direction and that is incident from the right. The solution is the same, except that k_1 and k_2 are interchanged,

$$r' = \frac{k_2 - k_1}{k_1 + k_2} \tag{9.82a}$$

$$t' = \frac{2k_2}{k_1 + k_2}. \tag{9.82b}$$

The results for this problem can be summarized compactly in terms of a scattering matrix (Fig. 9.13). If we have incident waves from the left, $\phi_{L\to R}(0^-)$ and from the right, $\phi_{R\to L}(0^+)$, then we can relate the emerging waves to the incident waves by a scattering matrix and write the result as

$$\begin{bmatrix} \phi_{L\to R}(0^+) \\ \phi_{R\to L}(0^-) \end{bmatrix} = \begin{bmatrix} t & r' \\ r & t' \end{bmatrix} \begin{bmatrix} \phi_{L\to R}(0^-) \\ \phi_{R\to L}(0^+) \end{bmatrix}. \tag{9.83}$$

Note that the elements of this scattering matrix related the amplitudes of the incident and emerging waves; in general, they are complex numbers, and $t + r$ need not sum to unity.

To relate the fluxes on the two sides of the barrier, recall the definition of the quantum mechanical probability current,

$$J = \frac{\hbar}{2m^* i}[\phi^*(\nabla\phi) - \phi(\nabla\phi^*)]. \tag{9.84}$$

Using the wavefunction, eq. (9.78), we find

$$J = \frac{\hbar k_1}{m^*}(1 - |r|^2) \equiv J_{inc} - J_{refl} \qquad z < 0 \tag{9.85a}$$

$$J = \frac{\hbar k_2}{m^*}|t|^2 \equiv J_{trans} \qquad z > 0. \tag{9.85b}$$

Current reflection and transmission coefficients can be defined as

$$R \equiv J_{refl}/J_{inc} = |r|^2 \tag{9.86a}$$

and

$$T \equiv J_{trans}/J_{inc} = |t|^2(k_2/k_1). \tag{9.86b}$$

By the same techniques, we can show that $R' = R$ and $T' = T$. The final result is that we can relate the emerging currents to the incident currents by another scattering matrix,

$$\begin{bmatrix} J_{L\to R}(0^+) \\ J_{R\to L}(0^-) \end{bmatrix} = \begin{bmatrix} T & R' \\ R & T' \end{bmatrix} \begin{bmatrix} J_{L\to R}(0^-) \\ J_{R\to L}(0^+) \end{bmatrix}. \tag{9.87}$$

When relating currents, the elements of the scattering matrix are real numbers, and $T + R = 1$ expresses current conservation. This scattering matrix is much like those we computed for semiclassical transport, but T and T' are evaluated by solving the wave equation and ignoring inelastic, phase randomizing scattering. Figure 9.14 is a sketch of T from eq. (9.86) for the potential step. Below the top of the energy step, $T = 0$. When the energy is just above the step, T begins to increase, but quantum mechanical reflections are strong. As the energy increases, T approaches unity as for the classical calculation.

9.8.1 Cascading amplitude scattering matrices

We consider now the scattering matrix for the more complex structure sketched in Fig. 9.15. For this structure, we expect resonances because the reflections

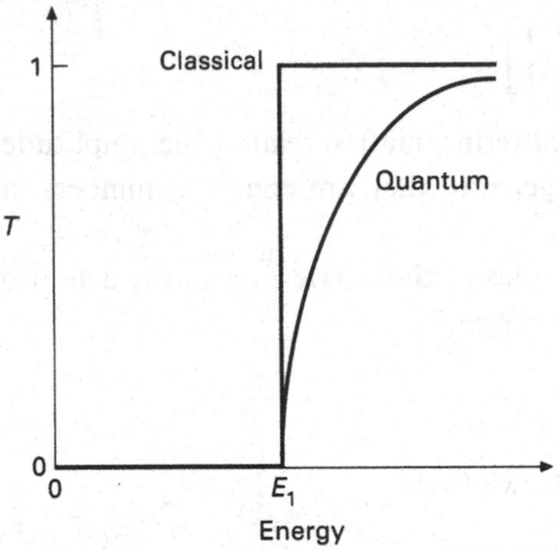

Fig. 9.14 The expected transmission coefficient versus energy for the potential step of Fig. 9.13. Classical and quantum mechanical treatments are shown. (The height of the potential energy step is E_1.)

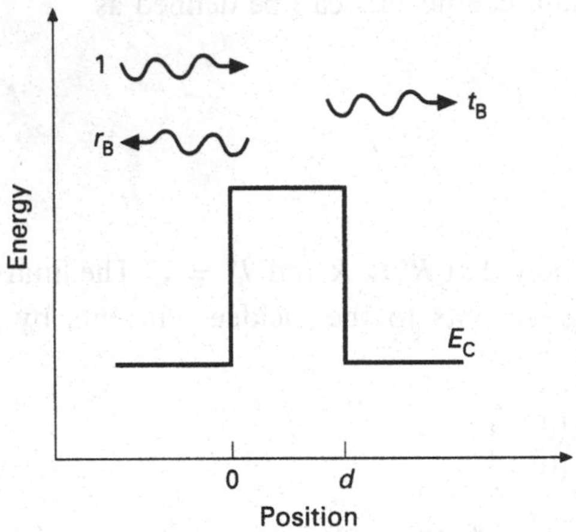

Fig. 9.15 A barrier with an electron with energy above the barrier incident.

occurring at $z = 0$ and at $z = d$ will interfere. We already know the amplitude scattering matrices for the potential step at $z = 0$, and a similar matrix describes the step at $z = d$. It's easy to show that for the region in between where the potential energy is constant,

$$\begin{bmatrix} \phi_{L\to R}(d^-) \\ \phi_{R\to L}(0^+) \end{bmatrix} = \begin{bmatrix} e^{ik_2 d} & 0 \\ 0 & e^{-ik_2 d} \end{bmatrix} \begin{bmatrix} \phi_{L\to R}(0^+) \\ \phi_{R\to L}(d^-) \end{bmatrix}. \tag{9.88}$$

We can combine these three scattering matrices using the cascading rules given in eqs. (9.9); the result is [9.2]

$$\begin{bmatrix} \phi_{\mathrm{L}\to\mathrm{R}}(d^+) \\ \phi_{\mathrm{R}\to\mathrm{L}}(0^-) \end{bmatrix} = \begin{bmatrix} t_{\mathrm{B}} & r'_{\mathrm{B}} \\ r_{\mathrm{B}} & t'_{\mathrm{B}} \end{bmatrix} \begin{bmatrix} \phi_{\mathrm{L}\to\mathrm{R}}(0^-) \\ \phi_{\mathrm{R}\to\mathrm{L}}(d^+) \end{bmatrix}, \tag{9.89}$$

where

$$t_{\mathrm{B}} = t'_{\mathrm{B}} = \frac{tt'}{1 - r'^2 \mathrm{e}^{\mathrm{i}k_2 d}} \tag{9.90a}$$

and

$$r_{\mathrm{B}} = r'_{\mathrm{B}} = \frac{r(1 - \mathrm{e}^{\mathrm{i}2k_2 L})}{1 - r^2 \mathrm{e}^{\mathrm{i}2k_2 L}}. \tag{9.90b}$$

The current transmission coefficients for the structure can also be evaluated to find

$$T_{\mathrm{B}} = |t_{\mathrm{B}}|^2 = \frac{T^2}{1 + R^2 - 2R\cos 2k_2 d} \tag{9.91a}$$

and

$$R_{\mathrm{B}} = |r_{\mathrm{B}}|^2 = \frac{2R - 2R\cos 2k_2 d}{1 + R^2 - 2R\cos 2k_2 d}. \tag{9.91b}$$

As expected by current conservation, $T_{\mathrm{B}} + R_{\mathrm{B}} = 1$.

This example shows the resonances that can occur when quantum mechanical interference effects are not washed out by phase randomizing scattering. According to eq. (9.91a), $T_{\mathrm{B}} = 1$ when $2k_2 d = n(2\pi)$, or when d is a multiple of $\lambda_{\mathrm{B}}/2$ where λ_{B} is the wavelength of electrons in the middle region. When this occurs, reflections at $z = 0$ and at $z = d$ add in phase. To evaluate these interference effects, we need to cascade *amplitude* scattering matrices then determine the current transmission matrices from the composite scattering matrix. If we cascade the *current* scattering matrices, then all phase information is lost and no interference effects occur. *This is the essential difference between quantum and classical transport. In quantum transport we multiply transmission amplitudes: in classical transport we multiply transmission probabilities.*

Interference effects in these kinds of structures tend to be weak at room temperature for two reasons. One is that phonon scattering imparts random phases to the electron waves, which washes out interference patterns. Another is that a thermal distribution of electrons is incident on the barrier. Each electron in the distribution has a different wavelength, and the overall effect is the sum of many interference patterns with different wavelengths. If the measurements are done at low temperatures and biases, however, interference effects are readily observed.

9.8.2 Tunneling

In the previous example, we assumed that the electron energy was above the barrier; it is also interesting to see what happens when the electron energy is below the barrier as illustrated in Fig. 9.16. We still have

$$k_1 = \sqrt{\frac{2m^*(E - E_{C1})}{\hbar^2}} \tag{9.92}$$

but now k_2, which will be imaginary, is written as

$$k_2 = i\gamma = i\sqrt{\frac{2m^*(E_{C2} - E)}{\hbar^2}}. \tag{9.93}$$

The scattering matrix for the barrier is found by cascading three scattering matrices as before to find [9.2]

$$t_B = \frac{i4k_1^2\gamma^2}{(k_1 + i\gamma)^2 - (k_1 - i\gamma)^2 e^{-2\gamma d}} \tag{9.94a}$$

and

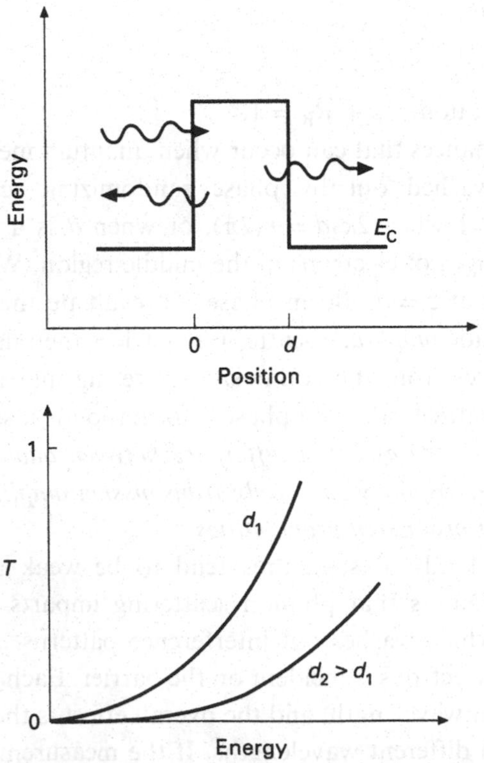

Fig. 9.16 (a) A barrier with an incident electron with energy below the barrier incident. (b) The expected transmission coefficient versus energy.

$$r_B = \frac{(k_1^2+\gamma^2)(1-e^{-2\gamma d})}{(k_1+i\gamma)^2-(k_1-i\gamma)^2 e^{-2\gamma d}}. \tag{9.94b}$$

The current transmission coefficient for this barrier can be written as

$$T_B = \frac{1}{1+\dfrac{(E_{C2}-E_{C1})^2\sinh^2\gamma d}{4(E-E_{C1})(E_{C2}-E)}}. \tag{9.95}$$

If the energy is well below the top of the barrier, then $\gamma d \gg 1$, and eq. (9.95) can be simplified as

$$T_B = \frac{4(E-E_{C1})(E_{C2}-E)}{(E_{C2}-E_{C1})^2} e^{-2\gamma d}. \tag{9.96}$$

Equation (9.96) is an interesting result because classically we would expect that if the electron's energy were below the barrier, it could not transmit across. Quantum mechanically, there is some probability that the electron can *tunnel* through the barrier. The tunneling probability depends exponentially on the thickness and height of the barrier.

9.8.3 Resonant tunneling

Finally, let's consider one more problem. Figure 9.17 shows two tunneling barriers in series separated by a distance L. Equations (9.94) give the amplitude scattering matrices for the two identical tunneling barriers, and there is a uniform region described by an amplitude scattering matrix like eq. (9.88) in between. By cascading the three scattering matrices, we can show that [9.2]

$$t_{2B} = \frac{t_B^2 e^{ik_1 L}}{1-r_B^2 e^{i2k_1 L}}. \tag{9.97}$$

The corresponding current transmission coefficient for the double barrier can be written as [9.2]

$$T_{2B} = |t_{2B}|^2 = \left[1+\frac{4R_B}{T_B^2}\sin^2(k_1 L-\theta)\right]^{-1} \tag{9.98}$$

where $r_B^2 = R_B e^{-2i\theta}$.

Figure 9.17b compares the current transmission coefficients, T_B and T_{2B} for the single and double barrier structures. The striking feature is the strong resonances and unity transmission that occur at certain critical energies. An electron that may have a small probability of tunneling through a single barrier can go through two barriers in series with no attenuation at all. This *resonant tunneling*

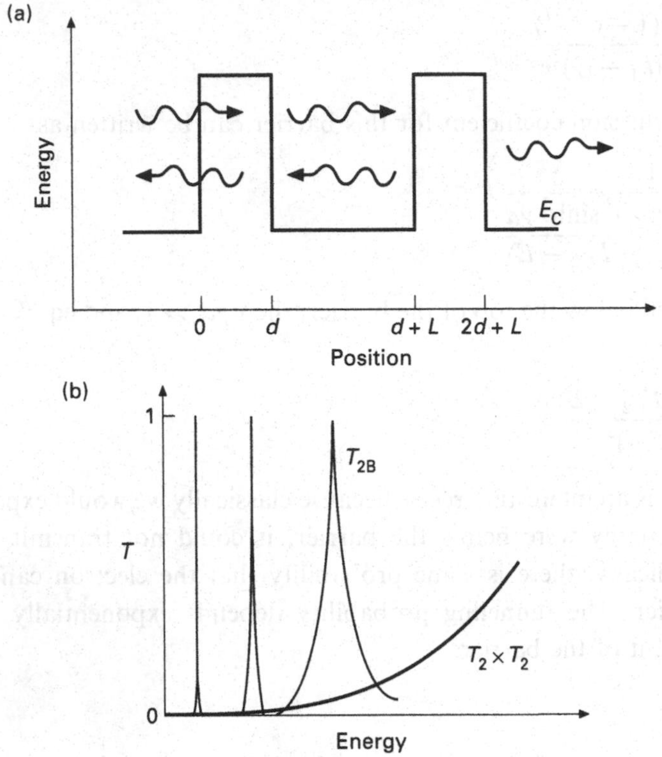

Fig. 9.17 A double-barrier structure with an electron with energy above the barrier incident. (a) The structure and (b) the current transmission coefficient versus energy.

is a consequence of quantum mechanical interference and would have been completely missed had we cascaded the current transmission matrices for the three regions.

9.8.4 Resonant tunneling diodes

Resonant tunneling is a phenomenon that can be used to realize useful electronic devices. Figure 9.18a shows a simplified energy band diagram for a resonant tunneling diode under four different bias conditions. Figure 9.18b shows the corresponding current versus voltage characteristic. Under zero bias, the currents carried by electrons injected from the left contact cancel with those injected from the right contact, and no current flows. The resonant energies for transmission through the double barrier correspond to the energy of the quasi-bound states in the quantum well between the two barriers. (These are referred to as *quasi-bound states* because the barriers are thin enough that electrons in the well eventually tunnel out.) A bias on the anode lowers the energy of electrons in the anode so that there are fewer available to tunnel through the double barrier. A current,

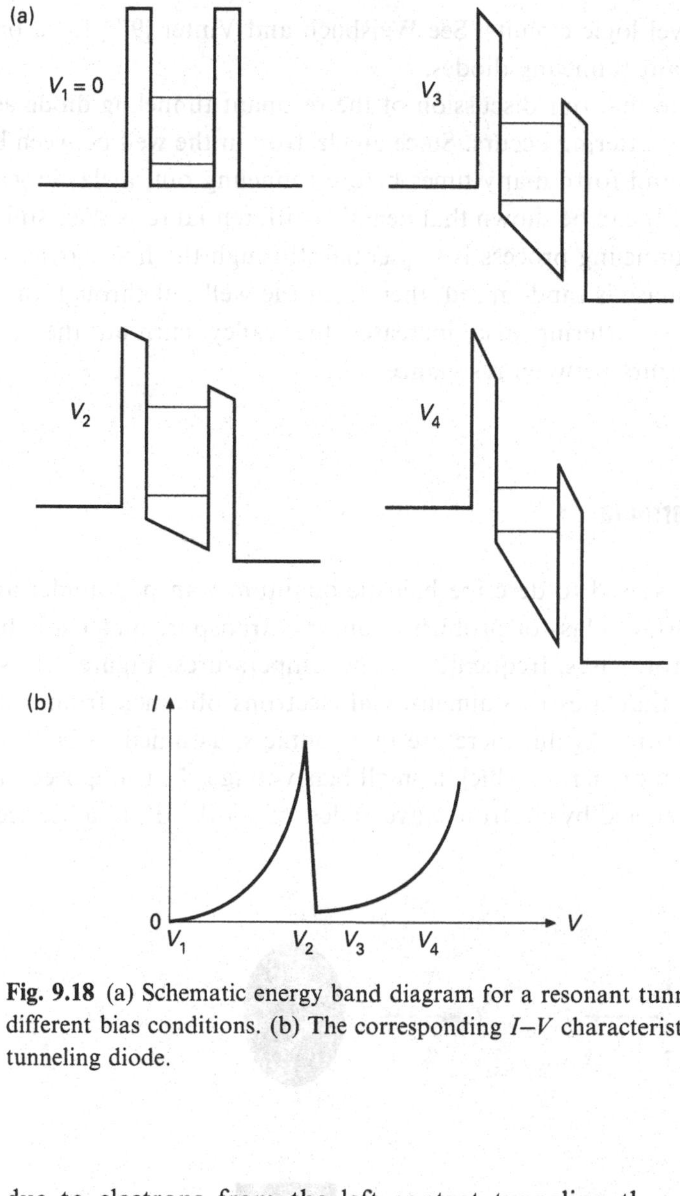

Fig. 9.18 (a) Schematic energy band diagram for a resonant tunneling diode under four different bias conditions. (b) The corresponding *I*–*V* characteristic for the resonant tunneling diode.

due to electrons from the left contact tunneling through the double barrier, begins to flow.

As the bias increases, the resonant tunneling energy aligns with the average carrier energy in the cathode, and the current becomes very large. A further increase in bias, however, removes the resonant tunneling condition, and the current plummets. Further increases in energy increase the current slowly again, until a second resonant energy is aligned with the cathode and a large current flows again. The key feature of the device is that a region of negative differential resistance occurs which can be exploited to produce oscillators. Alternatively, digital circuits with multiple stable points can be realized to produce multiple level logic circuits. See Weisbuch and Vinter [9.8] for a brief discussion of resonant tunneling diodes.

We should note that our discussion of the resonant tunneling diode assumed that no inelastic scattering occurs. Since an electron in the well between barriers can reflect back and forth many times before tunneling out, inelastic scattering events can occur. It can be shown that negative differential resistance still occurs, even when the tunneling process is sequential, through the first barrier into the well where the phase is randomized, then from the well out through the second barrier. Inelastic scattering also increases the valley current, the minimum current which occurs between resonances.

9.9 The Landauer formula

Equation (9.71) is used to describe ballistic quantum transport under arbitrary biases. An important class of problems concerns transport under low biases in degenerate electron gases, frequently at low temperatures. Figure 9.19 shows a quantum device that uses two-dimensional electrons obtained from a modulation-doped structure. Again, there are two contacts, assumed to be in thermodynamic equilibrium, across which a small bias voltage, V, is imposed. The two contacts are connected by electron 'waveguides' of width, W, to a 'device' struc-

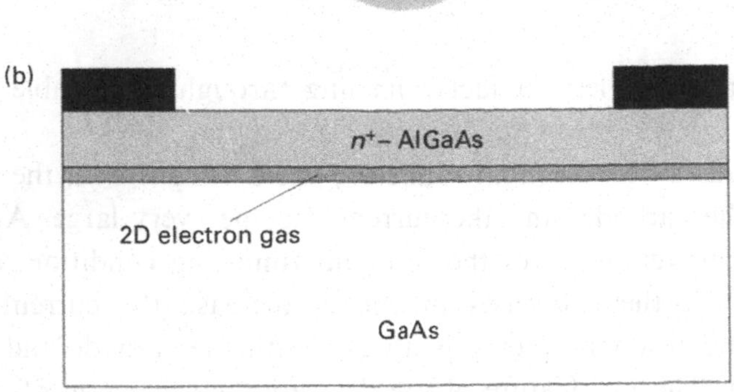

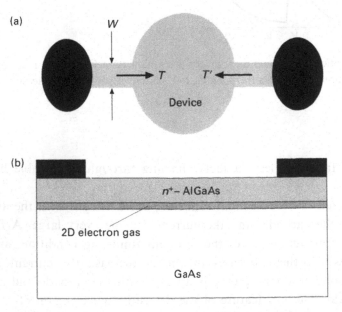

Fig. 9.19 Illustration of a 2D quantum device. (a) Top view and (b) cross-section showing how it is constructed in a modulation-doped structure using 2D, confined electrons.

ture described by a current transmission coefficient, τ_{mn}. (We will assume that only elastic scattering exists, so that $\tau_{mn} = \tau'_{mn}$.) Equation (9.71) can be used to describe the current versus voltage characteristics of such a quantum device.

This bias voltage across the quantum device lowers the Fermi energy in the right contact; since the bias is small,

$$f_{\mathrm{L}} - f_{\mathrm{R}} \approx -\frac{\partial f}{\partial E}(qV). \tag{9.99}$$

In a degenerate electron system, the Fermi function changes from one to zero over a range of a few $k_{\mathrm{B}}T$ about the Fermi level. At low temperatures,

$$\frac{\partial f}{\partial E} \approx -\delta(E_{\mathrm{F}}). \tag{9.100}$$

The result is that eq. (9.72) becomes

$$I = \frac{2q^2}{h}\sum_{n=1}^{M} T_n(E_{\mathrm{F}})V_{\mathrm{D}} = \frac{2q^2}{h} M\bar{T}(E_{\mathrm{F}})V_{\mathrm{D}}, \tag{9.101}$$

where M is the number of transverse modes in the electron waveguides connecting the contacts to the device itself and $\bar{T}(E_{\mathrm{F}})$ is the average mode-dependent transmission coefficient at the Fermi energy. For resonant tunneling devices, the cross-sectional area, A, is typically large, and the sum over transverse modes in eq. (9.72) can be converted to an integral (see homework problem 9.7). For mesoscopic devices like that sketched in Fig. 9.19, however, the number of modes as determined by the width of the electron waveguides is small, so we count them rather than integrating over them.

Equation (9.101) gives the conductance of a quantum device as

$$\boxed{G = \frac{2q^2}{h} M\bar{T}(E_{\mathrm{F}})}, \tag{9.102}$$

which has become known as the *Landauer formula* and has been widely used to interpret experiments and guide thinking in the exploration of mesoscopic transport physics [9.1, 9.9]. According to the Landauer formula, the conductance of a mesoscopic device is the product of some fundamental constants, the number of propagating transverse modes, and the transmission coefficient of the device. It is interesting to note that even in the ballistic limit where $T = 1$, there is still a finite conductance (or resistance) for the device. Each mode has a resistance of 12.9 kΩ, which is interpreted as a contact resistance due to the rearrangement of charge between the contacts and the device. For a large device, there are many transverse modes, and this fundamental resistance becomes negligibly small, but as devices shrink, it may become important.

9.10 Quantized conductance

To actually compute the conductance of a quantum device, we need to compute its transmission coefficient and the number of transverse modes. Consider the number of modes and recall from Section 1.4.1 that each state occupies a space, $2\pi/W$ in **k**-space. The number of transverse modes is, therefore,

$$M = \frac{2k_F}{(2\pi/W)}, \tag{9.103}$$

where k_F is the maximum wavevector in the degenerate electron gas system, and the factor of two arises because states from $-k_F$ to $+k_F$ are occupied. Recall from Section 1.5.2 that in a degenerate 2D system,

$$n_S = \frac{k_F^2}{2\pi} \tag{9.104}$$

(Section 1.5.2). We find, therefore, that

$$M = \sqrt{\frac{2n_S}{\pi}}W \tag{9.105}$$

so the number of transverse modes, and therefore the conductance, is proportional to the width of the electron waveguide and to the square root of the electron density.

Figure 9.20a shows a device in which these features of quantum transport are displayed clearly. The active portion of the structure consists of a region with metal gates that can electrostatically control the width of the conducting region. A classical theory would predict that the conductance of the device would increase linearly as the width, W, increases. According to the Landauer formula, however, the conductance should increase in steps as M increases in steps. The measured results, shown in Fig. 9.20b, clearly show that G increases in steps. Since the length of the active region is so short, there is little scattering, and $T \approx 1$. For this reason, each step increase in G corresponds very closely to the quantum of conductance, $(2q^2/h)$.

Quantized conductance very clearly shows that non-classical transport can occur in small structures at low temperatures. To observe these effects, the device needs to be small, so that a few transverse modes are involved. The temperature also needs to be low and the length of the critical region short so that $T \approx 1$.

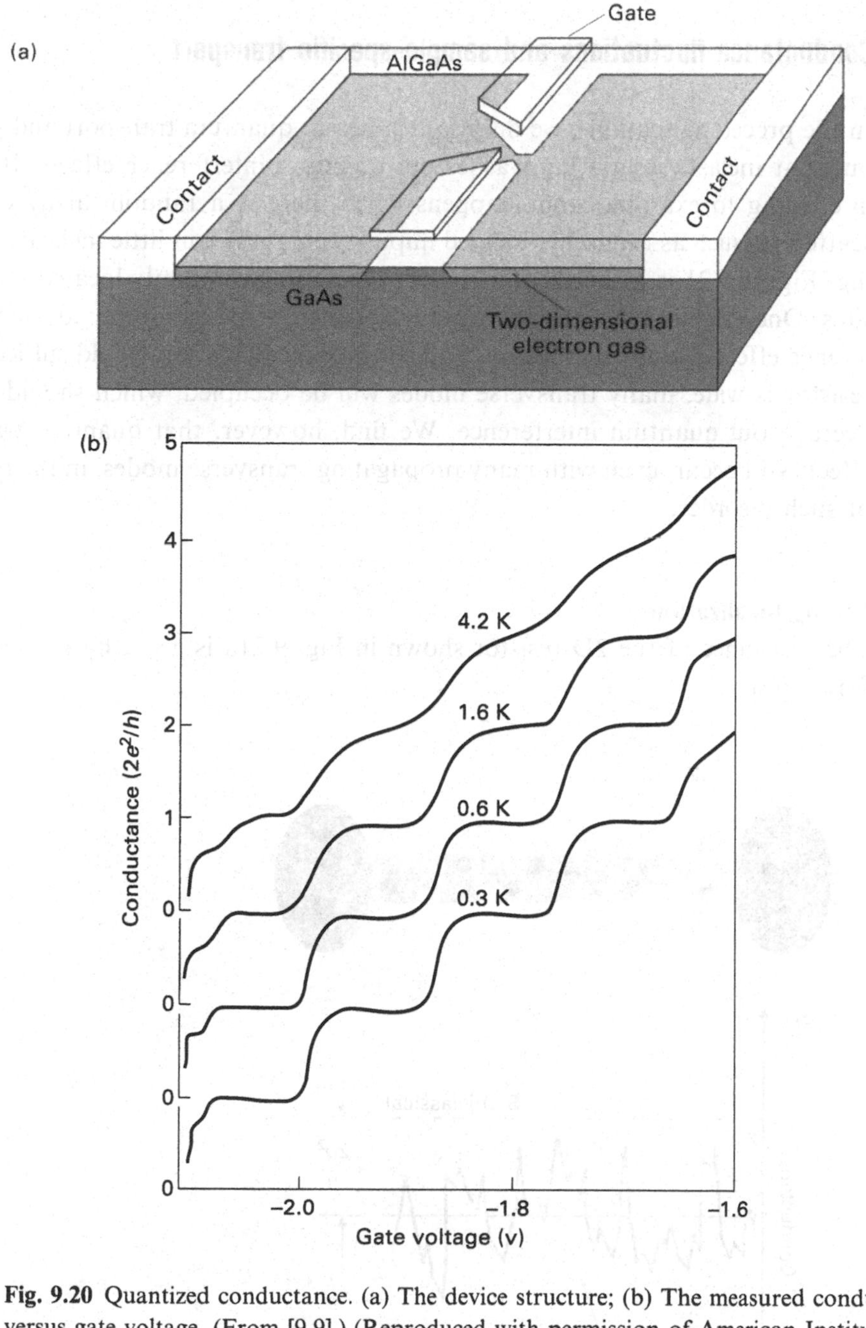

Fig. 9.20 Quantized conductance. (a) The device structure; (b) The measured conductance versus gate voltage. (From [9.9].) (Reproduced with permission of American Institute of Physics.)

9.11 Conductance fluctuations and sample-specific transport

In the preceding sections, we discussed coherent quantum transport and pointed out that inelastic scattering washes out quantum interference effects. It is also interesting to examine what happens when there is a random array of static scatterers (such as randomly located impurity charges), but little inelastic scattering. Figure 9.21 is a sketch of a resistor containing randomly located scattering sites. One might expect that because the scatterers are randomly located, interference effects would average out and a classical calculation would suffice. If the resistor is wide, many transverse modes will be occupied, which should further average out quantum interference. We find, however, that quantum transport effects still occur, even with many propagating transverse modes, in the presence of such disorder.

Strong localization

The resistance of the 2D resistor shown in Fig. 9.21a is given by the Landauer formula as

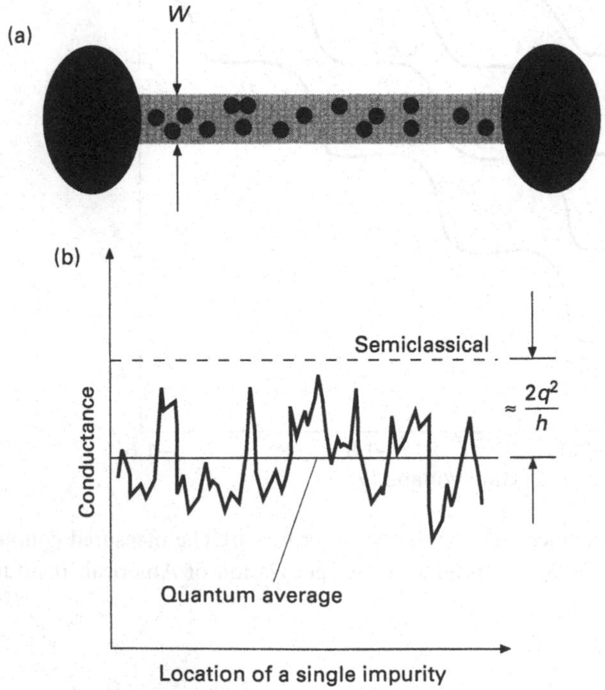

Fig. 9.21 Conductance fluctuations. (a) A 2D resistor with randomly located scattering sites. (b) The expected conductance as a single impurity is moved. (After [9.1].)

$$\mathcal{R} = \left(\frac{h}{2q^2}\right)\frac{1}{MT}, \tag{9.106a}$$

where we use the symbol, $\mathcal{R}$, for resistance to differentiate it from the R used to denote reflection. If we were to calculate the resistance classically, we would use eq. (9.16) for T to find

$$\mathcal{R} = \left(\frac{h}{2q^2 M}\right) + \left(\frac{h}{2q^2}\right)\frac{L}{M\lambda}. \tag{9.106b}$$

The first term in eq. (9.106b) is independent of the length of the resistor and is identified as a contact resistance (it is the limiting resistance that occurs when $T = 1$). The second term, which is proportional to the L, is the resistance of the resistor itself. The resistance is determined by the mean-free-path for scattering, which in turn is determined by the averaged density of the scattering sites. Quantum mechanically the transmission coefficient is determined by reflection from the scattering sites and by interference between the various reflection paths. The resistance should depend not only on the number of scatters, but also on exactly how they are arranged. For a given number of scattering centers, each sample will have a different resistance depending on how the scattering sites are arranged. It only makes sense, therefore, to talk about the average resistance of an ensemble of resistors.

Classically, the resistance is proportional to the length of a resistor (i.e. to the number of scattering sites). If phase randomizing scattering is weak, however, the resistance will increase exponentially if the resistor is long enough so that its resistance is $\sim (h/2q^2)$. According to eq. (9.106b), this effect, known as strong localization, occurs when the resistor is longer than

$$L_C = M\lambda, \tag{9.107}$$

which is known as the *localization length*. Practical wires or resistors are composed of phase coherent segments which are short compared to the localization length so that strong localization does not occur, and the resistance scales with length.

Weak localization

When the resistance is less than $\sim (h/2q^2)$, quantum mechanical effects can still influence the performance of a resistor. Since the resistance is determined by an interference pattern set up by multiple reflections from the impurities, if a single impurity is moved, the interference pattern will change, which will change $T(E_F)$ and, therefore, the conductance. Alternatively, if the electron density is changed (perhaps by a gate above the resistor), then k_F will change which, by changing the electron wavelength, will change the interference pattern. (The same effect also occurs if a small magnetic field is applied.) Figure 9.21b shows what happens when a single impurity is moved across the width of the resistor. Two effects are worth noting. First, the conductance fluctuates as the interference pattern changes, and the magnitude of the fluctuations is about $2q^2/h$. The second effect to note is that the average value of the quantum mechanical conductance is $2q^2/h$ less than the classical value.

The surprising thing about the fluctuations is not that they occur but that they have a universal value of $\sim 2q^2/h$ independent of the size of the average conductance itself. One might have expected that the size of the fluctuations would vary as $1/M$, where M is the number of modes, so that if the conductance were high, it would imply many transverse modes, so that the fluctuations would be small. In samples that are large compared to a phase relaxation length, however, these *universal conductance fluctuations* are not observed. Fluctuations occur in sub-elements of the resistor with dimensions of a phase relaxation length, but these sub-elements are uncorrelated so the fluctuations average out in large samples. A discussion of why the magnitude of the conductance fluctuations has a universal value is beyond the scope of this discussion; the interested reader should consult Datta [9.1].

Intuitively, we might have expected fluctuations in the presence of disorder, but the lowering of the average conductance due to quantum mechanical interference is something of a surprise. The following simple argument helps explain why this occurs [9.1]. We begin with the Landauer expression for the conductance of the resistor,

$$G = \frac{2q^2}{h} M(1 - R) \tag{9.108}$$

and recall from eq. (9.16) that in semiclassical transport,

$$R_{\mathrm{CL}} = \frac{L}{L + \lambda}, \tag{9.109}$$

where the subscript, CL, denotes semiclassical. If the scattering is isotropic, then backscattering should occur with equal probability into each of the M transverse modes. We expect, therefore, that the probability of backscattering from mode m into mode n is

$$R_{nm} = \frac{1}{M} \frac{L}{L + \lambda}. \tag{9.110}$$

When we compute the backscattering from the random array of scatterers by quantum mechanics, eq. (9.109) still holds on average when $n \neq m$

$$\langle R_{nm} \rangle = \frac{1}{M} \frac{L}{L + \lambda} \qquad n \neq m \tag{9.111a}$$

but when $n = m$, the probability for backscattering doubles,

$$\langle R_{nm} \rangle = \frac{2}{M} \frac{L}{L + \lambda} \cdot \qquad n = m. \tag{9.111b}$$

To find the total probability for backscattering, we sum the contributions for all modes to find

$$\langle R \rangle = \frac{L}{L + \lambda} + \frac{1}{M} \frac{L}{L + \lambda} \approx R_{\text{CL}} + \frac{1}{M} \frac{L}{L + \lambda} \approx R_{\text{CL}} + \frac{1}{M}, \tag{9.112}$$

where we have assumed that $L \gg \lambda$, the resistor is many mean-free-paths long. When eq. (9.112) is inserted into eq. (9.108), we find

$$\langle G \rangle = G_{\text{CL}} - \frac{2q^2}{h}. \tag{9.113}$$

Equation (9.113) explains why the quantum mechanically evaluated conductance is about $2q^2/h$ lower than the value from a classical calculation. It arises because the probability for backscattering increases by a factor of two when quantum mechanics is considered. Why does quantum mechanics double the probability for backscattering?

Consider an electron injected from the resistor's left contact in mode m, that subsequently reflects and emerges back into the left contact in mode n. There are many different reflection and transmission paths (so-called *Feynman paths*) within the resistor that contribute to this process. For example, we can write the amplitude for one specific path as

$$r_{nm}^{(1)} = t_{m_2 m} \times t_{m_5 m_2} \times r_{m_1 m_5} \times t_{n m_1}, \tag{9.114}$$

which in words says that the electron first transmits from the injected mode, m, into another transverse mode, m_2. Then it transmits to another mode, m_5, after which it reverses direction by reflecting into yet another mode, m_1. Finally it transmits from mode m_1 into mode, n, and exists at the left contact. Note that there are an infinite number of possible reflection paths within the resistors and that these transmission and reflection coefficients are complex numbers. To compute the real number, R_{nm}, which gives the probability that an incident current in mode m reflects into a current emerging in mode, n, we should add up the contributions from all possible paths. Because of the disorder, there is no phase relationship between the different paths, so

$$R_{nm} \neq |r_{nm}^{(1)} + r_{nm}^{(2)} + r_{nm}^{(3)} + \ldots|^2, \tag{9.115a}$$

rather,

$$R_{nm} = |r_{nm}^{(1)}|^2 + |r_{nm}^{(2)}|^2 + |r_{nm}^{(3)}|^2 + \ldots. \tag{9.115b}$$

Because of the disorder, there are no phase relationships between different paths, so we add probabilities as in semiclassical transport rather than adding amplitudes.

The case for which an electron incident in mode m reflects back into the same mode, m, is, however, special. Consider one possible path for such a process,

$$r_{nm}^{(1)} = t_{m_2 m} \times t_{m_5 m_2} \times r_{m_1 m_5} \times t_{m m_1}. \tag{9.116a}$$

This process could also occur in the opposite order,

$$\hat{r}_{mm}^{(1)} = t_{m_1 m} \times r_{m_5 m_1} \times t_{m_2 m_5} \times t_{m m_2} \tag{9.116b}$$

This pair of paths (so-called *time-reversed paths*) is special. Because the scattering is elastic, we can show that the two paths have the same amplitude (recall the discussion in Section 9.7 about T and T'). They also experience exactly the same phase shift because they traverse exactly the same path, just in opposite orders. Since there is a definite phase relation between these two paths, their amplitudes should be added rather than their probabilities. Consequently,

$$R_{mm} = |r_{mm}^{(1)} + \hat{r}_{mm}^{(1)}|^2 + |r_{mm}^{(2)} + \hat{r}_{mm}^{(2)}|^2 + |r_{mm}^{(3)} + \hat{r}_{mm}^{(3)}|^2 + \ldots. \tag{9.117a}$$

Since $r_{mm}^{(1)} = \hat{r}_{mm}^{(1)}$,

$$R_{mm} = 4|r_{mm}^{(1)}|^2 + 4|r_{mm}^{(2)}|^2 + 4|r_{mm}^{(3)}|^2 + \ldots, \tag{9.117b}$$

which is twice the value we would obtain if we had added the probabilities for each path. Interference between time-reverse paths doubles the probability of reflecting back into the same mode, and therefore lowers the conductance of the resistor by $2q^2/h$ as shown in eq. (9.113).

9.12 Discussion: quantum transport

Quantum transport is most readily understood and treated when inelastic scattering is weak so that phase coherence can be assumed. In that case we see effects such as resonant tunneling, quantized conductance, strong and weak localization, and conductance fluctuations. We find that the properties of a device can be sensitive to the precise, microscopic arrangement of scattering sites. At the opposite extreme is the phase incoherent limit of semiclassical transport, which we treated in Sections 9.2–9.6. Room temperature electronic devices have historically operated in the incoherent regime, but as devices get smaller, quantum transport effects may become important, even at room temperature, in electronic devices. In that regime, both quantum interference and inelastic scattering processes will have to be treated on a more or less equal basis.

We saw in Section 9.6 that the proper generalization of the simple, one-flux scattering theory is the Boltzmann Transport Equation. Similarly, the generalization of the transmission approach to quantum transport is a quantum kinetic equation. For the Boltzmann equation, the unknown being solved for is the distribution function, $f(\mathbf{r}, \mathbf{k}, t)$. For the quantum kinetic equation, the corresponding quantity is $-iG^{<}(\mathbf{r}, \mathbf{r}', t, t')$, which gives the correlation between the amplitude in a state $\mathbf{r}$ at rime t and another in state $\mathbf{r}'$ at time t'. See Datta (Chapter 8 of [9.1]) for an introduction to quantum kinetic equations.

9.13 Summary

Scattering (or transmission) theory provides a simple, conceptual way to think about transport in small structures. It is useful in short devices dominate by semiclassical transport as well as in small devices controlled by quantum transport. For semiclassical transport, the scattering approach provides an alternative to the more traditional methods of transport theory that have been the subject of Chapters 1–8. It is especially useful as quasi-ballistic transport begins to become important. The analogous approach, transmission theory, has provided the most widely-used conceptual tools for understanding a variety of quantum transport effects that occur primarily at low temperatures. As devices continue to shrink, however, quantum effects may eventually become important at room temperatures. This chapter has been a brief introduction to the kinds of effects that can occur in small structures. For a thorough introduction to quantum transport, the reader is advised to consult Datta's book [9.1].

References and further reading

For an introduction to quantum transport in mesoscopic structures, see

9.1 Datta, S. *Electronic Transport in Mesoscopic Systems.* Cambridge University Press, Cambridge, UK, 1995.

9.2 Datta, S. *Quantum Phenomena.* Addison-Wesley, Reading, MA, 1989.

For a discussion of the scattering approach to semiclassical transport, including the treatment of recombination–generation processes, time dependence, and applications to devices, consult

9.3 Tanaka, S.-I. and Lundstrom, M. S. A compact HBT device model based on a one-flux treatment of carrier transport. *Solid-State Electronics*, **37**, 401–10, 1994.

9.4 Alam, M. A. and Lundstrom, M. S. A small-signal, one-flux analysis of short-base transport. *Solid-State Electronics*, **38**, 177–82, 1995.

9.5 Tanaka, S.-I. and Lundstrom, M. S. A flux-based study of carrier transport in thin-base diodes and transistors. *IEEE Transactions on Electron Devices*, **42**, 1806–15, 1995.

9.6 Lundstrom, M. S. Elementary scattering theory of the Si MOSFET. *IEEE Electron Devices Letters*, **18**, 361–3, 1997.

9.7 Alam, M. A., Stettler, M. S. and Lundstrom, M. S. Formulation of the Boltzmann equation in terms of scattering matrices. *Solid-State Electronics*, **36**, 263–71, 1993.

For discussion of resonant tunneling devices and quantum transport effects, see

9.8 Weisbuch, C. and Vinter, B. *Quantum Semiconductor Structures*. Academic Press, New York, 1991.

9.9 van Hauten, H. and Beenakker, C. Quantum point contacts. *Physics Today*, **July**, 22–7, 1996.

Problems

9.1 The diffusion coefficient is often equated to the product of the thermal velocity and the carrier mean-free-path for scattering, λ.

(a) Derive an expression relating D and λ when $\lambda = \lambda_0$ is constant. Begin with

$$D = \frac{k_B T_L}{m^*} \langle\langle \tau \rangle\rangle.$$

Next, relate the scattering time to the mean-free-path by

$$\tau = \frac{\lambda_0}{\upsilon},$$

where υ is the carrier velocity,

$$(1/2)m^* \upsilon^2 = E.$$

Show that τ can be written in power law form with an exponent of $s = -1/2$. Finally, use this result to evaluate D and show that it can be written as

$$D = \left(\frac{4}{3}\right)\lambda_0 \upsilon_R,$$

where

$$\upsilon_R = \sqrt{\frac{k_B T_L}{2\pi m^*}}$$

is the so-called Richardson velocity.

(b) Derive an expression relating D and λ when the scattering time is constant. Begin with

$$D = \frac{k_B T_L}{m^*} \tau_0$$

and define the average mean-free-path as

$$\langle \lambda \rangle = \langle \upsilon \rangle \tau_0.$$

For the average carrier velocity, use the rms thermal velocity,

$$\langle v \rangle = v_{\text{rms}} = \sqrt{\frac{3k_B T_L}{m^*}}.$$

Use these assumptions to show that when scattering time is constant,

$$D = \left(\sqrt{\frac{2\pi}{3}}\right)\langle \lambda \rangle v_R.$$

9.2 Figure 9.4 shows a set of two interconnected scattering matrices. The effect of these two scattering matrices can be described by the single, composite scattering matrix. Derive expressions for the elements of the composite scattering matrix, T and T', and show that the results are eqs. (9.9a)–(9.9d). Hint: begin by writing expressions for each of the emerging fluxes, then eliminate the interior fluxes, a_1 and b_1.

9.3 A scattering matrix relates the fluxes emerging from a slab to those incident upon it. For the slab shown in Fig. P9.3, the scattering matrix gives

$$\begin{pmatrix} a_1 \\ b_0 \end{pmatrix} = \begin{bmatrix} T & R' \\ R & T' \end{bmatrix} \begin{pmatrix} a_0 \\ b_1 \end{pmatrix}.$$

A transmission matrix relates the fluxes on the right side of a slab to those on the left,

$$\begin{pmatrix} a_1 \\ b_1 \end{pmatrix} = \begin{bmatrix} W_{11} & W_{12} \\ W_{21} & W_{22} \end{bmatrix} \begin{pmatrix} a_0 \\ b_0 \end{pmatrix}.$$

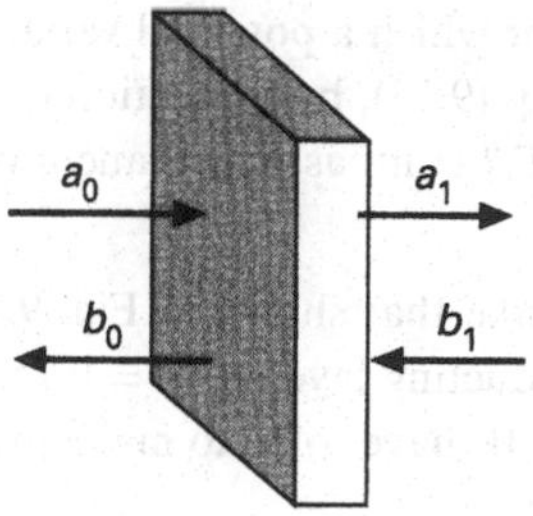

Fig. P9.3

Show that

$$W_{11} = T - R'[T']^{-1}R$$
$$W_{12} = R'[T']^{-1}$$
$$W_{21} = -[T']^{-1}R$$
$$W_{22} = [T']^{-1}.$$

9.4 Use Fick's Law of Diffusion, $F = -D\dfrac{dn}{dz}$, to solve for the net flux in a slab of thickness L. Assume that $n(0)$ is given and that $F = Sn(L)$ at the right (S is a 'surface recombination velocity' in cm/s). Show that the results are equivalent to the flux analysis if we assume that

$S = \upsilon_T$. This example shows that diffusion can be described by Fick's Law even when the region is thin compared to a mean-free-path.

9.5 Consider a flux of carriers into a surface as illustrated in Fig. P9.5. Conventionally, we write, $F = Sn(0)$, where F is the carrier flux, S the surface recombination velocity, and $n(0)$ the carrier density at the surface. If the flux comprises two near-equilibrium (thermal Maxwellian) halves, show that the surface recombination velocity is related to the surface backscattering coefficient, R_S, by

$$S = \upsilon_T\left(\frac{1 - R_S}{1 + R_S}\right).$$

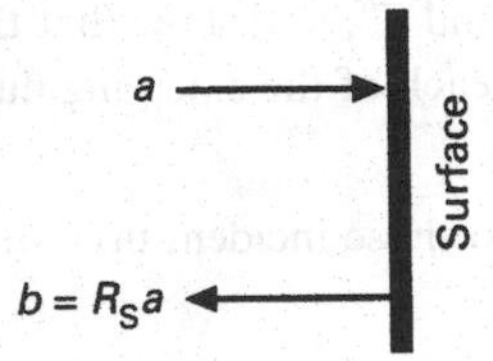

Fig. P9.5

9.6 Derive eq. (9.68) from eq. 9.67.

9.7 In eq. (9.72), the sum over n represents the sum over the k-states in the x–y plane. Assuming that the cross-sectional area the device is large, work out this sum.

9.8 Insert the $(1 - f)$ factors, which account for the fact that the final state may be filled, in eq. (9.71). Show that these factors cancel out under the assumed conditions of elastic scattering.

9.9 Consider a device which is uniform in the x–y plane but for which a potential variation exists in the z-direction. Solve the steady-state wave equation, eq. (9.73), by separation of variables and show that the result is eqs. (9.75) and (9.76). What is E_t? Hint: assume plane waves in the x- and y-directions.

9.10 Consider an AlGaAs/GaAs modulation-doped structure like that shown in Fig. 9.19. If the electron density is $5 \times 10^{11}\ \text{cm}^{-2}$ and the width of the conducting layer is $W = 0.25\ \mu\text{m}$, then how many transverse modes are occupied? How thin does W have to be to make the electron waveguide single-moded?

9.11 Consider an ultrasmall silicon MOSFET with $W = 150\ \text{nm}$. If the inversion layer density is $1 \times 10^{12}\ \text{cm}^{-2}$, what is the minimum resistance of the MOSFET?

Appendices

Appendix 1: summary of indicial notation and tensors

Scalars, vectors, and tensors appear in transport theory. For our purposes, only a few definitions relating to tensors and an acquaintance with indicial notation are necessary. We deal with *Cartesian tensors*, those referred to a right-handed Cartesian coordinate system. A tensor of order zero is a scalar, which is completely defined by its magnitude. A tensor of order one is a vector, which is defined by its magnitude and direction. A tensor of order two is a matrix. Higher order tensors also appear in the transport equations. Tensors may be written in *symbolic notation* or in *indicial notation.* Indicial notation displays the indices explicitly and often simplifies the manipulation of equations involving tensors. In this appendix, we present a brief synopsis of indicial notation for Cartesian tensors.

In symbolic notation, we write a vector as **V** or $\vec{V}$. Alternatively, we can write a vector in terms of its three components with respect to a right-handed Cartesian coordinate system,

$$V_i \text{ for } i = 1, 2, 3 \text{ or } i = \hat{x}, \hat{y}, \hat{z}. \tag{A1.1a}$$

The vector is the sum of its three components,

$$\mathbf{V} = V_1\hat{x}_1 + V_2\hat{x}_2 + V_3\hat{x}_3, \tag{A1.1b}$$

where $(\hat{x}_1, \hat{x}_2, \hat{x}_3)$ are the three orthogonal unit vectors. We can write eq. (A1.1b) more compactly by adopting the *summation convention*, which states that a repeated index is to be summed over its allowed values. With this convention, eq. (A1.1b) becomes

$$\mathbf{V} = V_i\hat{x}_i. \tag{A1.1c}$$

We often speak of V_i as a vector, but it should be kept in mind that V_i just represents the three components of the Cartesian vector, **V**.

The convention of summing over all repeated indices is used in all indicial equations. For example,

$$A_iB_jC_j \equiv A_iB_1C_1 + A_iB_2C_2 + A_iB_3C_3. \tag{A1.2}$$

Note that this expression is a vector; the $\hat{x}_1$ component of the vector is $A_1(B_1C_1 + B_2C_2 + B_3C_3)$. The repeated indices in the expression get *summed out* and are called *dummy indices*. In eq. (A1.2) we could replace the dummy index, j, with any other letter (except i) without changing the meaning of the expression. The indices that remain after summing out the dummy indices are call *floating subscripts*.

Occasionally, it is necessary to refer to one specific term of the sum rather than to the complete sum, for example, V_{zz}. To do this, we adopt the convention that repeated Greek subscripts refer to a specific term and do not imply a sum. For example, $V_{\alpha\alpha}$ with $\alpha = z$ refers to V_{zz}.

The dot product of two vectors is written as

$$\mathbf{V} \cdot \mathbf{W} = \sum_{j=1}^{3} V_j W_j \equiv V_j W_j, \tag{A1.3}$$

where the expression on the left-hand side is the dot product written in symbolic notation and the others in indicial notation. The form on the right employs the summation convention.

A vector can be formed by taking the gradient of a scalar. In symbolic notation, we write this as

$$\mathbf{V} = \nabla v, \tag{A1.4}$$

or in indicial notation as

$$V_j = \partial_j(v) \equiv \frac{\partial}{\partial x_j}(v) \qquad j = 1, 2, \text{ or } 3 \quad (\text{or } j = x, y, \text{ or } z), \tag{A1.5}$$

where j ranges over the three coordinate axes. Similarly, the divergence of a vector, a scalar, can be written as

$$\nabla \cdot \mathbf{V} = \frac{\partial V_1}{\partial x_1} + \frac{\partial V_2}{\partial x_2} + \frac{\partial V_3}{\partial x_3} = \partial_j V_j. \tag{A1.6}$$

We write a 3 × 3 matrix, a Cartesian tensor of order two, symbolically as [A] or in indicial notation as

$$A_{ij} \quad \text{for} \quad i \text{ and } j = 1, 2, \text{ or } 3 \text{ (or } x, y, \text{ or } z). \tag{A1.7}$$

A second order tensor represents nine numbers, which can be written as a matrix

$$A_{ij} = \begin{bmatrix} A_{11} & A_{12} & A_{13} \\ A_{21} & A_{22} & A_{23} \\ A_{31} & A_{32} & A_{33} \end{bmatrix}. \tag{A1.8}$$

The product of a matrix times a vector (a second order tensor times a first) can be written in either symbolic or in indicial notation,

$$[A]\mathbf{V} = \sum_{j=1}^{3} A_{ij} V_j = A_{ij} V_j, \tag{A1.9}$$

where the last expression on the right-hand side makes use of the summation convention. Equation (A1.9) is readily identified as the usual matrix–vector multiplication. The trace of a matrix is the sum of its diagonal components,

$$Tr([A]) = A_{ii}. \tag{A1.10}$$

The product of two second order tensors is

$$C_{ik} = A_{ij} B_{jk}, \tag{A1.11}$$

which can be written in matrix notation as

$$\begin{bmatrix} C_{11} & C_{12} & C_{13} \\ C_{21} & C_{22} & C_{23} \\ C_{31} & C_{32} & C_{33} \end{bmatrix} = \begin{bmatrix} A_{11} & A_{12} & A_{13} \\ A_{21} & A_{22} & A_{23} \\ A_{31} & A_{32} & A_{33} \end{bmatrix} \begin{bmatrix} B_{11} & B_{12} & B_{13} \\ B_{21} & B_{22} & B_{23} \\ B_{31} & B_{32} & B_{33} \end{bmatrix} \tag{A1.12}$$

and evaluated by the normal rules of matrix multiplication.

A commonly used second rank tensor is the *Kronecker delta*,

$$\delta_{ij} = 0 \quad \text{if} \quad i \neq j \tag{A1.13a}$$

$$\delta_{ij} = 1 \quad \text{if} \quad i = j. \tag{A1.13b}$$

One can readily verify that

$$V_j \delta_{ij} = V_i. \tag{A1.14}$$

Another useful tensor, ε_{ijk}, the *alternating unit tensor*, is defined as

$$\begin{aligned} \varepsilon_{ijk} &= +1 \quad \text{for } i, j, k \text{ in cyclic order (1,2,3 2,3,1, etc)} \\ &= -1 \quad \text{for } i, j, k \text{ in anti-cyclic order (3,2,1 1,2,3, etc)} \\ &= 0 \quad \text{otherwise.} \end{aligned} \tag{A1.15}$$

Note that interchanging two adjacent subscripts changes the sign of the alternating unit tensor. For example,

$$\varepsilon_{ijk} = -\varepsilon_{jik} = \varepsilon_{jki} - \varepsilon_{kji}. \tag{A1.16}$$

The alternating unit tensor allows us to write cross-products in indicial notation. To do so, we express the cross-product as

$$\mathbf{A} \times \mathbf{B} = \varepsilon_{1jk} A_j B_k \hat{x}_1 + \varepsilon_{2jk} A_j B_k \hat{x}_2 + \varepsilon_{3jk} A_j B_k \hat{x}_3. \tag{A1.17}$$

The *i*th component of the cross-product is written as

$$(\mathbf{A} \times \mathbf{B}) \cdot \hat{x}_1 = \varepsilon_{ijk} A_j B_k. \tag{A1.18}$$

You can verify that these expressions are correct by simply working out the cross-product and comparing it component by component to eq. (A1.18). The triple product $\mathbf{A} \cdot \mathbf{B} \times \mathbf{C}$ is written in indicial notation as $\varepsilon_{ijk} A_i B_j C_k$.

There is an important relationship between the alternating unity tensor and the Kronecker delta. The relationship is

$$\varepsilon_{ijk} \varepsilon_{irs} = \delta_{jr} \delta_{ks} - \delta_{js} \delta_{kr}. \tag{A1.19}$$

Since each index runs over the three coordinate axes, eq. (A1.19) actually represents 81 separate equations.

We have described the dot product of two vectors and their cross-product. A third vector product is the *dyadic product*, which is written as **VW**. The nine components of this product can be written as

$$V_i W_j = \begin{bmatrix} V_1 W_1 & V_1 W_2 & V_1 W_3 \\ V_2 W_1 & V_2 W_2 & V_2 W_3 \\ V_3 W_1 & V_3 W_2 & V_3 W_3 \end{bmatrix}. \tag{A1.20}$$

If a tensor is a function of the spatial coordinates, then the tensor may be differentiated. For example, consider a second-order tensor, $M_{ij}(x_1, x_2, x_3)$. Differentiation produces a third-order tensor, M_{ijk}, where

$$M_{ijk} = \frac{\partial}{\partial x_k} M_{ij}. \tag{A1.21}$$

We can also generalize the concept of the gradient. Note that the gradient of a scalar (a tensor of order zero) produces a vector (a tensor of order one). Similarly, the gradient of a vector, V, produces a *dyad* $\nabla\mathbf{V}$, a tensor of order two. The nine components of the dyad are:

$$\frac{\partial V_i}{\partial x_j} \equiv V_{ij}. \tag{A1.22}$$

Tensors are classified according to their order, and the order of a tensor can be easily determined by identifying the number of unsummed indices. If there are no unsummed indices, the quantity is a tensor of order zero, a scalar. If there is one unsummed index, the quantity is a tensor of order one, a vector, which is described by three numbers. If two indices are unsummed, the tensor has order two, etc. For example, $C_{ij} = A_{ik}B_{kj}$ is a second-order tensor, with nine components, formed by 'matrix multiplication' of the two second-order tensors, A_{ik} and B_{kj}, but $C_{ijkl} = A_{ij}B_{kl}$ is a fourth-order tensor, with 81 components, formed by the dyadic product of two second-order tensors.

This short discussion provides an adequate background to follow the discussion in this text. For a thorough introduction to tensor analysis, the reader should consult a text on applied mathematics or on tensor analysis.

Appendix 2: some useful integrals

$$\int_0^\infty e^{-\alpha x^2}dx = \frac{1}{2}\sqrt{\frac{\pi}{\alpha}} \qquad \int_0^\infty x\,e^{-\alpha x^2}dx = \frac{1}{2\alpha}$$

$$\int_0^\infty x^2e^{-\alpha x^2}dx = \frac{1}{4\alpha}\sqrt{\frac{\pi}{\alpha}} \qquad \int_0^\infty x^3e^{-\alpha x^2}dx = \frac{1}{2\alpha^2}$$

$$\int_0^\infty x^4e^{-\alpha x^2}dx = \frac{3}{8\alpha^2}\sqrt{\frac{\pi}{\alpha}} \qquad \int_0^\infty x^5e^{-\alpha x^2}dx = \frac{1}{\alpha^3}$$

$$\int_0^\infty x^6e^{-\alpha x^2}dx = \frac{15}{16\alpha^3}\sqrt{\frac{\pi}{\alpha}} \qquad \int_0^\infty x^7e^{-\alpha x^2}dx = \frac{3}{\alpha^4}$$

$$\int_0^\infty \frac{dx}{1+ae^x} = -\ln(1+a^{-1}e^{-x})|_0^\infty = \ln(1+a^{-1})$$

Fermi–Dirac integral: (typically, $\eta = (E_F - E_C)/k_BT$)

order 1/2:

$$\mathfrak{F}_{1/2}(\eta) = \frac{2}{\sqrt{\pi}}\int_0^\infty \frac{x^{1/2}dx}{1+\exp(x-\eta)}$$

order j:

$$\mathfrak{F}_j(\eta) = \frac{1}{\Gamma(j+1)}\int_0^\infty \frac{x^j dx}{1+\exp(x-\eta)}$$

$$\frac{d}{d\eta}\mathfrak{F}_j(\eta) = \mathfrak{F}_{j-1}(\eta)$$

nondegenerate limit:

$$\lim_{\eta\to-\infty}\mathfrak{F}_j(\eta) = \exp(\eta)$$

degenerate limit:

$$\lim_{\eta\to+\infty}\mathfrak{F}_{1/2}(\eta) = \frac{4\eta^{3/2}}{3\sqrt{\pi}}$$

Index